Flight Dynamics and Control of Aero and Space Vehicles

Aerospace Series

Stability and Control of Aircraft Systems: Introduction to Classical Feedback Control
Roy Langton

Aerospace Propulsion
T. W. Lee

Civil Avionics Systems, 2nd Edition
Ian Moir, Allan Seabridge, Malcolm Jukes

Aircraft Flight Dynamics and Control
Wayne Durham

Modelling and Managing Airport Performance
Konstantinos Zografos, Giovanni Andreatta, Amedeo Odoni

Advanced Aircraft Design: Conceptual Design, Analysis and Optimization of Subsonic Civil
Airplanes
Egbert Torenbeek

Design and Analysis of Composite Structures: With Applications to Aerospace Structures, 2nd
Edition
Christos Kassapoglou

Aircraft Systems Integration of Air-Launched Weapons
Keith A. Rigby

Understanding Aerodynamics: Arguing from the Real Physics
Doug McLean

Design and Development of Aircraft Systems, 2nd Edition
Ian Moir, Allan Seabridge

Aircraft Design: A Systems Engineering Approach
Mohammad H. Sadraey

Introduction to UAV Systems, 4th Edition
Paul Fahlstrom, Thomas Gleason

Theory of Lift: Introductory Computational Aerodynamics in MATLAB/Octave
G. D. McBain

Sense and Avoid in UAS: Research and Applications
Plamen Angelov

Morphing Aerospace Vehicles and Structures
John Valasek

Spacecraft Systems Engineering, 4th Edition
Peter Fortescue, Graham Swinerd, John Stark

Unmanned Aircraft Systems: UAVS Design, Development and Deployment
Reg Austin

Gas Turbine Propulsion Systems
Bernie MacIsaac, Roy Langton

Aircraft Systems: Mechanical, Electrical, and Avionics Subsystems Integration, 3rd Edition
Ian Moir, Allan Seabridge

Basic Helicopter Aerodynamics, 3rd Edition
John M. Seddon, Simon Newman

System Health Management: with Aerospace Applications
Stephen B Johnson, Thomas Gormley, Seth Kessler, Charles Mott, Ann Patterson-Hine, Karl Reichard, Philip Scandura Jr.

Advanced Control of Aircraft, Spacecraft and Rockets
Ashish Tewari

Air Travel and Health: A Systems Perspective
Allan Seabridge, Shirley Morgan

Principles of Flight for Pilots
Peter J. Swatton

Handbook of Space Technology
Wilfried Ley, Klaus Wittmann, Willi Hallmann

Cooperative Path Planning of Unmanned Aerial Vehicles
Antonios Tsourdos, Brian White, Madhavan Shanmugavel

Design and Analysis of Composite Structures: With Applications to Aerospace Structures
Christos Kassapoglou

Introduction to Antenna Placement and Installation
Thereza Macnamara

Principles of Flight Simulation
David Allerton

Aircraft Fuel Systems
Roy Langton, Chuck Clark, Martin Hewitt, Lonnie Richards

Computational Modelling and Simulation of Aircraft and the Environment, Volume 1: Platform Kinematics and Synthetic Environment
Dominic J. Diston

Aircraft Performance Theory and Practice for Pilots, 2nd Edition
Peter J. Swatton

Military Avionics Systems
Ian Moir, Allan Seabridge, Malcolm Jukes

Aircraft Conceptual Design Synthesis
Denis Howe

Flight Dynamics and Control of Aero and Space Vehicles

Rama K. Yedavalli
The Ohio State University
USA

Registered Offices
John Wiley & Sons, Inc., 111 River Street, Hoboken, NJ 07030, USA
John Wiley & Sons Ltd, The Atrium, Southern Gate, Chichester, West Sussex, PO19 8SQ, UK

Editorial Office
John Wiley & Sons Ltd, The Atrium, Southern Gate, Chichester, West Sussex, PO19 8SQ, UK

For details of our global editorial offices, customer services, and more information about Wiley products visit us at
www.wiley.com.

Wiley also publishes its books in a variety of electronic formats and by print-on-demand. Some content that appears in
standard print versions of this book may not be available in other formats.

Library of Congress Cataloging-in-Publication Data

Names: Yedavalli, Rama K., author.
Title: Flight dynamics and control of aero and space vehicles / Professor
    Rama K. Yedavalli, The Ohio State University.
Description: First edition. | Hoboken, NJ : John Wiley & Sons, Inc., 2020.
    | Series: Aerospace series | Includes bibliographical references and
    index.
Identifiers: LCCN 2019024450 (print) | LCCN 2019024451 (ebook) | ISBN
    9781118934456 (cloth) | ISBN 9781118934425 (adobe pdf) | ISBN
    9781118934432 (epub)
Subjects: LCSH: Flight control. | Airplanes–Control systems. | Space
    vehicles–Dynamics. | Space vehicles–Control systems.
Classification: LCC TL589.4 .Y444 2020 (print) | LCC TL589.4 (ebook) |
    DDC 629.132/6–dc23
LC record available at https://lccn.loc.gov/2019024450
LC ebook record available at https://lccn.loc.gov/2019024451

Cover image: © Dima Zel/Shutterstock, © Egorych/Getty Images
Cover design: Wiley

Set in 10/12pt WarnockPro by SPi Global, Chennai, India

10 9 8 7 6 5 4 3 2 1

Contents

Preface

The subject of flight dynamics and control is an important and integral part of any quality aerospace education curriculum. With the affinity and bias I have for this subject, I even go to the extent of saying that this discipline is an *essential* part of aerospace education. If we make the analogy of a flight vehicle to a human body, I liken this subject as the "brain" of the human body. After all, a flight vehicle on its own is of no use (or lifeless) without its ability to maneuver from one point to another point. In the first half of the 20th century, flight essentially meant atmospheric flight, while the second half of the 20th century expanded that notion to space vehicles as well. Thus, it is only fitting that the students of the 21st century be conversant with both aero and space flight vehicle dynamics and control. It is indeed accepted that there are plenty of excellent textbooks available on this general subject area. However, a close examination of the contents of the currently available textbooks reveals that the majority of those textbooks are exclusively aimed at either aircraft flight dynamics and control or at spacecraft flight dynamics and control, as can be seen from the references given at the end of this preface.

While universities and academic institutions with large and separate aerospace engineering departments can afford the luxury of teaching flight vehicle dynamics and control at the undergraduate level separately for air vehicles and space vehicles, in general, the most likely scenario in majority of the undergraduate curricula across major higher education institutions across the globe is that this type of separate, exclusive treatment for both of these types of vehicles within the available undergraduate curriculum became constrained by faculty/staff resources, the academic institution's mission as well as student body interests and the local job market. Hence, of late, it is felt that the undergraduate student body is better served if it is introduced to few basics of *both* aircraft flight vehicle dynamics and of spacecraft (satellite) dynamics and control to conform to the ABET guidelines of a satisfactory and adequate dynamics and control discipline coverage in a typical *aerospace* engineering department. Our then aerospace engineering department at the Ohio State University embraced this viewpoint. Within this viewpoint, it became increasingly clear that there is a need for an undergraduate level textbook that provides the needed exposure to the fundamentals of *both* air vehicles as well as space vehicles, adjacent to one another, so that the undergraduate student has the option to specialize in either of those two application areas for their advanced learning. For a long period of time, two separate textbooks (expensive) were prescribed; one catering to aircraft dynamics and control and another catering to spacecraft dynamics and control. The practicality of the coverage of the subject in a limited time (of either in a semester or in a quarter, depending on the academic institution's calendar) dictated that only a very minor part of each of those books was used in the entire course, leaving the students somewhat dissatisfied with a feeling that they did not get "value" for the money they spent on the dual set of textbooks. This observation solidified my desire to author a textbook that covers both topics in a single volume, which in turn would serve as a single textbook for the entire core/elective sequence of courses in the undergraduate curriculum. Hence the resulting title of this book, namely "Flight Dynamics and Control of Aero and Space Vehicles". Even though there are few books that treat both of these vehicles together, the covered material is too advanced and not suitable for the standard undergraduate population.

Typically, in a semester system, the Fall semester of the junior year starts with a flight vehicle dynamics (AAE 3520, at OSU) core course, then the Spring semester has a core course on the fundamentals of flight vehicle control (AAE 3521, at OSU), which deals with basic transfer function based linear control systems theory with applications to flight vehicles. A more advanced state space based time domain modern control theory based course with flight vehicle applications is offered as an elective at the senior level. Thus the entire flight vehicle dynamics and control *at the undergraduate level* consists of a year long sequence of three courses. This scenario at OSU, that existed for a long time and continues to exist even now, provided this author the needed incentive to serve the undergraduate student body at a place like OSU by offering them a single textbook that they can use throughout their undergraduate days at OSU. This type of textbook has to have contents such that it provides sufficiently strong coverage of both aircraft and spacecraft dynamics and control areas simultaneously, thereby preparing them to embark on pursuing higher learning in either of those two areas of their choice and passion. It turns out that while writing this book with this viewpoint, many intellectually stimulating and rewarding insights surfaced that clearly highlighted the similarities as well as the differences in the subject matter between these two types of flight vehicles. As an educator, this author believes that this type of overview on the treatment of the subject between these two types of vehicles is much more valuable than mastering the subject matter related to either of those two types of vehicles individually. This in itself provided sufficient impetus for the author to complete this textbook with a unified and integrated treatment given to these types of highly important flight vehicles that form the backbone of the aerospace education and practice.

As such, by its scope and intent, this book does not promise elaborate discussion and exposure to a variety of topics within each of these two types of vehicles, namely aircraft and spacecraft. Instead, it offers the minimum needed, yet sufficiently strong exposure, to the basic topics in dynamics and control of each of these two types of vehicles. Thus, it is hoped that the content of this book is evaluated and appreciated more from the appropriate balance between breadth and depth in the coverage on each of these flight vehicles. The overall objective of the book is to achieve a reasonably satisfactory balance between the coverage on each of these two types of vehicles. In that sense, this book does not conflict or replace the contributions of the many excellent textbooks available on each of these individual types of vehicles, but instead gets inspired by them and makes that type of subject matter available to the student in a single volume, but with only the needed degree of emphasis each type of vehicle warrants, in an undergraduate curriculum. Thus the interested student is left with the option of learning additional advanced material in any single discipline from those textbooks specialized in either aircraft or spacecraft.

The material covered in this book is essentially divided into four parts; Parts I (flight vehicle dynamics), II (flight vehicle control via classical transfer function based methods), III (flight vehicle control via modern, state space based methods), and IV (other related flight vehicles). It also contains four Appendices (A,B,C, and D), where Appendix A presents useful data related to aircraft and satellites (needed for Part I), Appendix B summarizes a brief review of Laplace transform theory (needed for Part II), Appendix C summarizes a brief review of matrix theory and linear algebra (needed for Part III), and finally Appendix D, which summarizes all the MATLAB commands used or needed along with author supplied MATLAB subroutines (for forming the Fuller matrices). The suggested options for use of the entire material in the book are as follows.

Courses	Suggested parts in textbook
Flight vehicle dynamics	Part I + Appendices A, C, and D
Flight vehicle control using transfer function based control theory	Part II + Part IV + Appendices A, B, C, and D
Flight vehicle control using time domain state space based control theory	Part III + Part IV + Appendices A, B, C, and D

Each of the above suggested courses is suitable for a complete semester long course on the said subject matter in the undergraduate curriculum at a standard American (possibly worldwide) university. For example, at OSU, the first course content is taught as a core course AAE3520 in the Fall semester of the junior year, the second course content is taught as as a core course AAE3521 in the Spring semester of the junior year and finally the third course content is taught as a technical elective at the senior year. Thus this book is intended to serve as a single textbook for the undergraduate to cover entire the flight dynamics and control course sequence at OSU, covering the needed material in both the aeronautical as well as space vehicles in a single volume for each of the courses mentioned above. It is believed that this feature is indeed the strength of this book that would serve the undergraduate education in a standard aerospace engineering department at any university.

While the above arrangement of the usage of the book in its entirety portrays the situation at The Ohio State University, it is possible that the contents of the book can also be used in various different combinations, tailored to the situation of any specific dynamics and control sequence of courses at a given academic institution. For example, within the flight vehicle dynamics course, the chapters are arranged in such a way that all the chapters dealing with aircraft dynamics can be gathered together and that content can be taught continually focusing only on aircraft and then do the same with all the chapters dealing with satellite/spacecraft. However, this author still believes that treating both these types of vehicles together and highlighting the similarities and differences at appropriate junctures in the course would be more rewarding for the student.

Each chapter starts with an overview of its highlights and then ends with a chapter summary, underscoring the utility of that chapter's contents in the overall scheme of things. The body of the chapter is illustrated with a generous number of solved examples and the exercises at the end of each chapter would help the student evaluate his/her progress in understanding the material within that chapter. References relevant to that chapter are provided at the the end of that chapter.

An undertaking of this magnitude would not have been possible without the help and support of an amazing group of friends, colleagues and students with like minded zeal and enthusiasm for this project. I would like to take this opportunity to convey my heartfelt thanks to all the many individuals who helped either directly or indirectly. In particular, the following individuals deserve a special mention. It is indeed a pleasure to get a chance to acknowledge their support and contributions. They include undergraduate students Gabe Scherer, Apoorva Chaubal, and Ilyas Wan Mohd Zani, who not only helped with the word processing but also provided valuable feedback on the organization of the content having learnt the subject taught by me. Their "student perspective" helped immensely in my presentation of the content of the book. Among them, Ilyas deserves additional appreciation for his passion and efforts towards this project. I would also like to thank the administration of the Department of Mechanical and Aerospace Engineering at the Ohio State University for providing the needed resources for this project. I would also like to thank the Aerospace Engineering department at the Indian Institute of Science, Bangalore, India, for honoring me with the Satish Dhawan Visiting Chaired Professorship during my sabbatical there and providing me with the much needed solitude and intellectual atmosphere that was essential in completing this project. It is also a pleasure to thank the staff at Wiley, especially Paul Petralia, who initiated this process, and Sherlin Benjamin, Hannah Lee, Shalisha Sukanya and Ashwani Veejai Raj who helped bring this book production process to completion. Finally, it is indeed a pleasure to formally thank my wife Sreerama, who put up with my absence for long periods of time immersed in this task, and supported me by making me focus on my goal with minimal distractions. In addition, it is also a pleasure to thank my other family members, my eldest son Vivek, his wife Aditi, my younger son Pavan and his wife Sasha. They, based on their interaction with their peer undergraduate student community at Cornell, Columbia, Yale, Berkeley and Stanford (during their collective, undergraduate (and graduate) education at those institutions), felt that my authorship of an undergraduate level textbook with this type of unified treatment to both air and space vehicles would be of appreciable value to the students and thus they not only encouraged me to stay on course but also supported me by taking an active role in shaping the final product. It is my sincere hope and wish that this contribution to undergraduate education is deemed successful and useful by the student community, not only at OSU but also at all other universities as well.

Rama K. Yedavalli
Columbus, OH, USA

Bibliography

1 M.J. Abzug and E.E Larrabee. *Airplane Stability and Control: A history of the Technologies that made Aviation possible*. Cambridge University Press, Cambridge, U.K, 2 edition, 2002.

2 R.J. Adams. *Multivariable Flight Control*. Springer-Verlag, New York, NY, 1995.

3 B.N. Agrawal. *Design of geosynchronous spacecraft*. Prentice-Hall, Englewood Cliffs, NJ, 1986.

4 D. Allerton. *Principles of flight simulation.* Wiley-Blackwell, Chichester, U.K., 2009.

5 J. Anderson. *Aircraft Performance and Design.* McGraw Hill, New York, 1999.

6 J. Anderson. *Introduction to Flight.* McGraw-Hill, New York, 4 edition, 2000.

7 A. W. Babister. *Aircraft Dynamic Stability and Response.* Pergamon, 1980.

8 R.H. Battin. *An introduction to the mathematics and methods of astrodynamics.* Washington, DC: AIAA, 1990.

9 J.H. Blakelock. *Automatic Control of Aircraft and Missiles.* Wiley Interscience, New York, 1991.

10 Jean-Luc Boiffier. *The Dynamics of Flight, The Equations.* Wiley, 1998.

11 J.V. Breakwell and R.E. Roberson. *Orbital and Attitude Dynamics.* American Institute of Aeronautics and Astronautics, 1969.

12 L.S. Brian and L.L. Frank. *Aircraft control and simulation.* John Wiley & Sons, Inc., 2003.

13 A. E Bryson. *Control of Spacecraft and Aircraft.* Princeton University Press, Princeton, 1994.

14 Chris Carpenter. *Flightwise: Principles of Aircraft Flight.* Airlife Pub Ltd, 2002.

15 P.R.K Chetty. *Satellite Technology and Its Application.* Tab Books, 1988.

16 Vladimir A. Chobotov. *Spacecraft Attitude Dynamics and Control.* Krieger Publishing Company, 1991.

17 Michael V Cook. *Flight dynamics principles.* Arnold, London, U.K., 1997.

18 Anton H. de Ruiter, Christopher Damaren, and James R. Forbes. *Spacecraft Attitude Dynamics and Control: An Introduction.* Wiley, 2013.

19 R. Deutsch. *Orbital dynamics of space vehicles.* Prentice-Hall, Englewood Cliffs, NJ, 1963.

20 Wayne Durham. *Aircraft flight dynamics and control.* John Wiley & Sons, Inc., 2013.

21 Pedro Ramon Escobal. *Methods of orbit determination.* Krieger, 1965.

22 B Etkin and L.D. Reid. *Dynamics of flight: stability and control,* volume 3. John Wiley & Sons, New York, 1998.

23 E.T. Falangas. *Performance Evaluation and Design of Flight Vehicle Control Systems.* Wiley and IEEE Press, Hoboken, N.J, 2016.

24 A.L. Greensite. *Analysis and design of space vehicle flight control systems.* spartan Books, New York, 1970.

25 T. Hacker. *Flight Stability and Control.* Elsevier Science Ltd, 1970.

26 G.J. Hancock. *An Introduction to the Flight Dynamics of Rigid Aeroplanes.* Number Section III.5-III.6. Ellis Hornwood, New York, 1995.

27 P.C. Hughes. *Spacecraft attitude dynamics.* Wiley, New York, 1988.

28 D.G. Hull. *Fundamentals of Airplane Flight Mechanics.* Springer International, Berlin, Germany, 2007.

29 John L Junkins and James D Turner. *Optimal spacecraft rotational maneuvers,* volume 3. Elsevier, Amsterdam, 1986.

30 M.H. Kaplan. *Modern spacecraft dynamics and control.* Wiley, New York, 1976.

31 D. McLean. Automatic flight control systems. *Prentice Hall International Series in System and Control Engineering,* 1990.

32 Duane T McRuer, Dunstan Graham, and Irving Ashkenas. *Aircraft dynamics and automatic control.* Princeton University Press, Princeton, NJ, 1973.

33 A. Miele. *Flight Mechanics: Theory of Flight Paths.* Addison-Wesley, New York, 1962.

34 Ian Moir and Allan Seabridge. *Aicraft Systems: Mechanical, Electrical and Avionics Subsytems Integration*, 3rd Edition. John Wiley & sons, 2008.

35 Ian Moir and Allan Seabridge. *Design and Development of Aircraft Systems*. John Wiley & sons, 2012.

36 Robert C Nelson. *Flight stability and automatic control*. McGraw Hill, New York, 2 edition, 1998.

37 C. Perkins and R. Hage. *Aircraft Performance, Stability and Control*. John Wiley & Sons, London, U.K., 1949.

38 John Stark Peter Fortescue, Graham Swinerd. *Spacecraft Systems Engineering, 4th Edition*. Wiley Publications, 2011.

39 J.J. Pocha. *An Introduction to Mission Design for Geostationary Satellites*. Springer Science & Business Media, 1 edition, 2012.

40 L.W. Pritchard and A.J. Sciulli. *Satellite communication systems engineering*. Prentice-Hall, Englewood Cliffs, NJ, 1986.

41 J.M. Rolfe and K.J. Staples. *Flight simulation*. Number 1. Cambridge University Press, Cambridge, U.K., 1988.

42 Jan Roskam. *Airplane Flight Dynamics and Automatic Flight Controls: Part I*. Roskam Aviation and Engineering Corp, 1979.

43 Jan Roskam. *Airplane Flight Dynamics and Automatic Flight Controls: Part II*. Roskam Aviation and Engineering Corp, 1979.

44 J.B. Russell. *Performance and Stability of Aircraft*. Arnold, London, U.K., 1996.

45 Hanspeter Schaub and John L. Junkins. *Analytical Mechanics of Space Systems*. AIAA, 2014.

46 D.K. Schmidt. *Modern Flight Dynamics*. McGraw Hill, New York, 2012.

47 Louis V Schmidt. *Introduction to Aircraft Flight Dynamics*. AIAA Education Series, Reston, VA, 1998.

48 D. Seckel. *Stability and control of airplanes and helicopters*. Academic Press, 2014.

49 R. Shevell. *Fundamentals of Flight*. Prentice Hall, Englewood Cliffs, NJ, 2 edition, 1989.

50 M.J. Sidi. *Spacecraft Dynamics and Control: A Practical Engineering Approach*. Cambridge University Press, New York, N.Y., 1997.

51 Frederick O Smetana. *Computer assisted analysis of aircraft performance, stability, and control*. McGraw-Hill College., New York, 1984.

52 R.E. Stengel. *Flight Dynamics*. Princeton University Press, Princeton, 2004.

53 Brian L Stevens, Frank L Lewis, and Eric N Johnson. *Aircraft control and simulation*. Interscience, New York, 1 edition, 1992.

54 P. Swatton. *Aircraft Performance Theory and Practice for Pilots*. Wiley Publications, 2 edition, 2008.

55 A. Tewari. *Atmospheric and Space Flight Dynamics*. Birkhauser, Boston, 2006.

56 A. Tewari. *Advanced Control of Aircraft, Spacecraft and Rockets*. Wiley, Chichester, UK, 2011.

57 W.T. Thomson. *Introduction to space dynamics*. Dover, 1986.

58 Ranjan Vepa. *Flight Dynamics, Simulation and Control for Rigid and Flexible Aircraft*. CRC Press, New York, 1 edition, 2015.

59 N. Vinh. *Flight Mechanics of High Performance Aircraft*. Cambridge University Press, New York, 1993.

60 J.R. Wertz. *Spacecraft attitude determination and control*. Reidel, Dordrecht, the Netherlands, 1978.

61 Bong Wie. *Space Vehicle Dynamics and Control*. AIAA Education Series, Reston, 2008.

62 Wili Hallmann Wilfried Ley, Klaus Wittmann. *Handbook of Space Technology*. Wiley Publications, 2009.

63 P.H. Zipfel. *Modeling and simulation of aerospace vehicle dynamics*. American Institute of Aeronautics and Astronautics Inc., 2000.

Perspective of the Book

The overarching objective of this book is to present the fundamentals of the analysis (mathematical modeling) and synthesis (designing control systems) in a conceptual way within a systems level framework, for a dynamic system so that it behaves the way we desire and then show the application of these concepts to the specific application area, namely dynamics and control of flight vehicles, including both aero (atmospheric) as well as space vehicles.

Dynamic systems are those whose behavior varies as a function of time. As such, their dynamic behavior is described by differential equations whereas static systems are those whose behavior is described by algebraic equations with no dependence on time. Since most physical systems engineers are interested in controlling are dynamic systems, in this book, we focus our attention on dynamic systems. Depending on the underlying basic principles used in the investigation, dynamic systems can be categorized as mechanical, electrical, aerospace, chemical, biological, and various other systems, as shown in Figure 1. The approach taken in this book is to be as generic as possible by treating the subject at a "systems" level in the preliminary stages and then specializing the details to a specific application system. In our case, the field considered for application is the dynamics and control of flight vehicles, which include atmospheric vehicles such as aircraft, as well as space vehicles such as satellites.

Once the focus of book is determined to be aero and space vehicles, it is acknowledged that for a really comprehensive coverage of this subject, one needs to consider the four pillars of this subject matter, namely dynamics, navigation, guidance, and control. However, realizing that in an undergraduate curriculum there is not sufficient time to cover all these four aspects, we focus our attention to two pillars, which we consider to be the most important, namely dynamics and control. The reason for this viewpoint is that navigation (position information of the vehicle) tends to demand knowledge that is centered around sensors and avionics with a considerable electrical engineering flavor and fortunately the fundamental concepts behind navigation are somewhat embedded in the dynamics pillar (in the form of Euler angles and coordinate transformations, etc.) and thus we assume the navigation discussion is indirectly absorbed in the dynamics pillar. Similarly the guidance pillar revolves around the determination of the "desired behavior" for the vehicle and typically guidance commands act as reference inputs in a control system. Thus we design a control system assuming the desired or reference behavior is given by the guidance system (guidance computer). Thus in the end, we believe that emphasizing the dynamics part and the control part of the subject matter, with the understanding of the important role played by navigation and guidance in

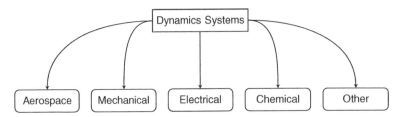

Figure 1 Dynamics systems in various fields.

Figure 2 Boeing 747-400. Credit: Delta.

the overall scheme of things, would do complete justice to the Undergraduate education objective and hence the book completely focuses on these two pillars, justifying the title of the book being "Flight Dynamics and Control of Aero and Space Vehicles".

Perspective on the Contents of the Book with its Unique Features

In Part I of the book, our focus is on obtaining mathematical models, namely the equations of motion, for aero and space flight vehicles. Since model based control systems analysis and design is the overarching objective of the subject matter to be grasped by the student, naturally the first step is to develop these mathematical models (i.e. getting the equations of motion) for use later in the control system analysis and design. The mathematical modeling for flight vehicles is carried out in this book by applying primarily Newton's laws of motion for a rigid body with specialization to aircraft motion as well as spacecraft motion. The Lagrangian energy approach to modeling

Figure 3 International Space Station. Courtesy of NASA.

is also briefly used in spacecraft dynamics to introduce the student to the many other, albeit more advanced methods of mathematical modeling of dynamic systems (such as Lagrange and Hamiltonian approaches) but it is felt that these advanced methods can be further pursued after mastering the Newtonian approach at the basic undergraduate level.

The unique features of Part I of the book include a thorough discussion of the conceptual similarities and differences in the state space representation of the equations of motion between atmospheric flight vehicles and space flight vehicles. This was possible because of the simultaneous treatment given to these two types of vehicles in this book in a unified framework.

In Part II of the book, our focus is on gathering the mathematical tools needed to solve and analyze the mathematical models (the equations of motion) developed in Part I. In accordance with the chronological progress of the subject matter, in this part we learn the methods to solve the linear, constant coefficient ordinary differential equations by the Laplace transform method. This in turn leads to the concept of the transfer function in the frequency domain. Hence, this part emphasizes the control systems analysis and design from the transfer function based frequency domain viewpoint. Accordingly, application to aircraft and spacecraft dynamics and control from frequency domain, transfer function based methods are covered and emphasized in this part.

The unique features of Part II of the book include the conceptual treatment of the systems level classical control design methods to aircraft control applications with heightened exposure to automatic landing control systems (within longitudinal aircraft dynamics) and a conceptual discussion on steady coordinated turn control methods (within lateral/directional aircraft dynamics). In addition, the classical control theory application to spacecraft includes an application problem that is non-routine, with specific focus on the control design for a specific satellite, with significant similarity to the "Aryabhata" satellite, which happens to be the first satellite built by the Indian Space Research Organization, with which this author had interaction with in his graduate student days. This is deemed to be of good educational value to the undergraduate students as they grasp all the control design steps and appreciate the importance of all the theory they have learnt so far.

In Part III of the book, our focus continues to be on developing the ability to solve and analyze the mathematical models (the equations of motion) developed in Part I, from a different point of view. In accordance with the chronological progress of the subject matter, in this part, we gather and learn other methods to solve the linear, constant coefficient ordinary differential equations, directly in the time domain using the state space system description. This in turn leads to the use of many matrix theory related methods such as the state transition matrix and the evolution of the state and output variables trajectories as a function of time. Topics such as controllability, observability, and dynamic stability via eigenvalues become very relevant and form the bulk of the content in this part. Control systems analysis and design via time domain, state space based methods, such as pole placement controllers and linear quadratic regulation (LQR) based controllers are covered along with their applications to aircraft and spacecraft flight control systems.

The unique features of Part III of the book include a thorough treatment on the necessary and sufficient conditions for the Hurwitz stability of a real matrix, which happens to be the plant matrix for a linear state space system. This author believes that this is the first time in a textbook that the Fuller matrices and the resulting conditions for Hurwitz stability of a real matrix are highlighted along with a discussion of the interrelationship between the Routh–Hurwitz criterion, Fuller's criterion and the popular Lyapunov matrix equation criterion. In addition, in the popular LQR optimal control method, the importance of getting the "trade-off" curve between the state regulation cost and control regulation cost is given heightened exposure. Also, the determination of multiple control gains (via Brogan's method) that all place the closed loop system eigenvalues at the same given "desired" locations is illustrated with examples. In this connection, the inadequacy of the currently existing MATLAB routine out-putting *only a single gain* as part of the solution to this problem formulation is pointed out. A few other inadequacies of currently existing MATLAB routines in a few other problem formulations are also pointed out, with the hope that in the future MATLAB will expand its algorithms to alleviate these inadequacies for the benefit of future research in state space based control methods. Finally, the application of the LQR design method in the spacecraft applications presented in this book is deemed unique, and is not discussed in other spacecraft control books as it involves a rewarding, insightful application to the satellite formation flying problem. Another unique feature is the introduction of concepts such as "strong stability" in observer based feedback control systems and "spillover instabilities" in two model (a higher order evaluation model being driven by lower order control design model) control theory.

In Part IV of the book, in addition to summarizing some overview type coverage on aircraft dynamics and control from an industry viewpoint as well as on satellite dynamics and control in a tutorial fashion, our focus extends to covering some other related fight vehicles such as helicopters, quadcopters (representing rotorcraft flight vehicles), missiles and hypersonic flight vehicles in the form of "mini" tutorials.

The unique features of Part IV of the book include bringing awareness to the conceptual similarities and differences in the dynamics and control issues related to these other types of flight vehicle. This coverage is thus deemed to offer "completeness" to the overall subject of dynamics and control of aero and space flight vehicles, making the student put the entire contents of the other three parts (I, II, and III) of the book in proper "perspective".

Part I

Flight Vehicle Dynamics

Roadmap to Part I

"Absence of Evidence is not Evidence of Absence"

– Carl Sagan *Astronomer*

Part I covers the dynamics (mathematical modeling of equations of motion) of both aero as well as space vehicles. This part in turn consists of chapters 1 to 7. Chapter 1 treats the subject in a generic way and introduces the basic concepts needed to be able to write down the elaborate equations of motion in later chapters. Chapter 2 then develops the detailed development of the equations of motion, first for a general rigid body in three dimensional space. Then one section specializes these general equations to the specific case of aircraft with the relevant assumptions specific to the aircraft situation.

Then in the next section, they are specialized to the case of spacecraft (satellites) with the assumptions and approaches followed in the spacecraft situation. Along the way, the student is made to clearly understand the similarities as well as differences in the developed equations of motion for these two distinct situations. Then Chapter 3 presents a generic viewpoint on how to linearize a set of nonlinear differential equations about a given operating point and then briefly presents the generic concepts of stability of a dynamic system for the general case and then thoroughly covers the stability conditions for a linear, time invariant system. Finally a thorough response analysis of a simple linear second order system is presented. Chapter 4 then focuses on aircraft static stability and control issues. Within that context, all the concepts of static stability and control for an aircraft in a trimmed condition considering both the longitudinal case as well as the lateral/directional case are presented. Then Chapter 5 considers the dynamic stability concept and develops the linearized perturbed motion equations needed later in the control design process, along with the longitudinal modes of approximation as well as the lateral/directional modes of approximations. Chapter 6 deals with spacecraft passive control and stabilization issues. It covers various passive and semi-active control methods for satellite attitude stabilization and control such as spin, dual-spin, gravity gradient stabilization, etc. Finally Chapter 7 covers spacecraft dynamic stability and control. It carries out the analysis of dynamic behavior for an axi-symmetric satellite in uncontrolled, torque free situation along with the linearized models for use in the three axis control design later, along with a discussion on the modeling of various disturbance torques in the space environment.

1

An Overview of the Fundamental Concepts of Modeling of a Dynamic System

1.1 Chapter Highlights

In this chapter, we discuss the fundamental concepts and steps needed to carry out the analysis of a generic dynamic system. The importance of making realistic, engineering judgment based approximations to develop tractable, simple mathematical models for the specific system at hand is highlighted. Then two very important basic mathematical tools, namely the time rate of change of vectors in absolute and relative frames as well as the role of coordinate transformations in deriving the equations of motion are explained.

1.2 Stages of a Dynamic System Investigation and Approximations

Dynamic systems, as mentioned earlier, are systems that change as a function of time. Our objective at this early stage will be to understand and predict the dynamic behavior of a given system. Then, in the later parts of this book, we expand our focus to controlling and improving the behavior of dynamic systems to the way we desire. Typically, there are few basic stages one needs to follow to make this investigation more systematic and logical. We now outline those fundamental stages of investigation. Clearly this list of stages is not exhaustive but these are the stages we focus on in this book.

Stage 1. Specify the system to be studied and imagine a physical model whose behavior will match sufficiently closely the behavior of the actual system.

Stage 2. Derive a mathematical model to represent the physical model, i.e. write the equations of motion (usually represented by differential equations).

Stage 3. Study and analyze the dynamical behavior by solving these differential equations.

Stage 4. Make design decisions, that is, design a control system so that the system behaves the way we want.

More advanced stages would include a thorough simulation of the fully designed system (both in a computational framework as well as in an experimental setup) and then possibly building a scalable prototype. This book emphasizes only the basic four stages mentioned above.

Flight Dynamics and Control of Aero and Space Vehicles, First Edition. Rama K. Yedavalli.
© 2020 John Wiley & Sons Ltd. Published 2020 by John Wiley & Sons Ltd.

A more detailed and enlightening discussion of these stages is given in classical textbooks such as [4]. An aspect that deserves special mention in following the above orderly steps is the need for diligence in deriving tractable, simple mathematical models that capture the essential features of the system under consideration without undue, and unnecessary details. This is accomplished by making reasonably valid approximations supported by common sense and engineering judgment. Let us elaborate on this. These engineering approximations can be categorized as follows.

1. **Neglecting small effects.** Neglecting small effects results in considerable savings in the number of variables in a differential equation as well as the complexity of the mathematical model. For example, when we attempt to write down the mathematical model of an aircraft in motion, we can neglect the effect of solar radiation pressure on the wings as that effect is much smaller and thus negligible compared to the major aerodynamic forces acting on the aircraft. Of course, what is a small effect is a relative concept, because the torque generated by same solar radiation pressure could be a significant effect when we are writing down the equations of motion for a satellite in a geosynchronous orbit. Thus extreme care needs to be exercised in making this small effect approximation. The underlying thought process is that there is nothing wrong in making approximations as long as they are reasonably valid for the specific system under consideration.

2. **Assuming the environment surrounding the system is unaltered by the system (independent environment).** This assumption is predicated by the need for us to be more precise about what constitutes the system under consideration. For example, when we model the automobile dynamics on a paved road, it is fairly reasonable to assume that we can study the automobile dynamics without worrying about modeling of the paved road. However, if all terrain vehicle dynamics is being modeled, it is only reasonable to include the characteristics of the terrain and its parameters (such as icy roads, soil properties of the road, etc.) as part of the system being considered. Thus being cognizant of what constitutes a system and its surroundings helps us to simplify the number and complexity of the equations, just as in the case of the small effects approximation.

3. **Replacing distributed characteristics by lumped characteristics.** This is a very important assumption one needs to be rigorous about. This is due to the fact that this assumption, when reasonably valid, helps not only lessen the complexity of the equations but even leads to simplicity in the method employed to solve the resulting equations. For example, for any body, the mass is distributed along the area (or volume) of the body with possible mass density variations along the body. So if we account for these, strictly speaking, our system becomes a distributed parameter system and we may end up with partial differential equations, which are obviously difficult to solve. However, if a body is fairly rigid with constant mass density, a lumped parameter approximation is reasonable, and in this case we can get away with ordinary differential equations, which are relatively easier to solve. Thus unless it is really necessary, there is no need to resort to distributed parameter modeling of structures (bodies). When the structure is highly flexible, it may be necessary to resort to distributed parameter modeling because only then can one account for or capture the structural flexibility and locations of actuators and sensor effects in the control system design. For this reason, in the early 1980s, the dynamics and control of large

flexible space structures became an active topic of research [1, 5–8] resulting in new insights such as control and observation spillover instabilities. Similarly, by the same token, neglecting structural flexibility and the inherent interaction between aeroelasticity and control system components resulted in the failure of flexible boom satellites such as Voyager, which were modeled as rigid bodies.

4. **Assuming simple linear cause and effect relationships between physical variables.** This assumption is also one that has far reaching consequences in simplifying the nature of resulting differential equations and the methodology needed to solve those equations. For tractability and gaining physical insight, this is a good approximation to start with. For example, when we assume a linear relationship between various physical variables, it is likely that the resulting differential equations turn out to be linear differential equations. In this connection, it is important for the student to have complete clarity in deciding when an ordinary differential equation is a linear and when it is a nonlinear equation, as there seems to be still some confusion in the student's mind. For example, the following differential equation

$$\frac{d^3x}{dt^3} + 4\frac{d^2x}{dt^2} + t^3\frac{dx}{dt} + 3x = \sin 5t \tag{1.1}$$

is a linear differential equation. Note that one of the coefficients (namely t^3) is nonlinear in nature but that nonlinearity is in the independent variable t. Thus this equation is still a linear differential equation, albeit with time varying coefficients. Clearly, by this logic

$$\frac{d^3x}{dt^3} + 4\frac{d^2x}{dt^2} + x^2\frac{dx}{dt} + 3x = \sin 5t \tag{1.2}$$

is a nonlinear differential equation because this time the coefficient x^2 is nonlinear in the dependent variable x, making it a nonlinear differential equation, which is obviously more difficult to solve.

Thus this linearity approximation makes it easy to come up with a linear differential equation (possibly with time varying coefficients), which are relatively easier to solve. In fact, it is true that even among linear differential equations, the ones with time varying coefficients (namely linear, time varying systems) are relatively more difficult to solve than those with constant coefficients [namely linear time invariant (LTI) systems]. This leads us to another following approximation that simplifies the analysis even further.

5. **Assume constant (time invariant) parameters.** As mentioned above, this approximation leads us to linear constant coefficient ordinary differential equations, which are the simplest to solve. In fact, in the entire rest of this book, we focus on LTI systems, which allows us to take advantage of the vast body of solution techniques, giving rise to amazingly rich theory for designing automatic control systems for linear dynamic systems.

6. **Neglecting uncertainty and noise.** Obviously, assuming fixed, known nominal values for all the parameters encountered in the system equations allows us to obtain a complete solution to that particular nominal system, which forms the basis for the content of this book. Assuming uncertainty (or perturbations) in the parameters and the solution techniques that account for the accommodation of this uncertainty in the analysis and design stage forms the new subject area of robust control of linear dynamic systems, which is beyond the scope this book. Interested students in this

Table 1.1 Approximations and their mathematical simplifications.

Approximation	Mathematical simplification
Neglect small effects	Reduces number and complexity of differential equation
Independent environment	Reduces number and complexity of differential equation
Replacing distributed characteristic by lumped characteristic	Leads to ordinary differential equations instead of partial differential equations
Assume linear relationships	Make equations linear
Constant parameters	Leads to constant coefficient in differential equations
Neglect uncertainty and noise	Avoids statistical treatment

research can consult various books written on this subject [2, 3, 9] including the one by this author [10].

Similarly, incorporating noise into the modeling process leads to other branches of control theory. For example, modeling noise as a stochastic process leads to stochastic control theory, which in itself is vast, based on probability theory concepts. Other areas like fuzzy control and adaptive control also address the effect of including the effects of noise and other disturbances.

In summary, the mathematical ramifications of these simplifications are shown in Table 1.1.

In this chapter, our aim is to derive the equations of motion for a body in three dimensional space. For this, we need to first gather a few fundamental concepts; just like when building a component or a device, we need few tools like a hammer and a screwdriver. In what follows, we attempt to gather those basic tools (concepts).

1.3 Concepts Needed to Derive Equations of Motion

There are two basic concepts we discuss here. One is the ability to derive an expression for the time rate of change of a vector, based on the coordinate frame in which that vector is expressed, and the other is the ability to make coordinate transformations so that one can easily transform the equations from one coordinate frame to another coordinate frame. These two concepts together allow us to derive the entire set of equations of motion for a body that undergoes translational as well as rotational motion. Since a body's motion is always expressed relative to a given coordinate frame, we need to introduce the notion of using different coordinate frames for a given purpose and then be able to transform from one frame to another. While in the most general framework we can visualize various frames and routinely go from one frame to another frame, it is important to start with some basic concepts related to coordinate frames and their transformations. In that spirit, we first consider the most basic situation of having two frames. One is the inertial coordinate frame, in which the frame stays fixed with neither (rectilinear) translation nor any rotation. Thus the inertial coordinate frame is a fixed coordinate frame. The other coordinate frame we need to introduce is the body (or vehicle) fixed coordinate frame. Since this frame is fixed to the body, as the body undergoes translational and rotational motion, so does this body fixed frame. In that sense, the body fixed frame is a moving frame (in contrast to the inertial frame, which is a stationary frame). It is important to recall here that Newton's laws of motion, which are basic laws of motion we invoke for writing down the equations of motion for a body in motion, are always stated with respect to the inertial coordinate frame. Thus, the time rate of change of a vector needs to be derived taking into account the coordinate frame in which the vector is expressed. This leads us to the first fundamental concept, explained below.

1.3.1 Time Rate of Change Vectors in a Moving (Body Fixed) Frame and a Stationary (Non-rotating, Inertial) Frame

When considering a body in motion, first attach a body fixed reference frame within the body. Quantities in the body fixed frame are indicated using lower case characters *xyz*. The term *xyz* frame or *ijk* frame may be used interchangeably with body frame. We also consider an inertial frame, which we fix (perhaps at the center of the earth, or at a point on the surface of the earth). The inertial (stationary)) frame is indicated using the upper case characters *XYZ*, and the term *XYZ* frame or *IJK* frame may be used interchangeably with "inertial frame. Thus we have a moving frame and an inertial reference frame (with respect to which Newton's laws are stated).

Sign convention for reference frames. We assume all the coordinate frame triads (with three axes) to follow the right hand rule. Once the first axis (X axis) is designated, all the other axes follow the right hand rule. Thus if north is taken as the X axis, then west (to its left) is taken as the Y axis and accordingly the Z axis will be pointing upwards and the counterclockwise rotation is taken as the positive angle of rotation. We designate this triad as the NWU (north, west, up) frame. By the same token, if we draw the Y axis to the right of the X axis, then the Z axis is downwards, and we label this as the NED (north, east, down) frame and in this the clockwise rotation is taken as the positive rotation. When we draw the triad on paper, in the NWU frame, the Z axis is pointing away from the page while in the NED frame, the Z axis is pointing into the page. It is better to be very clear from the beginning that these guidelines are being followed so that there is no confusion in visualizing the frames and being consistent with the sign convention. Adherence to this sign convention results in uniformity and uniqueness (whenever it exists) in the solutions to posed problems.

Any vector can then be expressed in terms of either the moving frame components or the inertial frame components. For example, consider an arbitrary vector $\underline{A}$ expressed in moving frame coordinates whose unit vectors are $\underline{i}, \underline{j}, \underline{k}$ as:

$$\underline{A} = A_x \underline{i} + A_y \underline{j} + A_z \underline{k} \tag{1.3}$$

where A_x, A_y and A_z are the components of $\underline{A}$ expressed in the moving *xyz* frame. By the same token, it is also possible to express $\underline{A}$ in terms of the inertial frame components, as:

$$\underline{A} = A_X \underline{I} + A_Y \underline{J} + A_Z \underline{K} \tag{1.4}$$

where A_X, A_Y and A_Z are the components in the XYZ inertial frame, and $\underline{I}, \underline{J}$, and $\underline{K}$ are the unit vectors in the XYZ frame.

Then, if we are interested in the time rate of change of vector $\underline{A}$, in the inertial frame coordinates, we have:

$$\frac{d}{dt}\underline{A} = \frac{d}{dt}A_X\underline{I} + \frac{d}{dt}A_Y\underline{J} + \frac{d}{dt}A_Z\underline{K} + A_x\overset{0}{\cancel{\frac{d\underline{I}}{dt}}} + A_y\overset{0}{\cancel{\frac{d\underline{J}}{dt}}} + A_z\overset{0}{\cancel{\frac{d\underline{K}}{dt}}}$$

$$= \dot{A}_X\underline{I} + \dot{A}_Y\underline{J} + \dot{A}_Z\underline{K}. \tag{1.5}$$

The unit vectors $\underline{I}, \underline{J}, \underline{K}$ in Equation 1.5 by definition have a fixed magnitude. It is important to note that these particular unit vectors, by virtue of our definition of the inertial frame, also have a fixed direction. Therefore they are constants with respect to time; their derivative with respect to time is zero.

Recall we defined the body frame with respect to our body (an aircraft or spacecraft); when the orientation of the body changes, so does the direction of the unit vectors $\underline{i}, \underline{j}, \underline{k}$. Therefore, when $\underline{A}$ is expressed in the body frame coordinates we have to take into consideration the change in the directions of the unit vectors $\underline{i}, \underline{j}, \underline{k}$. Mathematically, this means we have to consider the translation and rotation of the body frame with respect to the inertial frame.

If the body is in pure translation with respect to the inertial frame (i.e. the body frame is not rotating with respect to the inertial frame), then:

$$\frac{d}{dt}\underline{A} = \dot{A}_X\underline{I} + \dot{A}_Y\underline{J} + \dot{A}_Z\underline{K}$$

$$= \dot{A}_x\underline{i} + \dot{A}_y\underline{j} + \dot{A}_z\underline{k} + A_x\overset{0}{\cancel{\frac{d\underline{i}}{dt}}} + A_y\overset{0}{\cancel{\frac{d\underline{j}}{dt}}} + A_z\overset{0}{\cancel{\frac{d\underline{k}}{dt}}}$$

$$= \dot{A}_x\underline{i} + \dot{A}_y\underline{j} + \dot{A}_z\underline{k}. \tag{1.6}$$

The body frame maintains the same direction throughout the motion and thus the time rate of change of unit vectors $(\underline{i}, \underline{j}, \underline{k})$ is zero. However, this motion is a very restricted motion and does not represent the most common general motion, in which the body undergoes general translational as well as rotational motion.

In the most general case of the body undergoing both translational as well as rotational motion, where the moving frame rotates with an angular velocity ω, expressed in the moving frame components, then the time rate of change of $\underline{A}$ (expressed in body frame components) is given by:

$$\frac{d}{dt}\underline{A} = \dot{A}_x\underline{i} + \dot{A}_y\underline{j} + \dot{A}_z\underline{k} + A_x\frac{d\underline{i}}{dt} + A_y\frac{d\underline{j}}{dt} + A_z\frac{d\underline{k}}{dt}. \tag{1.7}$$

The above can in turn be written as,

$$\left(\frac{d}{dt}\underline{A}\right)_{XYZ} = \left(\frac{d}{dt}\underline{A}\right)_{xyz} + \underline{\omega} \times \underline{A}. \tag{1.8}$$

Here, $\underline{\omega}$ is expressed as,

$$\underline{\omega}(t) = \omega_x(t)\underline{i} + \omega_y(t)\underline{j} + \omega_z(t)\underline{k} \tag{1.9}$$

and $(\frac{dA}{dt})_{xyz}$ as,

$$\left(\frac{dA}{dt}\right)_{xyz} = \dot{A}_x(t)\underline{i} + \dot{A}_y(t)\underline{j} + \dot{A}_z(t)\underline{k}. \tag{1.10}$$

This is an extremely important relationship that gets applied again and again in the derivation of equations of motion for a body in a three dimensional space. In some literature, this equation is referred to as Charle's theorem and we follow the same nomenclature.

1.3.2 Coordinate Transformations

Another important concept needed is the concept of coordinate transformations through the rotation matrices. Whenever we have two reference frames, obviously the orientation of one reference frame with respect to the other can be related by a series of rotations through appropriate angles. For example, consider the simple situation below where we consider two reference frames differing in orientation by an arbitrary angle, say θ. For consistency in notation, as explained in the sign convention discussion before, we assume the positive angles are interpreted by following the right hand rule convention for all the triads of the coordinate frames. In turn, we follow the NED frame convention, wherein all positive angles are in the clockwise direction.

Clearly the components x and y are related to components X and Y as in Figure 1.1 (note $Z = z$, and both point into the plane of the page):

$$X = x\cos\theta - y\sin\theta$$
$$Y = y\cos\theta + x\sin\theta.$$

In other words, we can relate the (X, Y, Z) components (i.e. the XYZ frame) to the (x, y, z) components (i.e. the xyz frame) by the following matrix form:

$$\begin{bmatrix} X \\ Y \\ Z \end{bmatrix} = \begin{bmatrix} \cos\theta & -\sin\theta & 0 \\ \sin\theta & \cos\theta & 0 \\ 0 & 0 & 1 \end{bmatrix} \begin{bmatrix} x \\ y \\ z \end{bmatrix} \tag{1.11}$$

$$\begin{bmatrix} X \\ Y \\ Z \end{bmatrix} = \begin{bmatrix} R_\theta \end{bmatrix} \begin{bmatrix} x \\ y \\ z \end{bmatrix}. \tag{1.12}$$

Figure 1.1 A simple coordinate transformation.

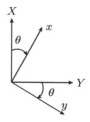

Obviously if the *xyz* frame components are known and rotation angle θ is known, we can determine the *XYZ* components and vice versa, i.e.:

$$\begin{bmatrix} x \\ y \\ z \end{bmatrix} = \begin{bmatrix} \text{Rot } \theta \end{bmatrix}^{-1} \begin{bmatrix} X \\ Y \\ Z \end{bmatrix}. \tag{1.13}$$

The rotation matrix (known as R_θ) is also called the direction cosine matrix.

You can observe that the above direction cosine matrix (DCM) is basically relating the unit vectors in one frame to the unit vectors in another frame, i.e.

$$\begin{bmatrix} i \\ j \\ k \end{bmatrix} = \begin{bmatrix} \text{DCM} \end{bmatrix} \begin{bmatrix} I \\ J \\ K \end{bmatrix}. \tag{1.14}$$

In general, if we have two reference frames with unit vectors $\underline{I}_1, \underline{J}_1, \underline{K}_1$ and $\underline{I}_2, \underline{J}_2, \underline{K}_2$, then it can be seen that we can write

$$\underline{I}_1 = \cos\alpha_{11}\underline{I}_2 + \cos\alpha_{12}\underline{J}_2 + \cos\alpha_{13}\underline{K}_2 \tag{1.15}$$

$$\underline{J}_1 = \cos\alpha_{21}\underline{I}_2 + \cos\alpha_{22}\underline{J}_2 + \cos\alpha_{23}\underline{K}_2 \tag{1.16}$$

$$\underline{K}_1 = \cos\alpha_{31}\underline{I}_2 + \cos\alpha_{32}\underline{J}_2 + \cos\alpha_{33}\underline{K}_2. \tag{1.17}$$

Thus

$$\begin{bmatrix} \underline{I}_1 \\ \underline{J}_1 \\ \underline{K}_1 \end{bmatrix} = \begin{bmatrix} \cos\alpha_{11} & \cos\alpha_{12} & \cos\alpha_{13} \\ \cos\alpha_{21} & \cos\alpha_{22} & \cos\alpha_{23} \\ \cos\alpha_{31} & \cos\alpha_{32} & \cos\alpha_{33} \end{bmatrix} \begin{bmatrix} \underline{I}_2 \\ \underline{J}_2 \\ \underline{K}_2 \end{bmatrix}. \tag{1.18}$$

Now you see why this matrix is called the direction cosine matrix. Obviously, knowing the direction cosine matrix elements we can also visualize the angle(s) between those two reference frames by taking inverse cosine angles corresponding to the entries of the direction cosine matrix.

A pictorial representation of this is depicted in Figure 1.2.

It is interesting and important to know that the direction cosine matrix has the following important features:

- It is an orthogonal matrix, meaning $[\text{DCM}]^{-1} = [\text{DCM}]^T$.
- All elements of this matrix have numerical values ≤ 1. Thus the length of each of its column vector is 1.
- The three column vectors of this matrix are mutually orthogonal. Note that two vectors x and y are orthogonal to each other when $x^T y = y^T x = 0$.
- The determinant of this matrix is ± 1. If the reference frame vectors follow the right hand rule, the determinant is $+1$.
- There is only one real eigenvalue for this matrix, which happens to be equal to ± 1, again $+1$ for right handed frames.

In the above simple illustration of the coordinate transformation concept we showed two frames, differing from each other by only one angular rotation, and how they can be related to each other. However, for a general body in motion in a three dimensional space, it is shown that a minimum of three angular rotations are needed to go from

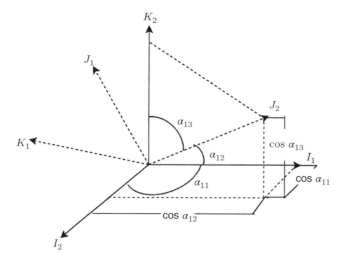

Figure 1.2 Direction cosines.

the stationary inertial reference frame to any arbitrary orientation of the moving body frame. While there are myriad ways of going from the XYZ frame to an arbitrary xyz frame, we follow a particular orderly sequence of three rotations labeled as Euler angles. Again, carefully following the clockwise as the positive angle rotation, the orderly sequence of rotations for bringing the XYZ frame to an arbitrary orientation of an xyz frame is as follows:

1. Rotation Ψ. *First rotation*: rotate the XYZ frame about the Z axis by an angle Ψ to go to an intermediate frame $x_1 y_1 z_1$. Thus $Z = z_1$ and the relationship between the XYZ frame and the $x_1 y_1 z_1$ frame is given in Figure 1.3.

$$\begin{bmatrix} X \\ Y \\ Z \end{bmatrix} = \begin{bmatrix} \cos \Psi & -\sin \Psi & 0 \\ \sin \Psi & \cos \Psi & 0 \\ 0 & 0 & 1 \end{bmatrix} \begin{bmatrix} x_1 \\ y_1 \\ z_1 \end{bmatrix} \tag{1.19}$$

$$\begin{bmatrix} X \\ Y \\ Z \end{bmatrix} = \begin{bmatrix} \text{Rot } \Psi \end{bmatrix} \begin{bmatrix} x_1 \\ y_1 \\ z_1 \end{bmatrix} \tag{1.20}$$

Figure 1.3 First rotation Ψ where the (Z, z_1) axes point into the page.

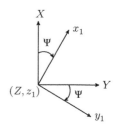

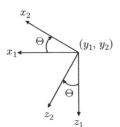

Figure 1.4 Second rotation Θ where the (y_1, y_2) axes point into the page.

2. Rotation Θ. *Second rotation*: rotate the $x_1 y_1 z_1$ frame about the y_1 axis by an angle Θ to go to an intermediate frame $x_2 y_2 z_2$. Thus $y_2 = y_1$ and we have (see Figure 1.4),

$$
\begin{bmatrix} x_1 \\ y_1 \\ z_1 \end{bmatrix} = \begin{bmatrix} \cos\Theta & 0 & \sin\Theta \\ 0 & 1 & 0 \\ -\sin\Theta & 0 & \cos\Theta \end{bmatrix} \begin{bmatrix} x_2 \\ y_2 \\ z_2 \end{bmatrix}
\tag{1.21}
$$

$$
\begin{bmatrix} x_1 \\ y_1 \\ z_1 \end{bmatrix} = \begin{bmatrix} \text{Rot } \Theta \end{bmatrix} \begin{bmatrix} x_2 \\ y_2 \\ z_2 \end{bmatrix}
\tag{1.22}
$$

3. Rotation Φ. *Third and final rotation*: rotate the $x_2 y_2 z_2$ frame about the x_2 axis and by angle Φ to go to the xyz frame. Thus we have (see Figure 1.5),

$$
\begin{bmatrix} x_2 \\ y_2 \\ z_2 \end{bmatrix} = \begin{bmatrix} 1 & 0 & 0 \\ 0 & \cos\Phi & -\sin\Phi \\ 0 & \sin\Phi & \cos\Phi \end{bmatrix} \begin{bmatrix} x \\ y \\ z \end{bmatrix}
\tag{1.23}
$$

$$
\begin{bmatrix} x_2 \\ y_2 \\ z_2 \end{bmatrix} = \begin{bmatrix} \text{Rot } \Phi \end{bmatrix} \begin{bmatrix} x \\ y \\ z \end{bmatrix}.
\tag{1.24}
$$

Thus finally the relationship between the XYZ frame and the moving frame xyz is given by:

$$
\begin{bmatrix} X \\ Y \\ Z \end{bmatrix} = \begin{bmatrix} \text{Rot } \Psi \end{bmatrix} \begin{bmatrix} \text{Rot } \Theta \end{bmatrix} \begin{bmatrix} \text{Rot } \Phi \end{bmatrix} \begin{bmatrix} x \\ y \\ z \end{bmatrix}.
\tag{1.25}
$$

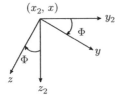

Figure 1.5 Third and final rotation Φ where the (x_2, x) axes point into the page.

Carrying out the above multiplication, we obtain

$$
\begin{bmatrix} X \\ Y \\ Z \end{bmatrix} = \begin{bmatrix} \cos\Theta\cos\Psi & \sin\Phi\sin\Theta\cos\Psi - \cos\Phi\sin\Psi & \cos\Phi\sin\Theta\cos\Psi + \sin\Phi\sin\Psi \\ \cos\Theta\sin\Psi & \sin\Phi\sin\Theta\sin\Psi + \cos\Phi\cos\Psi & \cos\Phi\sin\Theta\sin\Psi - \sin\Phi\cos\Psi \\ -\sin\Theta & \sin\Phi\cos\Theta & \cos\Phi\cos\Theta \end{bmatrix}
$$
$$
\times \begin{bmatrix} x \\ y \\ z \end{bmatrix}.
$$

We denote this composite direction cosine matrix as the S matrix. Just as each rotation matrix is an orthogonal matrix, meaning its inverse is simply its transpose, this composite rotation matrix S is also an orthogonal matrix. Thus $S^{-1} = S^T$.

- It may be noted that this orderly sequence of rotations to go from the XYZ to the xyz frame is not unique. However, for consistency, we choose to follow this particular orderly sequence so that all of us will get the same answer when solving problems involving these coordinate transformations. It is also interesting to point out that, for sufficiently small angles, this orderly sequence does not matter.
- It can be seen that the above rotation matrices are essentially functions of trigonometric terms involving sines and cosines of angles. As such, there exists the possibility of singularities occurring in these matrices when the angles reach those values that make the the determinants of those matrices zero. However, to keep the concepts (at the undergraduate level) simple and straightforward without deviating too much from the main ideas, we assume that these angles are such that they keep the rotation matrices non-singular.
- Also, when we resolve a vector into components in two different coordinate frames, it is implicitly assumed that the origin of the two reference frames are coincident. In other words, we translate (without any rotation) one coordinate frame to the other until the origins of both frames coincide.

We now conclude this chapter with an example through which we can illustrate the application of the above two fundamental concepts. The example we consider helps us to understand why it is beneficial to launch a satellite from a launching station located at the equator.

1.4 Illustrative Example

A vehicle is at rest on the surface of the Earth at latitude λ and longitude μ. Find the absolute velocity and acceleration of the body (i.e. velocity and acceleration of the body with respect to the initial frame S, which is as shown in Figure 1.6). The Earth's rotation rate is denoted by ω_e.

In Figure 1.6, we included a few coordinate frames, S, E, and V. We take S frame to be the inertial (fixed) reference frame. The E frame is the one fixed at a point on the surface of the Earth, where that point is at a latitude λ and a longitude μ. The V frame is the vehicle fixed frame, fixed at a point within the vehicle. For this problem at hand, we do not require this V frame. As usual, in each of these frames, we define the X, Y, Z vectors

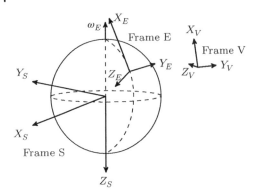

Figure 1.6 Motion of a body on the surface of the Earth.

using the right hand rule (positive) sign convention, giving rise to the NED frame (where north is the positive X direction, East is the positive Y direction and down is the positive Z direction). Clearly, the vehicle's angular velocity with respect to the S frame is given by $\omega_{E,S} = -\omega_e k_S$ (because the body is at rest, this is the only angular velocity component we have). Notice the negative sign in this expression, which is important.

We now need to establish the relationship between the S frame and the E frame. In this example, since the E frame has a specific orientation, it can be seen that we can start from the S frame and reach the E frame orientation by only two angular rotations. In other words, among the three Euler angle rotations, we need only two Euler angle rotations, so that the X_S, Y_S, Z_S frame can reach the X_E, Y_E, Z_E frame in two angular rotations, namely the longitude angle μ and the latitude angle λ. It is clear that the first Euler angle rotation, namely Ψ is same as the longitude angle μ. So, we can simply replace Ψ with the angle μ in the composite rotation matrix. Then, we observe that the other two Euler angle rotations Θ and Φ should satisfy the constraint that

$$\Theta + \Phi = 90 - \lambda \tag{1.26}$$

because λ varies from zero (Equator) to 90 degrees. This constraint can be satisfied with two options, namely either (i) $\Theta = 90 - \lambda$ and $\Phi = 0$ or (ii) $\Theta = 0$ and $\Phi = 90 - \lambda$. Thus we can enforce these options into the composite rotation matrix S to get the final relationship between the X_S, Y_S, Z_S frame and the X_E, Y_E, Z_E frame in terms of the two angles μ and λ.

Let us select the first option. So we substitute $(90 - \lambda)$ for Θ and select $\Phi = 0$ in the composite rotation matrix S. Noting that $\cos(90 - \lambda) = \sin \lambda$ and $\sin(90 - \lambda) = \cos \lambda$ we then get the relationship between X_S, Y_S, Z_S and X_E, Y_E, Z_E frames as given by

$$\begin{bmatrix} X_S \\ Y_S \\ Z_S \end{bmatrix} = \begin{bmatrix} \cos \mu \sin \lambda & -\sin \mu & \cos \mu \cos \lambda \\ \sin \mu \sin \lambda & \cos \mu & \sin \mu \cos \lambda \\ -\cos \lambda & 0 & \sin \lambda \end{bmatrix} \begin{bmatrix} X_E \\ Y_E \\ Z_E \end{bmatrix}.$$

Another way to arrive at this conclusion is as follows: *First rotation*: rotate the X_S, Y_S, Z_S axis by the longitude angle μ about the Z_S axis to reach an intermediate frame X_1, Y_1, Z_1. So from Figure 1.7

$$X_S = X_1 \cos \mu - Y_1 \sin \mu$$
$$Y_S = X_1 \sin \mu + Y_1 \cos \mu$$
$$Z_S = Z_1.$$

Figure 1.7 First rotation μ where the (Z_S, Z_1) axes point into the page.

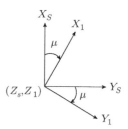

Thus,

$$\begin{bmatrix} X_S \\ Y_S \\ Z_S \end{bmatrix} = \begin{bmatrix} \cos\mu & -\sin\mu & 0 \\ \sin\mu & \cos\mu & 0 \\ 0 & 0 & 1 \end{bmatrix} \begin{bmatrix} X_1 \\ Y_1 \\ Z_1 \end{bmatrix}.$$

Now we do the second rotation with the Euler angle Θ.

Second rotation: rotate X_1, Y_1, Z_1 about Y_1 axis by the second Euler angle Θ to reach X_E, Y_E, Z_E.

Thus the relationship between X_1, Y_1, Z_1 and X_E, Y_E, Z_E is given by

$$\begin{bmatrix} X_1 \\ Y_1 \\ Z_1 \end{bmatrix} = \begin{bmatrix} \cos\Theta & 0 & \sin\Theta \\ 0 & 1 & 0 \\ -\sin\Theta & 0 & \cos\Theta \end{bmatrix} \begin{bmatrix} X_E \\ Y_E \\ Z_E \end{bmatrix}.$$

However from the definition of the latitude angle λ, we observe that Θ is nothing but $(90 - \lambda)$. Hence, we substitute $(90 - \lambda)$ for Θ. Noting that $\cos(90 - \lambda) = \sin\lambda$ and $\sin(90 - \lambda) = \cos\lambda$ we then get the relationship between X_1, Y_1, Z_1 and X_E, Y_E, Z_E as given by (see Figure 1.8)

$$\begin{bmatrix} X_1 \\ Y_1 \\ Z_1 \end{bmatrix} = \begin{bmatrix} \sin\lambda & 0 & \cos\lambda \\ 0 & 1 & 0 \\ -\cos\lambda & 0 & \sin\lambda \end{bmatrix} \begin{bmatrix} X_E \\ Y_E \\ Z_E \end{bmatrix}.$$

Then the final relationship between X_S, Y_S, Z_S and X_E, Y_E, Z_E is given by

$$\begin{bmatrix} X_S \\ Y_S \\ Z_S \end{bmatrix} = \begin{bmatrix} \cos\mu & -\sin\mu & 0 \\ \sin\mu & \cos\mu & 0 \\ 0 & 0 & 1 \end{bmatrix} \begin{bmatrix} \sin\lambda & 0 & \cos\lambda \\ 0 & 1 & 0 \\ -\cos\lambda & 0 & \sin\lambda \end{bmatrix} \begin{bmatrix} X_E \\ Y_E \\ Z_E \end{bmatrix}.$$

So,

$$\begin{bmatrix} X_S \\ Y_S \\ Z_S \end{bmatrix} = \begin{bmatrix} \cos\mu\sin\lambda & -\sin\mu & \cos\mu\cos\lambda \\ \sin\mu\sin\lambda & \cos\mu & \sin\mu\cos\lambda \\ -\cos\lambda & 0 & \sin\lambda \end{bmatrix} \begin{bmatrix} X_E \\ Y_E \\ Z_E \end{bmatrix}.$$

Figure 1.8 Second rotation θ where the (Y_1, Y_E) axes point into the page.

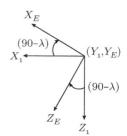

which is exactly the same as the one we obtained before. What is important to realize from the above exercises is that the Euler angle Θ itself is not same as the latitude angle λ, because Θ is an angle measured from an intermediate axis frame, whereas the latitude angle is an angle measured from either the equitorial plane or from the polar plane. Thus, in general, care needs to be taken in getting the relationship between two frames with constraints as a special case of the general three Euler angle orderly rotation sequence. It is left to the reader to check that we obtain the same relationship as the one we derived using option two as well, where we could substitute $\Theta = 0$ and make $\Phi = 90 - \lambda$ in the composite rotation matrix S.

Finally, at any rate, from the above relationship between the X_S, Y_S, Z_S and X_E, Y_E, Z_E frames, we note that

$$Z_S = -\cos \lambda X_E + \sin \lambda Z_E$$

i.e.

$$\underline{k}_S = -\cos \lambda \underline{i}_E + \sin \lambda \underline{k}_E.$$

Now the angular velocity $\omega_{E,S}$ can be expressed in terms of the components of the E frame as

$$
\begin{aligned}
\underline{\omega}_{E,S} &= -\omega_e \underline{k}_S \\
&= -\omega_e(-\cos \lambda \underline{i}_E + \sin \lambda \underline{k}_E) \\
&= \omega_e \cos \lambda \underline{i}_E - \omega_e \sin \lambda \underline{k}_E.
\end{aligned}
\tag{1.27}
$$

Now, the absolute velocity of the vehicle is given by

$$\left(\frac{d\underline{r}}{dt}\right)_S = \left(\frac{d\underline{r}}{dt}\right)_E + \underline{\omega}_{E,S} \times \underline{r} \tag{1.28}$$

where $\underline{r}$ is the position vector of the vehicle, which can be expressed as $\underline{r} = -R_e \underline{k}_E$ where R_e is the radius of the Earth (≈ 4000 miles). Since R_e is constant, obviously $\left(\frac{d\underline{r}}{dt}\right)_E = 0$. Therefore, the absolute velocity $\underline{u}_{abs}$ is

$$
\begin{aligned}
\underline{u}_{abs} &= \left(\frac{d\underline{r}}{dt}\right)_S \\
&= 0 + \underline{\omega}_{E,S} \times \underline{r} \\
&= \begin{vmatrix} \underline{i}_E & \underline{j}_E & \underline{k}_E \\ \omega_e \cos \lambda & 0 & -\omega_e \sin \lambda \\ 0 & 0 & -R_e \end{vmatrix} \\
&= R_e \omega_e \cos \lambda \underline{j}_E
\end{aligned}
\tag{1.29}
$$

i.e. the absolute velocity is of the magnitude $R_e \omega_e \cos \lambda$ in the local easterly direction.

Now you see why it is advisable to launch a satellite or any object in the equatorial plane in the easterly direction! Because even at rest, the object possesses a velocity of

$R_e\omega_e$ (almost close to $1500\,\text{ft s}^{-1}$) at a point on the equator ($\lambda = 0$, $R_e = 4000$ miles, $\omega_e = 0.728 \times 10^{-4}\,\text{rad s}^{-1}$ with respect to the inertial S frame).

Notice how the two fundamental concepts we learnt have helped us to rigorously establish that it is indeed advantageous to launch a satellite from a launching station located at the equator. For this reason, there are nations that have built launching stations on the equator (for example, France). Obviously, for locations with high latitude, this almost becomes a necessity, because we know from the orbital mechanics course that plane changes in orbits are prohibitively fuel expensive. However, countries like Russia, which is at a very high latitude, choose to use different ways to launch their satellites without requiring plane changes (for example, via Molinya orbits). The US launches most of its satellites from the Cape Caneveral station in Florida, which is at a latitude of about 28°. This concludes the message given by this illustrative example.

We now present another example, which further illustrates the importance of coordinate transformations.

Example 1.1   In the previous example, we obtained

$$
\begin{aligned}
\underline{u}_{abs} &= \left(\frac{d\underline{r}}{dt}\right)_{XYZ} \\
&= \left(\frac{d\underline{r}}{dt}\right)_{xyz} + \underline{\omega} \times \underline{r} \\
&= 0 + \underline{\omega} \times \underline{r} \\
&= R_e\omega_e \cos \lambda \underline{j}_{-E}.
\end{aligned}
\tag{1.30}
$$

Our aim now is to write $\underline{u}_{abs}$ in the inertial frame components $\underline{i}_S, \underline{j}_S$, and $\underline{k}_S$.

Solution

We know the relationship between $(\underline{i}_S, \underline{j}_S, \underline{k}_S)$ in terms of $(\underline{i}_E, \underline{j}_E, \underline{k}_E)$ via the transformation matrix given in the previous problem. Since this rotation matrix is orthogonal, it has the property that its inverse is equal to its transpose.

Therefore we have

$$
\begin{bmatrix} \underline{i}_E \\ \underline{j}_E \\ \underline{k}_E \end{bmatrix} = \begin{bmatrix} \cos\mu \sin\lambda & \sin\mu \sin\lambda & -\cos\lambda \\ -\sin\mu & \cos\mu & 0 \\ \cos\mu \cos\lambda & \sin\mu \cos\lambda & \sin\lambda \end{bmatrix} \begin{bmatrix} \underline{i}_S \\ \underline{j}_S \\ \underline{k}_S \end{bmatrix}.
\tag{1.31}
$$

From the above matrix, we observe that

$$
\underline{j}_E = -\sin\mu \underline{i}_S + \cos\mu \underline{j}_S.
\tag{1.32}
$$

And so

$$
\underline{u}_{abs} = -R_e\omega_e \cos\lambda \sin\mu \underline{i}_S + R_e\omega_e \cos\lambda \cos\mu \underline{j}_S.
\tag{1.33}
$$

Thus we now have the absolute velocity in terms of the inertial frame components.

1.5 Further Insight into Absolute Acceleration

Now going back to the analysis, we can now calculate the absolute acceleration of the body by applying Charle's theorem:

$$
\begin{aligned}
\underline{a}_{abs} &= (\frac{d^2 r}{dt^2})_S \\
&= (\frac{d\underline{u}_{abs}}{dt})_S \\
&= (\frac{d\underline{u}_{abs}}{dt})_E + \omega_{E,S} \times \underline{u}_{abs}
\end{aligned}
\tag{1.34}
$$

where u_{abs} is expressed in the E frame components, i.e.

$$
u_{abs} = R_e \omega_e \cos \lambda \underline{j}_{-E}.
\tag{1.35}
$$

Now

$$
\begin{aligned}
(\frac{d\underline{u}_{abs}}{dt})_E &= \frac{d}{dt}(R_e \omega_e \cos \lambda \underline{j}_{-E}) \\
&= -R_e \omega_e \dot{\lambda} \sin \lambda \underline{j}_{-E}
\end{aligned}
\tag{1.36}
$$

and the cross product term, $\omega_{E,S} \times \underline{u}_{abs}$, is

$$
\begin{vmatrix}
\underline{i}_E & \underline{j}_{-E} & \underline{k}_E \\
\omega_e \cos \lambda & 0 & -\omega_e \sin \lambda \\
0 & R_e \omega_e \cos \lambda & 0
\end{vmatrix}.
\tag{1.37}
$$

Since the vehicle is at rest at the position considered, $\dot{\lambda} = 0$. Therefore, $\underline{a}_{abs} = \underline{\omega}_{E,S} \times \underline{u}_{abs}$. Note that if the body was not at rest in the problem to start with, we would get many more terms in the absolute velocity and acceleration expressions. You may recognize that these extra terms in the acceleration to be of Coriolis in nature. In Exercise 1.3, you will get a chance to do a similar problem with $\dot{\lambda} \neq 0$.

1.6 Chapter Summary

- In this chapter, we presented an overview of the standard practice of dynamics investigation along with the need to make approximations to keep the modeling process as simple and tractable as possible without compromising the salient points of the dynamical behavior.
- We have learnt that making approximations is valid as long as there is a reasonable justification for making that approximation.
- We have learnt two fundamental concepts that are useful for later use in deriving the detailed equations of motion. The first one is Charle's theorem, which governs the rate of change of a vector, when the vector is expressed in body fixed (moving) coordinate frame. The second one is the concept of coordinate transformations in which we learnt that the components in different coordinate frames can be related to each other with the help of the Euler angles. Their usefulness is demonstrated by

application to a practical situation wherein it is established that launching a satellite from a station located on the equator has an advantage in saving fuel.

1.7 Exercises

- **Exercise 1.1.** As part of the solution to the example problem above, we obtained the relationship between the components of the S frame in terms of the components of the E frame. Now obtain the components of the E frame in terms of the components of the S frame analytically. Then obtain the corresponding numerical values for $\mu = 30°$ and $\lambda = 45°$.
- **Exercise 1.2.** Complete the expansion of absolute acceleration $\underline{a}_{abs} = \underline{\omega}_{E,S} \times \underline{u}_{abs}$ for a body at rest on the surface of the Earth.
- **Exercise 1.3.** An expression for the absolute velocity and absolute acceleration of a vehicle on the surface of the spherical Earth was derived under the assumption that the vehicle was at rest in a previous worked out example. Now assume a non-zero rate of change for the latitude angle $\dot{\lambda}$, where $\dot{\lambda}$ is defined as

$$\dot{\lambda} \equiv \frac{u}{R_e} \tag{1.38}$$

where u is the velocity of the vehicle, on the equator, headed on a highway due north. Also note

$$\ddot{\lambda} \equiv \frac{a}{R_e} \tag{1.39}$$

where a is the acceleration of the vehicle. Find the expressions for $\underline{u}_{abs}$ and $\underline{a}_{abs}$ given $\mu = 0$,

$$u = 60 \text{ mph} \tag{1.40}$$

$$a = 0.05 \quad \text{ft} \quad \text{s}^{-2} \tag{1.41}$$

$$\omega_e = 0.728 \times 10^{-4} \quad \text{rad/s}^{-1}. \tag{1.42}$$

- **Exercise 1.4.** Obtain the composite rotation matrix S making a small angle approximation. Note that when the angles (in radians) are assumed to be very small, $sin\theta$ is approximately equal to θ and $cos\theta$ is approximately equal to 1.
- **Exercise 1.5.** Investigate which of the following matrices can be qualified to be a proper, rotation matrix between two frames that obey the right hand rule.

$$R_1 = \begin{bmatrix} -0.6 & 0 & -0.8 \\ 0.8 & 0 & -0.6 \\ 0 & -1 & 0 \end{bmatrix} \tag{1.43}$$

$$R_2 = \begin{bmatrix} 0.6 & 0 & -0.8 \\ -0.8 & 0 & -0.6 \\ 0 & -1 & 0 \end{bmatrix} \tag{1.44}$$

$$R_3 = \begin{bmatrix} 0 & -0.8 & -0.6 \\ 0 & -0.6 & 0.8 \\ -1.1 & 0 & 0 \end{bmatrix}. \tag{1.45}$$

Bibliography

1 M. Balas. Feedback control of flexible systems. *IEEE Transactions on Automatic Control*, 23(4).

2 B.R. Barmish. *New tools for robustness of linear systems*. Macmillan Publishing company, Riverside, NJ, 1994.

3 S.P Bhattacharyya, H Chappellat, and L.H Keel. *Robust control: The Parametric approach*. Prentice Hall, Englewood cliffs, NJ, 1995.

4 R.H. Cannon. *Dynamics of Physical Systems*. Dover Publications, New York, 1953.

5 L. Meirovitch and I. Tuzcu. Unified theory for the dynamics and control of maneuvering flexible aircraft. *AIAA Journal*, 42(4):714–727, 2004.

6 D.K. Schmidt and D.L. Raney. Modeling and simulation of flexible flight vehicles. *Journal of Guidance, Control, and Dynamics*, 24(3):539–546, 2001.

7 R.E. Skelton. *Dynamic Systems Control: Linear Systems Analysis and Synthesis*. Wiley, New York, 1 edition, 1988.

8 M.R. Waszak and D.K. Schmidt. Flight dynamics of aeroelastic vehicles. *Journal of Aircraft*, 25(6):563–571, 1988.

9 K.A. Wise. Comparison of six robustness tests evaluating missile autopilot robustness to uncertain aerodynamics. *AIAA Journal of Guidance Control and Dynamics*, 15(4):861–870, 1992.

10 Rama K Yedavalli. *Robust Control of Uncertain Dynamic Systems: A Linear State Space Approach*. Springer, 2014.

2

Basic Nonlinear Equations of Motion in Three Dimensional Space

2.1 Chapter Highlights

In this chapter, we embark on the task of deriving the equations of motion for a rigid body in three dimensional motion. This general motion can be viewed as consisting of the *translational motion* of the center of mass of the rigid body and the rotational motion of the rigid body about its center of mass. Note that for a general motion of the rigid body in three dimensions, we thus have three translational degrees of freedom and three rotational degrees of freedom. We plan to derive these six degrees of freedom equations by careful application of Newton's law of motion. The translational motion is governed by the rate of change of linear momentum and the rotational motion is governed by the rate of change of angular momentum. Note that the rotational motion about the center of mass is also referred to as attitude dynamics. The following development is valid for any rigid body in motion (with respect to an inertial reference frame) and thus up to some point is applicable to deriving equations of motion for an aircraft in atmosphere as well as for a satellite in Earth's orbit. By recognizing the appropriate external forces acting on the body, we eventually specialize these equations of motion separately for aircraft and satellites in Earth's orbit. The equations of motion for these two situations (aircraft/spacecraft) are also specialized in accordance with the body frame (moving frame) selection, as we shall see later. It is also important to observe as to at which juncture in the derivation of these equations of motion the body is assumed to be rigid. This helps us to figure out the needed modifications for deriving the equations of motion for a flexible body. For extension of these equations for a flexible body, the reader is referred to other excellent books such as [4, 8]. Since this book primarily caters to an undergraduate student body, an advanced topic such as flexible body equations of motion is considered beyond the scope of this book.

2.2 Derivation of Equations of Motion for a General Rigid Body

Since we are attempting to derive the equations of motion, the starting point is the application of Newton's laws of motion to the body. Let us start with the two fundamental laws of motion, namely:

Flight Dynamics and Control of Aero and Space Vehicles, First Edition. Rama K. Yedavalli.
© 2020 John Wiley & Sons Ltd. Published 2020 by John Wiley & Sons Ltd.

(a) Rate of change of linear momentum:

$$\underline{\dot{L}} = \frac{d}{dt}\underline{L} = \sum \underline{F}_{\text{applied}} \rightarrow \text{translational dynamics.}$$

(b) Rate of change of angular momentum:

$$\underline{\dot{H}} = \frac{d}{dt}\underline{H} = \sum \underline{M}_{\text{applied}} \rightarrow \text{rotational/attitude dynamics.}$$

We start by assuming two coordinate frames, one is a body-fixed (moving) reference frame and the other is an inertial (fixed) reference frame. The inertial reference frame can be conveniently selected to be an Earth centered reference frame (ideally with one axis passing through the poles and the other two axes in the equatorial plane), see Figure 2.1.

Consider a mass particle within the body denoted by *dm*. Let the position vectors of the origin of the body frame *C*, the mass particle *dm* be as shown (with respect to the inertial reference frame) in Figure 2.1.

Thus we have:

$$\frac{d}{dt}\int \frac{d\underline{r}'}{dt}dm = \sum \underline{F}_{\text{applied}} \tag{2.1}$$

$$\frac{d}{dt}\int \underline{r}' \times \frac{d\underline{r}'}{dt}dm = \sum \underline{M}_{\text{applied}}. \tag{2.2}$$

2.2.1 Translational Motion: Force Equations for a General Rigid Body

Note that Newton's laws are always stated with respect to the fixed inertial reference frame. Note also that

$$\underline{r}' = \underline{r}_c + \underline{r}. \tag{2.3}$$

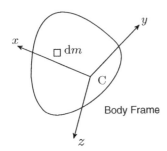

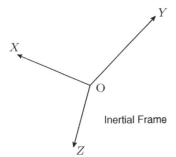

Figure 2.1 The body frame relative to the inertial frame.

Figure 2.2 The position vectors of the origin of the body frame C and the mass particle dm.

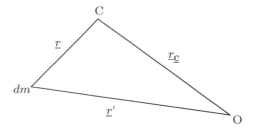

Refer to Figure 2.2 above. If cm is the center of mass of the body, then by the definition of the center of mass of the body, we have:

$$\int \underline{r} dm = m\underline{r}_{cm/c} \tag{2.4}$$

where $\underline{r}_{cm/c}$ is the position vector of the center of mass of the body from the origin of the body fixed moving reference frame. Obviously if we select the origin of the body frame as the center of mass of the body, we get the simplification

$$\int \underline{r} dm = 0. \tag{2.5}$$

So we select C as the center of mass of the body. Now consider the linear momentum equation

$$\frac{d}{dt} \int \frac{d}{dt} \underline{r}' dm = \frac{d}{dt} \int \frac{d}{dt} (\underline{r}_c + \underline{r}) dm$$

$$= \frac{d}{dt} \frac{d}{dt} \int (\underline{r}_c + \underline{r}) dm. \tag{2.6}$$

Then, at this stage, we make another assumption, namely that the mass is constant (i.e. $\frac{dm}{dt} = 0$). Under this assumption, we observe that we can interchange the integral and derivative operations in the right hand side of the above equation and thus equation 2.6 can be further simplified as

$$\frac{d}{dt} \frac{d}{dt} \left[\int \underline{r}_c dm + \int \underline{r} dm \right]^{0} = \frac{d}{dt} \frac{d}{dt} \int \underline{r}_c dm$$

$$= \frac{d}{dt} \frac{d}{dt} \underline{r}_c \int dm \tag{2.7}$$

$$= \frac{d}{dt} \frac{d}{dt} \underline{r}_c m.$$

Note $\underline{r}_c$ is independent of dm. Therefore

$$m\frac{d}{dt} \frac{d}{dt} \underline{r}_c = m\frac{d}{dt} \underline{v}_c = \sum \underline{F}_{applied} \tag{2.8}$$

where $\frac{d}{dt} \underline{r}_c = \underline{v}_c$ the velocity of the center of mass of the body.

This equation is the iconic Newton's second law of motion ($\underline{F} = m\underline{a}$); but it can be seen that it is too simplistic and somewhat casual. It is very important to understand how many assumptions and approximations were made to get to this point.

Note that when the force $\underline{F}$ vector in Equation 2.1 is taken to be the inverse square law of the Earth's gravitational field and is $|\underline{F}| = \frac{GMm}{r^2}$ acting along the vector, $\underline{r_c}$, where M is the mass of the Earth and m is the mass of the orbiting body, we can write:

$$\frac{d^2 \underline{r_c}}{dt^2} + \frac{\mu}{r_c^3}\underline{r_c} = 0 \tag{2.9}$$

where $\mu = GM$ is the gravitational constant. This is the so-called orbit equation. Thus the translational motion of a rigid body in Earth's central force field becomes the orbital motion.

However for any rigid body flying in the Earth's atmosphere, the gravitational acceleration can be taken to be fairly constant and thus, from now on, for a rigid body in the Earth's atmosphere, we take the fixed inertial reference frame to be simply the coordinate frame fixed on the surface of the Earth with one axis going local north, one axis going east, and the local vertical down as the third axis. Equations of motion derived under this simple assumption are the so-called flat Earth approximation equations. Thus the translational motion for a rigid body under flat Earth approximation (i.e. acceleration due to gravity is taken as a constant) is:

$$
\begin{aligned}
m\frac{d\underline{v_c}}{dt} &= m\frac{d^2\underline{r_c}}{dt^2} \\
&= mg\underline{K} + \underline{F}_{aero} + \underline{F}_{thrust} \\
&= \underline{F}_{gravity} + \underline{F}_{aero} + \underline{F}_{thrust}
\end{aligned}
\tag{2.10}
$$

i.e.

$$\boxed{m\frac{d\underline{v_c}}{dt} = m\frac{d^2\underline{r_c}}{dt^2} = \underline{F}_g + \underline{F}_{aero} + \underline{F}_{thrust}} \tag{2.11}$$

where $\underline{F}_g$ is the weight acting vertically downwards.

2.2.2 Rotational Motion: moment equations for a General Rigid Body

Now consider moment equations. We have:

$$\frac{d}{dt}\int \frac{d\underline{r'}}{dt}dm = \sum \underline{F}_{applied} \tag{2.12}$$

$$\frac{d}{dt}\int \underline{r'} \times \frac{d\underline{r'}}{dt}dm = \underline{r'} \times \sum \underline{F}_{applied}. \tag{2.13}$$

Substituting $\underline{r'} = \underline{r_c} + \underline{r}$ into the above equations, we have:

$$\frac{d}{dt}\int (\underline{r_c} + \underline{r}) \times \frac{d\underline{r'}}{dt}dm = (\underline{r_c} + \underline{r}) \times \sum \underline{F}_{applied} \tag{2.14}$$

$$\frac{d}{dt}\left[\int \underline{r_c} \times \frac{d\underline{r'}}{dt}dm + \int \underline{r} \times \frac{d\underline{r'}}{dt}dm\right] = \underline{r_c} \times \sum \underline{F}_{applied} + \underline{r} \times \sum \underline{F}_{applied}. \tag{2.15}$$

Now cross product equation 2.12 with $\underline{r_c}$ to get:

$$\frac{d}{dt}\int \underline{r_c} \times \frac{d\underline{r'}}{dt}dm = \underline{r_c} \times \sum \underline{F}_{applied} \tag{2.16}$$

i.e.

$$\underline{r}_c \times \sum \underline{F}_{\text{applied}} + \frac{\mathrm{d}}{\mathrm{d}t} \int \underline{r} \times \frac{\mathrm{d}\underline{r}'}{\mathrm{d}t} \mathrm{d}m = \underline{r}_c \times \sum \underline{F}_{\text{applied}} + \underline{r} \times \sum \underline{F}_{\text{applied}}. \tag{2.17}$$

From this, we get

$$
\begin{aligned}
\underline{r} \times \sum \underline{F}_{\text{applied}} &= \frac{\mathrm{d}}{\mathrm{d}t} \int \underline{r} \times \frac{\mathrm{d}}{\mathrm{d}t} \underline{r}' \mathrm{d}m \\
&= \frac{\mathrm{d}}{\mathrm{d}t} \int \underline{r} \times \frac{\mathrm{d}}{\mathrm{d}t} (\underline{r}_c + \underline{r}) \mathrm{d}m \\
&= \frac{\mathrm{d}}{\mathrm{d}t} \int \underline{r} \times \frac{\mathrm{d}\underline{r}_c}{\mathrm{d}t} \mathrm{d}m + \frac{\mathrm{d}}{\mathrm{d}t} \int \underline{r} \times \frac{\mathrm{d}\underline{r}}{\mathrm{d}t} \mathrm{d}m.
\end{aligned}
\tag{2.18}
$$

In the first term above, since $\underline{r}_c$ is independent of mass particle $\mathrm{d}m$, we can make it come out of that time derivative term. Then we have the right hand side as

$$= \frac{\mathrm{d}}{\mathrm{d}t} \int \underline{r} \mathrm{d}m^{\nearrow 0} \times \frac{\mathrm{d}\underline{r}_c}{\mathrm{d}t} + \frac{\mathrm{d}}{\mathrm{d}t} \int \underline{r} \times \frac{\mathrm{d}\underline{r}}{\mathrm{d}t} \mathrm{d}m. \tag{2.19}$$

Thus we finally have

$$\boxed{\frac{\mathrm{d}}{\mathrm{d}t} \int \underline{r} \times \frac{\mathrm{d}\underline{r}}{\mathrm{d}t} \mathrm{d}m = \underline{r} \times \sum \underline{F}_{\text{applied}} = \sum \underline{M}_{\text{applied}}.} \tag{2.20}$$

The interesting thing here is that we finally have the moment equation expressed in terms of $\underline{r}$ rather than $\underline{r}'$. This is important to note, because $\underline{r}$ is a quantity one can measure or physically visualize. So, now we have the translational (force) equations and the rotational (moment) equations, both in vector form.

$$m \frac{\mathrm{d}\underline{v}_c}{\mathrm{d}t} = \underline{F}_{\text{applied}} \qquad \text{general translational motion} \tag{2.21a}$$

$$\frac{\mathrm{d}}{\mathrm{d}t} \int \underline{r} \times \frac{\mathrm{d}\underline{r}}{\mathrm{d}t} \mathrm{d}m = \sum \underline{M}_{\text{applied}} \qquad \begin{array}{l}\text{general rotational motion about the}\\ \text{center of mass of the body.}\end{array} \tag{2.21b}$$

When we assume acceleration due to gravity is constant, they can be modified to

$$m \frac{\mathrm{d}\underline{v}_c}{\mathrm{d}t} = \underline{W} + \underline{F}_A + \underline{F}_T \qquad \begin{array}{l}\text{flat Earth approximation}\\ \text{translational motion}\end{array} \tag{2.22a}$$

$$\frac{\mathrm{d}}{\mathrm{d}t} \int \underline{r} \times \frac{\mathrm{d}\underline{r}}{\mathrm{d}t} \mathrm{d}m = \sum \underline{M}_{\text{applied}} \qquad \begin{array}{l}\text{rotational motion, describing attitude}\\ \text{dynamics about the center of gravity.}\end{array} \tag{2.22b}$$

Here $\underline{W}$ is the weight of the rigid body, which is nothing but the force of gravity under constant acceleration assumption.

2.2.3 Scalar Motion Equations for a General Rigid Body

Unless these two vector relations are resolved into scalar equations (three translational and three rotational), they are not useful. So we need to convert them to scalar equations. Thus we need to express vectors $\underline{v}_c$, $\underline{r}_c$, $\underline{r}$ and $\underline{F}$ and $\underline{M}$, etc., into component form. We can resolve all these vectors in question either into components along the body reference frame or along the inertial frame. For reasons that will become clear soon, we choose to express these vectors along the components in the body frame. Thus let,

$$\underline{v}_c = v_{cx}(t)\underline{i} + v_{cy}(t)\underline{j} + v_{cz}(t)\underline{k}$$
$$\underline{\omega}(t) = \omega_x(t)\underline{i} + \omega_y(t)\underline{j} + \omega_z(t)\underline{k} \tag{2.23}$$
$$\underline{M}_{ext}(t) = M_x(t)\underline{i} + M_y(t)\underline{j} + M_z(t)\underline{k}.$$

Since all these vectors are expressed in body frame components and this body is translating and rotating with respect to the inertial frame, we have to use Charle's theorem to take into account the moving nature of the $\underline{ijk}$ frame whenever we encounter any term containing a derivative of a vector (i.e. whenever we see $\frac{d}{dt}\underline{A}$, where $\underline{A}$ is any arbitrary vector). Once that vector is expressed in the body frame components, we use Charle's theorem, given by

$$\left(\frac{d\underline{A}}{dt}\right)_{XYZ\ \text{frame}} = \left(\frac{d\underline{A}}{dt}\right)_{xyz\ \text{frame}} + \underline{\omega} \times \underline{A}. \tag{2.24}$$

Now consider the force equation again:

$$m\frac{d\underline{v}_c}{dt} = \underline{F}_{applied}. \tag{2.25}$$

Since $\underline{v}_c$ is expressed in body frame components and the derivation in the above equation is with respect to the inertial frame, we have to apply equation 2.24. Doing this we get:

$$m\left[\left(\frac{d\underline{v}_c}{dt}\right) + \underline{\omega} \times \underline{v}_c\right] = \underline{F}_{applied} \tag{2.26}$$

where

$$\frac{d\underline{v}_c}{dt} = \dot{v}_{cx}\underline{i} + \dot{v}_{cy}\underline{j} + \dot{v}_{cz}\underline{k} \tag{2.27}$$

and

$$\underline{\omega} \times \underline{v}_c = \begin{vmatrix} \underline{i} & \underline{j} & \underline{k} \\ \omega_x & \omega_y & \omega_z \\ v_{cx} & v_{cy} & v_{cz} \end{vmatrix}. \tag{2.28}$$

Carrying out the cross multiplication and gathering the $\underline{i}, \underline{j},$ and $\underline{k}$ components and we get (omitting the time arguments):

force equations
$$\begin{aligned} m(\dot{v}_{cx} - v_{cy}\omega_z + v_{cz}\omega_y) &= F_x \\ m(\dot{v}_{cy} + v_{cx}\omega_z - v_{cz}\omega_x) &= F_y . \\ m(\dot{v}_{cz} - v_{cx}\omega_y + v_{cy}\omega_x) &= F_z \end{aligned} \tag{2.29}$$

Thus, the above three scalar equations describe the general translational motion of a rigid body in three dimensional space. These are commonly referred to as the three degrees of freedom (3DOF) translational motion equations.

Now consider the moment equation:

$$\frac{d}{dt}\int \underline{r} \times \frac{d}{dt}\underline{r}\,dm = \sum \underline{M}_{applied} \tag{2.30}$$

i.e.

$$\underline{H} = \int \underline{r} \times \frac{d\underline{r}}{dt} dm$$

angular momentum

$$= \int \underline{r} \times (\underline{\dot{r}} + \underline{\omega} \times \underline{r}) dm \tag{2.31}$$

However, from the rigid body assumption $\underline{\dot{r}} = 0$ (i.e. the mass particle does not have any relative motion within the body). Mathematically, this is expressed as

$$\frac{dx(t)}{dt} = \frac{dy(t)}{dt} = \frac{dz(t)}{dt} = 0. \tag{2.32}$$

Thus, under the rigid body assumption, position components x, y, and z are not functions of time and can be taken as constants. Therefore

$$\underline{H} = \int \underline{r} \times (\underline{\omega} \times \underline{r}) dm. \tag{2.33}$$

Now,

$$\begin{aligned}
\underline{r} \times (\underline{\omega} \times \underline{r}) &= [\omega_x(y^2 + z^2) - \omega_y xy - \omega_z xz]\underline{i} \\
&\quad + [-\omega_x xy + \omega_y(x^2 + z^2) - \omega_z yz]\underline{j} \\
&\quad + [-\omega_x xz - \omega_y yz + \omega_z(y^2 + x^2)]\underline{k}.
\end{aligned} \tag{2.34}$$

Now define moments and products of inertia:

$$\int (y^2 + z^2) dm = I_{xx} \tag{2.35a}$$

$$\int (x^2 + z^2) dm = I_{yy} \tag{2.35b}$$

$$\int (x^2 + y^2) dm = I_{zz} \tag{2.35c}$$

$$\int xy dm = I_{xy} \tag{2.36a}$$

$$\int xz dm = I_{xz} \tag{2.36b}$$

$$\int yz dm = I_{yz}. \tag{2.36c}$$

Note that these are constant for a given mass distribution of the body. Keep in mind that moments of inertia are always positive, whereas products of inertia can be positive or negative. Now you realize why we choose to express $\underline{r}(t)$ in terms of body frame components because this choice gives moments and products of inertia to be constant and they can be computed a priori, without any consideration to the position of the body in the three dimensional space. Thus,

$$\underline{H} = H_x \underline{i} + H_y \underline{j} + H_z \underline{k} \tag{2.37}$$

where,

$$H_x = I_{xx}\omega_x - I_{xy}\omega_y - I_{xz}\omega_z \tag{2.38}$$

$$H_y = -I_{xy}\omega_x + I_{yy}\omega_y - I_{yz}\omega_z \tag{2.39}$$

$$H_z = -I_{xz}\omega_x - I_{yz}\omega_y + I_{zz}\omega_z \tag{2.40}$$

or, in matrix form,

$$\begin{bmatrix} H_x \\ H_y \\ H_z \end{bmatrix} = \begin{bmatrix} I_{xx} & -I_{xy} & -I_{xz} \\ -I_{xy} & I_{yy} & -I_{yz} \\ -I_{xz} & -I_{yz} & I_{zz} \end{bmatrix} \begin{bmatrix} \omega_x \\ \omega_y \\ \omega_z \end{bmatrix} \tag{2.41}$$

i.e.

$$\underline{H} = \underline{I} \ \underline{\omega} \quad \text{(where I is a symmetric matrix with constant entries.)} \tag{2.42}$$

$$\underline{H} = H_x \underline{i} + H_y \underline{j} + H_z \underline{k} \tag{2.43}$$

i.e.

$$H_x = I_{xx}\omega_x - I_{xy}\omega_y - I_{xz}\omega_z \tag{2.44}$$

$$H_y = -I_{xy}\omega_x + I_{yy}\omega_y - I_{yz}\omega_z \tag{2.45}$$

$$H_x = -I_{xz}\omega_x - I_{yz}\omega_y + I_{zz}\omega_z. \tag{2.46}$$

This is expressed more conveniently in matrix form

$$\underline{H} = \underline{I}\underline{\omega} \tag{2.47}$$

where,

$$\begin{bmatrix} H_x \\ H_y \\ H_z \end{bmatrix} = \begin{bmatrix} I_{xx} & -I_{xy} & -I_{xz} \\ -I_{xy} & I_{yy} & -I_{yz} \\ -I_{xz} & -I_{yz} & I_{zz} \end{bmatrix} \begin{bmatrix} \omega_x \\ \omega_y \\ \omega_z \end{bmatrix}. \tag{2.48}$$

Note the moment of inertia matrix $\underline{I}$ is a symmetric, positive definite matrix. Here the body's angular velocity $\underline{\omega}$ expressed in body fixed coordinate frame is

$$\underline{\omega} = \omega_x \underline{i} + \omega_y \underline{j} + \omega_z \underline{k}. \tag{2.49}$$

The external applied moments are then given by the rate of change of angular momentum, namely:

$$\sum \underline{M}_{\text{applied}} = M_x \underline{i} + M_y \underline{j} + M_z \underline{k} \tag{2.50}$$
$$= \underline{\dot{H}} + \underline{\omega} \times \underline{H}.$$

Thus we have:

$$M_x = \dot{H}_x + (\omega_y H_z - \omega_z H_y) \tag{2.51}$$

$$M_y = \dot{H}_y + (\omega_z H_x - \omega_x H_z) \tag{2.52}$$

$$M_z = \dot{H}_z + (\omega_x H_y - \omega_y H_x). \tag{2.53}$$

Carrying out the cross multiplication and gathering all the $\underline{i}, \underline{j}, \underline{k}$ components we finally get the scalar moment equations as

$$\dot{H}_x + (\omega_y H_z - \omega_z H_y) = M_{\text{roll}} \quad \text{rolling moment equation} \tag{2.54a}$$

$$\dot{H}_y + (\omega_z H_x - \omega_x H_z) = M_{\text{pitch}} \quad \text{pitching moment equation} \tag{2.54b}$$

$$\dot{H}_z + (\omega_x H_y - \omega_y H_x) = M_{\text{yaw}} \quad \text{yawing moment equation.} \tag{2.54c}$$

The above three scalar equations describe the general rotational motion about the center of mass for a rigid body.

This completes the derivation of the force and moment equations for a general rigid body in three dimensional space.

Example 2.1 The angular momentum components of a rigid body about its mass center is observed (in units of $N \cdot m \cdot s$) to be:

$$\underline{H} = \begin{bmatrix} 1000 \\ 300 \\ 500 \end{bmatrix}$$

and the body's moment of inertia is given in units of N m s^2 as

$$I = \begin{bmatrix} 30 & -10 & 0 \\ -10 & 40 & 0 \\ 0 & 0 & 20 \end{bmatrix}.$$

Obtain the angular velocity components for the body.

Solution
We know $\underline{H} = I\underline{\omega}$; therefore,

$$\underline{\omega} = I^{-1}\underline{H}$$
$$= \begin{bmatrix} 0.0364 & 0.0091 & 0 \\ 0.0091 & 0.0273 & 0 \\ 0 & 0 & 0.05 \end{bmatrix} \begin{bmatrix} 1000 \\ 300 \\ 500 \end{bmatrix}$$
$$= \begin{bmatrix} 6.36 \\ 9.09 \\ 25 \end{bmatrix}.$$

Example 2.2 Suppose we are given incomplete data characterizing the rotation of a rigid body about its center of mass:

$$\begin{bmatrix} H_x \\ 475 \\ 2250 \end{bmatrix} = \begin{bmatrix} 50 & -15 & -I_{xz} \\ -15 & 30 & -I_{yz} \\ -I_{xz} & -I_{yz} & 60 \end{bmatrix} \begin{bmatrix} 15 \\ 30 \\ 40 \end{bmatrix}.$$

Find the angular momentum component H_x.

Solution
Recognize that this is an algebraic system of equations in three equations and three unknowns, I_{xz}, I_{yz}, and H_x. Taking advantage of the fact that moment of inertia matrix is a symmetric matrix and then solving those three simultaneous equations, we obtain

$$I_{xz} = 0$$
$$I_{yz} = 5$$
$$H_x = 300 \quad N\ m\ s.$$

Depending on the external applied forces and moments, these equations can be specialized to different scenarios. In the next two sections, we specialize these equations of motion for an aircraft and and a spacecraft (satellite) respectively, taking into account the external applied forces and moments specific to their situations.

2.3 Specialization of Equations of Motion to Aero (Atmospheric) Vehicles

In this section, we specialize the above equations of motion, taking into account the the nature of applied external forces and moments that govern an aircraft in motion. In this connection, the first step in tailoring these equations to the aircraft dynamics is to change the nomenclature to that is widely used and embraced by the aircraft dynamics community. Accordingly, we express the velocity of the center of mass v_c, the angular velocity vector $\underline{\omega}$, the moment vector $\underline{M}$ in the body fixed frame as follows:

$$
\begin{aligned}
\underline{r}(t) &= x(t)\underline{i} + y(t)\underline{j} + z(t)\underline{k} \\
\underline{v}_c(t) &= U(t)\underline{i} + V(t)\underline{j} + W(t)\underline{k} \\
\underline{\omega}(t) &= P(t)\underline{i} + Q(t)\underline{j} + R(t)\underline{k} \\
\underline{M}_{\text{aero}}(t) &= L(t)\underline{i} + M(t)\underline{j} + N(t)\underline{k}.
\end{aligned}
\tag{2.55}
$$

In other words, we use letters U, V, and W for denoting linear velocity components; P, Q, and R for angular velocity components; and L, M, and N for aerodynamic moment components, which happens to be a well established standard notation for writing down aircraft specific dynamics. Here the letter L used for rolling moment should not be confused with the lift force, which is sometimes also denoted as L. We assume caution on the part of the reader and believe that we are able to distinguish between them based on the context in which this letter is used. From now on, we follow this specialized notation for aircraft specific dynamics. Thus, we have

$$
\frac{d\underline{v}_c}{dt} = \dot{U}\underline{i} + \dot{V}\underline{j} + \dot{W}\underline{k}
\tag{2.56}
$$

and

$$
\underline{\omega} \times \underline{v}_c = \begin{vmatrix} \underline{i} & \underline{j} & \underline{k} \\ P & Q & R \\ U & V & W \end{vmatrix}.
\tag{2.57}
$$

Carrying out the cross multiplication and gathering the $\underline{i}$, $\underline{j}$, and $\underline{k}$ components we get the force equations (omitting the time arguments)

$$
m(\dot{U} - VR + WQ) = F_x
\tag{2.58}
$$
$$
m(\dot{V} + UR - WP) = F_y
\tag{2.59}
$$
$$
m(\dot{W} - UQ + VP) = F_z
\tag{2.60}
$$

where

$$
F_x = F_{gx} + F_{Ax} + F_{Tx}
\tag{2.61}
$$

$$F_y = F_{gy} + F_{Ay} + F_{Ty} \qquad (2.62)$$

$$F_z = F_{gz} + F_{Az} + F_{Tz} \qquad (2.63)$$

where F_g is the gravitational force, F_A represents the aerodynamic forces, and F_T is the thrust force(s). Note that F_g, under the flat earth approximation, is nothing but the weight. We now want to express the gravitational force (weight), which always acts vertically downwards, in terms of the body frame axes. We denote the weight by $\underline{W}_t$ to differentiate it from the letter W we used for the vertical velocity.

2.3.1 Components of the Weight in Body Frame

Since the gravitational acceleration always acts vertically downwards, we have the gravitational force (weight) vector given by:

$$\underline{W}_t = mg\underline{K} \qquad (2.64)$$

where $\underline{K}$ is the unit vector in the fixed frame vertically downwards. However, we want to express this force in terms of body frame components. So using the Euler angle relationships, we can get:

$$\begin{aligned} \underline{W}_t &= mg\underline{K} \\ &= m(g_x\underline{i} + g_y\underline{j} + g_z\underline{k}). \end{aligned} \qquad (2.65)$$

However, through the composite rotation matrix S, we know the relationship between the unit vector $\underline{K}$ and the body frame unit vectors $\underline{i}, \underline{j}$ and $\underline{k}$, which is given by

$$\underline{K} = -\sin\Theta\underline{i} + \cos\Theta(\sin\Phi\underline{j} + \cos\Phi\underline{k}) = -\sin\Theta\underline{i} + \cos\Theta\sin\Phi\underline{j} + \cos\Theta\cos\Phi\underline{k}.$$

Substituting for $\underline{K}$ in terms of $\underline{i}, \underline{j}, \underline{k}$, we get:

$$g_x = -g\sin\Theta \qquad (2.66)$$

$$g_y = g\sin\Phi\cos\Theta \qquad (2.67)$$

$$g_z = g\cos\Phi\cos\Theta. \qquad (2.68)$$

Until now, we have focused on the force equations describing the translational motion of the aircraft. Now consider the moment equation:

$$\frac{d}{dt}\int \underline{r} \times \frac{d}{dt}\underline{r}dm = \sum \underline{M}_{\text{applied}} \qquad (2.69)$$

i.e. the angular momentum is

$$\begin{aligned} \underline{H} &= \int \underline{r} \times \frac{d\underline{r}}{dt}dm \\ &= \int \underline{r} \times (\dot{\underline{r}} + \underline{\omega} \times \underline{r})dm. \end{aligned} \qquad (2.70)$$

However, from the rigid body assumption $\dot{\underline{r}} = 0$ (i.e. the mass particle does not have any relative motion within the body). Mathematically, this is expressed as

$$\frac{dx(t)}{dt} = \frac{dy(t)}{dt} = \frac{dz(t)}{dt} = 0. \qquad (2.71)$$

Therefore

$$H = \int r \times (\omega \times r) dm \tag{2.72}$$

Now,

$$r \times (\omega \times r) = [P(y^2 + z^2) - Qxy - Rxz]\underline{i} + [-Pxy + Q(x^2 + z^2) - Ryz]\underline{j}$$
$$+ [-Pxz - Qyz + R(y^2 + x^2)]\underline{k}. \tag{2.73}$$

As mentioned before, moments and products of inertia are given by:

$$\int (y^2 + z^2) dm = I_{xx} \tag{2.74a}$$

$$\int (x^2 + z^2) dm = I_{yy} \tag{2.74b}$$

$$\int (x^2 + y^2) dm = I_{zz} \tag{2.74c}$$

$$\int xy dm = I_{xy} \tag{2.75a}$$

$$\int xz dm = I_{xz} \tag{2.75b}$$

$$\int yz dm = I_{yz}. \tag{2.75c}$$

Thus,

$$\underline{H} = H_x \underline{i} + H_y \underline{j} + H_z \underline{k} \tag{2.76}$$

where

$$H_x = I_{xx}P - I_{xy}Q - I_{xz}R \tag{2.77}$$
$$H_y = -I_{xy}P + I_{yy}Q - I_{yz}R \tag{2.78}$$
$$H_z = -I_{xz}P - I_{yz}Q + I_{zz}R \tag{2.79}$$

or, in matrix form

$$\begin{bmatrix} H_x \\ H_y \\ H_z \end{bmatrix} = \begin{bmatrix} I_{xx} & -I_{xy} & -I_{xz} \\ -I_{xy} & I_{yy} & -I_{yz} \\ -I_{xz} & -I_{yz} & I_{zz} \end{bmatrix} \begin{bmatrix} P \\ Q \\ R \end{bmatrix} \tag{2.80}$$

i.e.

$$\underline{H} = I \; \underline{\omega} \tag{2.81}$$

where I is a symmetric (constant) matrix.

Note that these moments of inertia can be computed a priori once we select the body fixed frame within the body. For example, for an aircraft, the $(+)x$ axis can be taken along a fuselage reference line and the $(+)y$ axis along the right wing and the moments of inertia can be calculated based on these axes. However there is a special axes frame called a principal axes frame that one can use in computing the moments of inertia. This special frame has the property that the moment of inertia matrix $\underline{I}$ is pure diagonal and thus in the principal axes frame, the products of inertia are all zero and the diagonal moments of inertia are called principal moments of inertia'. Note that every arbitrary shaped body has a principal axes frame in it. However, for an arbitrarily shaped body, it is difficult to identify how the principal axes are oriented within the body. Of course, for some specially shaped bodies, we can easily identify the principal axes. For example, for the majority of the time if a body has symmetry about an axis, that axis qualifies as being a principal axis. Generally, in practice, a convenient general body frame is selected and the principal axes are identified in relation to this original body frame.

Recall that

$$\frac{d}{dt}\underline{H} = \frac{d\underline{H}}{dt} + \underline{\omega} \times \underline{H} = \underline{\dot{H}} + \underline{\omega} \times \underline{H} = \sum \underline{M}_{\text{applied}}. \tag{2.82}$$

Carrying out the cross multiplication and gathering all the $\underline{i}$, $\underline{j}$, and $\underline{k}$ components we finally get the scalar moment equations as

$$\dot{H}_x + (QH_z - RH_y) = L \quad \text{rolling moment equation} \tag{2.83a}$$

$$\dot{H}_y + (RH_x - PH_z) = M \quad \text{pitching moment equation} \tag{2.83b}$$

$$\dot{H}_z + (PH_y - QH_x) = N \quad \text{yawing moment equation.} \tag{2.83c}$$

Recalling that

$$\frac{d}{dt}\underline{H} = \frac{d\underline{H}}{dt} + \underline{\omega} \times \underline{H} = \underline{\dot{H}} + \underline{\omega} \times \underline{H} = \sum \underline{M}_{\text{applied}} \tag{2.84}$$

where

$$\underline{\dot{H}} = \dot{H}_x \underline{i} + \dot{H}_y \underline{j} + \dot{H}_z \underline{k} \tag{2.85a}$$

$$\dot{H}_x = I_{xx}\dot{P} - I_{xy}\dot{Q} - I_{xz}\dot{R} \tag{2.85b}$$

$$\dot{H}_y = -I_{xy}\dot{P} + I_{yy}\dot{Q} - I_{yz}\dot{R} \tag{2.85c}$$

$$\dot{H}_z = -I_{xz}\dot{P} - I_{yz}\dot{Q} + I_{zz}\dot{R}. \tag{2.85d}$$

Since an aircraft is symmetric about the xz plane, $I_{xy} = I_{yz} = 0$. Thus we finally get the moment equations, which describe the rotational motion of the aircraft:

$$I_{xx}\dot{P} - I_{xz}\dot{R} + QR(I_{zz} - I_{yy}) - I_{xz}PQ = L \tag{2.86a}$$

$$I_{yy}\dot{Q} + RP(I_{xx} - I_{zz}) + I_{xz}(P^2 - R^2) = M \tag{2.86b}$$

$$I_{zz}\dot{R} - I_{xz}\dot{P} + PQ(I_{yy} - I_{xx}) + I_{xz}QR = N. \tag{2.86c}$$

This completes the discussion of aircraft equations of motion.

Note that these equations are nonlinear, highly coupled first order differential equations, which are obviously very difficult to solve analytically.

Having reached an important juncture in the derivation of equations of motion for an aircraft, it is time to pause and review and summarize all these equations of motion in a compact way in one place.

2.3.2 Review of the Equations of Motion for Aircraft

XYZ or $X_f Y_f Z_f$ → inertial fixed eeference frame($\underline{I}, \underline{J}, \underline{K}$ or $\underline{i}_f, \underline{j}_f, \underline{k}_f$ are unit vectors)

xyz → body axes reference frame($\underline{i}, \underline{j}, \underline{k}$ are unit vectors)

$\underline{v}_c$ = velocity of the mass center of mass of the body (i.e. a/c) $= U\underline{i} + V\underline{j} + W\underline{k}$

$\underline{\omega}$ = angular velocity of the body $= P\underline{i} + Q\underline{j} + R\underline{k}$

$\underline{F}_A$ = aerodynamic forces $= F_{A_x}\underline{i} + F_{A_y}\underline{j} + F_{A_z}\underline{k}$

$\underline{g}$ = gravitational acceleration $= g_x\underline{i} + g_y\underline{j} + g_z\underline{k}$

$\underline{M}_A$ = aerodynamic moments $= L_A\underline{i} + M_A\underline{j} + N_A\underline{k}$

$\underline{M}_T$ = propulsive moments $= L_T\underline{i} + M_T\underline{j} + N_T\underline{k}$

$\underline{F}_T$ = thrust forces $= F_{T_x}\underline{i} + F_{T_y}\underline{j} + F_{T_z}\underline{k}$

Ψ, Θ, and Φ → Euler angles

$$m(\dot{U} - VR + WQ) = -mg \sin \Theta + F_x$$
$$m(\dot{V} + UR - WP) = mg \sin \Phi \quad \cos \Theta + F_y \tag{2.87}$$
$$m(\dot{W} - UQ + VP) = mg \cos\Phi \quad \cos \Theta + F_z$$

$$I_{xx}\dot{P} - I_{xz}\dot{R} - I_{xz}PQ + (I_{zz} - I_{yy})RQ = L$$
$$I_{yy}\dot{Q} + (I_{xx} - I_{zz})PR - I_{xz}(P^2 - R^2) = M \tag{2.88}$$
$$I_{zz}\dot{R} - I_{xz}\dot{P} + (I_{yy} - I_{xx})PQ + I_{xz}QR = N$$

$$P = \dot{\Phi} - \dot{\Psi} \sin \Theta$$
$$Q = \dot{\Theta} \cos \Phi + \dot{\Psi} \cos \Theta \; \sin \Phi \tag{2.89a}$$
$$R = \dot{\Psi} \cos \Theta \; \cos \Phi - \dot{\Theta} \sin \Phi \quad .$$

It is helpful to view this relationship between body angular rates and the Euler angle rates in a matrix format as follows:

$$\begin{bmatrix} P \\ Q \\ R \end{bmatrix} = \begin{bmatrix} 1 & 0 & -\sin \Theta \\ 0 & \cos \Phi & \cos \Theta \sin \Phi \\ 0 & -\sin \Phi & \cos \Theta \cos \Phi \end{bmatrix} \begin{bmatrix} \dot{\Phi} \\ \dot{\Theta} \\ \dot{\Psi} \end{bmatrix} . \tag{2.89b}$$

Note that, in general,

$$P \neq \dot{\Phi} \quad Q \neq \dot{\Theta} \quad R \neq \dot{\Psi}. \tag{2.89c}$$

In an exercise at the end of this chapter, the student is asked to derive the above relationship between body angular rates and the Euler angle rates.

Alternatively, it is insightful to look at this same relationship linking Euler angle rates to the body angular rates as follows:

$$\dot{\Phi} = P + Q \sin \Phi \tan \Theta + R \cos \Phi \tan \Theta \tag{2.89d}$$

$$\dot{\Theta} = Q \cos \Phi - R \sin \Phi \tag{2.89e}$$

$$\dot{\Psi} = (Q \sin \Phi + R \cos \Phi) \sec \Theta. \tag{2.89f}$$

Finally,

$$\begin{bmatrix} \underline{I} \\ \underline{J} \\ \underline{K} \end{bmatrix} = \begin{bmatrix} \cos\Theta\cos\Psi & \sin\Phi\sin\Theta\cos\Psi - \cos\Phi\sin\Psi & \cos\Phi\sin\Theta\cos\Psi + \sin\Phi\sin\Psi \\ \cos\Theta\sin\Psi & \sin\Phi\sin\Theta\sin\Psi + \cos\Phi\cos\Psi & \cos\Phi\sin\Theta\sin\Psi - \sin\Phi\cos\Phi \\ -\sin\Theta & \sin\Phi\cos\Theta & \cos\Phi\cos\Theta \end{bmatrix}$$

$$\times \begin{bmatrix} \underline{i} \\ \underline{j} \\ \underline{k} \end{bmatrix}. \tag{2.90}$$

At this point in time, we observe that we have a set of nonlinear, highly coupled, first order constant coefficient ordinary differential equations completely describing the translational and rotational motion of an aircraft in a flat earth situation. However, these equations still do not help us to visualize the flight path of an aircraft with respect to a ground station on Earth (i.e. with respect to the inertial frame, which for our current purposes is a frame fixed at a point on the surface of the Earth). The reason the above equations do not serve that purpose is that all these equations are expressed in the body fixed (moving) coordinate frame. In order to get the orientation and flight path of the aircraft with respect to the inertial frame, we need to relate these body fixed frame components to the inertial frame components. Recall that this was done with the help of Euler angles. In what follows, we now present a procedure to visualize the orientation and flight path of an aircraft.

2.3.3 Orientation and Flight Path of the Aircraft Relative to a Fixed Frame

Recall that we chose to express:

$$\underline{v}_c(t) = U(t)\underline{i} + V(t)\underline{j} + W(t)\underline{k} \tag{2.91}$$

where $\underline{v}_c$ is the velocity of the aircraft in the body frame components, with respect to the inertial frame. Note that the components of velocity $U(t)$, $V(t)$ and $W(t)$ are known quantities, measured from aircraft sensors. However, what we really want is the plane's position and velocity in the inertial frame components (i.e. an airport control tower):

$$\underline{v}_c(t) = U_f(t)\underline{I} + V_f(t)\underline{J} + W_f(t)\underline{K}$$
$$= \dot{p}_N(t)\underline{I} + \dot{p}_E(t)\underline{J} + [-\dot{h}(t)]\underline{K} \tag{2.92}$$

Here $\dot{p}_N$, $\dot{p}_E$ and $\dot{h}$ denote the time derivatives of the position coordinates (i.e. the velocity components) in the inertial frame, which has $(+)X$ towards the north, $(+)Y$ towards the east, and $(+)Z$ pointing downwards (the NED frame might be used interchangeably with inertial frame). Keep in mind that the height h in equation 2.92 has a negative sign in front of it because $h(t)$ is height (measured above the ground) where as $\underline{K}$ is taken positive vertically downwards. In order to get the position of the aircraft with respect to the inertial frame, we first get the velocity components with respect to the inertial frame and then integrate, i.e.:

$$\underline{r}_c(t) = \left[\int U_f(t)dt\right]\underline{I} + \left[\int V_f(t)dt\right]\underline{J} + \left[\int W_f(t)dt\right]\underline{K} \tag{2.93}$$
$$= p_N(t)\underline{I} + p_E(t)\underline{J} + [-h(t)]\underline{K}$$

where $\underline{r}_c(t)$ is the flight path of aircraft with respect to the inertial frame.

Fortunately, it is easy to see how to we get U_f, V_f and W_f, the components of velocity in the inertial frame. Remember that $U(t)$, $V(t)$ and $W(t)$ are quantities that can be measured by instrumentation on the aircraft. So knowing $U(t)$, $V(t)$ and $W(t)$, we have to compute U_f, V_f and W_f. This can be done easily through the Euler angles. We have already established the relationship between body frame components and the inertial frame components via the composite rotation matrix S. Thus

$$\begin{bmatrix} U_f \\ V_f \\ W_f \end{bmatrix} = [\text{the composite rotation matrix } S] \begin{bmatrix} U \\ V \\ W \end{bmatrix} \qquad (2.94)$$

or

$$\begin{bmatrix} \underline{I} \\ \underline{J} \\ \underline{K} \end{bmatrix} = [\text{the composite rotation matrix } S] \begin{bmatrix} \underline{i} \\ \underline{j} \\ \underline{k} \end{bmatrix}. \qquad (2.95)$$

This composite rotation matrix S was derived in the previous chapter. Recall that these Euler angles allow us to relate the body frame components to the inertial frame components and vice versa. Thus:

$$\begin{bmatrix} U_f \\ V_f \\ W_f \end{bmatrix} = [(\text{Rot}\Psi)(\text{Rot}\Theta)(\text{Rot}\Phi)] \begin{bmatrix} U \\ V \\ W \end{bmatrix} \qquad (2.96)$$

i.e.

$$\begin{bmatrix} \underline{I} \\ \underline{J} \\ \underline{K} \end{bmatrix} = [(Rot\Psi)(Rot\Theta)(Rot\Phi)] \begin{bmatrix} \underline{i} \\ \underline{j} \\ \underline{k} \end{bmatrix} \qquad (2.97)$$

$$\begin{bmatrix} \underline{I} \\ \underline{J} \\ \underline{K} \end{bmatrix} = \begin{bmatrix} \cos\Theta\cos\Psi & \sin\Phi\sin\Theta\cos\Psi - \cos\Phi\sin\Psi & \cos\Phi\sin\Theta\cos\Psi + \sin\Phi\sin\Psi \\ \cos\Theta\sin\Psi & \sin\Phi\sin\Theta\sin\Psi + \cos\Phi\cos\Psi & \cos\Phi\sin\Theta\sin\Psi - \sin\Phi\cos\Psi \\ -\sin\Theta & \sin\Phi\cos\Theta & \cos\Phi\cos\Theta \end{bmatrix}$$
$$\times \begin{bmatrix} \underline{i} \\ \underline{j} \\ \underline{k} \end{bmatrix}.$$

2.3.4 Procedure to get the Flight Path with Respect to a Fixed Frame

1. For given torques L, M, and N (they are zero if uncontrolled motion is of interest) and inertia distribution I_{xx}, I_{yy}, I_{zz}, etc. integrate equations 2.88 to P, Q, R.
2. Substituting these P, Q, R functions, integrate the equations 2.89a to get the Euler angles, Ψ, Θ, Φ, as functions of time.
3. Knowing P, Q, R and Ψ, Θ, Φ as functions of time as well as the applied forces F_x, F_y and F_z, integrate equations 2.87 to get U, V, and W.
4. Knowing these components (in the i, j, k frame), obtain the corresponding components of the velocity vector in the inertial frame (I, J, K frame) using the direction cosine matrix (equation 2.90).

Once we get U_f, V_f, and W_f, the position coordinates in the inertial frame are obtained by:

$$p_N(t) = \int U_f dt; \quad p_E(t) = \int V_f dt; \quad -h(t) = \int W_f dt. \tag{2.98}$$

Obviously, in a real life situation, all the above steps are carried out by online computations for which extensive simulation codes with sophisticated numerical analysis tools being employed are made available by industries like Boeing and others. However, to convey the basic idea of visualizing the flight path determination in a conceptual way, in what follows, we present a very simple example in which appropriate assumptions are made to simplify the above equations and steps.

Example 2.3   An aircraft is flying straight and level at a constant velocity of 337.77 ft s^{-1}, and then performs a symmetric pull up such that $\dot{\Theta} = 0.05$ rads^{-1} = constant. Assume the aircraft's x axis is aligned with the flight path throughout the motion and that at $t = 0$, $p_N = 0$, $p_E = 0$ and $h = 5000$ ft. Find the position coordinates p_N, p_E and h at $t = 5$ s. Assume $\Psi = 0$.

Solution

This is a very special case of the general motion discussed in the equations of motion. Given that:

$U = $ constant $= 337.77$ ft s^{-1}, $V = 0$ (motion only in the vertical plane).

Since the body x axis is aligned with the free stream velocity vector throughout the motion, we see that $W = 0$.

Also, as far as rotational motion is concerned, $P = 0$ (no roll), $R = 0$ (no yaw), $\Psi = 0$, $\Phi = 0$ (because of symmetrical pull up) and $\dot{\Theta} = Q = $ constant $= 0.05$ rad s^{-1}. $\dot{\Theta} = Q$ because in this motion throughout, the body y axis and the inertial Y axis are parallel to each other.

Now

$$\begin{bmatrix} \dot{p}_N \\ \dot{p}_E \\ -\dot{h} \end{bmatrix} = \begin{bmatrix} \cos\Theta\cos\Psi & * & * \\ \cos\Theta\sin\Psi & * & * \\ -\sin\Theta & * & * \end{bmatrix} \begin{bmatrix} U \\ V \\ W \end{bmatrix} \tag{2.99}$$

where $*$ denotes entries we do not care or worry about. For the above given situation, the navigation equations are:

$$\begin{bmatrix} \dot{p}_N \\ \dot{p}_E \\ \dot{h} \end{bmatrix} = \begin{bmatrix} \cos\Theta & * & * \\ 0 & * & * \\ \sin\Theta & * & * \end{bmatrix} \begin{bmatrix} U \\ 0 \\ 0 \end{bmatrix}. \tag{2.100}$$

Thus $\dot{p}_N = U\cos\Theta$, $\dot{p}_E = 0$, $\dot{h} = U\sin\Theta$.

Given that $\dot{\Theta} = 0.05$ rads^{-1} = constant; $\Theta(t) = 0.05t$

$$\therefore \dot{p}_N = 337.77\cos(0.05t)$$

$$p_N(t)|_{t_0=0, t=5} = \int_0^5 337.77\cos(0.05t)dt = 1671.3 \text{ ft}$$

$$p_E(t) = 0$$

$$\dot{h} = 337.77 \sin(0.05t) \qquad h(0) = 5000 \ \text{ft}$$

$$h(t) = \int_0^5 337.77 \sin(0.05t) \quad dt = 5210 \ \text{ft}.$$

Then the aircraft's initial position and its position after five seconds are shown in Figures 2.3 and 2.4.

2.3.5 Point Mass Performance Equations

Up until now, we have focused on the derivation of equations of motion that consisted of all the six degrees of freedom, namely three equations governing the translational degrees of freedom and three equations governing the attitude (rotational) degrees of

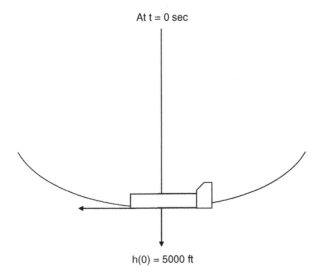

Figure 2.3 Position of aircraft at $t = 0$.

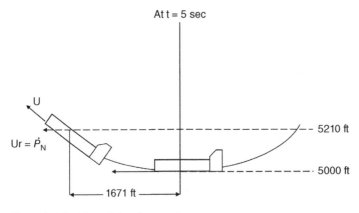

Figure 2.4 Position of aircraft at $t = 5$.

freedom. However, sometimes in trajectory planning, performance analysis (like range, endurance, etc.), and optimization related issues, there is no need for any emphasis on the attitude rotational degrees of freedom, but instead the aircraft can be treated as a particle, i.e. as a point of fixed mass. However, in this situation, the emphasis is more on how this particle is traveling in an aerodynamic environment, that is we attempt to follow the wind. As such, we employ the so-called vehicle carried frame (as [4] denotes it), rather than a vehicle fixed frame, which can still be present. The flat Earth approximation still holds. Therefore we now need to concern ourselves with two body frames.

1. The vehicle-carried frame, or the wind frame, indicated by the subscript W. The origin of this frame is fixed to the center of mass of the vehicle. Its orientation moves with the body as such: its X_W axis aligned with the relative wind v_∞; the Z_W axis is in the longitudinal plane of the aircraft; and the Y_W axis is directed along the right wing of the aircraft normal to the X_W axis.
2. The inertial frame, indicated by subscript I. As you recall, the inertial frame is fixed on a flat earth defined such that its X_I axis is positive North; Y_I axis positive east; and Z_I axis positive downwards.

Of course, the body-fixed frame, indicated in this section with the subscript V can still be present. This frame, as you recall, has its origin at the center of mass; the X_V axis pointing out the nose of the aircraft; the Y_V axis pointing out the right wing of the aircraft; and the Z_V mutually perpendicular.

The angle of attack α and sideslip β relate the wind frame to the body fixed frame. Here, α lies in the $X_V Z_V$ plane, and β lies in the $X_W Y_W$ plane. To relate the wind frame and inertial frame, we once again enlist the aid of Euler angles ψ_W (labeled velocity heading angle), γ (labeled the flight path angle) and ϕ_W (labeled velocity bank angle). The subscript W is to distinguish these Euler angles from those relating the body fixed frame to the inertial frame. Then the direction cosine matrix becomes:

$$
\begin{bmatrix} \overline{i_W} \\ \overline{j_W} \\ \overline{k_W} \end{bmatrix} = \begin{bmatrix} 1 & 0 & 0 \\ 0 & \cos\phi_W & \sin\phi_W \\ 0 & -\sin\phi_W & \cos\phi_W \end{bmatrix} \begin{bmatrix} \cos\gamma & 0 & -\sin\gamma \\ 0 & 1 & 0 \\ \sin\gamma & 0 & \cos\gamma \end{bmatrix} \begin{bmatrix} \cos\psi_W & \sin\psi_W & 0 \\ -\sin\psi_W & \cos\psi_W & 0 \\ 0 & 0 & 1 \end{bmatrix} \begin{bmatrix} \overline{i_I} \\ \overline{j_I} \\ \overline{k_I} \end{bmatrix}.
$$

$$(2.101)$$

Notice that in the wind frame velocity has a component only along the X_W axis, as such:

$$\overline{u_V} = u_V \overline{i_W} + 0\overline{j_W} + 0\overline{k_W}. \tag{2.102}$$

Likewise, the aerodynamic force vector can be decomposed into the drag, sideforce, and lift

$$\overline{F_A} = -D\overline{i_W} + S\overline{j_W} - L\overline{k_W}. \tag{2.103}$$

Here the letter L is used to denote the lift force, letter D to denote the drag force and letter S to denote the side force. The thrust forces in the wind frame become:

$$\overline{F_T} = T_x \overline{i_W} + T_y \overline{j_W} + T_z \overline{k_W}. \tag{2.104}$$

The main advantage of the wind frame is the simplification of the velocity and aerodynamic force vectors.

We now derive the equations of motion in the wind axes frame. We start with Newton's second law of motion, stated here again for convenience:

$$m\frac{d}{dt}\overline{u_V} = \sum \overline{F}. \tag{2.105}$$

Recall Charles's theorem, also restated here for convenience

$$\left(\frac{d\overline{u_V}}{dt}\right)_I = \left(\frac{d\overline{u_V}}{dt}\right)_W + \overline{\omega} \times \overline{u_V}. \tag{2.106}$$

Defining angular velocity ω as

$$\overline{\omega} \equiv P_W \overline{i_W} + Q_W \overline{j_W} + R_W \overline{k_W} \tag{2.107}$$

and carrying out the cross product, we arrive at the following:

$$\left(\frac{d\overline{u_V}}{dt}\right)_I = \dot{u}_V \overline{i_W} + u_V R_W \overline{j_W} - u_V Q_W \overline{k_W}. \tag{2.108}$$

Substituting Equations 2.108, 2.102, 2.103, and 2.104 into Newton's second law, we arrive at the three scalar equations of motion:

$$m\dot{u}_V = mg_{xW} + F_{Px} - D \tag{2.109}$$

$$mu R_W = mg_{yW} + F_{Py} + S \tag{2.110}$$

$$-mu_V Q_W = mg_{zW} + F_{Pz} - L. \tag{2.111}$$

Notice that the components of gravitational force in the wind frames $g_{xW}, g_{yW},$ and g_{zW} need to be related to the inertial frame using Euler angles. Analogous to the gravitational force in the vehicle-fixed frame, the gravitational forces in the vehicle-carried, or wind frame, are:

$$g_{xW} = -g \sin \gamma \tag{2.112}$$

$$g_{yW} = g \sin \phi_W \cos \gamma \tag{2.113}$$

$$g_{zW} = g \cos \phi_W \cos \gamma. \tag{2.114}$$

It can be shown that the Euler rates relating the wind frame to the inertial frame are expressed:

$$P_W = \dot{\phi}_W - \dot{\psi}_W \sin \gamma \tag{2.115}$$

$$Q_W = \dot{\gamma} \cos \phi_W + \dot{\psi}_W \sin \phi_W \cos \gamma \tag{2.116}$$

$$R_W = \dot{\psi}_W \cos \phi_W \cos \gamma - \dot{\gamma} \sin \phi_W. \tag{2.117}$$

Substituting Equations 2.112 through 2.117 into equations 2.109 through 2.111, we arrive at the scalar equations of translational motion in the wind frame

$$m\dot{u}_V = -mg \sin \gamma + F'_{Px} - D \tag{2.118}$$

$$mu_V(\dot{\psi}_W \cos \phi_W \cos \gamma - \dot{\gamma} \sin \phi_W) = mg \sin \phi_W \cos \gamma + F'_{PY} + S \tag{2.119}$$

$$mu_V(\dot{\gamma} \cos \phi_W + \dot{\psi}_W \sin \phi_W \cos \gamma) = -mg \cos \phi_W \cos \gamma - F'_{PZ} + L. \tag{2.120}$$

Now, supposing the thrust acts along the X_V axis, the translational equations become:

$$m\dot{u}_V = -mg \sin \gamma + F'_{Px} - D + T \cos \alpha \cos \beta \tag{2.121}$$

$$mu_V(\dot{\psi}_W \cos \phi_W \cos \gamma - \dot{\gamma} \sin \phi_W) = mg \sin \phi_W \cos \gamma + T \cos \alpha \sin -\beta + S \tag{2.122}$$

$$mu_V(\dot{\gamma} \cos \phi_W + \dot{\psi}_W \sin \phi_W \cos \gamma) = -mg \cos \phi_W \cos \gamma + T \sin \alpha + L. \tag{2.123}$$

It can also be shown that the position equations in the wind frame are:

$$\dot{X}_I = u_V \cos \gamma \cos \psi_W \tag{2.124}$$

$$\dot{Y}_I = u_V \cos \gamma \sin \psi_W \tag{2.125}$$

$$\dot{h} = u_V \sin \gamma. \tag{2.126}$$

These above equations together completely describe the point mass performance equations, which are helpful in trajectory planning, performance analysis, and any optimization exercises in which taking the aircraft as a point mass is justified.

Note that the above vehicle carried frame is strictly used only for the translational motion of the aircraft and that too for simple performance analysis wherein treating the aircraft as a point mass is justified. It is important to realize that there is a significant conceptual difference between a vehicle-carried frame and a vehicle-fixed frame. It is left as an exercise for the student to argue as to why this vehicle-carried frame analysis cannot be used for describing the rotational motion equations (see Exercise 2.6).

In the next section, we focus on the task of specializing the general rigid body equations of motion to the spacecraft specific dynamics.

2.4 Specialization of Equations of Motion to Spacecraft

2.4.1 Translational Motion: Orbit Equation

Recall that the dominant force in space is gravity force governed by Newton's law of gravitation which states that the gravitational force between the Earth and the (smaller mass) body is inversely proportional to the square of the distance between them acting along the line joining the centers of the two bodies. Thus

$$\underline{F}_g = m\frac{d\underline{v}_c}{dt} \tag{2.127}$$

$$= -\frac{GMm}{r_c^3}\underline{r}_c. \tag{2.128}$$

Rearranging,

$$m\frac{d^2\underline{r}_c}{dt^2} + \frac{GMm}{r_c^3}\underline{r}_c = 0 \tag{2.129}$$

or

$$m\frac{d^2\underline{r}_c}{dt^2} + m\frac{\mu}{r_c^3}\underline{r}_c = 0 \tag{2.130}$$

where $\mu = GM$ is a known, physical constant. G is the universal constant of gravitation with a value 6.6695×10^{-11} m^3 kg^{-1} s^{-2}. The combined constant $\mu = GM$ with Earth taken as the main body exerting the gravitational force (with satellite being the other small body) has a value in the MKS units as $\mu = 3.986 \times 10^{14}$ m^3 s^{-2} and is referred to as the Earth gravitational constant. Keep in mind that this physical constant has a numerical value that depends on the units chosen as well as on the issue of which celestial body is the main body of gravitational attraction in the two-body problem we are considering. Clearly the two-body problem we are looking into in this chapter is the one with Earth and the satellite together.

This is the famous orbit equation, the study of which leads to a separate course on orbital mechanics, which is out of scope of this book. The interested student needs to take a full length course, typically offered as a technical elective at the undergraduate senior level.

However, the orbit equation can be simplified using the concept of the two-body problem in orbital mechanics where the satellite and the Earth are the only two bodies considered with the satellite experiencing only the Earth's gravitational field. In this problem formulation, the satellite is taken as point mass. In what follows, we present the translational equations of motion for a satellite, taken as a point mass in Earth's gravitational field.

2.4.2 Point Mass Satellite Translational Motion Equations in Earth's Gravitational Field

This two-body problem formulation is pictorially represented in Figure 2.5. Going forward, we omit the subscript C for the position vector of the point mass satellite (with mass m) and simply denote it as r. These equations of motion are derived using the Lagrangian method, which is based on the concept of conservation of energy. We use the polar coordinates, namely radial direction along r, the transverse planar direction via the angle θ and the azimuthal direction via the angle ϕ. Note that the polar coordinates are also sometimes called spherical coordinates and are thus different from the standard Cartesian coordinates. The relationships between these different coordinates are discussed in [3].

To derive the equations of motion in this framework, we use the Lagrangian method as opposed to the Newtonian method we adopted before. Note that Newton's method

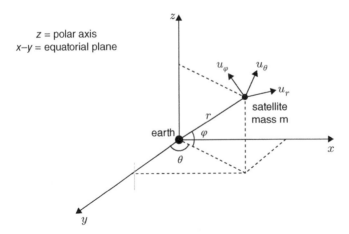

Figure 2.5 Satellite in orbit.

is based on the momentum (linear and angualr momentum) concept, whereas the Lagrangian method is based on the concept of energy. Accordingly, we obtain the kinetic energy and potential energy of the point mass and derive the equations of motion by evaluating the rate of change of the these energies via the Lagrangian formula.

The kinetic energy is basically $K = \frac{1}{2}mv^2$ (where v is the magnitude of the satellite's velocity) and is given by

$$K = \frac{m}{2}(\dot{r}^2 + (r\dot{\phi})^2 + (r\dot{\theta}\cos\phi)^2). \tag{2.131}$$

The potential energy P is given by $P = -\mu m/r$ where μ is the gravitational constant mentioned before.

The Lagrangian function is defined as the difference between the kinetic and potential energies and is denoted by $\mathcal{L} = K - P$. Then the dynamics of the system is specified by the famous Lagrange equations, namely that

$$\frac{d}{dt}\left(\frac{\partial L}{\partial \dot{r}}\right) - \frac{\partial L}{\partial r} = u_r \tag{2.132}$$

$$\frac{d}{dt}\left(\frac{\partial L}{\partial \dot{\theta}}\right) - \frac{\partial L}{\partial \theta} = u_\theta \tag{2.133}$$

$$\frac{d}{dt}\left(\frac{\partial L}{\partial \dot{\phi}}\right) - \frac{\partial L}{\partial \phi} = u_\phi. \tag{2.134}$$

Taking derivatives as needed, and assembling all the terms, we finally obtain the following equations.

$$m\left(\ddot{r} - r\dot{\theta}^2\cos^2\phi - r\dot{\phi}^2 + \frac{\mu}{r^2}\right) = u_r \tag{2.135}$$

$$m(\ddot{\theta}r^2\cos^2\phi + 2r\dot{r}\dot{\theta}\cos^2\phi - 2r^2\dot{\theta}\dot{\phi}\cos\phi\sin\phi) = (r\cos\phi)u_\theta \tag{2.136}$$

$$m(\ddot{\phi}r^2 + r^2\dot{\theta}^2\cos\phi\sin\phi + 2r\dot{r}\dot{\phi}) = ru_\phi. \tag{2.137}$$

As before, it can be seen that these equations are also highly nonlinear coupled ordinary differential equations. At a later stage, we linearize these equations about an equilibrium solution, as will be discussed later in Chapter 3, and use the resulting linearized equations for control design purposes.

One solution of these equations turns out to be the situation that corresponds to a circular, equatorial orbit, given by the values

$$\underline{x}^*(t) = \begin{bmatrix} r^*(t) \\ \dot{r}^*(t) \\ \theta^*(t) \\ \dot{\theta}^*(t) \\ \phi^*(t) \\ \dot{\phi}^*(t) \end{bmatrix} = \begin{bmatrix} r_0 \\ 0 \\ \omega t \\ \omega \\ 0 \\ 0 \end{bmatrix} \tag{2.138}$$

where r_0 is the radius of the circular orbit. Note that for this nominal situation, the control variables are all zero, i.e

$$\underline{u}^*(t) = \underline{0}. \tag{2.139}$$

Thus, we linearize the above nonlinear equations about this nominal circular, equatorial orbit, following the linearization procedure discussed later in Chapter 3, and attempt to design control systems to maintain this nominal scenario under the presence of disturbances and perturbations.

We continue the discussion of the equations of motion development for satellites/spacecraft by now switching our attention to the rotational (attitude) motion of the vehicle around its center of mass. Note that in a previous section, we already obtained the rotational motion equations for a general rigid body in a three dimensional space. However, in this section, we extend those equations specifically for a satellite acknowledging the fact that the body is orbiting around the Earth. In other words, the coupling between orbital motion and the satellite's rotational (attitude) motion is taken into consideration. For simplicity, we focus only on the simple case of an ideal circular Keplerian orbit without any orbital perturbation taken into account.

2.4.3 Rotational (Attitude) Motion Equations for a Satellite in a Circular Orbit

For this, we first define an orbit reference frame and a body fixed reference frame (just as before for the aircraft case) and relate these two frames by again the Euler angles. The orbit reference frame is selected such that its x axis, denoted by x_o, is along the tangential direction to the orbit plane. Let it be the northerly axis. Rotation about this x_o axis becomes the roll rotation. Then following the right hand rule, the orbit y axis, denoted by y_o, is perpendicular to the orbital plane, in the easterly direction. Rotation about this y_o axis is becomes the pitch axis rotation. Finally the downward local vertical axis is the yaw axis, denoted by z_o. Thus when the satellite body fixed frame axes, i.e. the body xyz frame, coincides with the orbital $x_o y_o z_o$ frame, which is labeled as the nominal situation, the Euler angles are zero. When these two frames differ by the standard Euler angles with Ψ as the first rotation about the z_o axis, Θ as the second rotation about the intermediate y axis and finally the last rotation Φ about the x axis, they become the error angles in the satellite body roll, pitch, and yaw angles. Thus the relationship between satellite body axes and the orbital frame axes is as usual given by the composite rotation matrix as follows:

$$\begin{bmatrix} x \\ y \\ z \end{bmatrix} = \begin{bmatrix} \cos\Theta\cos\Psi & \cos\Theta\sin\Psi & -\sin\Theta \\ \sin\Phi\sin\Theta\cos\Psi - \cos\Phi\sin\Psi & \sin\Phi\sin\Theta\sin\Psi + \cos\Phi\cos\Psi & \sin\Phi\cos\Theta \\ \cos\Phi\sin\Theta\cos\Psi + \sin\Phi\sin\Psi & \cos\Phi\sin\Theta\sin\Psi - \sin\Phi\cos\Psi & \cos\Phi\cos\Theta \end{bmatrix}$$
$$\times \begin{bmatrix} x_o \\ y_o \\ z_o \end{bmatrix}.$$

Note that satellite body kinematics relative to the orbiting frame (i.e. the relationship between body angular rates to the Euler angle rates) is as before given by

$$\begin{bmatrix} \omega_x \\ \omega_y \\ \omega_z \end{bmatrix} = \begin{bmatrix} 1 & 0 & -\sin\Theta \\ 0 & \cos\Phi & \cos\Theta\sin\Phi \\ 0 & -\sin\Phi & \cos\Theta\cos\Phi \end{bmatrix} \begin{bmatrix} \dot\Phi \\ \dot\Theta \\ \dot\Psi \end{bmatrix}. \tag{2.140}$$

The overall attitude dynamics of the spacecraft is given by the dynamics of the spacecraft body frame relative to an inertial frame. For that, we relate the body frame to the orbit frame and then relate the orbit frame to the inertial frame. For a general orbit, this is a quite a formidable task. To simplify matters, this is where we assume the orbit to be a simple circular, planar, ideal Keplerian orbit. With this assumption, we realize that in this special case, the orbit y_o axis is inertially fixed since it is parallel to the orbit angular momentum vector, and the orbiting frame rotates once per orbit about y_o with a constant angular velocity given by the orbital rate, denoted by

$$\omega_o = \sqrt{\frac{\mu}{r_o^3}} \tag{2.141}$$

where μ is the gravitational constant of the earth and r_o is the radius of the circular orbit. Note that components of the angular velocity vector of the orbit frame with respect to the inertial frame, namely, $\overline{\omega_{ol}^T}$, are $[0 - \omega_o \quad 0]$. The minus sign is due to the realization that the direction of y_o is the negative of the direction of the orbital angular momentum. Thus, the body angular velocity components with respect to the inertial frame, are finally given by

$$\omega_x = \dot{\Phi} - \dot{\Psi} \sin \Theta \tag{2.142}$$

$$\omega_y = (\dot{\Theta} - \omega_o) \cos \Phi + \dot{\Psi} \sin \Phi \cos \Theta \tag{2.143}$$

$$\omega_z = \dot{\Psi} \cos \Theta \cos \Phi - \dot{\Theta} \sin \Phi. \tag{2.144}$$

For now, the above nonlinear relationships relate the body angular rates with respect to the inertial frame (expressed in body frame components) through the Euler angles. Since it is difficult to analyze these nonlinear equations, we eventually linearize these about a given nominal scenario, which is the subject matter in a later chapter. The main purpose of the development of nonlinear equations up to this point is to parallel the treatment of the subject we have carried out in the aircraft case, so that the student can see the similarities and differences between the those two cases.

With that spirit, let us now complete the discussion on the development of the rest of the set of equations, namely the rotational attitude motion equations of a rigid body (which in our current case is the spacecraft). For that, let us recall the general rigid body rotational motion equations we derived in the first few sections. We had, for angular momentum:

$$\underline{H} = H_x \underline{i} + H_y \underline{j} + H_z \underline{k} \tag{2.145}$$

i.e.

$$H_x = I_{xx}\omega_x - I_{xy}\omega_y - I_{xz}\omega_z \tag{2.146}$$

$$H_y = -I_{xy}\omega_x + I_{yy}\omega_y - I_{yz}\omega_z \tag{2.147}$$

$$H_x = -I_{xz}\omega_x - I_{yz}\omega_y + I_{zz}\omega_z. \tag{2.148}$$

This is expressed more conveniently in matrix form

$$\underline{H} = \underline{I}\underline{\omega} \tag{2.149}$$

where

$$\begin{bmatrix} H_x \\ H_y \\ H_z \end{bmatrix} = \begin{bmatrix} I_{xx} & -I_{xy} & -I_{xz} \\ -I_{xy} & I_{yy} & -I_{yz} \\ -I_{xz} & -I_{yz} & I_{zz} \end{bmatrix} \begin{bmatrix} \omega_x \\ \omega_y \\ \omega_z \end{bmatrix}. \tag{2.150}$$

Note the moment of inertia matrix $\underline{I}$ is a symmetric, positive definite matrix. Here the body's angular velocity $\underline{\omega}$ expressed in a body-fixed coordinate frame is

$$\underline{\omega} = \omega_x \underline{i} + \omega_y \underline{j} + \omega_z \underline{k}. \tag{2.151}$$

The external applied moments are then given by the rate of change of angular momentum, namely:

$$\sum \underline{M}_{\text{applied}} = M_x \underline{i} + M_y \underline{j} + M_z \underline{k} \tag{2.152}$$
$$= \underline{\dot{H}} + \underline{\omega} \times \underline{H}.$$

Thus we have:

$$M_x = \dot{H}_x + (\omega_y H_z - \omega_z H_y) \tag{2.153}$$

$$M_y = \dot{H}_y + (\omega_z H_x - \omega_x H_z) \tag{2.154}$$

$$M_z = \dot{H}_z + (\omega_x H_y - \omega_y H_x). \tag{2.155}$$

In the original body-fixed axes, we have the corresponding rotational energy as

$$2T_{\text{rot}} = I_{xx}\omega_x^2 + I_{yy}\omega_y^2 + I_{zz}\omega_z^2 - 2\omega_x\omega_z I_{xz} - 2\omega_y\omega_z I_{yz} - 2\omega_x\omega_y I_{xy}. \tag{2.156}$$

Accordingly, note that if a rigid body has a rotational energy given by

$$2T_{\text{rot}} = 20\omega_x^2 + 30\omega_y^2 + 15\omega_z^2 - 20\omega_x\omega_y - 30\omega_x\omega_z \tag{2.157}$$

the corresponding moment of inertia matrix is given by

$$I = \begin{bmatrix} 20 & -10 & -15 \\ -10 & 30 & 0 \\ -15 & 0 & 15 \end{bmatrix}. \tag{2.158}$$

At this point, we introduce the concept of principal axes. The principal axes frame is a specific body fixed frame in which the products of inertia, when calculated along these axes, becomes zero. Thus the moment of inertia matrix is a pure diagonal matrix with the diagonal elements representing the principal moments of inertia. We get much more simplified equations of motion in the principal axes frame: observe Equation 2.149 simplify to:

$$H_p = \underline{I}_p \underline{\omega}_p \tag{2.159}$$

where the subscript p indicates the relevant quantities are expressed in the principle axis frame.

Note the angular velocity ω_p in the principle axes frame is expressed as

$$\omega_p = \omega_1 \underline{e}_1 + \omega_2 \underline{e}_2 + \omega_3 \underline{e}_3 \tag{2.160}$$

and $\underline{e}_1, \underline{e}_2, \underline{e}_3$ are the unit vectors along the principal axes. In other words,

$$\underline{H}_p = H_1 \underline{e}_1 + H_2 \underline{e}_2 + H_3 \underline{e}_3. \tag{2.161}$$

Again, this is expressed best in matrix form as

$$\begin{bmatrix} H_1 \\ H_2 \\ H_3 \end{bmatrix} = \begin{bmatrix} I_1 & 0 & 0 \\ 0 & I_2 & 0 \\ 0 & 0 & I_3 \end{bmatrix} \begin{bmatrix} \omega_1 \\ \omega_2 \\ \omega_3 \end{bmatrix}. \tag{2.162}$$

Thus we get:

$$M_1 = I_1\dot{\omega}_1 + I_3\omega_2\omega_3 - I_2\omega_2\omega_3 \qquad (2.163a)$$

$$M_2 = I_2\dot{\omega}_2 + I_1\omega_1\omega_3 - I_3\omega_1\omega_3 \qquad (2.163b)$$

$$M_3 = I_3\dot{\omega}_3 + I_2\omega_1\omega_2 - I_1\omega_1\omega_2 \qquad (2.163c)$$

or

$$M_1 = I_1\dot{\omega}_1 + (I_3 - I_2)\omega_2\omega_3 \qquad (2.164a)$$

$$M_2 = I_2\dot{\omega}_2 + (I_1 - I_3)\omega_1\omega_3 \qquad (2.164b)$$

$$M_3 = I_3\dot{\omega}_3 + (I_2 - I_1)\omega_1\omega_2. \qquad (2.164c)$$

Again you can see that these differential equations are nonlinear and coupled. We will linearize these equations about some operating trajectory or nominal trajectory later and then analyze the behavior of the motion under a small perturbation assumption. Before we do this, it is important to understand the relationship between an arbitrary body fixed reference frame (denoted by subscripts x, y, z) and the principal axes reference frame (denoted by subscripts $1, 2, 3$).

If we obtain the eigenvalues and normalized eigenvectors of the 3×3 matrix $\underline{I}$, then because the $\underline{I}$ matrix is symmetric, positive, and definite, we obtain three real positive eigenvalues and the corresponding three real eigenvectors. These three real, positive eigenvalues are nothing but the three principal moments of inertia, I_1, I_2, and I_3 from Equation 2.162.

Let the corresponding three real normalized eigenvectors e^1, e^2, e^3 respectively be:

$$e^1 = \begin{bmatrix} e_{1x} \\ e_{1y} \\ e_{1z} \end{bmatrix}, \quad e^2 = \begin{bmatrix} e_{2x} \\ e_{2y} \\ e_{2z} \end{bmatrix}, \quad e^3 = \begin{bmatrix} e_{3x} \\ e_{3y} \\ e_{3z} \end{bmatrix}. \qquad (2.165)$$

Then E, the 3×3 matrix formed by stacking these eigenvectors together is

$$E = \begin{bmatrix} e^1 & e^2 & e^3 \end{bmatrix} = \begin{bmatrix} e_{1x}^1 & e_{2x}^2 & e_{3x}^3 \\ e_{1y}^1 & e_{2y}^2 & e_{3y}^3 \\ e_{1z}^1 & e_{2z}^2 & e_{3z}^3 \end{bmatrix}. \qquad (2.166)$$

Note the matrix E is an orthogonal matrix, i.e. $E^{-1} = E^T$. Therefore,

$$E^{-1}\underline{I}E = E^T\underline{I}E = I_p \qquad (2.167)$$

The elements of the E matrix can be thought of as the direction cosines matrix elements of the angles that relate the body frame to the principle axes frame. Thus we have,

$$\begin{bmatrix} \omega_x \\ \omega_y \\ \omega_z \end{bmatrix} = \begin{bmatrix} E \end{bmatrix} \begin{bmatrix} \omega_1 \\ \omega_2 \\ \omega_3 \end{bmatrix} \qquad (2.168)$$

and similarly,

$$\begin{bmatrix} \omega_1 \\ \omega_2 \\ \omega_3 \end{bmatrix} = \begin{bmatrix} E^T \end{bmatrix} \begin{bmatrix} \omega_x \\ \omega_y \\ \omega_z \end{bmatrix}. \qquad (2.169)$$

The advantage of using principal axes and the principal moments of inertia is that it significantly simplifies the resulting equations. For example, the angular momentum components are simply $H_1 = I_1 \omega_1$, $H_2 = I_2 \omega_2$, and $H_3 = I_3 \omega_3$. Similarly, the rotational energy, denoted by T_{rot}, can simply be written as

$$2T_{rot} = I_1 \omega_1^2 + I_2 \omega_2^2 + I_3 \omega_3^2. \tag{2.170}$$

Note that the angular momentum vector is given by

$$\underline{H} = I_1 \omega_1 \underline{e}_1 + I_2 \omega_2 \underline{e}_2 + I_3 \omega_3 \underline{e}_3. \tag{2.171}$$

We have the magnitude of the angular momentum given by

$$H^2 = I_1^2 \omega_1^2 + I_2^2 \omega_2^2 + I_3^2 \omega_3^2. \tag{2.172}$$

2.4.4 Torque-Free Motion of an Axi-symmetric Spacecraft

Recalling the rotation motion equations, with no externally applied moments,

$$I_1 \dot{\omega}_1 + (I_3 - I_2)\omega_2 \omega_3 = 0 \tag{2.173a}$$

$$I_2 \dot{\omega}_2 + (I_1 - I_3)\omega_1 \omega_3 = 0 \tag{2.173b}$$

$$I_3 \dot{\omega}_3 + (I_2 - I_1)\omega_1 \omega_2 = 0. \tag{2.173c}$$

Now suppose we consider an axi-symmetric satellite, say with $I_1 = I_2 = I_T$; the subscript T denotes the transverse moment of inertia. Also denote $I_3 = I_S$; the subscript S indicates the spin axis. Then we see that:

$$I_T \dot{\omega}_1 + (I_3 - I_T)\omega_2 \omega_3 = 0 \tag{2.174a}$$

$$I_T \dot{\omega}_2 + (I_T - I_S)\omega_1 \omega_3 = 0 \tag{2.174b}$$

$$I_S \dot{\omega}_3 = 0. \tag{2.174c}$$

From (2.174c) we observe that ω_3 is a constant. Let $\omega_3 = \Omega$. Emphasizing the earlier point, the axis 3 is the spin axis and the axes 1 and 2 are the transverse' axes. Differentiating Equation (2.174a) and substituting $\dot{\omega}_2$ from Equation (2.174b) into the resulting equation, we get,

$$\ddot{\omega}_1 + \lambda^2 \omega_1 = 0 \tag{2.175}$$

where λ is the constant:

$$\lambda = \left[\frac{(I_S - I_T)}{I_T} \right] \Omega. \tag{2.176}$$

The solution to (2.175) is of the form

$$\omega_1(t) = a \cos \lambda t + b \sin \lambda t \tag{2.177}$$

where again a and b are constants and λ is the angular frequency. Multiplying (2.174a) by ω_1 and (2.174b) by ω_2 and adding the two, we get

$$\frac{d}{dt}(\omega_1^2 + \omega_2^2) = 0 \tag{2.178}$$

which indicates the quantity $(\omega_1^2 + \omega_2^2)$ is a constant. We now take a closer look at the transverse angular velocity resultant ω_T

$$\omega_T = \sqrt{\omega_1^2 + \omega_2^2}. \tag{2.179}$$

Then the magnitude of the total angular velocity ω can be expressed as

$$\omega = \sqrt{\omega_1^2 + \omega_2^2 + \Omega^2} = \sqrt{\omega_T^2 + \Omega^2} = \text{constant}. \tag{2.180}$$

Since the angular momentum (in inertial space) $\underline{H}$

$$\frac{d}{dt}\underline{H} = 0 \Longrightarrow \underline{H} = \text{constant} \tag{2.181}$$

thus the angular momentum in inertial space has constant magnitude and direction and is thus fixed in inertial space. Let $\underline{I}, \underline{J}, \underline{K}$ be the unit vectors in inertial space and since the total angular momentum vector is fixed in space, let us select, say vector $\underline{K}$ to coincide with this angular momentum vector. This situation is pictorially represented in Figure 2.6.

The spacecraft motion consists of the spacecraft rotation about its spin axis rotating about the angular momentum vector. This latter motion is called nutational motion. This situation is very similar to a rotating top or a spinning top.

This nutational motion can be represented in terms of the body cone and space cone, as shown in Figures 2.7 and 2.8. The body cone is fixed in the body and its axis coincides with the spin axis. The space cone is fixed in space and its axis is along the direction of the angular momentum $\underline{H}$. The total angular velocity is along the line of contact between the two cones.

For a disk shaped body, $I_S > I_T$, the inside surface of the body cone rolls on the outside surface of the space cone as shown in Figure 2.7.

For a rod shaped body, $I_S < I_T$, the outside surface of the body cone rolls on the outside surface of the space cone as shown in Figure 2.8.

Figure 2.6 Torque-free motion.

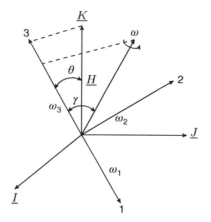

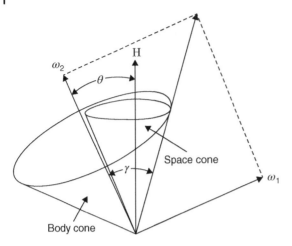

Figure 2.7 Motion for a disk shaped body.

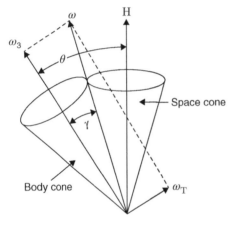

Figure 2.8 Motion for a rod shaped body.

The nutation half-angle θ and the whole wobble half angle γ are also constant and are given by:

$$\tan\theta = \frac{H_T}{H_3} = \frac{I_T\omega_T}{I_S\omega_3} = \frac{I_T\omega_T}{I_S\Omega} = \text{constant} \tag{2.182}$$

$$\tan\gamma = \frac{\omega_T}{\omega_3} = \frac{\omega_T}{\Omega} = \text{constant.} \tag{2.183}$$

This completes the discussion of the development of the basic motion specific to the attitude dynamics of a satellite in a circular orbit.

2.5 Flight Vehicle Dynamic Models in State Space Representation

Now that we have the complete set of aircraft equations of motion as well as spacecraft equations of motion , we observe a common feature for both of these sets of equations. That common feature is that they all turn out to be a set of simultaneous first order ordinary differential equations. Hence, from now on, to treat these equations in a unified framework from a systems level point of view, we express these equations in the form widely known as the state space representation of a dynamic system. This is elaborated next. Accordingly, let us now summarize the flight dynamic models of both aircraft as well as spacecraft in this state space representation, which would be useful to analyze in later chapters. State space representation of dynamic systems is a powerful concept and we review those fundamentals in the next chapter. For that reason, viewing all the equations of motion we derived up until now, whether it is an aircraft dynamics or satellite attitude dynamics, from a state space representation point of view is very beneficial.

In general, the state space description of any dynamic system described by nonlinear, ordinary differential equations is given by

$$\left.\begin{array}{l} \dot{\vec{x}} = \vec{f}(\vec{x}, \vec{u}, t) \\ \vec{y} = \vec{h}(\vec{x}, \vec{u}, t) \end{array}\right\} \tag{2.184}$$

where $\vec{x}$ is the vector of state variables, $\vec{u}$ is the vector of control variables, and $\vec{y}$ is the vector of output variables. $\vec{f}$ and $\vec{h}$ are vectors of nonlinear functions in x_i and u_i. We refer to this model as the state space representation of a dynamic system.

From the equations of motion we have derived, it can be seen that these equations can be cast in the above form of a state variable representation. We now formally summarize those equations (both for aircraft as well as for spacecraft) in the state variable form.

2.5.1 Aircraft Dynamics from the State Space Representation Point of View

Let us now summarize the final set of aircraft equations of motion we derived as follows.
The force equations describing translational motion:

$$\dot{U} = VR - WQ - g\sin\Theta + \frac{F_x}{m}$$

$$\dot{V} = WP - UR + g\sin\Phi\cos\Theta + \frac{F_y}{m} \tag{2.185}$$

$$\dot{W} = UQ - VP + g\cos\Phi\cos\Theta + \frac{F_z}{m}.$$

The moment equations describing rotational motion:

$$\gamma\dot{P} = I_{xz}[I_{xx} - I_{yy} + I_{zz}]PQ - [I_{zz}(I_{zz} - I_{yy}) + I_{xz}{}^2]RQ + I_{zz}L + I_{xz}N$$

$$I_{yy}\dot{Q} = (I_{zz} - I_{xx})PR - I_{xz}(P^2 - R^2) + M$$

$$\gamma\dot{R} = [I_{xx}((I_{xx} - I_{yy}) + I_{xz}{}^2]PQ - I_{xz}[I_{xx} - I_{yy} + I_{xz}]QR + I_{xz}L + I_{xx}N \tag{2.186}$$

$$\gamma = I_{xx}I_{zz} - I_{xz}{}^2.$$

The kinematic equations:

$$\dot{\Phi} = P + Q\sin\Phi\tan\Theta + R\cos\Phi\tan\Theta \tag{2.187}$$

$$\dot{\Theta} = Q\cos\Phi - R\sin\Phi \tag{2.188}$$

$$\dot{\Psi} = (Q\sin\Phi + R\cos\Phi)\sec\Theta. \tag{2.189}$$

And the navigation equations:

$$
\begin{bmatrix} \dot{p}_N \\ \dot{p}_E \\ \dot{h} \end{bmatrix} = \begin{bmatrix} \cos\Theta\cos\Psi & \sin\Phi\sin\Theta\cos\Psi - \cos\Phi\sin\Psi & \cos\Phi\sin\Theta\cos\Psi + \sin\Phi\sin\Psi \\ \cos\Theta\sin\Psi & \sin\Phi\sin\Theta\sin\Psi + \cos\Phi\cos\Psi & \cos\Phi\sin\Theta\sin\Psi - \sin\Phi\cos\Psi \\ \sin\Theta & -\sin\Phi\cos\Theta & -\cos\Phi\cos\Theta \end{bmatrix}
$$

$$
\times \begin{bmatrix} U \\ V \\ W \end{bmatrix}. \tag{2.190}
$$

Be careful to notice that in the above navigation equations, the last variable in the position coordinates vector is taken to be h and so the last row elements of the above navigation rotation matrix entries have a negative sign in front of them, compared to the last row elements of the original composite rotation matrix S given in Chapter 1.

Note that the above moment equations are deliberately written in such a way that in each of those equations, the left hand side has the first order derivative of only one state variable, so that it fits into the definition of state variable representation. Thus considerable effort went into getting the original equations into the above state variable form. The reason for this extra effort is that this current form can be seen to be a set of first order differential equations in the most important twelve motion variables, namely U, V, W (linear velocity components); P, Q, R (angular velocity components); Ψ, Θ, Φ (the Euler angles); and finally p_N, p_E, h (the position coordinates). These critical motion variables, whose derivatives describe the entire motion of the aircraft are labeled, in a generic fashion, as the state variables. That is the reason this full set of equations is referred to as the state variable representation of the aircraft dynamics. More about the importance and utility of state space representation of dynamic systems is discussed in the next chapter. For a complete description of the state space representation, in addition to state variables, it is important to also identify the output variables and control variables. Output variables are the variables we wish to control, that is, they are the controlled variables. Typically, output variables are either a subset of state variables or they could be new variables but as a (possibly nonlinear) function of the state variables. In the above aircraft dynamics, an example for an output variable could be the flight path angle, which we wish to control. The flight path angle is obviously a function of the above state variables and it is assumed that we are able to express the output variable in terms of these state variables. In a later chapter dealing with linear models, we can demonstrate the process of getting an output equation in terms of state variables. Next, we discuss the role of control variables, sometimes referred to as input variables. As the name implies, control variables are those which are available for us to manipulate so that we can make the output variables behave the way we want or desire. Note the subtle difference between controlled variables and control variables. In our aircraft dynamics, for example, there are many control variables such as aerodynamic forces, and moments generated by the control surfaces (like elevators, ailerons and rudder, etc.), which are discussed in detail in later chapters. Loosely speaking, the control variables are those variables that appear as forcing functions, i.e inputs, that appear in the right hand side of the differential equation. That is the reason control variables are often referred to as input variables. In flight dynamics and control jargon, control variables provide actuation and thus we can think of them as actuators. Finally, we do have what we label as measurement (or sensor) variables. Obviously, these are variables that we measure with sensors. Clearly, measurement (sensors) variables are typically a small subset of state variables as it is highly unlikely that we measure all the state variables. Again, just like output variables, we assume that we know the functional relationship between measurement variables and state variables. Note that typically the output variables happen to be the measurement variables as well, but in general conceptually, output variables could be different from measurement variables.

2.5.2 Spacecraft Dynamics from a State Space Representation Point of View

We have seen that in the spacecraft dynamics case, we have two sets of equations. One set is for the translational motion of a point mass satellite in Earth's gravitational field in polar coordinates and the second set is the spacecraft attitude rotational dynamics about its mass center. We summarize these two sets of equations separately from state space representation framework.

2.5.2.1 Satellite Point Mass Translational Equations of Motion in Polar Coordinates
These equations can be written in terms of the state, input, and output vectors, defined by

$$\underline{x}(t) = \begin{bmatrix} r(t) \\ \dot{r}(t) \\ \theta(t) \\ \dot{\theta}(t) \\ \phi(t) \\ \dot{\phi}(t) \end{bmatrix} \tag{2.191}$$

$$\underline{u}(t) = \begin{bmatrix} u_r(t) \\ u_\theta(t) \\ u_\phi(t) \end{bmatrix} \tag{2.192}$$

$$\underline{y}(t) = \begin{bmatrix} r(t) \\ \theta(t) \\ \phi(t) \end{bmatrix}. \tag{2.193}$$

Thus, the state space representation of these equations is given by

$$\dot{\underline{x}} = \underline{f}(\underline{x}, \underline{u}) = \begin{bmatrix} \dot{r} \\ r\dot{\theta}^2\cos^2\phi + r\dot{\phi}^2 - \frac{\mu}{r^2} + \frac{u_r}{m} \\ \dot{\theta} \\ -2\dot{r}\dot{\theta}/r + 2\dot{\theta}\dot{\phi}\frac{\sin\phi}{\cos\phi} + \frac{u_\theta}{mr\cos\phi} \\ \dot{\phi} \\ -\dot{\theta}^2\cos\phi\sin\phi - 2\dot{r}\dot{\phi}/r + u_\phi/mr \end{bmatrix} \tag{2.194}$$

and

$$\underline{y} = \underline{C}\underline{x} = \begin{bmatrix} 1 & 0 & 0 & 0 & 0 & 0 \\ 0 & 0 & 1 & 0 & 0 & 0 \\ 1 & 0 & 0 & 0 & 1 & 0 \end{bmatrix} \underline{x}. \tag{2.195}$$

Thus, in these satellite point mass translational motion equations in polar coordinates, we observe that we have six state variables (basically the radial, tangential and azimuthal positions and their respective velocities) and three control variables, which are nothing but the accelerations in the radial, tangential, and azimuthal directions.

2.5.2.2 Spacecraft Attitude (Rotational) Motion about its Center of Mass

Now we focus our attention on the attitude (rotational) motion of the spacecraft (i.e. satellite) about its center of mass. We recall those equations of motion we derived in the principal axes frame, set in a state space form, as follows:

$$\dot{\omega}_1 = [(I_2 - I_3)/I_1]\omega_2\omega_3 + M_1/I_1 \tag{2.196a}$$

$$\dot{\omega}_2 = [(I_3 - I_1)/I_2]\omega_1\omega_3 + M_2/I_2 \tag{2.196b}$$

$$\dot{\omega}_3 = [(I_1 - I_2)/I_3]\omega_1\omega_2 + M_3/I_3 \tag{2.196c}$$

$$\omega_1 = \dot{\Phi} - \dot{\Psi}\sin\Theta \tag{2.197}$$

$$\omega_2 = (\dot{\Theta} - \omega_o)\cos\Phi + \dot{\Psi}\sin\Phi\cos\Theta \tag{2.198}$$

$$\omega_3 = \dot{\Psi}\cos\Theta\cos\Phi - \dot{\Theta}\sin\Phi. \tag{2.199}$$

Keep in mind that the above equations are also specialized to a satellite in a circular Keplerian orbit, wherein the circular orbit's constant angular velocity is given by ω_o. Note that in this situation, the Euler angles essentially take the role of roll angle error, pitch angle error and the yaw angle error of the satellite's deviation from the nominal scenario (in which the body frame coincides with the orbit frame).

It is also useful to recall that the principal angular velocities are related to the (arbitrary) body-fixed frame angular velocities by the normalized eigenvector matrix E, given by

$$\begin{bmatrix} \omega_1 \\ \omega_2 \\ \omega_3 \end{bmatrix} = \begin{bmatrix} E^T \end{bmatrix} \begin{bmatrix} \omega_x \\ \omega_y \\ \omega_z \end{bmatrix}. \tag{2.200}$$

In the spacecraft case, we realized that the translational motion in the Earth's gravity field became the orbit equation and in the idealistic case of perfect circular orbit with no orbital perturbations, the orbital motion is essentially decoupled from the body attitude motion. It needs to be kept in mind that in the presence of the realistic, practical situation of orbital motion, where there are always orbital perturbations, this decoupling assumption may not be valid. However, for a conceptual preliminary design, we ignore this coupling and focus on the pure rotational attitude motion given by the above equations. It is easy to see that the above equations are somewhat simplified (even though they are also highly coupled, nonlinear differential equations like in the aircraft case) because these equations are written in the special body axes frame, namely, the principal axes frame. Following the discussion given for the aircraft case, it is not difficult to observe that in this dynamics, the three angular velocities ω_1, ω_2, and ω_3 qualify to be the state variables. The moments M_1, M_2, and M_3 on the right hand side of the equations act as the forcing functions. These torques can be viewed as consisting of control torques as well as disturbance torques. Obviously control torques are the control variables in this case. In the spacecraft situation, it is easy to observe that considerable effort and care needs to be expended to model and estimate the various disturbance torques acting on the satellite and that information be used in designing the needed control torques to control the attitude of the spacecraft in the presence of those disturbance torques. The same comments we made regarding output variables and measurement variables in the aircraft dynamics case equally hold in this case as well. The main point to take away from this discussion is that describing the satellite attitude dynamics in the state variable representation makes its analysis and design follow almost the same conceptual path as in the case of aircraft dynamics. This is the reason, in the next chapter, we treat the state space representation of the flight vehicle dynamics from a generic, systems viewpoint and analyze it.

As mentioned earlier, in a later chapter, we linearize these nonlinear equations about their equilibrium solutions and use them for control design purposes.

2.5.3 Conceptual Differences Between Aircraft Dynamic Models and Spacecraft Dynamic Models

Even though we observe that, in the end, both aircraft dynamics and spacecraft dynamics lend themselves neatly to follow a rigorous mathematical framework through the state space representation, there are few subtle differences from a practical viewpoint. The first observation is that the above mathematical model we arrived at for aircraft is of very high fidelity in the sense that the real motion of aircraft is likely to follow the motion suggested by this model. In that sense, aircraft engineers need to think that it is somewhat of a luxury that they have a thoroughly rigorous and reasonably dependable mathematical model. In addition, it is always possible to estimate the values of the various parameters in the equations of motion and verify their validity in a ground based wind tunnel tests from their aerodynamics colleagues. The spacecraft engineers do not have that luxury. In that sense, they are heavily dependent on the fidelity of the available mathematical models generated by first principles, such as those we developed in this chapter. Thus robustness to perturbations is a necessity in the case of satellite attitude control systems design. Of course, robustness to perturbations is of importance in any control systems design exercise but even more important in satellite attitude control systems. This is justified even more by the fact that estimating the parameters and disturbance torques encountered in the spacecraft environment is a non-trivial task in view of the lack of any ground based tests that mimic the space environment. Another observation, at least in the current state variable representation point of view, is that the spacecraft attitude dynamics model has only the principal moments of inertia as the real parameters within the model whereas in the aircraft equations, mass, moments of inertia, and products of inertia become the real parameters within the dynamics. Notice that moments and products of inertia are very much functions of the shape and size of the rigid body. In the case of aircraft, in addition, many aerodynamic coefficients that determine the stability derivatives (as discussed in later chapters) become the real parameters. It is well recognized by now that robustness to perturbations in these real parameters is of paramount importance in designing a control system for these aerospace applications.

Fundamentals such as these covered in this chapter are also available in many other textbooks such as [1, 2, 4–8] and the reader is encouraged to consult these and other books to expand their horizons, not getting limited by the essential, necessary information provided in this book. This is intentional because the only major objective of this book is to cover the most basic subject that is expected at an undergraduate level, in both aero as well as space flight vehicle dynamics. As such, in each category (i.e. aircraft and spacecraft), many elaborate and advanced concepts have not been covered deliberately because within the time limitation inherent in an undergraduate curriculum, the total content of coverage is necessarily limited anyway.

2.6 Chapter Summary

In this chapter, we first derived the equations of motion for a general rigid body in three dimensional space. Then we specialized those equations to aircraft, in which the flat Earth approximation (acceleration due to gravity taken as a constant) is enforced, along with other features specific to an aircraft. We then carry out a similar exercise making the equations specific to spacecraft (or a satellite). We recognize that in this case, the translational motion of the center of mass becomes the orbit equation, while the rotational motion equations describe the attitude dynamics of the center of mass. Finally we represented all these equations of motion in a state variable representation, clearly identifying the state variables, control variables, output variables and measurement variables. The state variable representation of these equations of motion forms the starting point for analyzing the dynamics and synthesizing control systems in subsequent chapters. With this setup in mind, the next task of understanding the importance of linearizing these nonlinear models is discussed in the next chapter.

2.7 Exercises

Exercise 2.1. An aircraft in a steady, level right turn is flying at a constant altitude and a constant turning rate ω_{turn}. If P is the vehicle's roll rate, Q the pitch rate, and R the yaw rate, assuming a flat, non-rotating Earth, show that the vehicle's roll, pitch, and yaw rates will equal

$$P = -\omega_{turn} \sin \Theta$$

$$Q = \omega_{turn} \sin \Phi \cos \Theta$$

$$R = \omega_{turn} \cos \Phi \cos \Theta$$

where Φ is the roll Euler angle and Θ is the pitch Euler angle.

Exercise 2.2. Convert the angular velocity vector $\underline{\omega}$, given in terms of a mixture of intermediate and inertial frame components as follows

$$\underline{\omega} = \underline{\dot{\Psi}} + \underline{\dot{\Theta}} + \underline{\dot{\Phi}}$$
$$= \dot{\Psi}\underline{K} + \dot{\Theta}\underline{J}_1 + \dot{\Phi}\underline{I}_2$$

into body frame components

$$\underline{\omega} = P\underline{i} + Q\underline{j} + R\underline{k}.$$

In other words, get the relationship between the body angular velocity components and the Euler angles and their rates.

Exercise 2.3. Given that the rotational kinetic energy of a rigid body about its center of mass is

$$T_{rot} = \frac{1}{2}[25\omega_x^2 + 34\omega_y^2 + 41\omega_z^2 - 24\omega_y\omega_z]$$

where x, y, z is a specified body-fixed coordinate set.
1. Determine the principal moments of inertia.
2. Calculate the angles between x, y, z and the principal axes 1, 2, 3.
3. Determine the magnitude of angular momentum, H.

Exercise 2.4. Consider a rigid body with inertia tensor

$$\vec{I} = \begin{bmatrix} 30 & -I_{xy} & -I_{xz} \\ -10 & 20 & -I_{yz} \\ 0 & -I_{zy} & 30 \end{bmatrix} (N \text{ m s}^2)$$

and angular velocity

$$\vec{\omega} = 10\vec{i} + 10\vec{j} + 10\vec{k}(\text{rad s})^{-1}.$$

If

$$\vec{H} = 200\vec{i} + 200\vec{j} + 400\vec{k}(N \text{ m s}^2)$$

determine the following.
1. Values of I_{zy}, I_{xy}, I_{xx}, and I_{yz}.
2. Principal moments of inertia.
3. Rotational kinetic energy.

Exercise 2.5. Consider a rigid body with inertia tensor

$$\vec{I} = \begin{bmatrix} 20 & -10 & 0 \\ -10 & 30 & 0 \\ 0 & 0 & 40 \end{bmatrix} (N \text{ m s}^2)$$

and angular velocity

$$\vec{\omega} = 10\vec{i} + 20\vec{j} + 30\vec{k}(\text{rad s})^{-1}.$$

1. Find the angular momentum of this body about its center of mass
2. Find the principal moments of inertia.
3. Find the rotational kinetic energy.

Exercise 2.6. Given:

$$\underline{I} = \begin{bmatrix} 10 & -7 & -8 \\ -7 & 20 & -5 \\ -8 & -5 & 30 \end{bmatrix}$$

obtain the principal moments of inertia and the orientation of the (x, y, z) frame with respect to the $(1, 2, 3)$ frame. Recall the principle moments of inertia are nothing but the three eigenvalues of the $\underline{I}$ matrix. Likewise, the direction cosine matrix is the matrix composed of the three normalized eigenvectors. We can easily determine both the eigenvalues and the normalized eigenvectors using MATLAB's built-in functions.

Exercise 2.7. Explain what difficulties you would face if you used the vehicle carried frame for describing the rotational motion of an aircraft and thus to conclude that we only can use the standard vehicle fixed frame for describing the rotational motion of the aircraft (as we have done before the discussion of the vehicle carried frame based point mass performance equations).

Exercise 2.8. Consider a rigid body with axial symmetry and principal inertia

$$I_1 = I_2 \tag{2.201}$$

$$I_3 > I_1. \tag{2.202}$$

If it is acted upon by a small body-fixed, transverse torque, $M_1 = M$, with $M_2 = M_3 = 0$, then:

(a) Write the differential equations of motion for this situation. What special nature do you observe in the resulting equations for this situation?

(b) Is ω_3 affected M_1? Explain.

(c) How is the angular momentum vector affected by M_1? Explain the angular momentum property when (i) $M_1 = 0$, (ii) $M_1 = M$ = constant and finally when $M_1 = M_1(t)$, i.e it is a time varying torque.

Bibliography

1 B.N. Agrawal. *Design of geosynchronous spacecraft*. Prentice-Hall, Englewood Cliffs, NJ, 1986.

2 M.H. Kaplan. *Modern spacecraft dynamics and control*. Wiley, New York, 1976.

3 Hanspeter Schaub and John L. Junkins. *Analytical Mechanics of Space Systems*. AIAA, 2014.

4 D.K. Schmidt. *Modern Flight Dynamics*. McGraw Hill, New York, 2012.

5 Louis V Schmidt. *Introduction to Aircraft Flight Dynamics*. AIAA Education Series, Reston, VA, 1998.

6 R.E. Stengel. *Flight Dynamics*. Princeton University Press, Princeton, 2004.

7 Brian L Stevens, Frank L Lewis, and Eric N Johnson. *Aircraft control and simulation*. Interscience, New York, 1 edition, 1992.

8 Ranjan Vepa. *Flight Dynamics, Simulation and Control for Rigid and Flexible Aircraft*. CRC Press, New York, 1 edition, 2015.

3

Linearization and Stability of Linear Time Invariant Systems

3.1 Chapter Highlights

In this chapter, we focus on the process of linearizing a set of nonlinear differential equations about a given equilibrium (or steady state or nominal) point. This process of linearization becomes an important step for the material in later chapters because it is these linearized models that we use later for control design purposes. Once we decide to deal with linearized/linear models, in this chapter, we also cover briefly a few basic concepts about the stability of continuous time linear systems, with special emphasis on a simple second order linear system. A thorough understanding of the simple second order linear system stability is of high importance as it forms the fundamental, backbone concept to deal with higher order systems later, because typically the stability behavior of all higher order systems can be inferred or approximated from their dominant second order system stability behavior. This chapter necessarily treats this subject matter in a generic systems level framework, with the intent of applying these methods to aero and space flight vehicle dynamic models in later chapters.

3.2 State Space Representation of Dynamic Systems

The state of a dynamic system is the smallest set of linearly independent variables (called state variables) such that the knowledge of these variables at $t = t_0$ together with the input at $t \geq t_0$ completely determines the behavior of the system for any time $t \geq t_0$.

When a dynamic system is modeled by ordinary differential equations, it is relatively easy to identify the set of state variables. For example, if we have a differential equation

$$\frac{d^2\theta}{dt^2} + 5\frac{d\theta}{dt} + 6\theta = e^t \tag{3.1}$$

then it is easy to observe that we need $\theta(t_0)$ and $\frac{d\theta}{dt}(t_0)$ to completely determine the behavior of $\theta(t)$ for all $t \geq t_0$. Thus $\theta(t)$ and $\frac{d\theta}{dt}(t)$ become the two state variables.

One main feature of the state space representation is that the set of differential equations are expressed in first order form (in the state variables). Thus the state space

Flight Dynamics and Control of Aero and Space Vehicles, First Edition. Rama K. Yedavalli.
© 2020 John Wiley & Sons Ltd. Published 2020 by John Wiley & Sons Ltd.

representation of the above second order differential equation is obtained by first defining

$$\left.\begin{aligned} \theta(t) &= x_1(t) \\ \text{and } \frac{d\theta(t)}{dt} &= x_2(t) \end{aligned}\right\} \rightarrow 2 \text{ state variables}$$

and then rewriting the above equation as two first order equations in the state variables $x_1(t)$ and $x_2(t)$.

$$\dot{x}_1(t) = x_2(t)$$

and

$$\dot{x}_2(t) = \frac{d^2\theta}{dt^2}(t) = -5x_2(t) - 6x_1(t) + e^t$$

i.e.

$$\begin{bmatrix} \dot{x}_1(t) \\ \dot{x}_2(t) \end{bmatrix} = \begin{bmatrix} 0 & 1 \\ -6 & -5 \end{bmatrix} \begin{bmatrix} x_1(t) \\ x_2(t) \end{bmatrix} + \begin{bmatrix} 0 \\ 1 \end{bmatrix} e^t.$$

This is in the form

$$\dot{\vec{x}} = A\vec{x} + B\vec{u}$$

where $u(t) = e^t$. This is referred to as the state space representation of the dynamic system represented by the equation 3.1.

In the above example, the equation considered is a linear differential equation and thus the resulting state space description is a linear state space description, but this representation can also be generalized to nonlinear set of equations. In fact the mathematical models of many physical systems derived from first principles turn out to be nonlinear differential equations. In particular, we focus here on systems governed by nonlinear ordinary differential equations.

So in general, the state space description of any dynamic system described by nonlinear first order ordinary differential equations is given by

$$\dot{\vec{x}} = \vec{f}(\vec{x}, \vec{u}, t)$$

where $\vec{x}$ is the state vector, $\vec{u}$ is the control vector and $\vec{f}$ is a vector of nonlinear functions in x_i and u_i. Typically, we write

$$\vec{x} \in R^n \text{ i.e. } \vec{x} = \begin{bmatrix} x_1 \\ x_2 \\ \vdots \\ x_n \end{bmatrix}$$

$$\vec{u} \in R^m \text{ i.e. } \vec{u} = \begin{bmatrix} u_1 \\ u_2 \\ \vdots \\ u_m \end{bmatrix}.$$

Consider the following three classes of nonlinear systems:

1. $\dot{\vec{x}} = \vec{f}(\vec{x}, \vec{u}, t)$
2. $\dot{\vec{x}} = \vec{f}(\vec{x}, t)$
3. $\dot{\vec{x}} = \vec{f}(\vec{x})$.

Out of these, consider the second class of systems

$$\dot{\vec{x}} = \vec{f}(\vec{x}, t), \ \vec{x}(t_0) = \vec{x}_0.$$

We assume that the above equation, has a unique solution starting at the given initial condition, i.e. we have one single solution corresponding to each initial condition. Let us denote this solution as

$$\vec{x}(t; \vec{x}_0, t_0) \equiv \vec{x}(t) \text{ for simplicity}$$
$$\vec{x}(t_0; \vec{x}_0, t_0) \equiv \vec{x}_0$$

3.2.1 Equilibrium State

In the above class of systems, a state $\vec{x}_e$ where $\vec{f}(\vec{x}_e, t) = 0$ for all t is called an equilibrium state of the system, i.e. the equilibrium state corresponds to the constant solution of the system. If the system is linear time invariant (i.e. $\vec{f}(\vec{x}, t) = A\vec{x}$), then there exists only one equilibrium state if A is non-singular and many equilibrium states if A is singular. For nonlinear systems there may be one or more equilibrium states.

Any isolated equilibrium point can always be transferred to the origin of the coordinates

$$\text{i.e. } \vec{f}(0, t) = 0$$

by a proper coordinate transformation. So one can always take $\vec{x}_e = 0$ without any loss of generality. The origin of state space is always an equilibrium point for linear systems and for linear systems all equilibrium states behave the same way (because if $\vec{x}(t)$ is a solution $\overline{\vec{x}}(t)$ is also a solution; then $\vec{x}(t) \rightarrow \overline{\vec{x}}(t)$ is also a solution for the linear system).

In such cases, the nonlinear differential equations are linearized about an equilibrium to get a linear state space representation in small motions around the equilibrium. One such linearization process is labeled as the Jacobian method and the resulting linearized state space matrix is called the Jacobian matrix.

3.3 Linearizing a Nonlinear State Space Model

Consider the general nonlinear state variable model

$$\left. \begin{array}{l} \dot{\vec{x}} = \vec{f}(\vec{x}, \vec{u}, t) \\ \vec{y} = \vec{h}(\vec{x}, \vec{u}, t) \end{array} \right\}. \tag{3.2}$$

The above set of nonlinear differential equations can be linearized about a constant, equilibrium solution, which can also be called the steady state solution. This is the most common type of linearization process, i.e. linearization about a given steady state condition. In a slightly different viewpoint, the nonlinear system of equations can also be linearized about a given nominal trajectory, where the nominal trajectory satisfies the original nonlinear set of differential equations. It does not have to be a constant solution. This linearization process can be referred to as the Jacobian linearization. Of course, Jacobian linearization holds good for constant, steady state equilibrium conditions. In what follows, we consider a simple, brute force linearization process which involves expanding the original nonlinear equations in terms of the steady state plus some perturbation. Then assuming those perturbations to be small, we neglect second and higher order terms along with making a small angle approximation whenever there are trigonometric functions involved. We discuss the more general Jacobian linearization process later in Part III of the book where we devote the entire discussion to linear state space models.

3.3.1 Linearization about a Given Steady State Condition by Neglecting Higher Order Terms

Let us assume the steady state of the system, denoted by $\vec{x}_{ss}$, $\vec{u}_{ss}$, and $\vec{y}_{ss}$, is known or given. Note that this steady state is typically a constant solution of the nonlinear system, which is nothing but the equilibrium solution of the nonlinear equations.

Then we assume that the current state is a perturbation from this steady state (equilibrium) condition. Then the difference between these steady state vector functions and the current state functions $\vec{x}(t)$, $\vec{u}(t)$, and $\vec{y}(t)$ are the perturbation functions. Thus we have

$$\vec{x}(t) = \vec{x}_{ss} + \vec{\delta x}(t); \qquad \vec{u}(t) = \vec{u}_{ss} + \vec{\delta u}(t); \qquad \vec{y}(t) = \vec{y}_{ss} + \vec{\delta y}(t). \tag{3.3}$$

We then expand the original nonlinear equations in terms of the steady state plus perturbation terms. Note that the derivative of the steady state (constant) function is zero. Similarly, if there are any trigonometric functions, such as $\sin \theta(t)$, we write $\theta(t)$ as a summation of a steady state angle θ_{ss} plus a perturbation angle $\delta\theta(t)$, and expand $\sin \theta(t)$ in terms of the steady state angle θ_{ss} and the perturbation angle $\delta\theta(t)$. Until now, we have not made any assumptions on the perturbation functions and angles. Now the formal process of linearization involves the assumption that these perturbation terms are sufficiently small. Under this assumption, we can afford to neglect the second and higher order terms (the nonlinearity causing terms) because those terms would become even much smaller than the linear terms. This also means that in the case of angles, making small angle approximation entails writing $\sin \delta\theta(t)$ as $\delta\theta(t)$ and $\cos \delta\theta(t)$ as equal to 1. Note that in a small angle approximation, we are assuming the angles to be expressed in radians.

Thus, in summary, the process of linearization necessarily involves assuming small perturbations around the nominal (or steady state or equilibrium) motion and neglecting the second and higher order terms in the dependent variable (not in the independent variable, time) and making a small angle approximation. Thus a linearized model is valid, meaning it produces reasonably accurate motion behavior only for small perturbations about the constant steady state (equilibrium) condition. Since large motions have to go through these small motion phases anyway, understanding the behavior of linearized models becomes very important and forms the basis for understanding the more involved nonlinear system behavior. Because of this, in this book, our focus and emphasis will be on linear and linearized models.

Example 3.1   Consider a simple nonlinear differential equation given by:

$$\dot{h} = U \sin \Theta.$$

Linearize the equation about the steady state values h_{ss}, U_{ss}, and Θ_{ss}, where the perturbations are taken as δh, δU, and $\delta \Theta$. Write the reference (steady state) equation as well as the linearized, perturbation equation.

Solution
We expand each of the state variables into a steady state (static) term indicated by subscript ss and a perturbation (dynamic) term as explained before. Thus, we have:

$$\frac{d}{dt}(h_{ss} + \delta h) = (U_{ss} + \delta U) \sin(\Theta_o + \delta \Theta).$$

We can expand the trigonometric terms with the following identities:

$$\sin(A + B) = \sin A \cos B + \cos A \sin B$$
$$\cos(A + B) = \cos A \cos B - \sin A \sin B.$$

We get,

$$\dot{h}_{ss} + \delta \dot{h} = U_{ss} \sin \Theta_{ss} \cos \delta \Theta + U_{ss} \cos \Theta_{ss} \sin \delta \Theta + \delta U \sin \Theta_{ss} \cos \delta \Theta$$
$$+ \delta U \cos \Theta_{ss} \sin \delta \Theta.$$

Note that since the steady state is constant, we take $\dot{h}_{ss}$ as equal to zero.

Also make small angle approximation, namely that $\sin \delta x \approx \delta x$ and $\cos \delta x \approx 1$, x here being an arbitrary angle (usually less than ~15°). With this we get,

$$\delta \dot{h} = U_{ss} \sin \Theta_{ss} \times 1 + U_{ss} \cos \Theta_{ss} \delta \Theta + \delta U \sin \Theta_{ss} \times 1 + cos\Theta_{ss} \delta U \delta \Theta.$$

Notice that the above equation still contains a nonlinear term in the perturbation variables, namely the term $\delta U \delta \Theta$. Since we assume small perturbations for linearization purposes, we need to neglect and remove this nonlinear term. Thus we get

$$\delta \dot{h} = U_{ss} \sin \Theta_{ss} \times 1 + U_{ss} \cos \Theta_{ss} \delta \Theta + \sin \Theta_{ss} \delta U.$$

Now gather all those terms containing strictly only all pure static terms (those indicated by subscript ss) and designate that as the steady state reference equation and gather all the rest of the terms, the dynamic terms (indicated by the attached δ) into the perturbation equation. Following these steps, the steady state reference equation is as follows:

$$h_{ss} = U_{ss} \sin \Theta_{ss}.$$

Note that the reference equation's right hand side term mimics the right hand side of the original nonlinear function. The perturbation equation is as follows:

$$\delta \dot{h} = U_{ss} \cos \Theta_{ss} \delta \Theta + \delta U \sin \Theta_{ss}.$$

Suppose we know or are given the steady state values of U_{ss} and Θ_{ss}, then clearly the above linearized equation can be compactly written as

$$\delta \dot{h} = c_1 \delta \Theta + c_2 \delta U$$

where c_1 and c_2 are some known constants.

This is the final linearized perturbation equation. Note that, with this linearization procedure, the differential equation is always in the perturbation variable(s). It is important to keep in mind that it is the perturbation angle that is assumed to be small, not the steady state angle. The steady state angle Θ_{ss} does not have any restrictions and thus could be very large. Care needs to be taken in carefully expanding all the terms in the original nonlinear equation, keeping track of any nonlinear terms in the perturbation variables and then discarding them. Similarly vigilance is needed in making the small angle approximation. A common mistake that often occurs is to make the $\sin \delta x$ term zero instead of it being δx. Also, note that all the steady state variables, which take on constant values, become constant coefficients in the linearized perturbation differential equation. In that standard case, we label the resulting constant coefficient linear system (in a state space representation) as a linear time invariant (LTI) system. It is possible that in some cases, when the nominal (or the steady state or the equilibrium) motion contains coefficients which are functions of the independent variable time, t, (even non-linear terms in time, t), the dynamic system is still a linear system but with time varying coefficients. In that case, we label it as a linear time varying (LTV) system. In other words, in the generic state space system representation, the LTI models (in a vector, matrix notation) are given by

$$\dot{\vec{x}} = A\vec{x} + B\vec{u}, \qquad \vec{x}(0) = \vec{x}_0 \tag{3.4}$$

where matrices A and B are constant matrices, $\underline{x}$ and $\underline{u}$ denote vectors of dimension n and m respectively, whereas the LTV models are given by

$$\dot{\vec{x}} = A(t)\vec{x} + B(t)\vec{u}, \qquad \vec{x}(0) = \vec{x}_0 \tag{3.5}$$

where the matrices contain elements which could be time varying.

Once the linearized state space system is obtained about the equilibrium points, our interest would then shift to understanding the stability of the resulting linear systems, especially in the natural motion (i.e. uncontrolled, open loop situation). In the next section, we learn about some fundamental concepts of stability of linear state space dynamic systems,

3.4 Uncontrolled, Natural Dynamic Response and Stability of First and Second Order Linear Dynamic Systems with State Space Representation

In this section, we now focus our attention on the natural, uncontrolled (often referred to as open loop situation in controls jargon) response of an LTI system, i.e we investigate the nature of the response of

$$\dot{\vec{x}} = A\vec{x} \qquad \vec{x}(0) = \vec{x}_0. \tag{3.6}$$

3.4.1 Dynamic Response of a First Order Linear System

Obviously, the simplest case for analyzing the natural dynamic response of linear dynamic system would be a first order system, namely, when the constant matrix A is simply a 1×1 matrix, i.e. a scalar system ($n = 1$). In other words, a constant coefficient differential equation in a single dependent variable. Thus we consider

$$\dot{x} = ax \qquad x(0) = x_0. \tag{3.7}$$

Note that for this simple case, the response is given by

$$x(t) = e^{at} x_0. \tag{3.8}$$

Thus, the natural motion is simply an exponential motion. It is a stable (converging to zero as time tends to ∞) motion when a is negative, a constant when $a = 0$ and finally unstable (divergent motion) when a is positive. Thus it can be easily concluded that first order systems never produce oscillatory behavior, and instead are always of pure exponential in nature.

Next we turn our attention to second order systems. Since second order systems are simple and quite frequently encountered, it is extremely important for the student to be thoroughly familiar with a second order system response. An undergraduate student is expected to be thoroughly conversant with this material.

3.4.2 Dynamic Response of a Second Order Linear System

The state space model of second order LTI system is given by:

$$\begin{bmatrix} \dot{x}_1(t) \\ \dot{x}_2(t) \end{bmatrix} = \begin{bmatrix} a_{11} & a_{12} \\ a_{21} & a_{22} \end{bmatrix} \begin{bmatrix} x_1(t) \\ x_2(t) \end{bmatrix}. \tag{3.9}$$

It may be recalled from the basic ordinary differential equation course the reader may have taken in their undergraduate education that the solution to this simple system of a set of simultaneous first order constant coefficient ordinary differential equations is obtained by examining the eigenvalues and eigenvectors of the constant matrix A.

In other words, just as the first order linear system is of exponential nature, the second order system response is also of exponential nature, albeit in terms of the eigenvalues of the matrix A, namely

$$\underline{x}(t) = c_1 e^{\lambda_1 t} + c_2 e^{\lambda_2 t} \tag{3.10}$$

where the constants c_1 and c_2 determine the amplitude of the motion, which depend on the eigenvector elements and the initial condition information.

Recall from your linear algebra or matrix theory class that the eigenvalues of a matrix A are given by the roots of the characteristic equation $\det(\lambda I - A) = 0$ (or $\det(A - \lambda I) = 0$). Sometimes, this characteristic equation is also referred to as the characteristic polynomial. For an nth order matrix, this characteristic polynomial is an nth degree polynomial. The n roots of this polynomial are the eigenvalues of the matrix A, denoted by λ_1, $\lambda_2, \ldots \lambda_n$. For a second order system described above, we obtain a second degree polynomial as follows:

$$\lambda^2 + p\lambda + q = 0. \tag{3.11}$$

The roots of the characteristic polynomial, namely λ_1 and λ_2, determine the nature of the time response of the above linear second order system and the eigenvector and initial condition information determine the amplitude of the time response. For understanding only the stability nature of the dynamic response, it is sufficient for us to focus on the nature of the eigenvalues.

It is also known from basic linear algebra and matrix theory that

$$a_{11} + a_{22} = \text{trace of } A = \lambda_1 + \lambda_2 \tag{3.12}$$

and that

$$a_{11}a_{22} - a_{12}a_{21} = \text{determinant of } A = \lambda_1\lambda_2. \tag{3.13}$$

That is, the trace of A is equal to the summation of the eigenvalues and the determinant of A is equal to the product of eigenvalues.

Note that the two eigenvalues can be two real numbers or a complex conjugate pair with some real part and an imaginary part. The real part itself can be zero or non-zero.

A zero real part of a complex conjugate pair (i.e pure imaginary eigenvalues) would produce a pure simple harmonic motion type oscillatory motion because we know

$$e^{i\omega} = \sin\omega + i\cos\omega \tag{3.14}$$

and

$$e^{-i\omega} = \sin\omega - i\cos\omega \tag{3.15}$$

where $i = \sqrt{(-1)}$.

From the above observation, it is clear that for stability of the above second order system it is necessary that the eigenvalues have real parts that are negative. Clearly positive real part eigenvalues result in unstable (divergent) motion and zero real part (pure imaginary) eigenvalues produce pure simple harmonic motion (neutrally stable behavior).

This in turn implies that in the following characteristic polynomial

$$\lambda^2 + p\lambda + q = 0 \tag{3.16}$$

the constants, p and q can be represented as follows:

$$p = -\text{trace of } A$$
$$\text{and}$$
$$q = \text{determinant of } A.$$

So both coefficients p and q are required to be positive for stability of a second order system.

Assuming this necessary condition is satisfied, we now write this characteristic polynomial in a generic form as:

$$\lambda^2 + 2\xi\omega_n\lambda + \omega_n^2 = 0 \tag{3.17}$$

where ξ is the damping ratio (which is by default assumed to be positive) and ω_n is the undamped natural frequency. Note that this nomenclature in terms of damping ratio and undamped natural frequency is appropriate only when both coefficients of the above characteristic polynomial are positive. If any one of the coefficients p or q is negative, it can be seen that the real parts of the roots of this second degree polynomial are always positive indicating that time response becomes unbounded either in a non-oscillatory (pure exponential fashion) or in an oscillatory way. For those situations, we simply label

Figure 3.1 Second order system eigenvalue locations.

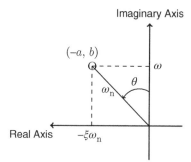

those systems as unstable systems and refer to them as systems having negative damping ratio. Thus the following discussion is specifically for those systems with both coefficients being positive. Thus, the roots of the characteristic equation, $\lambda_{1,2}$, are written in the form:

$$\lambda_{1,2} = \eta \pm i\omega \tag{3.18}$$

where η and ω are defined as such:

$$\eta = -\xi\omega_n \tag{3.19}$$

$$\omega = \omega_n \sqrt{1 - \xi^2}. \tag{3.20}$$

Here, η is related to the damping ratio and ω is the damped natural frequency. These eigenvalue locations are depicted in Figure 3.1.

In this notation, it is clear that the possibility of the two eigenvalues becoming pure real, is manifested when the (positive) damping ratio ξ becoming ≥ 1.

This in turn implies that whether the second degree characteristic polynomial of a second order system possesses pure real roots or a complex conjugate pair, very much depends on the numerical values of the two coefficients p and q of that characteristic polynomial. Note that when two roots are pure real, the motion is pure exponential and it does not make sense to refer that motion using the terms natural frequencies (whether undamped or damped). Those terms are used only when the motion is oscillatory, i.e when the (non-negative) damping ratio is such that $0 \leq \xi < 1$.

These concepts can also be illustrated graphically, as in Figure 3.1. Here, O marks one complex-conjugate eigenvalue ($\lambda_1 = -a + ib$). The other complex-conjugate eigenvalue in the pair ($\lambda_2 = -a - ib$), also exists in the third quadrant, but is not depicted explicitly in the diagram; looking at one eigenvalue is sufficient to generate meaningful insights. The reader should keep in mind though, that for real matrices the complex eigenvalues occur only as complex conjugate pairs.

From earlier discussions, we observe that this is an exponentially damped oscillatory motion.

Now, using the graphical depiction in Figure 3.1, we see the damping ratio can also be expressed as:

$$\xi = \sin \theta$$
$$= \frac{a}{\sqrt{a^2 + b^2}}. \tag{3.21}$$

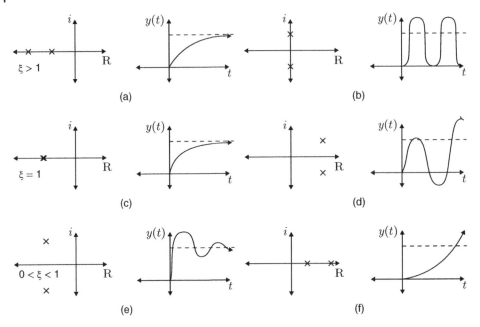

Figure 3.2 Time responses for various damping ratio.

We reiterate that whenever the second order system has a complex conjugate pair as its two eigenvalues, the resulting time response has oscillatory nature, where the imaginary part of the eigenvalues, namely the damped natural frequency ω dictates the oscillatory nature, while the real part dictates the nature of those oscillations. If the real part is negative, those oscillations decay over time and thus the resulting time response is a damped sinusoid with a rate of decay decided by the damping ratio ξ. The period of this oscillatory motion is given by

$$T = \frac{2\pi}{\omega}.$$ (3.22)

The time responses for various damping ratios are depicted in Figure 3.2.

As mentioned earlier, when the damping ratio ξ is zero, the time response is a pure sinusoidal motion (simple harmonic motion). When the real part is negative, then we refer the time response to be a stable time response, and when it is zero, it is referred to as neutrally stable response. The damping ratio ξ is of paramount importance to the control system designer; indeed, its value determines the speed of response of the system! For convenience, a summary of time responses for different values of damping ratio is given in Table 3.1. One can also see in Figure 3.1 that for complex conjugate roots, the diagonal distance from the origin to the root is the undamped natural frequency ω_n and the imaginary component of the root is the damped natural frequency ω.

We reiterate that for a second order linear system, both coefficients in the characteristic equation being positive is a necessary and sufficient condition for stability of the system. In other words, the moment one of the coefficients is negative or zero we can say that the second order system is not stable. As mentioned earlier, an unstable system with positive real part is labeled as having a negative damping ratio. Thus only for a stable system, the damping ratio ξ is always positive and the undamped natural frequency is a real (positive) quantity.

Table 3.1 Summary of time responses for various damping ratios.

Magnitude of damping ratio	Type of root	Time response
$\xi < -1$	Two distinct positive, purely real roots	Exponentially diverging
$0 > \xi > -1$	Complex conjugate pair, positive real part	Exponentially diverging sinusoid
$\xi = 0$	Purely imaginary	Purely sinusoidal
$0 < \xi < 1$	Complex conjugate pair, negative real part	Exponentially decaying sinusoid (i.e. underdamped)
$\xi = 1$	Two repeated negative, purely real	Exponentially decaying (i.e. critically damped)
$\xi > 1$	Two distinct negative, purely real	Exponentially decaying (i.e. overdamped)

We can generalize the above discussion to higher order linear state space systems as well where the plant matrix A is an $n \times n$ matrix. Given $\dot{\vec{x}} = A\vec{x}$ with $\vec{x}(0) = \vec{x}_0$, the solution to the above system of equations is given by

$$\vec{x}(t) = c_1 \vec{v}^1 e^{\lambda_1 t} + c_2 \vec{v}^2 e^{\lambda_2 t} + \ldots + c_n \vec{v}^n e^{\lambda_n t} \tag{3.23}$$

where λ_i is the n eigenvalues of the matrix A and $\vec{v}^i$ is the n normalized eigenvectors corresponding to each eigenvalue λ_i and c_i scalar constants to be determined from the initial condition vector $\vec{x}_0$. From this expression, it is clear that the eigenvalues determine the nature (i.e whether the trajectories $x_i(t)$ are converging to zero or diverging) of the time response whereas the eigenvectors determine the amplitude of the time response.

Thus, we could determine the stability of this higher order system by computing the eigenvalues of this higher order A matrix. Note that the eigenvalues of the matrix A are given by the roots of the characteristic equation $\det(\lambda I - A) = 0$. Thus the stability of the general linear time invariant system $\dot{x} = Ax$ (ignoring the vector notation) is completely determined by the real parts of the eigenvalues of the matrix A. Thus for continuous time systems, the Hurwitz stability of the system is thus given by the following criterion. Let $\lambda_i = \beta_i + i\omega_i$ where $i = \sqrt{(-1)}$.

(a) The system is unstable if β_i (real part of λ_i) > 0 for any distinct root or $\beta_i \geq 0$ for any repeated root.
(b) The system is stable in the sense of Lyapunov if $\beta_i \leq 0$ for all distinct roots and $\beta_i < 0$ for all repeated roots, i.e. there are no multiple poles on the imaginary axis and all distinct poles are in the left half of the complex plane.
(c) The system is asypmtotically stable if $\beta_i < 0$ for all roots.

Table 3.2 summarizes the Hurwitz stability conditions. A more thorough discussion of the stability conditions for linear time invariant systems is discussed in a later chapter on state space based modern control theory. For now, because of its simplicity as well as its importance, it is necessary and sufficient for the undergraduate student to be thoroughly familiar with the stability and dynamic response discussion we had in this chapter, especially for a simple second order linear state space system.

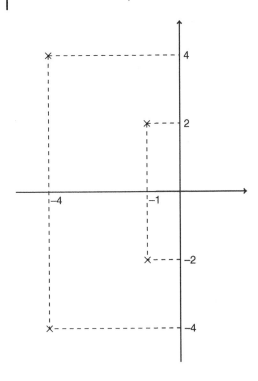

Figure 3.3 Fourth order system pole locations

Table 3.2 Hurwitz stability criteria for continuous time systems.

Unstable	If $\beta_i > 0$ for any single (or distinct) root or $\beta_i \geq 0$ for any repeated root
Stable in the sense of Lyapunov or neutrally stable	If $\beta_i \leq 0$ for all distinct roots and $\beta_i < 0$ for all repeated roots
Asymptotically stable	If $\beta_i < 0$ for all roots

Example 3.2   Consider a standard second order constant coefficient linear ordinary differential equation, say of the form:

$$\frac{d^2y(t)}{dt^2} + 2\frac{dy(t)}{dt} + 4y(t) = f(t) \tag{3.24}$$

where $f(t)$ is an external forcing function. For the homogeneous (no forcing function case), $f(t) = 0$. Thus the dynamics is initiated by non-zero initial conditions. Let us consider the homogeneous case.

Note that this second order single differential equation can also be written as a set of two first order differential equations as follows. First define:

$$y(t) = x_1(t) \tag{3.25}$$

$$\frac{dy(t)}{dt} = \dot{y}(t) = x_2(t). \tag{3.26}$$

Then, we have:

$$\dot{x}_1(t) = x_2(t) \tag{3.27}$$

$$\dot{x}_2(t) = \ddot{y}(t) = -2x_2(t) - 4x_1(t). \tag{3.28}$$

The equivalent state space model is:

$$\begin{bmatrix} \dot{x}_1(t) \\ \dot{x}_2(t) \end{bmatrix} = \begin{bmatrix} 0 & 1 \\ -4 & -2 \end{bmatrix} \begin{bmatrix} x_1(t) \\ x_2(t) \end{bmatrix}. \tag{3.29}$$

The characteristic polynomial is given by

$$\lambda^2 + 2\lambda + 4 = 0. \tag{3.30}$$

Comparing it with the generic characteristic polynomial

$$\lambda^2 + 2\xi\omega_n\lambda + \omega_n^2 = 0 \tag{3.31}$$

we observe that

$$\omega_n^2 = 4 \tag{3.32}$$

giving $\omega_n = 2$. Note that $2\xi\omega_n = 2$. From this, it is seen that the real part $-\xi\omega_n = -1$, so that $\xi\omega_n = 1$. Substituting $\omega_n = 2$, gives $\xi = 0.5$. Since ξ is such that $0 < \xi < 1$, we immediately conclude that the response is a stable, damped oscillatory motion, with the damping ratio and undamped natural frequency as determined above. Note that the damped natural frequency $\omega = 1.732$. This is nothing but the imaginary part of the complex conjugate pair. The real part is -1. Notice that the undamped natural frequency ω_n is nothing but the square root of the sum of the real part square and the imaginary part square. In other words, the two eigenvalues are $-1 \pm i1.732$. We obtained this result without actually computing the eigenvalues per se, but by first determining and knowing the damping ratio value.

Thus, in summary, for determining the stability of a second order system response, the damping ratio ξ (and thus the trace of A) plays a very important role. The trace of A has to be negative and the damping ratio ξ has to be positive for stability. The natural frequencies and the related determinant of A also play an important role. The determinant of A has to be positive for stability. A guideline for determining a priori whether a second degree characteristic polynomial possesses real or complex conjugate pairs is to examine whether the coefficient p is $> \sqrt{q}$ or not. If it is greater, it possesses pure real roots. If not, it possesses complex complex conjugate pairs. If $p = 0$ or negative, the system is not stable anyway.

The basic concepts covered in this chapter are also available in any of the excellent textbooks such as [1, 2].

Now that we have developed the theory for linear continuous time time invariant systems, in subsequent chapters, we apply these tools to study aircraft and spacecraft dynamics.

3.5 Chapter Summary

In this chapter, we learned how to linearize a set of nonlinear differential equations in state space representation about a given steady state (equilibrium) point so that the linearized state space models can be used later for control design purposes. We then thoroughly analyzed the dynamic response and stability behavior of a simple yet highly important linear second order state space system. In addition, we learnt that, also for higher order linear state space dynamic systems, the negativity of the real parts of the eigenvalues of the state space matrix A is a necessary and sufficient condition for Hurwitz stability of the dynamic system.

3.6 Exercises

Exercise 3.1: The relation describing the height of an aircraft is a nonlinear differential differential equation, as given here:

$$\dot{h} = U \sin \Theta - V \sin \Phi \cos \Theta - W \cos \Phi \cos \Theta.$$

Linearize the equation about the steady state values h_o, U_o, V_o, W_o, Θ_o, and Φ_o, where the perturbations are taken as δh, δU, δV, δW, $\delta \Theta$, and $\delta \Phi$. Write the reference equation as well as the perturbation equation.

Exercise 3.2: Linearize the composite rotation matrix S (we came across this in Chapter 1) and investigate if the linearized matrix is still orthogonal or not.

Exercise 3.3: Recall the rotational motion equations of a rigid body about its center of mass in the principal axis frame (reproduced here for your convenience):

$$\dot{\omega}_1 = [(I_2 - I_3)/I_1]\omega_2\omega_3 + M_1/I_1 \tag{3.33a}$$

$$\dot{\omega}_2 = [(I_3 - I_1)/I_2]\omega_1\omega_3 + M_2/I_2 \tag{3.33b}$$

$$\dot{\omega}_3 = [(I_1 - I_2)/I_3]\omega_1\omega_2 + M_3/I_3. \tag{3.33c}$$

Now assume, that the rigid body has axial symmetry so that principal inertia satisfy the following conditions:

$$I_1 = I_2 \tag{3.34}$$

$$I_3 > I_1. \tag{3.35}$$

If it is acted upon by a small body-fixed, transverse torque, $M_1 = M$, with $M_2 = M_3 = 0$, then

(a) Write the resulting differential equations of motion for this situation in state space form. Is it a nonlinear state space form or a linear state space form?
(b) Is ω_3 affected by M_1? Explain.
(c) Solve for the three angular velocities from the resulting state space equations and explain which of the angular velocities is constant and which ones are not constant for the above (given) external torque situation.

Exercise 3.4: The eigenvalues of a square 4 x 4 matrix are given in the complex plane as in Figure 3.3.
Calculate the natural frequency, damping ratio, the damped natural frequency and the period for each of these complex conjugate pairs.

Exercise 3.5: Investigate the nature of time response for the linear second order systems given by the following characteristic equations. Where appropriate, calculate the natural frequency, damping ratio, the damped natural frequency and the period.

1. $\lambda^2 + 2\lambda + 4 = 0$
2. $\lambda^2 + 5\lambda + 16 = 0$
3. $\lambda^2 + 4 = 0$
4. $\lambda^2 + 4\lambda + 4 = 0.$

Bibliography

1 K. Ogata and Y. Yang. *Modern control engineering*. Prentice-Hall, Englewood Cliffs, NJ, 1970.
2 W.L. Brogan. *Modern Control Theory*. Prentice Hall, 1974.

4

Aircraft Static Stability and Control

4.1 Chapter Highlights

In this chapter, we focus our attention on the analysis of dynamics specific to aircraft via the linearized equations of motion. It turns out that in aircraft dynamics, because of the safety considerations of a commercial aircraft, it becomes important to make sure that even in a steady state (static) situation, the aircraft possesses characteristics which make it tend to bring the attitude back to its steady state condition, when disturbed from the trim condition, without any external control being applied. This feature of tendency to come back to the trimmed state when disturbed from it, on its own without any external forcing function is labeled static stability and control of an aircraft. In dynamic stability and control we are concerned with the actual motion of the vehicle under various control inputs. Thus for an aircraft both static stability as well as dynamic stability are of importance. Hence in this chapter, first we thoroughly discuss static stability (both longitudinal as well as lateral/directional) issues and then shift the attention to dynamic stability analysis in the next chapter. In some sense, this type of discussion involving both static stability as well as dynamic stability analysis is somewhat unique for the case of aircraft dynamics.

4.2 Analysis of Equilibrium (Trim) Flight for Aircraft: Static Stability and Control

Assumptions:

- aircraft is a rigid body
- flat earth inertial frame (i.e. axes aligned with the local north, east, and vertical), see Figure 4.1.
  1. Nose up is a positive pitching moment.
  2. Right wing down is a positive rolling moment.
  3. Nose turning toward the right wing is a positive yawing moment.

Flight Dynamics and Control of Aero and Space Vehicles, First Edition. Rama K. Yedavalli.
© 2020 John Wiley & Sons Ltd. Published 2020 by John Wiley & Sons Ltd.

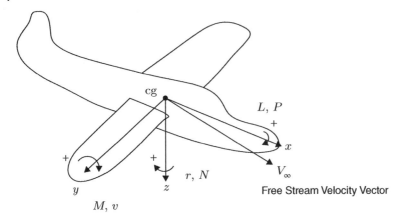

Figure 4.1 Body fixed frame.

Table 4.1 below lists the standard Aircraft variables and parameters. Now consider the steady state forces and moments:

$$F_x = C_x \overline{Q} S \qquad L = C_l \overline{Q} S b \qquad b = \text{wing span}$$
$$F_y = C_y \overline{Q} S \qquad M = C_m \overline{Q} S \overline{c} \qquad \overline{c} = \text{mean aerodynamic chord}$$
$$F_z = C_z \overline{Q} S \qquad N = C_n \overline{Q} S b \qquad S = \text{wing area}$$
$$\overline{Q} = \text{dynamic pressure} = \tfrac{1}{2} \rho V_\infty^2$$

$C_x, C_y, C_z, C_l, C_m, C_n$ are functions of Mach number, Reynolds number, $\alpha, \beta, \alpha', \beta'$. Decouple the motion of a/c into:

Longitudinal → pitch (motion in the xz plane)
Lateral/directional → roll/yaw (coupled).

(a) Equilibrium or trim condition. If an aircraft is to remain in steady, uniform flight

$$(\Sigma F)_{cg} = 0$$
$$(\Sigma M)_{cg} = 0.$$

Table 4.1 Aircraft variables and parameters.

Motion	Roll	Pitch	Yaw
Aerodynamic forces	F_x	F_y	F_z
Thrust forces	T_x	T_y	T_z
Gravitational	W_x	W_y	W_z
Angular velocity, ω	$p(P)$	$q(Q)$	$r(R)$
Linear velocity, V_∞	$u(U)$	$v(V)$	$w(W)$
Moments	$l(L)$	$m(M)$	$n(N)$
Moments of inertia	I_{xx}	I_{yy}	I_{zz}
Products of inertia	I_{yx}	I_{zx}	I_{xy}

Stable Eq. State   Unstable Eq. State   Neutrally State

Figure 4.2 State of stability of a system.

(b) Stability. Movement in returning or the tendency to return to a given state of equilibrium or trim condition, when perturbed from it.
(c) Static stability. Tendency to return to an equilibrium condition when disturbed from the trim condition.
(d) Dynamic stability. Actual time history of the resultant motion of a/c in response to disturbances in the form of either external unwanted inputs or pilot control actions.

From the above definitions, it is clear that the dynamic stability notion is of more importance and more encompassing than the pure static stability notion. However, in the aircraft case, the static stability issue is of prime importance in its own right because, in some sense, it is independent of any external excitation and simply brings out the tendency of the aircraft to return to its equilibrium position when disturbed from it. Put another way, an aircraft can be statically unstable, but yet can be made dynamically stable with the help of control actions. As we shall see later, this is the reason many military aircraft can be statically unstable, yet can be dynamically stable in the presence of an automatic flight control system, but it is clear that without the proper working of the automatic flight control system (i.e. an uncontrollable aircraft), the safety of the aircraft would be compromised. Thus for commercial aircraft, we demand not only dynamic stability, but a certain degree of static stability, so that on its own, in the uncontrolled situation, at least the aircraft has the tendency to return to its trim condition. It is clear that too much static stability interferes with aircraft's maneuverability (or controllability) and that is the reason, for highly maneuverable aircraft, it is common to sacrifice some degree of static stability to achieve better maneuverability. Thus, in summary, for safety critical systems like commercial aircraft, the issue of static stability becomes quite important, and in the next section we focus our attention on static stability analysis for aircraft. In this connection, it is convenient to analyze static longitudinal stability (namely, motion within the xz plane, i.e. essentially the pitching moment situation) separately from static lateral/directional stability case. With this backdrop, we now address the static longitudinal stability issue. The Figure 4.2 conceptually illustrates the concept of an equilibrium position to be stable, unstable, or neutrally stable.

4.3 Static Longitudinal Stability

Consider the pitching moment about the center of gravity of a given aerodynamic surface such as an airfoil as a function of the angle of attack α. To be able to trim (i.e. $C_m = 0$) at positive angles of attack is called balancing (see Figure 4.3).

For mathematical simplicity as well as to understand the basic concepts with high clarity, let us simply consider a simple basic linear relationship between the pitching

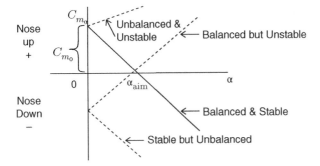

Figure 4.3 Static longitudinal Stability.

moment coefficient C_m and the angle of attack α, as follows:

$$C_m = C_{m_0} + \frac{dC_m}{d\alpha}\alpha$$
$$= C_{m_0} + C_{m_\alpha}\alpha. \tag{4.1}$$

Let us now consider different cases, where the slope C_{m_α} is of different signs. Consider the case when it is negative. Then, it can be seen that when there is a positive angle of attack change, there is a negative pitching moment generated, and vice versa. In other words, when $C_{m_\alpha} < 0$, the pitching moment generated is such that it tries to bring the airfoil back into the trim condition, i.e. the pitching moment generated is a restoring pitching moment; it has the tendency to bring it back to the trim condition, namely it is a statically stable condition. Thus to ensure static longitudinal stability, we require

$$C_{m_\alpha} < 0. \tag{4.2}$$

Since

$$C_{m_\alpha} \equiv \frac{dC_m}{d\alpha}$$
$$= \frac{dC_m}{dC_L}\frac{dC_L}{d\alpha} \tag{4.3}$$
$$= \frac{dC_m}{dC_L}C_{L\alpha}$$

and

$$C_{L\alpha} \equiv \frac{dC_L}{d\alpha} > 0. \tag{4.4}$$

Therefore we can establish another requirement for static longitudinal stability:

$$\frac{dC_m}{dC_L} < 0. \tag{4.5}$$

In the diagram above, we show four types of airfoils with different characteristics:

- $C_m0 > 0$ and $C_{m_\alpha} > 0$: this is an unbalanceable as well as a statically unstable situation.
- $C_m0 > 0$ and $C_{m_\alpha} < 0$: this is a balanceable as well as a statically stable situation.
- $C_m0 < 0$ and $C_{m_\alpha} < 0$: this is an unbalanceable but statically stable situation.
- $C_m0 < 0$ and $C_{m_\alpha} > 0$: This is a balanceable but statically unstable situation.

Note that for static stability as well as balancing, we need:

$$C_m 0 > 0 \text{ and } C_{m_\alpha} < 0.$$

Note that the pitching moment slope is heavily dependent on the center of gravity location about which we take moments, i.e. as the center of gravity moves, the C_m versus α curve changes.

Thus, basically C_{m_α} (or the slope $\frac{dC_m}{dC_L}$) essentially becomes a measure of the degree of static longitudinal stability.

In what follows, we attempt to analyze the contribution of each of the components of the aircraft such as the wing, fuselage, tail, etc. to the overall static longitudinal stability. However, since wing is expected to be the major contributor in relationship to the fuselage, we first analyze the contribution of the wing to static longitudinal stability and then later the contribution of the tail, neglecting the contribution of the fuselage altogether without sacrificing any conceptual understanding.

4.3.1 Contribution of Each Component to the Static Longitudinal Stability (i.e. to the Pitching Moment)

Wing Contribution (Figure 4.4)

$$\begin{aligned} M_{cg_w} &= M_{ac_w} + L\cos(\alpha_w - i_w)(x_{cg} - x_{ac_w}) + L_w\sin(\alpha_w - i_w)\ z_{cg} \\ &\quad + D_w\sin(\alpha_w - i_w)(x_{cg} - x_{ac_w}) - D_w\cos(\alpha_w - i_w)\ z_{cg} \qquad (4.6) \\ &= C_{m_{cg_w}}\overline{Q}S\bar{c} \end{aligned}$$

where

$$L_w = C_{L_w}\overline{Q}S, \quad D_w = C_{D_w}\overline{Q}S, \quad M_{ac_w} = C_{m_{ac_w}}\overline{Q}S\bar{c} \qquad (4.7)$$

$$\begin{aligned} C_{m_{cg_w}} &= C_{m_{ac_w}} + C_{L_w}\frac{1}{c}(x_{cg} - x_{ac_w})\cos(*) + C_{L_w}\frac{1}{c}z_{cg}\sin(*) \\ &\quad + C_{D_w}\frac{1}{c}(x_{cg} - x_{ac_w})\sin(*) - C_{D_w}\frac{1}{c}z_{cg}\sin(*) \qquad (4.8) \end{aligned}$$

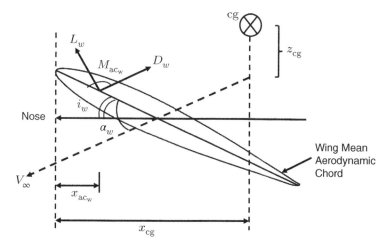

Figure 4.4 Wing contribution.

Assuming small angles and letting $C_L \gg C_D$ and $\bar{z}_{cg} \lll \bar{x}_{cg}$ where $\bar{x}_{cg} = \frac{x_{cg}}{\bar{c}}$ and $\bar{x}_{ac_w} = \frac{x_{ac_w}}{\bar{c}}$, we get

$$C_{m_{cg_w}} \approx C_{m_{ac_w}} + C_{L_w}(\bar{x}_{cg} - \bar{x}_{ac_w}). \tag{4.9}$$

Let us write

$$C_{L_w} = C_{L_{0_w}} + C_{L_{\alpha_w}} \alpha_w \tag{4.10}$$

and

$$C_{m_{cg_w}} = C_{m_{0_w}} + C_{m_{\alpha_w}} \alpha_w. \tag{4.11}$$

Then

$$C_{m_{cg_w}} = C_{m_{ac_w}} + C_{L_{0_w}}(\bar{x}_{cg} - \bar{x}_{ac_w}) + C_{L_{\alpha_w}} \alpha_w(\bar{x}_{cg} - \bar{x}_{ac_w}) \tag{4.12}$$

i.e.

$$C_{m_{0_w}} = C_{m_{ac_w}} + C_{L_{0_w}}(\bar{x}_{cg} - \bar{x}_{ac_w}) \tag{4.13}$$

and

$$C_{m_{\alpha_w}} = C_{L_{\alpha_w}}(\bar{x}_{cg} - \bar{x}_{ac_w}). \tag{4.14}$$

Since the lift curve slope $C_{L,\alpha,w}$ is always positive, for $C_{m_{\alpha_w}} < 0$ we require $\bar{x}_{cg} < \bar{x}_{ac_w}$. Physically, this means the aircraft's center of gravity has to be in front of the wing's aerodynamic center, but this is typically not the case.

Note that, if at zero lift, $C_{L_{0_w}} = 0$, then $C_{m_{0_w}} = C_{m_{ac_w}}$. Note that:

$$C_{m_{0_w}} = C_{m_{ac_w}} + C_{L_{0_w}}(\bar{x}_{cg} - \bar{x}_{ac_w}). \tag{4.15}$$

For $C_{m_{0_w}} > 0$, we have to ensure $C_{m_{ac_w}} \gg 0$. However, for the majority of low speed aircraft with positive camber, $C_{m_{ac_w}} < 0$ and the center of gravity is aft of the wing aerodynamic center. For example, look at Figure 4.5.

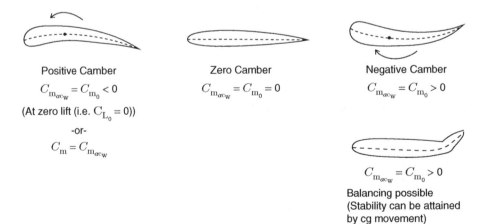

Positive Camber
$$C_{m_{ac_w}} = C_{m_0} < 0$$
(At zero lift (i.e. $C_{L_0} = 0$))
-or-
$$C_m = C_{m_{ac_w}}$$

Zero Camber
$$C_{m_{ac_w}} = C_{m_0} = 0$$

Negative Camber
$$C_{m_{ac_w}} = C_{m_0} > 0$$

$$C_{m_{ac_w}} = C_{m_0} > 0$$
Balancing possible
(Stability can be attained
by cg movement)

Figure 4.5 Cambers of wings.

So for all practical purposes, wing contribution to static longitudinal stability is mostly destabilizing.

This implies that there needs to be another means to make $C_{m_{0_w}} > 0$ and $C_{m_{\alpha_w}} < 0$ for standard positive camber airfoils. This is accomplished by introducing the horizontal tail.

Tail Contribution (Figure 4.6)

$$\alpha_t = (\alpha_w - i_w) - \varepsilon + i_t \tag{4.16}$$

Total lift $L = L_w + L_t$

$$C_L \overline{Q} S = C_{L_w} \overline{Q} S + C_{L_t} \overline{Q}_t S_t \tag{4.17}$$

$$C_L = C_{L_w} + \eta \frac{S_t}{S} C_{L_t} \tag{4.18}$$

where η is defined as the tail efficiency factor,

$$\eta = \frac{\overline{Q}_t}{\overline{Q}}$$
$$= \frac{\frac{1}{2}\rho V_t^2}{\frac{1}{2}\rho V_\infty^2}. \tag{4.19}$$

Values for η usually range from 0.8 to 1.2. In the wake region, $\eta < 1$, in the slip stream, $\eta > 1$.

Now the pitching moment due to the tail.

Assuming all the angles to be small so that we can make small angle approximations, and that the drag force is relatively negligible compared to the lift force and that the vertical distance between the tail and the center of gravity $z_{cg_t} \approx 0$, our equation simplifies considerably as follows:

$$M_t = -l_t[L_t \cos(*) + \cancel{D_t \sin(*)}] - \cancel{z_{cg_t}}[\cancel{D_t \cos(*)} - \cancel{L_t \sin(*)}] + \cancel{M_{ac_t}}$$
$$= -l_t L_t \tag{4.20}$$
$$= -l_t C_{L_t} \overline{Q}_t S_t.$$

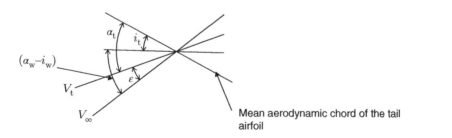

Mean aerodynamic chord of the tail airfoil

ε: Downwash angle
i_t: tail incidence angle
α: Effective angle of attack

Figure 4.6 Tail contributions.

Writing

$$M_t = C_{m_t} \overline{Q} S \bar{c} \tag{4.21}$$

we observe that

$$
\begin{aligned}
C_{m_t} &= \frac{M_t}{\overline{Q} S \bar{c}} \\
&= \frac{-l_t C_{L_t} \overline{Q}_t S_t}{\overline{Q} S \bar{c}} \tag{4.22} \\
&= -\frac{l_t S_t}{S \bar{c}} C_{L_t} \eta \\
&= -V_H \eta C_{L_t}
\end{aligned}
$$

where V_H is the horizontal tail volume ratio, defined

$$V_H = \frac{l_t S_t}{S \bar{c}} \tag{4.23}$$

Now

$$C_{L_t} = C_{L_{0_t}} + C_{L_{\alpha_t}} \alpha_t = C_{L_{\alpha_t}} [(\alpha_w - i_w) - \varepsilon + i_t] \tag{4.24}$$

Let

$$\varepsilon = \varepsilon_0 + \frac{d\varepsilon}{d\alpha} \alpha_w \tag{4.25}$$

where

$$\varepsilon = \frac{2 C_{L_w}}{\pi A R_w}, \tag{4.26}$$

$$\varepsilon_0 = \frac{2 C_{L_0}}{\pi A R_w} \tag{4.27}$$

and

$$\frac{d\varepsilon}{d\alpha} = \frac{2 C_{L_{\alpha_{cw}}}}{\pi A R_w} \tag{4.28}$$

$$C_{m_t} = -V_H \eta \left[C_{L_{\alpha_t}} \left\{ (\alpha_w - i_w) - \varepsilon_0 - \frac{d\varepsilon}{d\alpha} \alpha_w + i_t \right\} \right] \tag{4.29}$$

$$= V_H \eta C_{L_{\alpha_t}} (\varepsilon_0 + i_w - i_t) - V_H \eta C_{L_{\alpha_t}} \alpha_w \left(1 - \frac{d\varepsilon}{d\alpha} \right) \tag{4.30}$$

$$= C_{m_{0_t}} + C_{m_{\alpha_t}} \alpha_w \tag{4.31}$$

where the trail contribution to C_{m_0} is

$$C_{m_{0_t}} = V_H \eta C_{L_{\alpha_t}} (\varepsilon_0 + i_w - i_t). \tag{4.32}$$

The tail contribution to C_{m_α} is

$$C_{m_{\alpha_t}} = V_H \eta C_{L_{\alpha_t}} \left(1 - \frac{d\varepsilon}{d\alpha} \right). \tag{4.33}$$

So in total

$$C_{m_0} = C_{m_{0_w}} + C_{m_{0_t}}$$

$$= C_{m_{ac_w}} + C_{L_{0_w}}(\bar{x}_{cg} - \bar{x}_{ac_w}) + V_H \eta C_{L_{\alpha_t}}(\varepsilon_0 + i_w - i_t). \tag{4.34}$$

When $C_{L_{0_w}} = 0$ (which is the case most of the time), we observe that:

$$C_{m_0} = C_{m_{ac_w}} + V_H \eta C_{L_{\alpha_t}}(\varepsilon_0 + i_w - i_t) \tag{4.35}$$

From this we observe:

(i) C_{m_0} is independent of the CG location, $\bar{x}_{cg}$, and that
(ii) i_t should be such that it makes the second term highly positive (even when $\varepsilon_0 = o$ and $i_w = 0$).

Thus the tail should be placed as in Figure 4.7 (i.e. i_t is negative).
Similarly

$$C_{m_\alpha} = C_{m_{\alpha_w}} + C_{m_{\alpha_t}}$$

$$= C_{L_{\alpha_w}}(\bar{x}_{cg} - \bar{x}_{ac_w}) - V_H \eta C_{L_{\alpha_t}}\left(1 - \frac{d\varepsilon}{d\alpha}\right). \tag{4.36}$$

Notice how the second term in the above equation introduces a highly negative component into the C_{m_α} equation, so that the total C_{m_α} with the tail contribution becomes negative, thereby achieving static longitudinal stability for aircraft with positive camber airfoils.

Approximating $\alpha_{total} = \alpha_w$ write the total pitching moment coefficient $C_{m_{cg}}$ as:

$$C_{m_{cg}} = C_{m_0} + C_{m_\alpha}\alpha \tag{4.37}$$

where C_{m_0} is given by equation (4.35) and C_{m_α} is given by equation (4.36).
Note that

$$C_L = C_{L_w} + C_{L_t}\frac{\bar{Q}_t}{\bar{Q}}\frac{S_t}{S}$$

$$= C_{L_{0_w}} + C_{L_{\alpha_t}}\alpha\left\{C_{L_{\alpha_t}}\alpha\left(1 - \frac{d\varepsilon}{d\alpha}\right) - C_{L_{\alpha_t}}\frac{\bar{Q}_t}{\bar{Q}}\frac{S_t}{S}(\varepsilon_0 + i_w - i_t)\right\}. \tag{4.38}$$

Writing $C_L = C_{L_0} + C_{L_\alpha}\alpha$ we observe that for the total aircraft, which we approximate as the sum of wing and tail contributions:

$$C_{L_0} = C_{L_{0_w}} - C_{L_{\alpha_t}}\eta\frac{S_t}{S}(\varepsilon_0 + i_w - i_t) \tag{4.39}$$

$$C_{L_\alpha} = C_{L_{\alpha_w}} + C_{L_{\alpha_t}}\eta\frac{S_t}{S}\left(1 - \frac{d\varepsilon}{d\alpha}\right) \tag{4.40}$$

$$C_{m_\alpha} = C_{L_{\alpha_w}}(\bar{x}_{cg} - \bar{x}_{ac_w}) - C_{L_{\alpha_t}}\eta\frac{S_t}{S}\left(1 - \frac{d\varepsilon}{d\alpha}\right)(\bar{x}_{ac_t} - \bar{x}_{cg}) \tag{4.41}$$

$$C_{m_0} = C_{m_{ac_w}} + C_{L_{0_w}}(\bar{x}_{cg} - \bar{x}_{ac_w}) + C_{L_{\alpha_t}}\eta\frac{S_t}{S}(\varepsilon_0 + i_w - i_t)(\bar{x}_{ac_t} - \bar{x}_{cg}) \tag{4.42}$$

Figure 4.7 Placement of tail.

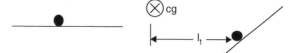

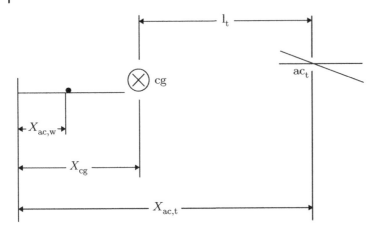

Figure 4.8 Pitching moment due to tail.

where we used the fact that

$$\frac{l_t}{\bar{c}} = \bar{x}_{ac_t} - \bar{x}_{cg} \tag{4.43}$$

In the majority of standard situations, we assume $C_{m_{ac_w}} < 0$ and $C_{L_0} \cong C_{L_{0_w}} = 0$.

An alternative to the wing/tail contribution for making $C_{m_0} > 0$ is the swept back wing with twisted tips. A variant to this is the delta wing. Either twist the tip or employ negative camber or upturn trailing edge flap.

Now that we have essentially gathered the contributions of the wing and tail together for the overall static longitudinal stability, we observe that there is a fictitious CG location point along the straight line segment extending from the wing leading edge to the tail cg, that has the property that if we take the pitching moment about that (special) CG location, the pitching moment would be independent of the angle of attack. In other words, this special CG location can be thought of as the aerodynamic center of the entire aircraft. That special CG location is labeled as the neutral point, or to be more precise in the current situation stick fixed neutral point (see Figure 4.8). It will become clear later as to why the adjective stick fixed is added here at this point, because, later we introduce the notion of a stick free neutral point at which time the difference between stick fixed and stick free becomes more clear.

4.4 Stick Fixed Neutral Point and CG Travel Limits

Stick fixed neutral point is that location of the center of gravity at which $C_{m_\alpha} = 0$, i.e. about this special center of gravity location the pitching moment is independent of the angle of attack, i.e. it is equivalent to the aerodynamic center for the entire aircraft. So solving equation (4.42) for the $\bar{x}_{cg}$ at which $C_{m_\alpha} = 0$, we get:

$$\bar{x}_{NP} = \bar{x}_{cg}|_{C_{m_\alpha}=0} = \frac{\left[x_{ac_w} + \eta \frac{S_t}{S} \frac{C_{L_{\alpha_t}}}{C_{L_{\alpha_w}}} \left(1 - \frac{d\varepsilon}{d\alpha} \right) \right]}{\left[1 + \eta \frac{S_t}{S} \frac{C_{L_{\alpha_t}}}{C_{L_{\alpha_w}}} \left(1 - \frac{d\varepsilon}{d\alpha} \right) \right]} \tag{4.44}$$

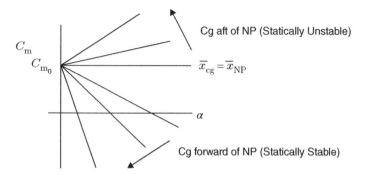

Figure 4.9 C_m versus α for low speed aircraft.

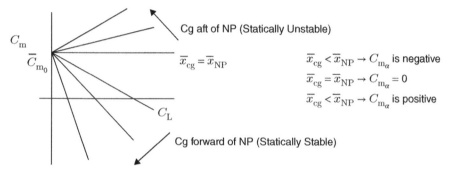

Figure 4.10 C_m versus C_L for low speed aircraft.

With this definition of $\bar{x}_{NP}$, we can show that:

$$C_{m_\alpha} = C_{L_\alpha}(\bar{x}_{cg} - \bar{x}_{NP})$$ (4.45)

i.e.

$$\frac{dC_m}{dC_L} = \frac{C_{m_\alpha}}{C_{L_\alpha}} = (\bar{x}_{cg} - \bar{x}_{NP}).$$ (4.46)

Of note is the quantity $\bar{x}_{cg} - \bar{x}_{NP}$, which we define as the Static Margin. See Figures 4.9 and 4.10.

Thus again, for static longitudinal stability, we see that the center of gravity should be in front of the neutral point, i.e. the stick fixed neutral point serves as the aft limit of center of gravity travel before loss of static longitudinal stability.

Example 4.1   For a flying wing configuration, the coefficient of moment about the aerodynamic center of the wing $C_{m_{ac_w}} = -0.1$; the zero-angle lift coefficient $C_{L_{0_{wing}}} = 0.3$; the location of the aerodynamic center is $\bar{x}_{ac_w} = 0.2$; the location of center of gravity is $\bar{x}_{cg} = 0.3$; and the rate of change of lift coefficient with respect to angle of attack $C_{L_\alpha} = 4.1$ per radian. Given this information,

1. Determine if the aircraft is balanceable and longitudinally stable. Plot $C_{m_{cg}}$ versus α, and if balanceable, find the trim angle of attack.

2. Repeat the above exercise with
   (a) $C_{m_{ac_w}} = 0.1$ and $\bar{x}_{cg} = 0.1$
   (b) $C_{m_{ac_w}} = 0.1$ and $\bar{x}_{cg} = 0.2$.

Solve this problem using the simplified equations:

$$C_L = C_{L_0} + C_{L_\alpha}\alpha$$

and

$$C_m = C_{m_0} + C_{m_\alpha}\alpha.$$

Solution

From the contribution of the wing to static stability,

$$C_{m_{cgw}} = C_{m_{ac_w}} + C_{L0_w}(\bar{x}_{cg} - \bar{x}_{ac_w}) + C_{L_{\alpha w}}(\bar{x}_{cg} - \bar{x}_{ac_w})\alpha_w$$
$$= C_{m0_w} + C_{m_\alpha}\alpha_w.$$

1. From the data

$$C_{m0_w} = -0.1 + (0.3)(0.3 - 0.2)$$
$$= -0.07.$$

Also, from the given data, we observe that,

$$C_{m_\alpha} = (4.1)(0.3 - 0.2)$$
$$= 0.41.$$

Therefore

$$C_{m_{cgw}} = -0.07 + 0.41\alpha_w.$$

From this it is clear that the aircraft is longitudinally unstable (because $C_{m_\alpha} > 0$ but balanceable because $C_m = 0$ when $\alpha_w = 9.9°$ (see Figure below).

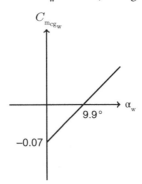

2. (a) When $C_{m_{ac_w}} = 0.1$ and $\bar{x}_{cg} = 0.1$, we have

$$C_{m0_w} = 0.1 + (0.3)(0.1 - 0.2)$$
$$= +0.07$$
$$C_{m_\alpha} = (4.1)(0.1 - 0.2)$$
$$= -0.41.$$

Therefore

$$C_{m_{cg_w}} = +0.07 - 0.41\alpha_w.$$

This aircraft is longitudinally statically stable because $C_{m_\alpha} < 0$ and also balanced, because $C_{m0_w} < 0$. The trim angle of attack is again $9.9°$

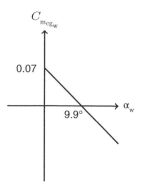

(b) When $C_{m_{ac_w}} = 0.1$ and $\bar{x}_{cg} = 0.2$

$$C_{m0_w} = 0.1 + (0.3)(0.2 - 0.2)$$
$$= 0.1$$

and

$$C_{m_\alpha} = (4.1)(0.2 - 0.2)$$
$$= 0.$$

This aircraft is neutrally stable, and unbalanceable

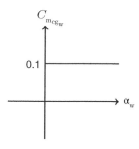

Example 4.2 Consider the Boeing 747-400 with the dimensions given in Figure 4.11. The wing area is $S = 5500$ ft^2; wingspan $b = 195$ ft; aerodynamic chord $\bar{c} = 27$ ft; and the center of gravity is located at $\bar{x}_{cg} = 0.25$. We are also given the following aerodynamic data:

$$\eta = 1.1$$
$$C_{L_{\alpha w}} = 5.7 \text{ rad}$$
$$C_{L_{\alpha t}} = 5.0 \text{ rad.}$$

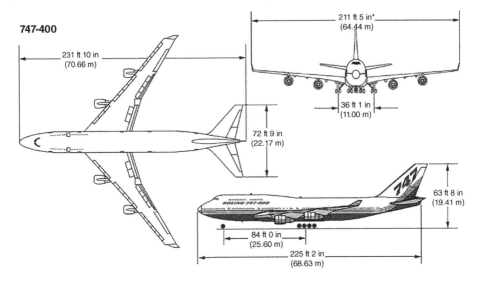

747-400

231 ft 10 in
(70.66 m)

211 ft 5 in*
(64.44 m)

36 ft 1 in
(11.00 m)

72 ft 9 in
(22.17 m)

63 ft 8 in
(19.41 m)

84 ft 0 in
(25.60 m)

225 ft 2 in
(68.63 m)

Figure 4.11 Dimensions of Boeing 747-400.

1. Without a tail, does this aircraft possess static longitudinal stability?
2. Find a set of different combinations of tail area S_t and the distance $\bar{x}_{ac_t}$, such that the total $C_{m\alpha} = -1.26$ is achieved corresponding to $\bar{x}_{cg} = 0.25$. Use

$$\frac{d\epsilon}{d\alpha} = \frac{2C_{L_{\alpha w}}}{\pi AR_w}.$$

3. For each of these configurations find the stick-fixed neutral point $\bar{x}_{NP}$ and give the neutral point location in physical units.
4. Now fix a particular S_t and $\bar{x}_{ac_t}$. For this particular choice, calculate C_{m_α} and static margin for different values of the center of gravity location $\bar{x}_{cg}$ where $\bar{x}_{cg}$ is selected to cover a wide range (aft and forward of $\bar{x}_{acw}$). For this problem, use the following simplified equations:

$$C_{m_{cg}} = C_m 0 + \left(C_{L_{\alpha w}} (\bar{x}_{cg} - \bar{x}_{acw}) - C_{L_{\alpha t}} \eta \frac{S_t}{S} \left(1 - \frac{d\epsilon}{d\alpha} \right) (\bar{x}_{ac_t} - \bar{x}_{cg}) \right) \alpha$$

where,

$$\eta = \frac{\bar{Q}_t}{\bar{Q}}$$

$$\frac{d\epsilon}{d\alpha} = \frac{2C_{L_{\alpha w}}}{\pi AR_w}$$

$$AR_w = \frac{b^2}{S}.$$

Solution

1. No. Since

$$\bar{x}_{cg} = 0.25$$

is same as

$$\bar{x}_{acw}$$

(which is assumed to be the quarter chord point), this aircraft's static margin, without tail, at this given CG is zero. Hence, it is not statically stable. It is neutrally stable.

2. We begin by evaluating the values of wing aspect ratio AR_w

$$AR_w = \frac{b^2}{S}$$
$$= \frac{(195)^2}{5500}$$
$$= 6.91$$

and the downwash $\frac{d\epsilon}{d\alpha}$ using the analytic expression provided

$$\frac{d\epsilon}{d\alpha} = \frac{2C_{L_{aw}}}{\pi AR_w}$$
$$= \frac{(2)(5.7)}{\pi * 6.91}$$
$$= 0.52$$

We have the rate of change of pitching moment with respect to angle of attack

$$C_{m_\alpha} = C_{L_{aw}}(\bar{x}_{cg} - \bar{x}_{ac_w}) - C_{L_{at}}\eta\frac{S_t}{S}(\bar{x}_{ac_t} - \bar{x}_{cg})\left(1 - \frac{d\epsilon}{d\alpha}\right).$$

Rearranging,

$$S_t = \frac{(C_{L_{aw}}(\bar{x}_{cg} - \bar{x}_{ac_w}) - C_{m_\alpha})S}{C_{L_{at}}\eta(\bar{x}_{ac_t} - \bar{x}_{cg})\left(1 - \frac{d\epsilon}{d\alpha}\right)}.$$

Plugging in known values and simplifying, we obtain a relation between the aerodynamic center location of the tail $\bar{x}_{ac_t}$ and the area of the tail S_t (see Figure 4.12) as such

$$S_t = \frac{2.65 \times 10^3}{\bar{x}_{ac_t} - 0.25}.$$

Taking different sets of $\bar{x}_{ac_t}$ and S_t, we have the values shown in Table below

$\bar{x}_{ac_t}$	$x_{ac_t} = \bar{x}_{ac_t}\bar{c}$	S_t	% of S
3.68	99.36 ft	772 ft^2	14%-appropriate
2.25	60.75 ft	1326 ft^2	24.1%
1.25	31.05 ft	2652 ft^2	47.72%

3. For each value of S_t and $\bar{x}_{ac_t}$, there is a corresponding value for the stick fixed neutral point $\bar{x}_{NP}$. In other words, tail design allows us to manipulate the stick fixed neutral point location, which in turn has an effect on static stability. The equation for neutral point is as follows:

$$\bar{x}_{NP} = \frac{\bar{x}_{ac_w} + \frac{C_{L_{\alpha t}}}{C_{L_{\alpha w}}} \eta \frac{S_t}{S} \bar{x}_{ac_t} \left(1 - \frac{d\epsilon}{d\alpha}\right)}{1 + \frac{C_{L_{\alpha t}}}{C_{L_{\alpha w}}} \eta \frac{S_t}{S} \left(1 - \frac{d\epsilon}{d\alpha}\right)}.$$

With $\bar{x}_{ac_t} = 3.68$ and $S_t = 772\,\text{ft}$, $\bar{x}_{NP}$ is calculated to be 0.46; neutral point locations can be found similarly for all other cases (See Figure 4.12).

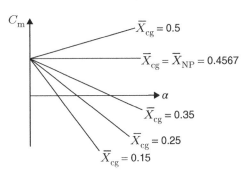

Figure 4.12

4. We fix $\bar{x}_{ac_t} = 3.68$ and $S_t = 772$ ft so that $\bar{x}_{NP} = 0.46$. We then vary the center of gravity location, finding the corresponding C_{m_α} and the static margin $(\bar{x}_{cg} - \bar{x}_{NP})$ (See table below).

$\bar{x}_{cg}$	$(\bar{x}_{cg} - \bar{x}_{NP})$	C_{m_α}
0.15	−0.3067	Negative
2.25	−0.2067	Negative
0.35	−0.1067	Negative
0.4567	0	0
0.5	0.0433	Positive

4.4.1 Canard Plus the Wing Combination

Another possible configuration to achieve static longitudinal stability is to use, instead of a tail, a control surface in front of the wing, which is called the Canard. In this case, the wing will have a stabilizing effect on the static longitudinal stability. An aircraft with Canard configuration, is shown in Figure 4.13.

4.5 Static Longitudinal Control with Elevator Deflection

In order to fly at different trim speeds (and angles of attack) without changing the static stability characteristics, we need some form of longitudinal control. This is provided by

Figure 4.13 Canard configuration. (Flickr.com: Peter Gronemann)

δe Up is
Negative

δe Down is
Positive

Figure 4.14 Elevator deflection.

the elevator (see Figure 4.14). Now we write

$$C_{\mathrm{L}} = C_{\mathrm{L}_0} + C_{\mathrm{L}_\alpha} \alpha + C_{\mathrm{L}_{\delta e}} \delta e \tag{4.47}$$

where

$$C_{\mathrm{L}_{\delta e}} = \frac{\mathrm{d}C_{\mathrm{L}}}{\mathrm{d}\delta e} \tag{4.48}$$

$$C_{\mathrm{m}} = C_{\mathrm{m}_0} + C_{\mathrm{m}_\alpha} \alpha + C_{\mathrm{m}_{\delta e}} \delta e$$
$$= \overline{C}_{\mathrm{m}_0} + \frac{\mathrm{d}C_{\mathrm{m}}}{\mathrm{d}C_{\mathrm{L}}} C_{\mathrm{L}} + \overline{C}_{\mathrm{m}_{\delta e}} \delta e \tag{4.49}$$

where

$$C_{\mathrm{m}_{\delta e}} \equiv \frac{\mathrm{d}C_{\mathrm{m}}}{\mathrm{d}\delta e}. \tag{4.50}$$

Note that, $C_{\mathrm{m}_{\delta e}}$ and $\overline{C}_{\mathrm{m}_{\delta e}}$ do not exist simultaneously. In other words, depending on how the aerodynamic data is presented, we use either the equation with C_{m_α} or the equation with

$$\frac{\mathrm{d}C_{\mathrm{m}}}{\mathrm{d}C_{\mathrm{L}}}$$

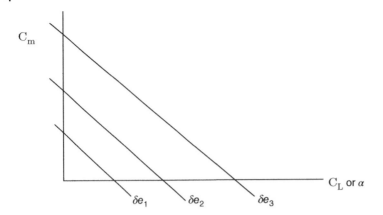

Figure 4.15 C_m versus c_L and/or alpha.

separately for C_m. There is no direct relationship between $C_{m_{\delta e}}$ and $\overline{C}_{m_{\delta e}}$ in this linear range. In other words, $C_{m_{\delta e}}$ is simply the slope of the C_m versus δ_e curve (line) when the data is given in terms the angle of attack α, whereas $\overline{C}_{m_{\delta e}}$ is the slope of the C_m versus δ_e curve (line) when the data is given in terms of C_L (Figure 4.15).

The change in lift due to the elevator is

$$\Delta L = \Delta L_t \tag{4.51}$$

$$\Delta C_L \overline{Q} S = \Delta C_{L_t} \overline{Q}_t S_t \tag{4.52}$$

$$\Delta C_L = \frac{\Delta C_{L_t} \overline{Q}_t S_t}{\overline{Q} S}$$

$$= \eta \Delta C_{L_t} \frac{S_t}{S}. \tag{4.53}$$

Let

$$\Delta C_{L_t} = \frac{dC_{L_t}}{d\delta e} \delta e. \tag{4.54}$$

However, as the elevator deflection changes, there is a change in the angle of attack of the tail:

$$\frac{dC_{L_t}}{d\delta e} = \frac{dC_{L_t}}{d\alpha_t} \frac{d\alpha_t}{d\delta e} \tag{4.55}$$

$$= C_{L_{\alpha t}} \tau$$

where τ is defined as

$$\tau = \frac{d\alpha_t}{d\delta e} \tag{4.56}$$

$$\therefore \boxed{C_{L_{\delta e}} = \frac{S_t}{S} \eta C_{L_{\alpha t}} \tau} \tag{4.57}$$

where τ is obtained empirically from wind tunnel tests. Similarly for pitching moment, recall that

$$C_{m_{tail}} = -V_H \eta C_{L_t}. \tag{4.58}$$

Due to the elevator

$$\Delta C_m = -V_H \eta \Delta C_{L_t}$$
$$= -V_H \eta C_{L_{\alpha t}} \tau \delta e. \tag{4.59}$$

If we write

$$\Delta C_m = C_{m_{\delta e}} \delta e$$
$$= \frac{dC_m}{d\delta e} \delta e. \tag{4.60}$$

Then

$$C_{m_{\delta e}} = -V_H \eta C_{L_{\alpha t}} \tau$$
$$= -\frac{S_t}{S} \eta C_{L_{\alpha t}} \tau \frac{l_t}{\bar{c}} \tag{4.61}$$

i.e.

$$\boxed{C_{m_{\delta e}} = -\frac{S_t}{S} \eta C_{L_{\alpha t}} \tau (\bar{x}_{ac_t} - \bar{x}_{cg}).} \tag{4.62}$$

From Equation (4.57) and Equation (4.62) we observe that,

$$C_{m_{\delta e}} = -C_{L_{\delta e}} (\bar{x}_{ac_t} - \bar{x}_{cg}). \tag{4.63}$$

We also have

$$C_{m_\alpha} = C_{L_\alpha} (\bar{x}_{cg} - \bar{x}_{NP}). \tag{4.64}$$

Now our interest is to obtain the trim angle of attack and elevator deflections.

4.5.1 Determination of Trim Angle of Attack and Trim Elevator Deflection

At trim, for small angles, from force distribution, we observe that the lift balances out the weight, i.e.

$$L = W = mg \tag{4.65}$$

$$C_{L_{trim}} \bar{Q} S = W \rightarrow C_{L_{trim}} = \frac{W}{\frac{1}{2}\rho S V_\infty^2}. \tag{4.66}$$

For moment trim, we need $C_m = 0$

$$\therefore C_{m_0} + C_{m_\alpha} \alpha_{trim} + C_{m_{\delta e}} \delta e_{trim} = 0 \tag{4.67}$$

Also

$$C_{L_{trim}} = \frac{W}{\frac{1}{2}\rho S V_\infty^2}$$
$$= C_{L_0} + C_{L_\alpha} \alpha_{trim} + C_{L_{\delta e}} \delta e_{trim}. \tag{4.68}$$

Thus from Equations (4.67) and (4.68), we have two equations (independent) and two unknowns α_{trim} and δe_{trim}. So, solving for them, we get

$$\alpha_{trim} = \frac{(C_{L_{trim}} - C_{L_0})C_{m_{\delta e}} + C_{m_0}C_{L_{\delta e}}}{C_{L_\alpha}C_{m_{\delta e}} - C_{m_\alpha}C_{L_{\delta e}}} \tag{4.69}$$

$$\delta e_{trim} = \frac{-[(C_{L_{trim}} - C_{L_0})C_{m_\alpha} + C_{m_0}C_{L_\alpha}]}{C_{m_{\delta e}}C_{L_\alpha} - C_{m_\alpha}C_{L_{\delta e}}}. \tag{4.70}$$

If we use the C_m versus C_L equation for trim,

$$0 = \overline{C}_{m_0} + (\overline{x}_{cg} - \overline{x}_{NP})C_{L_{trim}} + \overline{C}_{m_{\delta e}}\delta e_{trim}. \tag{4.71}$$

Solving for δe_{trim}, we get

$$\delta e_{trim} = \frac{-[\overline{C}_{m_0} + (\overline{x}_{cg} - \overline{x}_{NP})C_{L_{trim}}]}{\overline{C}_{m_{\delta e}}}. \tag{4.72}$$

Note that α_{trim} and δe_{trim} can be written as,

$$\alpha_{trim} = \alpha_{0_{trim}} + aC_{L_{trim}} \tag{4.73}$$

$$\delta e_{trim} = \delta e_{0_{trim}} + b'C_{L_{trim}} \tag{4.74}$$

with appropriate definitions of the coefficients a and b'.

Example 4.3 For a given aircraft, we have:

$$C_L = 0.1 + 0.1\alpha + 0.02\delta e$$
$$C_{m_{0.2\overline{c}}} = 0.025 - 0.3C_L - 0.01\delta e.$$

Find $\overline{x}_{NP}$, α_{trim} and δe_{trim} for a flight at $C_{L_{trim}} = 0.25$.

Solution
From the above equations and our notation, we see that:

$$\overline{C}_{m_0} = 0.025$$

$$\frac{dC_m}{dC_L} = (\overline{x}_{cg} - \overline{x}_{NP}) = -0.3$$

$$\overline{x}_{cg} = 0.2$$

$$\overline{C}_{m_{\delta e}} = -0.01$$

$$C_{L_0} = 0.1$$

$$C_{L_\alpha} = 0.1$$

$$C_{L_{\delta e}} = 0.02.$$

Since $\overline{x}_{cg} - \overline{x}_{NP} = 0.2 - \overline{x}_{NP} = -0.3 \rightarrow \overline{x}_{NP} = 0.5$.
Hence, for a flight at $C_{L_{trim}} = 0.25$, the α_{trim} and δe_{trim} are given by solving

$$0 = 0.025 - 0.3(0.25) - 0.01\delta e_{trim}$$
$$0.25 = 0.1 + 0.1\alpha_{trim} + 0.02\delta e_{trim}$$

from which we get

$$\delta e_{\text{trim}} = -5° \text{ (up elevator)}, \alpha_{\text{trim}} = 2.5°.$$

Now coming back to the δe_{trim} expressions given by either Equation (4.70) or Equation (4.72), we observe that δe_{trim} is a function of the $C_{L_{\text{trim}}}$ and $\bar{x}_{\text{cg}}$.

Note that

$$\delta e_{\text{trim}} = \frac{-(\bar{C}_{m_0} + (\bar{x}_{\text{cg}} - \bar{x}_{\text{NP}})C_{L_{\text{trim}}})}{\bar{C}_{m_{\delta e}}} \tag{4.75}$$

$$= \frac{-[(C_{L_{\text{trim}}} - C_{L_0})C_{m_\alpha} + C_{m_0}C_{L_\alpha}]}{C_{m_{\delta e}}C_{L_\alpha} - C_{m_\alpha}C_{L_{\delta e}}}. \tag{4.76}$$

From these two equations, we see that:

$$\frac{d\delta e_{\text{trim}}}{dC_{L_{\text{trim}}}} = \frac{-(\bar{x}_{\text{cg}} - \bar{x}_{\text{NP}})}{\bar{C}_{m_{\delta e}}} \tag{4.77}$$

$$= \frac{-C_{m_\alpha}}{C_{m_{\delta e}}C_{L_\alpha} - C_{m_\alpha}C_{L_{\delta e}}}.$$

Obviously, when $\bar{x}_{\text{cg}} = \bar{x}_{\text{NP}}$ or when $C_{m_\alpha} = 0$ we have

$$\frac{d\delta e_{\text{trim}}}{dC_{L_{\text{trim}}}} = 0. \tag{4.78}$$

This can be used as a new definition or interpretation of a neutral point: it is that center of gravity location at which $\frac{d\delta e_{\text{trim}}}{dC_{L_{\text{trim}}}} = 0$, i.e. at this center of gravity location, no change in elevator deflection is needed to get a change in lift coefficient, see Figures 4.16–4.17.

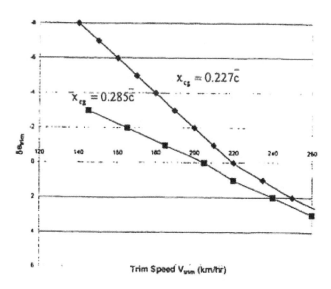

Figure 4.16 Trim speed versus δ_e trim.

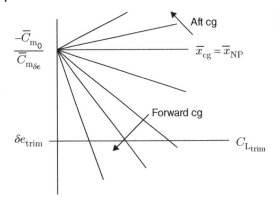

Aft cg

$\bar{x}_{cg} = \bar{x}_{NP}$

Forward cg

δe_{trim}

$C_{L_{trim}}$

$\dfrac{-\bar{C}_{m_0}}{\bar{C}_{m_{\delta e}}}$

Figure 4.17 Trim elevator angle versus $C_{L_{trim}}$.

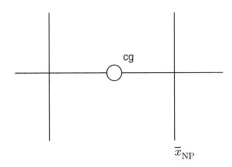

cg

$\bar{x}_{NP}$

Most forward cg location dictated by maximum up elevator deflection

Most aft cg location dictated by the Neutral point (Stick Fixed) (if static stability is desired)

Figure 4.18 Center of gravity travel.

From this plot, we observe that for a given trim lift coefficient (i.e. $C_{L_{trim}}$), as the center of gravity is more forward (of the neutral point), the more negative elevator deflection that is needed for trim. However, there is always a physical limitation on the elevator deflection. Thus we can conclude that the most forward center of gravity location is dictated by the maximum up elevator deflection allowed (Figure 4.18. We thus have a window for the center of gravity travel from the static longitudinal stability considerations.

This explains why in some small aircraft, the airline staff take note of the cargo and personnel distribution pattern during the flight envelope. It is important that the CG travel be restricted to this window as the mass distribution (cargo, personnel and fuel and other factors) keeps changing during the flight.

4.5.2 Practical Determination of Stick Fixed Neutral Point

Since

$$\frac{d\delta e_{trim}}{dC_{L_{trim}}} = 0 \quad \text{at} \quad \bar{x}_{cg} = \bar{x}_{NP} \tag{4.79}$$

Figure 4.19 Practical method of determining the neutral point.

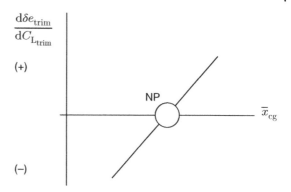

and

$$\frac{d\delta e_{\text{trim}}}{dC_{L_{\text{trim}}}} < 0 \quad \text{when} \quad \bar{x}_{\text{cg}} < \bar{x}_{\text{NP}} \tag{4.80}$$

and

$$\frac{d\delta e_{\text{trim}}}{dC_{L_{\text{trim}}}} > 0 \quad \text{when} \quad \bar{x}_{\text{cg}} > \bar{x}_{\text{NP}} \tag{4.81}$$

this is often used as another criterion for defining the stick fixed neutral point. Also this definition helps to find the stick fixed neutral point in a practical way, as in Figure 4.19:

For wind tunnel tests δe_{trim} versus $C_{L_{\text{trim}}}$ is plotted for different center of gravity locations. Then from that data, the slope $\frac{d\delta e_{\text{trim}}}{dC_{L_{\text{trim}}}}$ is obtained and this information is plotted as above (and interpolated if sufficient data is not available). The center of gravity location at which $\frac{d\delta e_{\text{trim}}}{dC_{L_{\text{trim}}}} = 0$ is obviously the neutral point.

Caution: All of the above analysis is valid for low Mach #s and altitudes because in this range, the values of coefficients such as C_{m_α}, C_{L_α}, C_{m_0}, C_{L_0} ... are almost constant. If these values change drastically during the flight conditions of interest, the above linear analysis is not accurate.

4.6 Reversible Flight Control Systems: Stick Free, Stick Force Considerations

In a reversible flight control system, the pilot stick movements get reflected in the horizontal stabilizer movements and vice versa through mechanical and other forms of linkages. In an irreversible control system, this transmission of control is one way, namely the pilot stick movements get reflected in the horizontal stabilizer movements but not the other way around. For this reason, in large aircraft like the Boeing 747, which uses irreversible control systems, an artificial feel is provided to the pilot for the horizontal stabilizer and other control surface movements, external to the cockpit.

An important consequence of the reversible control system is that the definition of neutral point (center of gravity location for which $C_{m_\alpha} = 0$) must be reviewed. In the case of a reversible control system, the stick is left free, and in the presence of the aerodynamic forces and moments, it is clear that the elevator will always float to such

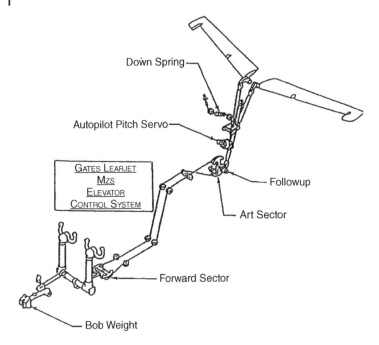

Figure 4.20 Example of a typical reversible flight control system.

a position that the hinge moment at the elevator control surface becomes equal to 0 (see Figure 4.20).

4.6.1 Stick Free Longitudinal Stability and Control

Hinge moment: $H_e = C_h \overline{Q}_t S_e \overline{c}_e$
  where

$$C_h = C_{h0} + C_{h_{\alpha_t}} \alpha_t + C_{h_{\delta e}} \delta e + C_{h_{\delta_t}} \delta_t \tag{4.82}$$

Here δ_t is the elevator tab deflection. Also $C_{h0} = 0$ for symmetric airfoils. Let $\delta_t = 0$. Thus

$$C_h = C_{h_{\alpha_t}} \alpha_t + C_{h_{\delta e}} \delta e. \tag{4.83}$$

Remember that when the control surface is left free, it will naturally float to an equilibrium position (i.e. where $H_e = 0$). Therefore Equation (4.83) becomes

$$C_{h_{\alpha_t}} \alpha_t + C_{h_{\delta e}} \delta e_{free} = 0 \tag{4.84}$$

i.e.

$$\delta e_{free} = -\frac{C_{h_{\alpha_t}}}{C_{h_{\delta e}}} \alpha_t. \tag{4.85}$$

Usually, $C_{h_{\alpha_t}}$ and $C_{h_{\delta e}}$ are negative. They are mostly determined empirically from wind tunnel tests. So for all positive α_t, the elevator floats into an up position in equilibrium[1]. Now we have

$$C_{L_t} = C_{L_0} + C_{L_{\alpha_t}} \alpha_t + C_{L_{\delta e}} \delta e_{\text{free}} \tag{4.86}$$

$$= C_{L_{\alpha_t}} \alpha_t - C_{L_{\delta e}} \frac{C_{h_{\alpha_t}}}{C_{h_{\delta e}}} \alpha_t \tag{4.87}$$

$$= C_{L_{\alpha_t}} \alpha_t \left(1 - \frac{C_{L_{\delta e}}}{C_{L_{\alpha_t}}} \frac{C_{h_{\alpha_t}}}{C_{h_{\delta e}}} \right) \tag{4.88}$$

$$= C'_{L_{\alpha_t}} \alpha_t \tag{4.89}$$

where

$$C'_{L_{\alpha_t}} = C_{L_{\alpha_t}} \left(1 - \frac{C_{L_{\delta e}}}{C_{L_{\alpha_t}}} \frac{C_{h_{\alpha_t}}}{C_{h_{\delta e}}} \right) \tag{4.90}$$

$$= C_{L_{\alpha_t}} f.$$

In order to determine the stick free neutral point, we follow the same procedure as for the stick fixed neutral point except that wherever $C_{L_{\alpha_t}}$ appears, we replace it by $C'_{L_{\alpha_t}}$. Carrying out this exercise (i.e. writing the C_m expression and finding C_{m_α} and then equating $C_{m_\alpha} = 0$), we get

$$\bar{x}'_{\text{NP}} = \frac{\bar{x}_{\text{ac}_w} + \frac{C'_{L_{\alpha_t}}}{C_{L_{\alpha w}}} \eta \frac{S_t}{S} \left(1 - \frac{d\varepsilon}{d\alpha} \right) \bar{x}_{\text{ac}_t}}{1 + \frac{C'_{L_{\alpha_t}}}{C_{L_{\alpha w}}} \eta \frac{S_t}{S} \left(1 - \frac{d\varepsilon}{d\alpha} \right)} \tag{4.91}$$

or

$$\bar{x}'_{\text{NP}} = \frac{\bar{x}_{\text{ac}_w} + \frac{C_{L_{\alpha_t}} f}{C_{L_{\alpha w}}} \eta \frac{S_t}{S} \left(1 - \frac{d\varepsilon}{d\alpha} \right) \bar{x}_{\text{ac}_t}}{1 + \frac{C_{L_{\alpha_t}} f}{C_{L_{\alpha w}}} \eta \frac{S_t}{S} \left(1 - \frac{d\varepsilon}{d\alpha} \right)}, \tag{4.92}$$

It is left as an exercise for the reader to get an expression for $\bar{x}_{\text{NP}} - \bar{x}'_{\text{NP}}$.

Generally, $\bar{x}'_{\text{NP}}$ is forward of $\bar{x}_{\text{NP}_{\text{fixed}}}$, i.e. the effect of the stick free situation is to reduce the static margin (Figure 4.21. Thus $\bar{x}'_{\text{NP}}$ (i.e. $\bar{x}_{\text{NP}_{\text{free}}}$) is very important because now $\bar{x}_{\text{NP}_{\text{free}}}$ becomes the most aft center of gravity location before we lose static longitudinal stability.

---

1 When this free floating elevator angle for trim δe_{free} differs from the δe_{trim}, obviously some extra force is required to hold the elevator. Obviously, this has to be provided by the pilot, which is very fatiguing. Trim tabs are provided to relieve the pilot of this load. Trim tabs make $\delta e_{\text{free}} = \delta e_{\text{trim}}$.

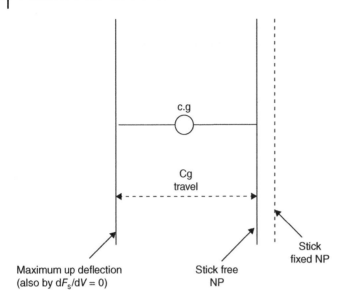

Figure 4.21 Cg travel under the stick free situation.

4.6.2 Stick Force

$$F_S = GH_e \tag{4.93}$$

$$= \overline{Q}_t S_e \overline{c}_e G \left[C_{h_0} + C_{h_{\alpha_t}} \left\{ \alpha \left(1 - \frac{d\varepsilon}{d\alpha} \right) + i_t - \varepsilon_0 + i_w \right\} + C_{h_{\delta e}} \delta e + C_{h_{\delta t}} \delta t \right] \tag{4.94}$$

where G =elevator gearing has units of rad ft^{-1} and is positive. However

$$\alpha_{\text{trim}} = \alpha_{0,\text{trim}} + aC_{L_{\text{trim}}} \tag{4.95}$$

$$\delta e_{\text{trim}} = \delta e_{0,\text{trim}} + b' C_{L_{\text{trim}}} \tag{4.96}$$

where

$$C_{L_{\text{trim}}} = \frac{W}{QS}. \tag{4.97}$$

Therefore

$$F_S = \eta \overline{Q} S_e \overline{c}_e G \left[C_{h_0} + C_{h_{\alpha_t}} \left\{ \alpha_{0,\text{trim}} \left(1 - \frac{d\varepsilon}{d\alpha} \right) + i_t - \varepsilon_0 + i_w \right\} \right.$$

$$\left. + C_{h_{\delta e}} \delta e_{0,\text{trim}} + C_{h_{\delta t}} \delta t \right]$$

$$+ \eta S_e \overline{c}_e G \left[a \frac{W}{S} C_{h_{\alpha_t}} \left(1 - \frac{d\varepsilon}{d\alpha} \right) + b' \frac{W}{S} C_{h_{\delta e}} \right] \tag{4.98}$$

i.e.

$$F_S = B \frac{1}{2} \rho V_{\text{trim}}^2 + A \tag{4.99}$$

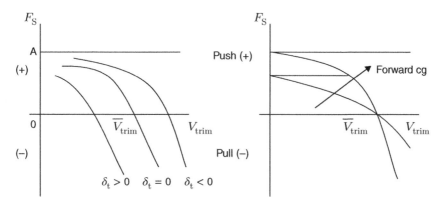

Figure 4.22 Stick force.

where A is a constant, independent of the speed V and B is a constant that serves as a coefficient of $\frac{1}{2}\rho V^2$. Note that $F_S = A$ when $V = 0$. By proper selection of i_t and δ_t one can make $F_S = 0$. The condition $F_S = 0$ is called the stick force trim. So essentially i_t and δ_t (tab angle) can be used to achieve stick force trim. Note that the constant A is a function of B, which in turn is a function of the center of gravity location. In other words, A is a linear function.

The stick force F_S is shown in Figure 4.22 and:

1. It is directly proportional to $S_e \bar{c}_e$ and thus very sensitive to the size of the tab.
2. It is proportional to the square of the speed.
3. The center of gravity position affects only the constant term A. A forward movement of the center of gravity produces an upward translation of the curve of F_S versus V.

4.6.3 Stick Force Gradient, $\frac{dF_S}{dV}$:

The stick force gradient (at the V_{trim}) plays an important role in speed stability analysis. For speed stability, we require $\frac{dF_S}{dV} < 0$. Of course, the higher the gradient, the better it is from the speed stability point of view, but at the same time, if the pilot does want to change the trim speed, he/she has to exert a lot of force to change the trim speed. Thus there is a trade-off between controllability and static speed stability. Consider

$$F_S = B\frac{1}{2}\rho V^2_{trim} + A \tag{4.100}$$

$$\frac{dF_S}{dV} = B\rho V. \tag{4.101}$$

At trim speed V_{trim}, $F_S = 0$ and $B = -\dfrac{A}{\frac{1}{2}\rho V^2_{trim}}$ (where it is assumed that a tab setting is chosen to achieve this trim speed) and thus:

$$\boxed{\frac{dF_S}{dV} = -\frac{2A}{V_{trim}}} \quad \text{(select } C_h \text{ terms such that } A \text{ is positive).} \tag{4.102}$$

Thus, we can observe the following (with respect to a fixed V_{trim}):

1. As the center of gravity moves forward, $\left|\frac{dF_s}{dV}\right|$ increases, i.e. the gradient becomes higher and higher as the center of gravity moves forward, but the stick force F_S also increases as the center of gravity moves forward.

   $\therefore$ Another criterion for most forward center of gravity location is the allowable stick force in the V_{trim} region of interest.

2. In the absence of compressibility effects, the elevator control will be heaviest at:
   a. sea level
   b. low speed
   c. forward center of gravity positions
   d. maximum weight.

3. At $\bar{x}_{cg} = \bar{x}'_{NP}$, $A = 0 \rightarrow$ can be used as an alternative way of defining the stick free neutral point. It is that center of gravity location at which $\frac{dF_s}{dV} = 0$, i.e. no stick force is needed to change the trim speed at this center of gravity location.

The stick force per gram is sometimes used as a measure of maneuvering stability by practicing engineers. Thus this stick force per gram is higher at a forward cg, regardless of altitude. This phenomenon has significant implications for steep turning maneuvers. For example, to perform a level turn at $60°$ of bank requires 2 g in any airplane. In general, the above analysis demonstrates that maneuvering at high altitude requires less column force than it does at low altitudes.

In summary, it is clear that static longitudinal stability and speed stability are important measures for the flight certification of an aircraft. Modern airplanes are evolving such that the emphasis on the aircraft possessing high static stability is being reduced. This is often termed as relaxed static stability (RSS). With RSS, it is possible to make the aircraft more fuel efficient by designing it to be more aerodynamically efficient. In such cases, the stability is augmented by active control systems, which will be discussed later in in chapters dealing with flight vehicle control, such as in Chapters 8 and 19. It is acknowledged by industries like Boeing that augmented stability provides better cruise performance with no increase in the workload of the pilot and no adverse effects from flying at an aft CG. Of course, the requirement on satisfactory handling qualities for continued safe flight and landing following an augmented system failure puts a limit on how far aft CG can go. As long as the CG travel is limited to the range where the most aft location is the neutral point, the handling qualities will be adequate with or without stability augmentation procedures.

This completes the discussion of the basic concepts needed in the topic of static longitudinal stability of aircraft. In a course that is explicitly dedicated to only aircraft dynamics and control, it is possible to continue further and discuss other concepts such as the maneuver point wherein the steady state considered is that of a steady vertical pull up. This special CG location is similar to the neutral point concept but the details are somewhat different. The interested student is encouraged to consult other excellent textbooks such as [35] to learn more about these advanced topics within static longitudinal stability.

Now that we have gathered few basic concepts needed in static longitudinal stability, we now shift our focus to briefly analyze similar concepts in the roll/yaw motion of the aircraft. Relative to static longitudinal stability, our coverage of static directional stability as well as the static lateral stability will be necessarily limited and brief.

4.7 Static Directional Stability and Control

The main, dominant variable in assessing static directional (yaw) stability is the sideslip angle, β. We can write the yawing moment coefficient C_n

$$C_n = C_{n\beta}\beta + C_{n\delta R}\delta R \tag{4.103}$$

where δR is the rudder deflection.

However, strictly speaking, physically the yaw motion and roll motion are coupled in the sense that roll motion creates yaw moments and yaw motion creates rolling moment. So we would really have

$$C_n = C_{n\beta}\beta + C_{n\delta R}\delta R + C_{n\delta A}\delta A \tag{4.104}$$

where δA is the aileron angle (mainly used for roll control) and $C_{n\delta A}$ is the cross coupling term. For the present discussion, let us neglect the cross coupling term.

$$C_n = C_{n\beta}\beta + C_{n\delta R}\delta R + \underset{\text{neglect}}{\cancel{C_{n\delta A}}} A \tag{4.105}$$

Static Directional Stability Criterion

When there is a positive sideslip angle disturbance (as shown in Figure 4.23), the restoring yawing moment (i.e. the moment which brings the aircraft back into the relative wind direction) has to be in the sense as shown (i.e. positive yawing moment). Therefore the criterion for static yaw stability is that:

$$C_{n\beta} > 0. \tag{4.106}$$

Static Roll Stability and Control

Again sideslip β is the main variable.

$$C_l = C_{l_\beta}\beta + C_{l_{\delta A}}\delta A. \tag{4.107}$$

Figure 4.23 Directional moments.

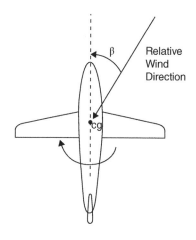

β

Relative
Wind
Direction

cg

Here

$$\delta A = \frac{1}{2}(\delta A_{L} + \delta A_{R}) \tag{4.108}$$

where δA_{L} is positive downwards and δA_{R} is positive upwards. So whenever the left aileron and right aileron are deflected in the opposite direction, there is always a net non-zero δA.

Rolling the right wing down (for positive β) is by convention a positive rolling moment. So the restoring moment (to put wings level) should be in the direction that makes the right wing go up (which is a negative rolling moment). Thus the static rolling stability criterion is that:

$$C_{l_{\beta}} < 0. \tag{4.109}$$

Roll stability is achieved by

1. Wing dihedral, Figure 4.24.
2. Wing position on the fuselage:
   (a) High wing–stabilizing, Figure 4.25.
   (b) Low wing–destabilizing, Figure 4.26.
3. Wing sweep–stabilizing, Figure 4.27. The higher the wing span the higher the stabilizing effect.

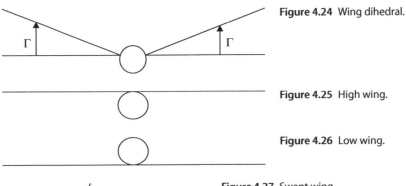

Figure 4.24 Wing dihedral.

Figure 4.25 High wing.

Figure 4.26 Low wing.

Figure 4.27 Swept wing.

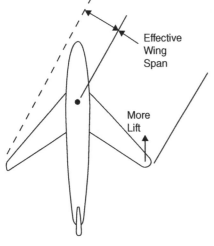

Figure 4.28 Forces in the event of an engine-out scenario.

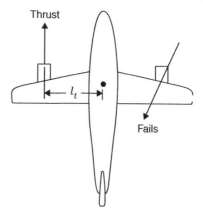

4.7.1 Static Lateral/Directional Control

Since roll and yaw are highly coupled, we discuss the lateral/directional control in a coupled way as well.

The main consideration in static lateral/directional control is the sizing of the lateral/directional motion control surfaces, namely rudder and aileron design. Major factors that influence this design are:

a) Adverse yaw: when the aircraft is rolled into a turn, there is a sideslip generated and this causes an unwanted yaw motion (called adverse yaw). So in order to have a coordinated turn, there should be a yawing moment generated through the rudder to compensate for adverse yaw.
b) Slip stream yaw motion: in propeller aircraft, if the vertical tail is in the slip stream, there will be some yawing motion.
c) Crosswinds at take-off and landing: self-explanatory.
d) Spinning: yaw control is necessary to come out of a spin.
e) Anti-symmetric power (engine out situation): for multi-engine aircraft, there is a heavy yawing moment that must be overcome by the rudder whenever one of the engines fails (Figure 4.28).

It turns out, that out of all these factors, the rudder design is most influenced by the engine out situation. Hence, in what follows, we analyze this particular situation and assume that a rudder design that is satisfactory for this important scenario would be adequate for the other scenarios considered above.

4.8 Engine Out Rudder/Aileron Power Determination: Minimum Control Speed, V_{MC}

Recall the yawing moment coefficient

$$C_n = C_{n_\beta}\beta + C_{n_{\delta R}}\delta R + \frac{N_T}{QSb} \qquad (4.110)$$

where

$$C_{n_{thrust}}\overline{Q}Sb = N_{thrust} = Tl_t. \qquad (4.111)$$

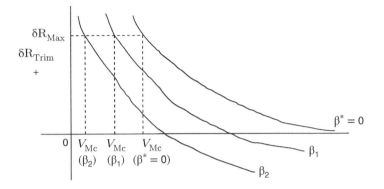

Figure 4.29 V_{MC} under rudder consideration.

In order to decide how much δR we need to trim after engine failure, for a given sideslip angle β^*

$$0 = C_{n\beta}\beta^* + C_{n\delta R}\delta R + \frac{N_T}{QSb}. \tag{4.112}$$

Rearranging the terms,

$$\delta R = \frac{-C_{n\beta}\beta^* - \frac{N_T}{QSb}}{C_{n\delta R}}. \tag{4.113}$$

For $\beta^* = 0$

$$\delta R = \frac{-N_T}{QSbC_{n\delta R}}$$

$$= -\frac{N_T}{\frac{1}{2}\rho V^2 SbC_{n\delta R}}. \tag{4.114}$$

From this expression, we see that as V decreases, δR increases (Figure 4.29).

The physical (i.e. mechanical or structural) limitation of δR gives rise to the concept of minimum control speed V_{Mc}. V_{Mc} is the speed corresponding to the maximum rudder deflection possible. Trying to trim at a speed lower than V_{Mc} would require a larger rudder deflection than structurally permitted.

4.8.1 V_{Mc} from β_{max} and Aileron Considerations

Since roll and yaw motion are coupled, it is possible to control yaw using the aileron as well as control the roll using the rudder. So let us examine if the yawing moment generated by an engine out situation can be controlled by ailerons. So consider the rolling moment equation:

$$C_l = C_{l\beta}\beta + C_{l\delta R}\delta R + C_{l\delta A}\delta A + \frac{\overset{0}{\cancel{L_T}}}{QSb} \tag{4.115}$$

where L_T is the rolling moment due to engine out situation, which we assume is negligible. Recall that from the yaw equation

$$0 = C_{n\beta}\beta + C_{n\delta R}\delta R + \frac{N_T}{\overline{Q}Sb}. \tag{4.116}$$

In order to assess if we have sufficient δA available, in case no δR is used (i.e. $\delta R = 0$), first approximately solve for the maximum β (sideslip) generated in engine out situations as such:

$$0 = C_{n\beta}\beta_{max} + \frac{N_T}{\overline{Q}Sb}. \tag{4.117}$$

In other other words,

$$\beta_{max} = -\frac{N_T}{C_{n\beta}\overline{Q}Sb}. \tag{4.118}$$

Then consider

$$0 = C_{l\beta}\beta_{max} + C_{l\delta A}\delta A \tag{4.119}$$

$$\delta A = -\frac{C_{l\beta}\beta_{max}}{C_{l\delta A}} \tag{4.120}$$

i.e.

$$\delta A = \frac{C_{l\beta}N_T}{C_{l\delta A}C_{n\beta}\overline{Q}Sb}. \tag{4.121}$$

If the δA determined from this approximate expression is below the maximum allowable δA, then we say the aircraft has sufficient lateral control power. Note that there is also a V_{Mc} possible from the maximum δA consideration (from the above expression). The eventual V_{Mc} is the one determined by the case that reaches $|\delta R|_{max}$ or $|\delta A|_{max}$ first.

Since roll/yaw are generally coupled, we now acknowledge this coupling between them and revisit the minimum control speed concept by analyzing it in the presence of this coupling between roll and yaw. We label this as V_{Mc} from exact equations.

V_{Mc} **from exact equations (for the engine out case):**

$$0 = C_{n\beta}\beta^* + C_{n\delta R}\delta R + C_{n\delta A}\delta A + \frac{N_T}{\overline{Q}Sb} \tag{4.122}$$

$$0 = C_{l\beta}\beta^* + C_{l\delta R}\delta R + C_{l\delta A}\delta A + \frac{L_T}{\overline{Q}Sb}, \quad \frac{L_T}{\overline{Q}Sb} = 0. \tag{4.123}$$

Thus:

$$\begin{bmatrix} C_{n\delta R} & C_{n\delta A} \\ C_{l\delta R} & C_{l\delta A} \end{bmatrix} \begin{bmatrix} \delta R \\ \delta A \end{bmatrix} = \begin{bmatrix} -C_{n\beta}\beta^* - \frac{N_T}{\overline{Q}Sb} \\ -C_{l\beta}\beta^* \end{bmatrix}. \tag{4.124}$$

This is in the form of $Ax = b$. So one can easily solve for δR and δA. Even if $\beta^* = 0$, you can see that both δR and δA are dependent on the term $\frac{N_T}{\overline{Q}Sb}$. So the minimum control speed is determined by the equations and the relative maximum deflections of δR and δA.

Table 4.2 Examples of flight control system types [27].

Airplane	Flaps	Longitudinal (pitch)	Lateral (roll)	Directional (yaw)
Cessna Skyhawk	Electric	Elevator + trim tab in left elevator, Reversible: cable driven	Frise ailerons, Reversible: Cable Driven	Rudder + ground adjustable trim tab, Reversible: cable driven
Cessna Cardinal	Electric	Stabilator with trim tab. Pilot controls tab, Reversible: cable driven	Frise ailerons, Reversible: Cable Driven	Rudder + trim tab, Reversible: cable driven
Rockwell Sabreliner	Electric trailing edge. Handley-Page slats	Dual electric variable incidence stabilizer, elevator + electric trim tab with manual override, Reversible: cable driven	Ailerons + electric trim tab in left aileron, Reversible: cable driven	Rudder + electric trim tab, Reversible: cable driven
Gates Learjet 35/36	Hydraulic	Dual electric variable incidence stabilizer + elevator, Reversible: cable and push-rod driven	Ailerons + servo tab + electric trim tab, Reversible: cable driven, Hydraulic spoilers: irreversible. Fly-by-wire	Rudder + electric trim tab, Reversible: cable driven
Boeing 737 (dual hydraulic control system)	Hydraulic	Elevator: hydraulic + manual override with servo tabs, cable driven, Variable incidence dual electric stabilizer with dual manual override	Hydraulic ailerons with manual reversion, Semi-reversible: cable driven, Hydraulic spoilers: reversible cable driven	Hydraulic rudder with manual reversion, Semi-reversible: cable driven
Boeing 747 (irreversible triple redundant + fourth back-up)	Hydraulic	Variable incidence tailplane with split elevators	Four ailerons + spoilers	Split rudder

The exercises at the end of this chapter will help the student test the grasp of the material related to V_{Mc} and other static lateral/directional stability concepts.

This completes the discussion of the basic exposure to static stability concepts related to aircraft. Now it is the appropriate juncture to shift our attention to the more important concept of dynamic stability and control. As mentioned earlier, in dynamic stability we analyze the actual time histories of the perturbed motion variables by developing linearized differential equations for the perturbed (small) motion variables where the perturbation is from the steady state trim (equilibrium) condition. Examples of flight control system types are shown in Table 4.2.

4.9 Chapter Summary

In this chapter, we discussed the static stability and control issues for an aircraft. From the dynamics point of view, the static case can be thought of as a snapshot at a given time, typically in the steady state (constant) situation. This type of static stability analysis takes on an important role, in particular for commercial aircraft, because we want to make sure, without any external control inputs, the aircraft possesses such characteristics that, even at the design stage, they make the aircraft have the tendency to come back to the equilibrium state when slightly disturbed from it. This analysis is carried out thoroughly for the longitudinal (pitching moment) as well as for lateral/directional motion. Typically for small motions the longitudinal analysis is decoupled from the lateral(roll)/directional (yaw) motion whereas roll and yaw motions are coupled to each other. The major concepts covered in the longitudinal analysis include concepts such as the importance of neutral point, stick force and stick force gradient. The importance of CG travel on static longitudinal stability was brought out. Also the concept of minimum control speed as a function of rudder and aileron angles is brought out, underscoring the importance of the engine out scenario that becomes important for rudder and aileron design. For more in depth treatment of any of the concepts presented in this chapter, the reader is encouraged to consult the many excellent textbooks, completely dedicated to the aircraft dynamics and control, [1–42], referenced at the end of this chapter.

4.10 Exercises

Exercise 4.1. If the slope of the C_{m} versus C_{L} plot is -0.15 and the pitching moment at zero lift is equal to 0.08 (per radian), determine the trim lift coefficient. If the center of gravity of the aircraft is located at $\bar{x}_{\text{cg}} = 0.3$, determine the stick-fixed neutral point. Assume $\delta_{\text{e}} = 0$ throughout the problem.

Exercise 4.2. This exercise is essentially the same as the worked out example we have done for the Boeing 747-400 jet transport aircraft. This time, consider an aircraft with slightly different data: let the wing area $S = 2600$ ft^2; wingspan $b = 125$ ft; aerodynamic chord $\bar{c} = 20$ ft; $\bar{x}_{\text{acw}} = 0.25$, and the center of gravity is located at $\bar{x}_{\text{cg}} = 0.35$.

We are also given the following aerodynamic data:

$$\eta = 1.1$$
$$C_{L_{aw}} = 4.81 \text{ rad}$$
$$C_{L_{at}} = 4.12 \text{ rad}.$$

1. Without a tail, does this aircraft possess static longitudinal stability?
2. Find a set of different combinations of tail area S_t and the distance $\bar{x}_{ac_t}$, such that total $C_{m\alpha} = -1.478$ is achieved corresponding to $\bar{x}_{cg} = 0.25$. Use

$$\frac{d\epsilon}{d\alpha} = \frac{2C_{L_{aw}}}{\pi AR_w}.$$

3. For each of these configurations find the stick-fixed neutral point $\bar{x}_{NP}$ and give the neutral point location in physical units.
4. Now fix a particular S_t and $\bar{x}_{ac_t}$. For this particular choice, calculate C_{m_a} and static margin for different values of the center of gravity location $\bar{x}_{cg}$ where $\bar{x}_{cg}$ is selected to cover a wide range (aft and forward of $\bar{x}_{acw}$). For this problem, use the following simplified equations:

$$C_{m_{cg}} = C_m 0 + \left(C_{L_{aw}}(\bar{x}_{cg} - \bar{x}_{acw}) - C_{L_{at}} \eta \frac{S_t}{S} \left(1 - \frac{d\epsilon}{d\alpha} \right) (\bar{x}_{ac_t} - \bar{x}_{cg}) \right) \alpha$$

where

$$\eta = \frac{\bar{Q}_t}{\bar{Q}}$$
$$\frac{d\epsilon}{d\alpha} = \frac{2C_{L_{aw}}}{\pi AR_w}$$
$$AR_w = \frac{b^2}{S}.$$

Exercise 4.3. An aircraft has a pitching moment coefficient at zero lift equal to 0.08 (per radian) and the stick-fixed neutral point is located at $\bar{x}_{NP} = 0.45$. If the aircraft is to be trimmed at $C_{L_{trim}} = 1.4$ and has the following elevator characteristics

$$\bar{C}_{m\delta_e} = -1.03 \quad \text{rad s}^{-1} \tag{4.125}$$

and the elevator deflection δ_e ranging from $10°$ to $-20°$, find the most forward CG location for this aircraft to possess static stability.

Exercise 4.4. For the data shown below, determine the following:
(a) Stick fixed neutral point
(b) If we wish to fly the aircraft at a velocity if 125 ft s^{-1} at sea level, what would be the trim lift coefficient and what would be the elevator angle.

$$W = 2750 \text{ lb}$$
$$S = 180 \text{ ft}^2$$
$$\bar{x}_{cg} = 0.25$$
$$\rho = 2.377 \times 10^{-3} \text{ slugs ft}^{-3}$$

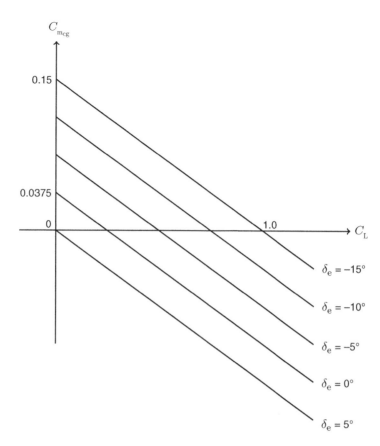

Exercise 4.5. For the aircraft data of the Boeing 747–400 jet transport, find the stick free Neutral point (for the three combinations of $\bar{x}_{act}$ and S_t, considered in the solution sheet) if $C_{h\delta_e} = -0.005$ per degree and $C_{h\alpha_t} = -0.003$ per degree and $\tau = 0.55 = \frac{d\alpha_t}{d\delta_e}$.

Exercise 4.6. Given the following data for a glider with an elevator of 20 ft^2 area and root mean square chord length of 2 ft. All aerodynamic coefficients given in the data are per degree.

$$C_{h_{\alpha_t}} = -0.003, \ C_{h_{\delta_e}} = -0.0055, \ C_{h_{\delta_t}} = -0.003$$

$$C_{h_e} = 0, \ \alpha_{0_{trim}} = 2°, \ i_\omega = 0, \ \epsilon = 0°, \ i_t = -1°$$

$$\delta_{e_{0_{trim}}} = 2°, \ \frac{d\epsilon}{d\alpha} = 0.5, \ \eta = 0.9, \ \frac{W}{S} = 30\text{lb ft}^{-2}$$

$$\bar{C}_{m\delta_e} = -0.008, \ \frac{1}{a} = 0.104, \ G = 10°/\text{ft}; \ \rho = 2.377 \times 10^{-3}\text{slugs/ft}^3$$

Calculate and plot the stick force required versus the speed for a stick fixed static margin $\frac{dC_m}{dC_L} = -0.15$ for different tab settings as follows:

a) $\delta_t = -5°$ (nose down, negative tab setting)
b) $\delta_t = -0°$

c) $\delta_t = +5°$ (nose up, positive tab setting)

Select a speed range of $0 \leq V \leq 3$ or 400 ft/sec. Investigate if you can achieve stick force trim with these tab settings or not.

$$\alpha_{trim} = \alpha_{0_{trim}} + aC_{L_{Itrim}}$$

$$\delta_{e_{trim}} = \delta_{e_{0_{trim}}} + b'C_{L_{trim}}$$

Exercise 4.7. A twin engine airplane has the following characteristics $W = 8000$ lbs, $S = 2000$ ft^2, $b = 40$ ft and $C_{L_{max}} = 1.8$. Each engine is capable of a velocity independent thrust (at low speeds) of 1000 lbs. The engine thrust lines are 10 ft to either side of the axis of symmetry. The aerodynamic data given by,

$$C_{l\beta} = -0.042 \text{ rad}^{-1} \qquad C_{n\beta} = 0.069 \text{ rad}^{-1}$$
$$C_{l\delta A} = 0.08 \text{ rad}^{-1} \qquad C_{n\delta A} = -0.008 \text{ rad}^{-1}$$
$$C_{l\delta R} = 0.008 \text{ rad}^{-1} \qquad C_{n\delta R} = -0.08 \text{ rad}^{-1}$$
$$|\delta A|_{max} = 25° \qquad |\delta R|_{max} = 30°$$

- Assuming that there is no rolling moment due to thrust ($L_T = 0$), determine, for the right engine out case, $\beta^* = 0$, the plots of δA versus speed V and δR versus speed V. Assume sea level conditions and use the uncoupled (approximate) equations. From these plots, determine the minimum control speed V_{MC}
- Also, using the coupled (exact) equations calculate the V_{MC} by drawing the plots of δA and δR versus V
- Compare the value of V_{m_C} you get from the coupled analysis and uncoupled analysis.

Bibliography

1 M.J. Abzug and E.E Larrabee. *Airplane Stability and Control: A history of the Technologies that made Aviation possible.* Cambridge University Press, Cambridge, U.K, 2 edition, 2002.
2 R.J. Adams. *Multivariable Flight Control.* Springer-Verlag, New York, NY, 1995.
3 D. Allerton. *Principles of flight simulation.* Wiley-Blackwell, Chichester, U.K., 2009.
4 J. Anderson. *Aircraft Performance and Design.* McGraw Hill, New York, 1999.
5 J. Anderson. *Introduction to Flight.* McGraw-Hill, New York, 4 edition, 2000.
6 A. W. Babister. *Aircraft Dynamic Stability and Response.* Pergamon, 1980.
7 J.H. Blakelock. *Automatic Control of Aircraft and Missiles.* Wiley Interscience, New York, 1991.
8 Jean-Luc Boiffier. *The Dynamics of Flight, The Equations.* Wiley, 1998.
9 L.S. Brian and L.L. Frank. *Aircraft control and simulation. John Wiley & Sons, Inc.,* 2003.
10 A. E. Bryson. *Control of Spacecraft and Aircraft.* Princeton University Press, Princeton, 1994.
11 Chris Carpenter. *Flightwise: Principles of Aircraft Flight.* Airlife Pub Ltd, 2002.
12 Michael V Cook. *Flight dynamics principles.* Arnold, London, U.K., 1997.
13 Wayne Durham. *Aircraft flight dynamics and control.* John Wiley & Sons, Inc., 2013.

14 B. Etkin and L.D. Reid. *Dynamics of flight: stability and control*, volume 3. John Wiley & Sons, New York, 1998.

15 E.T. Falangas. *Performance Evaluation and Design of Flight Vehicle Control Systems.* Wiley and IEEE Press, Hoboken, N.J, 2016.

16 T. Hacker. *Flight Stability and Control.* Elsevier Science Ltd, 1970.

17 G.J. Hancock. *An Introduction to the Flight Dynamics of Rigid Aeroplanes.* Number Section III.5-III.6. Ellis Hornwood, New York, 1995.

18 D.G. Hull. *Fundamentals of Airplane Flight Mechanics.* Springer International, Berlin, Germany, 2007.

19 D. McLean. Automatic flight control systems. *Prentice Hall International Series in System and Control Engineering*, 1990.

20 Duane T. McRuer, Dunstan Graham, and Irving Ashkenas. *Aircraft dynamics and automatic control.* Princeton University Press, Princeton, NJ, 1973.

21 A. Miele. *Flight Mechanics: Theory of Flight Paths.* Addison-Wesley, New York, 1962.

22 Ian Moir and Allan Seabridge. *Aicraft Systems: Mechanical, Electrical and Avionics Subsytems Integration, 3rd Edition.* John Wiley & sons, 2008.

23 Ian Moir and Allan Seabridge. *Design and Development of Aircraft Systems.* John Wiley & sons, 2012.

24 Robert C Nelson. *Flight stability and automatic control.* McGraw Hill, New York, 2 edition, 1998.

25 C. Perkins and R. Hage. *Aircraft Performance, Stability and Control.* John Wiley & Sons, London, U.K., 1949.

26 J.M. Rolfe and K.J. Staples. *Flight simulation.* Number 1. Cambridge University Press, Cambridge, U.K., 1988.

27 Jan Roskam. *Airplane Flight Dynamics and Automatic Flight Controls: Part I.* Roskam Aviation and Engineering Corp, 1979.

28 Jan Roskam. *Airplane Flight Dynamics and Automatic Flight Controls: Part II.* Roskam Aviation and Engineering Corp, 1979.

29 J.B. Russell. *Performance and Stability of Aircraft.* Arnold, London, U.K., 1996.

30 D.K. Schmidt. *Modern Flight Dynamics.* McGraw Hill, New York, 2012.

31 Louis V Schmidt. *Introduction to Aircraft Flight Dynamics.* AIAA Education Series, Reston, VA, 1998.

32 D. Seckel. *Stability and control of airplanes and helicopters.* Academic Press, 2014.

33 R. Shevell. *Fundamentals of Flight.* Prentice Hall, Englewood Cliffs, NJ, 2 edition, 1989.

34 Frederick O Smetana. *Computer assisted analysis of aircraft performance, stability, and control.* McGraw-Hill College., New York, 1984.

35 R.E. Stengel. *Flight Dynamics.* Princeton University Press, Princeton, 2004.

36 Brian L. Stevens, Frank L. Lewis, and Eric N. Johnson. *Aircraft control and simulation.* Interscience, New York, 1 edition, 1992.

37 P. Swatton. *Aircraft Performance Theory and Practice for Pilots.* Wiley Publications, 2 edition, 2008.

38 A. Tewari. *Atmospheric and Space Flight Dynamics.* Birkhauser, Boston, 2006.

39 A. Tewari. *Advanced Control of Aircraft, Spacecraft and Rockets.* Wiley, Chichester, UK, 2011.

40 Ranjan Vepa. *Flight Dynamics, Simulation and Control for Rigid and Flexible Aircraft*. CRC Press, New York, 1 edition, 2015.

41 N. Vinh. *Flight Mechanics of High Performance Aircraft*. Cambridge University Press, New York, 1993.

42 P.H. Zipfel. *Modeling and simulation of aerospace vehicle dynamics*. American Institute of Aeronautics and Astronautics Inc., 2000.

5

Aircraft Dynamic Stability and Control via Linearized Models

5.1 Chapter Highlights

In this chapter, as was the case with the previous chapter, we again focus our attention on the analysis of dynamics specific to aircraft, via the linearized equations of motion. In the previous chapter, we covered aircraft static stability and control. In this chapter, we turn our attention to dynamic stability and control. In dynamic stability and control we are concerned with the actual motion of the vehicle under various control inputs. Thus for an aircraft, in the end, it is the dynamic stability that is of utmost importance. Even if the aircraft is statically unstable (such as those military fighter aircraft which prefer high maneuverability, sacrificing static stability), keeping the aircraft dynamically stable is of paramount importance. Hence in this chapter, we thoroughly discuss dynamic stability using linearized models. In this connection, in this chapter, we focus on the analysis of natural motion (i.e. uncontrolled motion) and later in Parts II and III focus on designing feedback control systems to control the aircraft dynamics in a desired fashion.

5.2 Analysis of Perturbed Flight from Trim: Aircraft Dynamic Stability and Control

To develop the linearized equations of motion for the perturbed motion variables, we first express each motion variable as consisting of a steady state (equilibrium, trim) term, which is a constant plus a perturbation term. Then we expand the original nonlinear differential equations we developed in the previous chapters and then assume the perturbations to be small so as to make small angle approximations and neglect all the second and higher order terms in the perturbation variables, so that in the end, we come up with a set of linear differential equations in the (small) perturbation motion variables. Accordingly, the following substitutions are applied to all motion variables to derive the perturbed state equations of motion:

$$
\begin{aligned}
U &= U_0 + u_{sp} & V &= V_0 + v & W &= W_0 + w \\
P &= P_0 + p & Q &= Q_0 + q & R &= R_0 + r \\
\Psi &= \Psi_0 + \psi & \Theta &= \Theta_0 + \theta & \Phi &= \Phi_0 + \phi.
\end{aligned}
\tag{5.1}
$$

Equations (5.1) are known as the so-called perturbation equations: each variable is considered to be the sum of a steady state quantity (indicated by a subscript 0) and a

Flight Dynamics and Control of Aero and Space Vehicles, First Edition. Rama K. Yedavalli.
© 2020 John Wiley & Sons Ltd. Published 2020 by John Wiley & Sons Ltd.

perturbed state quantity (indicated by the lower case character). Similar substitutions are carried out for the aerodynamic and thrust force and moments. Forces:

$$
\begin{aligned}
F_{A_x} &= F_{A_{x_0}} + f_{A_x} \qquad F_{T_x} = F_{T_{x_0}} + f_{T_x} \\
F_{A_y} &= F_{A_{y_0}} + f_{A_y} \qquad F_{T_y} = F_{T_{y_0}} + f_{T_y} \\
F_{A_z} &= F_{A_{z_0}} + f_{A_z} \qquad F_{T_z} = F_{T_{z_0}} + f_{T_z}.
\end{aligned}
\tag{5.2}
$$

Moments:

$$
\begin{aligned}
L &= L_{A_0} + l_A \qquad L_T = L_{T_0} + l_T \\
M &= M_{A_0} + m_A \qquad M_T = M_{T_0} + m_T \\
N &= N_{A_0} + n_A \qquad N_T = N_{T_0} + n_T.
\end{aligned}
\tag{5.3}
$$

Carrying out the perturbation substitutions (5.1) through (5.3) in Equations 2.87 and 2.88:

$$
\begin{aligned}
m\{\dot{u}_{\rm sp} &- (V_0 + v)(R_0 + r) + (W_0 + w)(Q_0 + q)\} \\
&= -mg\sin(\Theta_0 + \theta) + F_{A_{x_0}} + f_{A_x} + F_{T_{x_0}} + f_{T_x}
\end{aligned}
\tag{5.4}
$$

$$
\begin{aligned}
m\{\dot{v} &+ (U_0 + u_{\rm sp})(R_0 + r) - (W_0 + w)(P_0 + p)\} \\
&= mg\sin(\Phi_0 + \phi)\cos(\Theta_0 + \theta) + F_{A_{y_0}} + f_{A_y} + F_{T_{y_0}} + f_{T_y}
\end{aligned}
\tag{5.5}
$$

$$
\begin{aligned}
m\{\dot{w} &- (U_0 + u_{\rm sp})(Q_0 + q) + (V_0 + v)(P_0 + p)\} \\
&= mg\cos(\Phi_0 + \phi)\cos(\Theta_0 + \theta) + F_{A_{z_0}} + f_{A_z} + F_{T_{z_0}} + f_{T_z}
\end{aligned}
\tag{5.6}
$$

$$
\begin{aligned}
I_{xx}\dot{p} &- I_{xz}\dot{r} - I_{xz}(P_0 + p)(Q_0 + q) + (I_{zz} - I_{yy})(R_0 + r)(Q_0 + q) \\
&= L_{A_0} + l_A + L_{T_0} + l_T
\end{aligned}
\tag{5.7}
$$

$$
\begin{aligned}
I_{yy}\dot{q} &+ (I_{xx} - I_{zz})(P_0 + p)(R_0 + r) + I_{xz}\{(P_0 + p)^2 - (R_0 + r)^2\} \\
&= M_{A_0} + m_A + M_{T_0} + m_T
\end{aligned}
\tag{5.8}
$$

$$
\begin{aligned}
I_{zz}\dot{r} &- I_{xz}\dot{p} + (I_{yy} - I_{xx})(P_0 + p)(Q_0 + q) + I_{xz}(R_0 + r)(Q_0 + q) \\
&= N_{A_0} + n_A + N_{T_0} + n_T.
\end{aligned}
\tag{5.9}
$$

Up to this point, our equations are still sufficiently general to be applicable to flight situations involving arbitrary perturbations. Now, we introduce a restriction on perturbations, and define the perturbations θ and ϕ such that:

$$
\cos\theta = \cos\phi \approx 1.0
\tag{5.10}
$$

$$
\sin\theta \approx \theta \quad \text{and} \quad \sin\phi \approx \phi.
\tag{5.11}
$$

This restricts the attitude and bank angle perturbations to roughly 15°, which is still sizable and therefore does not constitute any serious restriction from a practical point of view. Restrictions 5.10 and 5.11 allow the trigonometric terms in Equations 5.4 through 5.9 to be expanded as follows:

$$
\begin{aligned}
\sin(\Theta_0 + \theta) &= \sin\Theta_0\cos\theta + \cos\Theta_0\sin\theta \\
&= \sin\Theta_0 + \theta\cos\Theta_0
\end{aligned}
\tag{5.12}
$$

$$\sin(\Phi_0 + \phi)\cos(\Theta_0 + \theta) = (\sin\Theta_0\cos\phi + \cos\Phi_0\sin\phi)(\cos\Theta_0\cos\theta - \sin\Theta_0\sin\theta)$$
$$= (\sin\Phi_0 + \phi\cos\Phi_0)(\cos\Theta_0 - \theta\sin\Theta_0)$$
$$= \sin\Phi_0\cos\Theta_0 - \theta\sin\Phi_0\sin\Theta_0 + \phi\cos\Phi_0\cos\Theta_0$$
$$- \phi\theta\cos\Phi_0\sin\Theta_0 \tag{5.13}$$

$$\cos(\Theta_0 + \theta)\cos(\Psi_0 + \psi) = (\cos\Theta_0\cos\theta - \sin\Theta_0\sin\theta)(\cos\Psi_0\cos\psi - \sin\Psi_0\sin\psi)$$
$$= (\cos\Theta_0 - \theta\sin\Theta_0)(\cos\Phi_0 - \psi\sin\Psi_0)$$
$$= \cos\Theta_0\cos\Phi_0 - \psi\sin\Psi_0\cos\Theta_0 - \theta\sin\Theta_0\cos\Phi_0$$
$$+ \psi\theta\sin\Theta_0\sin\Psi_0 \tag{5.14}$$

Employing the trigonometric identities 5.12, 5.13, and 5.14 while expanding Equations 5.4 through 5.6 it is found that the equations of motion can be written as in Table 5.1.

Observe that parts of the equations in Table 5.1 are underlined with one line. These are steady-state terms. Note that the steady-state equations are embedded in the perturbed state equations of Table 5.1. Since the steady state equations are inherently satisfied, they can thus be eliminated from Table 5.1.

Observe that Table 5.1 also contains terms underlined with two lines. These items are all nonlinear in nature, that is they contain products or cross-products of the perturbation variables u, v, w, p, q, r, θ and ϕ.

At this point it is assumed that the perturbations are sufficiently small for products and cross-products of the perturbations to be negligible with respect to the perturbations themselves. With this assumption the nonlinear terms of Table 5.1 become negligible and the equations of perturbed motion simplify the following:

$$m(\dot{u}_{sp} - V_0 r - R_0 v + W_0 q + Q_0 w) = -mg\cos\Theta_0 + f_{A_x} + f_{T_x} \tag{5.15}$$
$$m(\dot{v} + U_0 r + R_0 u - W_0 p - P_0 w) = -mg\sin\Phi_0\sin\Theta_0 + mg\phi\cos\Phi_0\cos\Theta_0$$
$$+ f_{A_y} + f_{T_y} \tag{5.16}$$
$$m(\dot{w} - U_0 q - Q_0 u + V_0 p + P_0 v) = -mg\theta\cos\Phi_0\sin\Theta_0 - mg\phi\sin\Phi_0\cos\Theta_0$$
$$+ f_{A_z} + f_{T_z} \tag{5.17}$$
$$I_{xx}\dot{p} - I_{xz}\dot{r} - I_{xz}(P_0 q + Q_0 p) + (I_{zz} - I_{yy})(R_0 q + Q_0 r) = l_A + l_T \tag{5.18}$$
$$I_{yy}\dot{q} + (I_{xx} - I_{zz})(P_0 r + R_0 p) + I_{xz}(2P_0 p - 2R_0 r) = m_A + m_T \tag{5.19}$$
$$I_{zz}\dot{r} - I_{xz}\dot{p} + (I_{yy} - I_{xx})(P_0 q + Q_0 p) + I_{xz}(R_0 q + Q_0 r) = n_A + n_T \tag{5.20}$$

Provided the perturbed aerodynamic and thrust forces and moments are linear in the motion variables it is now clear that the equations of perturbed motion as reflected by 5.15 through 5.20 are linear in the variables u, v, w, p, q, r, ψ, θ and ϕ. Observe that we need to linearize the kinematic relations expressed by Equations 2.89. Carrying out the perturbation substitution in equations 2.89 yields:

$$P_0 + p = (\dot{\Phi}_0 + \dot{\phi}) - (\dot{\Psi}_0 + \dot{\psi})\sin(\Theta_0 + \theta) \tag{5.21}$$

$$Q_0 + q = (\dot{\Theta}_0 + \dot{\theta})\cos(\Phi_0 + \phi) + (\dot{\Psi}_0 + \dot{\psi})\cos(\Theta_0 + \theta)\sin(\Phi_0 + \phi) \tag{5.22}$$

$$R_0 + r = (\dot{\Psi}_0 + \dot{\psi})\cos(\Theta_0 + \theta)\cos(\Phi_0 + \phi) - (\dot{\Theta}_0 + \dot{\theta})\sin(\Phi_0 + \phi). \tag{5.23}$$

Table 5.1 Perturbation equations of motion with the restrictions $\cos\theta = \cos\phi \approx 1$; $\sin\theta \approx \theta$; $\sin\phi \approx \phi$.

$$m(-V_0 R_0 + W_0 Q_0) + m(\dot{u}_{sp} - V_0 v - R_0 v + W_0 q + Q_0 w) + m(-vr + uq) = -mg\sin\Theta_0 + F_{A_{x_0}} + F_{T_{x_0}} - mg\cos\Theta_0 + f_{A_x} + f_{T_x}$$

$$m(U_0 R_0 + W_0 P_0) + m(\dot{v} + U_0 r + R_0 u_{sp} - W_0 p - P_0 w) + m(u_{sp} r - wp) = mg\sin\Phi_0\cos\Theta_0 + F_{A_y}0 + F_{T_y}0 - mg\theta\sin\Phi_0\sin\Theta_0$$
$$+ mg\phi\cos\Phi_0\cos\Theta_0 + f_{A_y} + f_{T_y} - mg\phi\theta\cos\Phi_0\sin\Theta_0$$

$$m(-U_0 Q_0 + V_0 P_0) + m(\dot{w} - U_0 q - Q_0 u_{sp} + V_0 p + P_0 v) + m(-u_{sp}q + vp) = mg\cos\Phi_0\cos\Theta_0 + F_{A_z}0 + F_{T_z}0 - mg\theta\cos\Phi_0\sin\Theta_0$$
$$- mg\phi\sin\Phi_0\cos\Theta_0 + f_{A_z} + f_{T_z} + mg\phi\theta\sin\Phi_0\sin\Theta_0$$

$$-I_{xz}P_0 Q_0 + (I_{zz} - I_{yy})R_0 Q_0 + I_{xx}\dot{p} - I_{xz}\dot{r} - I_{xz}(P_0 q + Q_0 p) + (I_{zz} - I_{yy})(R_0 q + Q_0 r) - I_{xz}pq + (I_{zz} - I_{yy})rq = L_{A_0} + L_{T_0} + l_A + l_T$$

$$(I_{xx} - I_{zz})P_0 R_0 + I_{xz}(P_0^2 - Q_0^2) + I_{yy}\dot{q} + (I_{xx} - I_{zz})(P_0 r + R_0 p) + I_{xz}(2P_0 p - 2R_0 r) + (I_{xx} - I_{zz})pr + I_{xz}(p^2 - r^2) = M_{A0} + M_{T0} + m_A + m_T$$

$$(I_{yy} - I_{xx})P_0 Q_0 + I_{xz}Q_0 R_0 + I_{zz}\dot{r} - I_{xz}\dot{p} + (I_{yy} - I_{xx})(P_0 q + Q_0 p) + I_{xz}(R_0 q + Q_0 p) + (I_{yy} - I_{xx})pq + I_{xz}qr = N_{A_0} + N_{T_0} + n_A + n_T$$

Expanding these equations and using the restrictions 5.10 and 5.11 yields:

$$\underline{P_0} + p = \dot{\Phi}_0 + \dot{\phi} - \dot{\Psi}_0 \sin \Theta_0 - \underline{\dot{\Psi}_0 \theta \cos \Theta_0} - \underline{\underline{\dot{\psi} \sin \Theta_0 - \dot{\psi} \theta \cos \Theta_0}} \tag{5.24}$$

$$\underline{Q_0} + q = \dot{\Theta}_0 \cos \Phi_0 - \dot{\Theta}_0 \phi \sin \Phi_0 + \dot{\theta} \cos \Phi_0 - \underline{\underline{\dot{\theta} \phi \sin \Phi_0}} + \dot{\Psi}_0 \cos \Theta_0 \sin \Phi_0$$

$$+ \dot{\Psi}_0 \phi \cos \Theta_0 \sin \Phi_0 - \dot{\Psi}_0 \theta \sin \Theta_0 \sin \Phi_0 - \underline{\underline{\dot{\Psi}_0 \phi \theta \sin \Theta_0 \cos \Phi_0}}$$

$$+ \underline{\underline{\dot{\psi} \cos \Theta_0 \sin \Phi_0 + \dot{\psi} \phi \cos \Theta_0 \cos \Phi_0 - \dot{\psi} \theta \sin \Theta_0 \sin \Phi_0}}$$

$$- \underline{\underline{\dot{\psi} \theta \phi \sin \Theta_0 \cos \Phi_0}} \tag{5.25}$$

$$\underline{R_0} + r = \dot{\Psi}_0 \cos \Theta_0 \cos \Phi_0 - \dot{\Psi}_0 \phi \cos \Theta_0 \sin \Phi_0 - \dot{\Psi}_0 \theta \sin \Theta_0 \cos \Phi_0$$

$$- \underline{\underline{\dot{\Psi}_0 \phi \theta \sin \Theta_0 \sin \Phi_0}} + \underline{\underline{\dot{\psi} \cos \Theta_0 \cos \Phi_0 - \dot{\psi} \phi \cos \Theta_0 \sin \Phi_0}}$$

$$- \underline{\underline{\dot{\psi} \theta \sin \Theta_0 \cos \Phi_0 + \dot{\psi} \theta \phi \sin \Theta_0 \sin \Phi_0}} - \dot{\Theta}_0 \sin \Phi_0 - \dot{\Theta}_0 \phi \cos \Phi_0$$

$$- \underline{\underline{\dot{\theta} \sin \Phi_0 - \dot{\theta} \phi \cos \Phi_0}}. \tag{5.26}$$

By checking terms underlined with one line in equations 5.24 through 5.26 it is observed that the steady state equations 2.89 are embedded in equations 5.24 through 5.26. Terms underlined with two lines in equations 5.24 through 5.26 represent nonlinear terms. Eliminating the steady state parts and introducing the small perturbation assumption yields:

$$p = \dot{\phi} - \dot{\Psi}_0 \theta \cos \Theta_0 - \dot{\psi} \sin \Theta_0 \tag{5.27}$$

$$q = -\dot{\Theta}_0 \phi \sin \Phi_0 + \dot{\theta} \cos \Phi_0 + \dot{\Psi}_0 \cos \Theta_0 \cos \Phi_0 - \dot{\Psi}_0 \theta \sin \Theta_0 \sin \Phi_0$$

$$+ \dot{\psi} \cos \Theta_0 \sin_0 \tag{5.28}$$

$$r = -\dot{\Psi}_0 \phi \cos \Theta_0 \sin \Phi_0 - \dot{\Psi}_0 \theta \sin \Theta_0 \cos \Phi_0 + \dot{\psi} \cos \Theta_0 \cos \Phi_0 - \dot{\Theta}_0 \phi \cos \Phi_0$$

$$- \dot{\theta} \sin \Phi_0. \tag{5.29}$$

Equations 5.24 through 5.26 should be used in conjunction with Equations 5.15 through 5.20. Together they form nine equations in nine variables. The reader will observe that these equations are relative to an extremely general steady state. Once a specific steady state is assumed, then these linearized equations can be specialized to that particular steady state.

The majority of airplane dynamics problems are concerned with perturbations relative to a steady state for which

(a) no initial side velocity exists: $V_0 = 0$
(b) no initial bank angle exists: $\Phi_0 = 0$
(c) no initial angular velocities exist: $P_0 = Q_0 = R_0 = \dot{\Psi}_0 = \dot{\Theta}_0 = \dot{\Phi}_0 = 0$.

Imposing the above restrictions onto Equations 5.15 and 5.20, and Equations 5.24 through 5.26 yields:

$$m(\dot{u}_{sp} + W_0 q) = -mg \cos \Theta_0 + f_{A_x} + f_{T_x}$$

$$m(\dot{v} + U_0 r - W_0 p) = mg\phi \cos \Theta_0 + f_{A_y} + f_{T_y} \tag{5.30}$$

$$m(\dot{w} - U_0 q) = -mg\theta \sin \Theta_0 + f_{A_z} + f_{T_z}$$

$$I_{xx}\dot{p} - I_{xz}\dot{r} = l_A + l_T$$
$$I_{yy}\dot{q} = m_A + m_T \tag{5.31}$$
$$I_{zz}\dot{r} - I_{xz}\dot{p} = n_A + n_T$$

$$p = \dot{\phi} - \dot{\psi}\sin\Theta_0$$
$$q = \dot{\theta} \tag{5.32}$$
$$r = \dot{\psi}\cos\Theta_0.$$

Equations 5.30 through 5.32 form the basis for most studies of airplane dynamic stability, response-to-control and automatic flight control system studies.

5.3 Linearized Equations of Motion in Terms of Stability Derivatives For the Steady, Level Equilibrium Condition

In the stability axes frame, the aircraft body x axis is along the free stream velocity vector so that the steady state angle of attack is zero, which in turn means Θ_0 as well as W_0, in the above equations are zero. Thus the linearized equations of motion in stability axes are presented below. First the force equations (in the X, Y, and Z directions, respectively):

$$m\dot{u}_{sp} = -mg\theta + \frac{\partial F_{A_x}}{\partial u_{sp}}u_{sp} + \frac{\partial F_{A_x}}{\partial w}w + \frac{\partial F_{A_x}}{\partial q}q + \frac{\partial F_{A_x}}{\partial\theta}\theta + \frac{\partial F_{A_x}}{\partial\delta e}\delta e \tag{5.33}$$

$$m(\dot{v} + U_0 r) = mg\phi + \frac{\partial F_{A_y}}{\partial v}v + \frac{\partial F_{A_y}}{\partial p}p + \frac{\partial F_{A_y}}{\partial r}r + \frac{\partial F_{A_y}}{\partial\delta A}\delta A + \frac{\partial F_{A_y}}{\partial\delta R}\delta R \tag{5.34}$$

$$m(\dot{w} - U_0 q) = -mg\sin\Theta_0\theta + \frac{\partial F_{A_z}}{\partial u_{sp}}u_{sp} + \frac{\partial F_{A_z}}{\partial w}w + \frac{\partial F_{A_z}}{\partial q}q + \frac{\partial F_{A_z}}{\partial\theta}\theta + \frac{\partial F_{A_z}}{\partial\delta e}\delta e. \tag{5.35}$$

Then the moment equations (about the X, Y, and Z axes, respectively):

$$I_{xx}\dot{p} - I_{xz}\dot{r} = \frac{\partial L_A}{\partial v}v + \frac{\partial L_A}{\partial p}p + \frac{\partial L_A}{\partial r}r + \frac{\partial L_A}{\partial\delta A}\delta A + \frac{\partial L_A}{\partial\delta R}\delta R \tag{5.36}$$

$$I_{yy}\dot{q} = \frac{\partial M_A}{\partial u}u + \frac{\partial M_A}{\partial w}w + \frac{\partial M_A}{\partial q}q + \frac{\partial M_A}{\partial\theta}\theta + \frac{\partial M_A}{\partial\delta e}\delta e \tag{5.37}$$

$$I_{zz}\dot{r} - I_{xz}\dot{p} = \frac{\partial N_A}{\partial v}v + \frac{\partial N_A}{\partial p}p + \frac{\partial N_A}{\partial r}r + \frac{\partial N_A}{\partial\delta A}\delta A + \frac{\partial N_A}{\partial\delta R}\delta R. \tag{5.38}$$

And finally, the three kinematic equations:

$$\dot{\phi} = p \tag{5.39}$$

$$\dot{\theta} = q \tag{5.40}$$

$$\dot{\psi} = r. \tag{5.41}$$

Now denote

$$\frac{\partial F_{A_x}}{\partial u_{sp}}/m = X'_u \tag{5.42}$$

$$\frac{\partial F_{A_y}}{\partial v}/m = Y'_v \tag{5.43}$$

$$\frac{\partial F_{A_z}}{\partial u_{sp}}/m = Z'_u \tag{5.44}$$

$$\frac{\partial M_A}{\partial u_{sp}}/I_{yy} = M'_u \tag{5.45}$$

and so on.

We label the above right hand terms as dimensional stability derivatives. Gathering all the coefficients of each of the state variables, after appropriate manipulations and adjustment of terms on each side, it is possible to label all the stability derivatives in a concise form (as mentioned in [12]).

The linearized equations of motion, in the stability axes system, for an aircraft, are reproduced below where the steady state condition is taken to be a steady level, cruise flight and the control surface deflections are taken to be δe (elevator deflection), δA (aileron deflection) and δR (rudder deflection).

Out of these nine equations of motion, the four that describe motion in the (x, z) plane are called longitudinal equations of motion, and the other five out of plane equations are labeled as lateral/directional equations of motion.

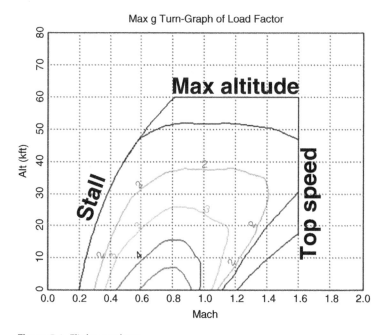

Figure 5.1 Flight envelope.

5.4 State Space Representation for Longitudinal Motion and Modes of Approximation

The forces in the X direction (longitudinal motion) can be expressed in a concise form [12] as:

$$\dot{u}_{sp} = X_u u_{sp} + X_w w + X_q q + X_\theta \theta + X_{\delta e} \delta e$$

$$\dot{w} = Z_u u_{sp} + Z_w w + Z_q q + Z_\theta \theta + Z_{\delta e} \delta e$$

$$\dot{q} = M_u u_{sp} + M_w w + M_q q + M_\theta \theta + M_{\delta e} \delta e$$

$$\dot{\theta} = q.$$

It is convention in some texts to replace the vertical velocity w with the angle of attack α in these equations. The reader should keep in mind that the variables w and α essentially convey the same information, and can therefore be used interchangeably. Note that, in the linear range, the vertical velocity w is often used interchangeably with the angle of attack α because in the linearized model w is related to α via the expression $\alpha = w/U_0$. In a vector-matrix format, they are:

$$
\begin{bmatrix} \dot{u}_{sp} \\ \dot{w} \\ \dot{q} \\ \dot{\theta} \end{bmatrix} =
\begin{bmatrix} X_u & X_w & X_q & X_\theta \\ Z_u & Z_w & Z_q & Z_\theta \\ M_u & M_w & M_q & M_\theta \\ 0 & 0 & 1 & 0 \end{bmatrix}
\begin{bmatrix} u_{sp} \\ w \\ q \\ \theta \end{bmatrix} +
\begin{bmatrix} X_{\delta e} \\ Z_{\delta e} \\ M_{\delta e} \\ 0 \end{bmatrix} \delta e
\tag{5.46}
$$

Here the motion variables u, w (or α), q, and θ are the state variables and the forcing function (input) variable δe is the control variable. We write the above linearized longitudinal equations of motion for an aircraft at a given steady state level flight condition (i.e. at a given speed and altitude) in the state variable representation or state space formulation mentioned in the previous chapters. Thus we have:

$$\dot{\underline{x}} = A\underline{x} + B\underline{u}$$

where $\underline{x}$ is the vector of state variables, A is the constant coefficient matrix that involves the stability derivatives, B is the constant coefficient matrix that involves the control surface stability derivatives, and $\underline{u}$ is the vector of control variable. In our case:

$$
\underline{x} = \begin{bmatrix} u_{sp} \\ w \\ q \\ \theta \end{bmatrix}
\tag{5.47}
$$

and

$$\underline{u} = \delta e.
\tag{5.48}$$

Note that in the longitudinal case the state vector $\underline{x}$ is of dimension 4 and the control vector $\underline{u}_c$ is of dimension 1 (single input) and thus A is a 4×4 matrix and B is a 4×1

matrix. Recall that in a general set of linear constant coefficient differential equations in state space representation, $\underline{x}$ is of dimension n and $\underline{u}_c$ is of dimension m and thus A is an $n \times n$ matrix and B is a $n \times m$ matrix.

We can get similar state variable representation for the other five equations involving lateral/directional motion with v, p, r, ϕ, and ψ as the state variables and the aileron deflection, δA, and the rudder deflection, δR, as the control variables. For now, we focus on the longitudinal case and provide a similar treatment in the lateral/directional case in Section 5.5. The elements in the matrices A and B in the above state space representation are labeled as the stability derivatives of the aircraft. Notice that these have specific numerical values based on the aircraft's geometric parameters (like wing area S, mean aerodynamic chord $\bar{c}$ and so on, mass and moment of inertia parameters, aerodynamic parameters such as C_L, C_D, $C_{L\alpha}$, $C_{m\alpha}$ and so on, and most importantly on the flight condition (for a given Mach number, altitude combination) itself. Since the linearization is done about a steady state condition, which in turn depends on the flight condition (cruise speed and altitude) all the rates of changes related to the aerodynamic forces (such as lift and drag) and the aerodynamic moments are evaluated at that flight condition and thus eventually take on some numerical values corresponding to that flight condition. In other words, we obtain a set of A and B matrices corresponding to a given flight condition within the flight envelope. The expressions for the entries of the A and B matrices could be the non-dimensional stability derivatives or dimensional stability derivatives depending on the nature of data used in the expressions for those stability derivatives. We assume the explicit expressions given for the entries to be the dimensional stability derivatives.

It is to be recognized that considerable effort goes into getting explicit expressions for these dimensional stability derivatives, mostly obtained from wind tunnel tests by aerodynamics engineers. In a course completely dedicated to aircraft dynamics and control, it is possible to go deeper into obtaining these explicit expressions. However, keeping the scope of this book in mind, we basically direct the reader to the many good textbooks available on pure aircraft dynamics and control such as [30, 31, 40] for those explicit formulae for the entries of the A matrix and B matrix in the above state space representation. Instead, we summarize the expressions for most of those stability derivatives and discuss their importance. The way to get final numerical values for each of the above stability derivatives (i.e. entries of the A and B matrices) for an aircraft at a given flight condition are elaborated in [24]. It suffices to emphasize the fact that the numerical value obtained for each of these stability derivatives embodies complete information about the mass, geometric and aerodynamic parameters of the aircraft as well as the specificity of the particular flight condition (Mach number and altitude). A typical flight envelope with various flight conditions is depicted in Figure 5.1. In that sense, it is worth reiterating that this simple state space equation $\dot{x} = Ax + Bu$ amazingly contains all the critical information that is specific to a particular aircraft at a particular flight condition. Looking at it from a mathematical point of view, it is a simple vector/matrix first order differential equation, but looking at from an aeronautical engineering point of view, that state space equation is indeed the aircraft itself! See Figure 5.2. This should provide sufficient motivation for the student to appreciate the use of mathematical modeling in capturing the dynamics of an engineering system, an aircraft in our case here. It underscores the close, powerful and useful connection between mathematics and engineering.

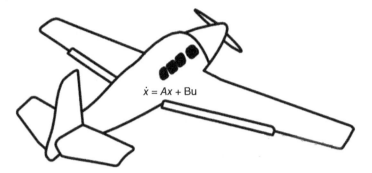

$$\dot{x} = Ax + Bu$$

Figure 5.2 Illustration of the importance of mathematical models.

5.4.1 Summary and Importance of the Stability Derivatives

Borrowing material from books such as [24], we list the expressions for the stability derivatives and when appropriate comment on the nature and importance of that derivative.

All stability derivatives are important but some are more important for flight control than others.

5.4.1.1 Importance of Various Stability Derivatives in Longitudinal Motion

We now focus on a few stability derivatives of importance in the longitudinal motion.

A number of parameters appear frequently in the equations defining stability derivatives. They are listed here for convenience (note that all the stability derivatives presented are dimensional): S is the surface area of the wing, $\bar{c}$ is the mean aerodynamic chord, ρ is the density, and b is the wing span.

5.4.2 Longitudinal Motion Stability Derivatives

5.4.2.1 Lift Related Stability Derivatives

$$Z_w = \frac{\rho S U_0}{2m}(C_{L_\alpha} + C_D).$$

The change in lift coefficient with a change in angle of attack, C_{L_α}, is often referred to as the lift curve slope. It is always positive for values of angle of attack below the stall value. The lift curve slope for the total airframe comprises components due to the wing, the fuselage and the tail. For most conventional aircraft it has been found to be generally true that the wing contributes 85-90 percent to the value of C_{L_α}. Consequently, any aeroelastic distortion of the wing can appreciably alter C_{L_α} and, hence, Z_w.

5.4.2.2 Pitching Moment Related Stability Derivatives

$$M_u \triangleq \frac{\rho S U_0 \bar{c}}{I_{yy}}(C_{m_u} + C_m).$$

The non-dimensional pitching moment coefficient, C_m, is usually zero in trimmed flight, except in cases of thrust asymmetry. M_u represents the change in pitching moment caused by a change in forward speed. Its magnitude can vary considerably and its sign can change with changes in Mach number and in dynamic pressure and also as a result of aeroelastic effects. In modern aircraft, the Mach number effects and the effects of aeroelasticity have become increasingly important.

$$M_w = \frac{\rho S U_0 \bar{c}}{2 I_{yy}} C_{m_\alpha}$$

The non-dimensional stability derivative, C_{m_α}, is the change in the pitching moment coefficient with angle of attack. It is referred to as the longitudinal static stability derivative. C_{m_α} is very much affected by any aeroelastic distortions of the wing, the tail and the fuselage. However, both sign and magnitude of C_{m_α} are principally affected by the location of the center of gravity of the aircraft. C_{m_α} is proportional to the distance, x_{AC}, between the center of gravity and the aerodynamic center of the whole aircraft (which often is also labeled as the static margin). x_{AC} is measured positive forwards. If x_{AC} is zero, C_{m_α} is zero. If $x_{AC} < 0$, C_{m_α} is negative and the aircraft is statically stable. If the center of gravity is aft of the aerodynamic center, $x_{AC} < 0$ and C_{m_α} is positive, with the consequence that the aircraft is statically unstable. In going from subsonic to supersonic flight the aerodynamic center generally moves aft, and, therefore, if the center of gravity remains fixed, C_{m_α} will tend to increase for a statically stable aircraft. Thus $M_w(M_\alpha)$ is closely related to the aircraft's static margin. Because of the significance of static stability, static margin, naturally M_w (or M_α) is one of the most important longitudinal stability derivatives.

$$M_q = \frac{\rho S U_0 \bar{c}^2}{4 I_{yy}} C_{m_q}$$

For conventional aircraft, M_q contributes a substantial part of the damping of the short period motion. This damping comes mostly from changes in the angle of attack of the tail and it is also proportional to the tail length, l_T.

M_q is a very significant stability derivative which has a primary effect on the handling qualities of the aircraft.

5.4.2.3 Control Related Stability Derivatives

$$Z_{\delta_E} = \frac{-\rho S U_0^2}{2m} C_{L_{\delta_E}}$$

Since $C_{L_{\delta_E}}$ is usually very small, Z_{δ_E} is normally unimportant except when an aircraft flight control system (AFCS) involving feedback of normal acceleration is used. Also, if a tailless aircraft is being considered, the effective lever arm for the elevator (or ailerons) is small, hence $C_{L_{\delta_E}}$ may be relatively large compared to $C_{m_{\delta_E}}$. In these cases, Z_{δ_E} cannot safely be neglected in any analysis.

$$M_{\delta_E} = \frac{\rho S U_0^2 \bar{c}}{2m} C_{m_{\delta_E}}.$$

$C_{m_{\delta_E}}$ is termed the elevator control effectiveness; it is very important in aircraft design and for AFCS work. When the elevator is located aft of the center of gravity, the normal location, $C_{m_{\delta_E}}$ is negative. Its value is determined chiefly by the maximum lift of the wing and also the range of center of gravity travel which can occur during flight.

Conceptually speaking, what is more important for the student to realize is that these stability derivatives are a function of the specific flight condition (at a given Mach number and altitude, i.e. a point in the flight envelope) and are evaluated at that specific flight condition. Thus we finally get numerical values for these stability derivatives as they are evaluated at given flight conditions within the flight envelope. In other words, this linearized model with a numerical set of entries in the A and B matrices is thought of as representing the aircraft's state at that particular flight condition. We then design a flight control system for the aircraft to stabilize and damp out any oscillations for small motions around that flight condition.

Note that we have linearized state space models of aircraft dynamics about a given flight condition, which turn out to produce a fourth order plant matrix A for the longitudinal motion (which is decoupled, in the linear range, from the lateral/directional motion). As mentioned in the discussion on stability of linear time invariant systems, the longitudinal dynamic stability and performance assessment involves finding the eigenvalues of a fourth order A matrix. Finding the eigenvalues of a fourth order system obviously involves solving for the roots of a fourth degree polynomial. For early aeronautical engineers working without the aid of computers, it was difficult, if not impossible, to solve for the roots of the polynomial. For this reason, they approximated the motion of the aircraft (both in longitudinal as well as lateral/directional cases) as smaller subsystems. This gives rise to the concept of modes of approximation, for longitudinal (as well lateral/directional motion). First we discuss the longitudinal approximations and then lateral/directional approximations.

5.4.3 Longitudinal Approximations

In the longitudinal case, there are two modes of approximation, namely the short period mode and the phugoid (long period) mode.

5.4.3.1 Phugoid Mode Approximation

We can think of the long-period or phugoid mode as a gradual interchange of potential and kinetic energy about the equilibrium altitude and airspeed. An approximation to the long period mode can be obtained by neglecting the pitching moment equation and assuming that the change in angle of attack is zero, i.e.

$$\alpha = \frac{w}{U_0} \quad \alpha = 0 \rightarrow w = 0. \tag{5.49}$$

Making these assumptions, the homogeneous longitudinal state equations reduce to the following:

$$\begin{bmatrix} \dot{u}_{sp} \\ \dot{\theta} \end{bmatrix} = \begin{bmatrix} X_u & -g \\ \frac{-Z_u}{U_0} & 0 \end{bmatrix} \begin{bmatrix} u_{sp} \\ \theta \end{bmatrix}. \tag{5.50}$$

Important remark: note that, since this second order system is an approximation of the large fourth order system, the individual element values of the A matrix of this approximate model do not necessarily coincide with the individual elemental values of the original fourth order matrix that correspond to the two state variables in the large order state space system. Hence stability derivative values that appear in the second order approximation model are obtained from the aerodynamic data with some approximations.

The eigenvalues of the long period approximation are obtained by solving the equation

$$|\lambda I - A| = 0 \tag{5.51}$$

or

$$\begin{vmatrix} \lambda - X_u & g \\ \dfrac{Z_u}{U_0} & \lambda \end{vmatrix} = 0. \tag{5.52}$$

Expanding the above determinant yields

$$\lambda^2 = X_u \lambda - \dfrac{Z_u g}{U_0} = 0 \tag{5.53}$$

or

$$\lambda_p = \left[X_u \pm \sqrt{X_u^2 + 4\dfrac{Z_u g}{U_0}} \right] / 2.0. \tag{5.54}$$

The frequency and damping ratio can be expressed as

$$\omega_{np} = \sqrt{\dfrac{-Z_u g}{U_0}} \tag{5.55}$$

$$\zeta_p = \dfrac{-X_u}{2\omega_{np}} \tag{5.56}$$

If we neglect compressibility effects, the frequency and damping ratios for the long-period motion can be approximated by the following equations:

$$\omega_{np} = \sqrt{2}\dfrac{g}{U_0} \tag{5.57}$$

$$\zeta_p = \dfrac{1}{\sqrt{2}}\dfrac{1}{L/D}. \tag{5.58}$$

Notice that the frequency of oscillation and the damping ratio are inversely proportional to the forward speed and the lift-to-drag ratio, respectively. We see from this approximation that the phugoid damping is degraded as the aerodynamic efficiency (L/D) is increased. When pilots are flying an airplane under visual flight rules, the phugoid damping and frequency can vary over a wide range and they will still find the airplane acceptable to fly. On the other hand, if they are flying the airplane under instrument flight rules, low phugoid damping will become very objectionable to them. To improve the damping of the phugoid motion, the designer would have to reduce the lift-to-drag ratio of the airplane. Because this would degrade the performance of the airplane, the designer would find such a choice unacceptable and would look for another alternative, such as an automatic stabilization system to provide the proper damping characteristics.

5.4.3.2 Short Period Approximation

We can think of the short period mode approximation as a motion with very little speed changes but appreciable angle of attack changes on relatively fast (short period) time scales. An approximation to the short period mode can be obtained by neglecting the speed changes but accounting for the changes in the angle of attack, i.e.

$$\alpha = \frac{w}{U_0}; \quad u_{sp} = 0 \rightarrow \theta = 0. \tag{5.59}$$

Making these assumptions, the homogeneous longitudinal state equations reduce to the following:

$$\begin{bmatrix} \dot{w} \\ \dot{q} \end{bmatrix} = \begin{bmatrix} Z_w & Z_q \\ M_w & M_q \end{bmatrix} \begin{bmatrix} w \\ q \end{bmatrix}. \tag{5.60}$$

Important remark: as mentioned in the case of the phugoid mode, the same remark holds for this case as well, namely that the individual element values of the A matrix of this approximate model do not necessarily coincide with the individual elemental values of the original fourth order matrix that correspond to the two state variables in the large order state space system. Hence stability derivative values that appear in the second order approximation model are obtained from the aerodynamic data with some approximations.

The eigenvalues of the short period approximation are obtained by solving the equation

$$|\lambda I - A| = 0 \tag{5.61}$$

and along the same lines as above for the phugoid mode, we can also obtain the short period mode natural frequency ω_{nSP} and its damping ratio ζ_{sp} in terms of the stability derivatives.

5.4.4 Summary of Longitudinal Approximation Modes

Summarizing, the short period approximation is arrived at by assuming motion at constant speed and the phugoid approximation is arrived at by assuming constant angle of attack motion. A pictorial representation of these modes of motion is shown in Figures 5.3 and 5.4.

1. Phugoid mode (long period mode):
   (a) long period, lightly damped
   (b) eigenvalues closer to the imaginary axis (dominant pair)
   (c) motion with significant u and e, and very small pitch rate and angle of attack
   (d) flight path: approximately one of constant total energy.

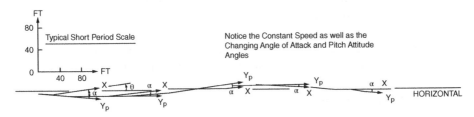

Figure 5.3 Short period as seen by an outside observer.

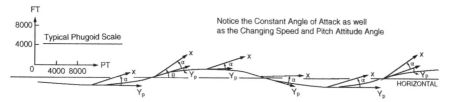

Figure 5.4 Phugoid as seen by an outside observer.

2. Short period mode:
   (a) short period, heavily damped
   (b) far from imaginary axis
   (c) motion with oscillation of angle of attack variation with negligible speed variation
   (d) flight path: damped rapidly

Important remark: it needs to be kept in mind that this type of mode separation is arrived at from physical considerations of the motion of a typical aircraft wherein lot of assumptions have been made about the stability derivative information. Thus it is not necessary that every aircraft exhibit a motion corresponding to these two types of modes. These two types modes can be identified in a distinct fashion only if the real parts and the imaginary parts of the two complex conjugate eigenvalue pairs of the larger fourth order matrix have appreciable separation among them. In other words, for an aircraft at a particular flight condition at which the stability derivatives take on such values that the overall longitudinal fourth order matrix has very closely spaced complex conjugate pair eigenvalues or possibly the eigenvalues do not even turn out to be two complex conjugate pairs at all (indicating no oscillatory behavior) then it is not possible to label them as strictly a short period mode and a phugoid mode. In the most common and standard data situations it may be relatively easy to identify one mode as a pure phugoid mode and another mode as a pure short period mode but this is not necessarily the case for each and every aircraft at each and every flight condition. This artifice of approximating a fourth order system dynamics into two smaller second order subsystem dynamics is very much dependent on the makeup of the eigenvalues of the larger fourth order system dynamics matrix and thus care needs to be exercised as to how to label the overall motion as consisting of a short period mode and a phugoid mode.

Next, we turn our attention to lateral/directional motion.

5.5 State Space Representation for Lateral/Directional Motion and Modes of Approximation

We can write expressions for the forces and moments in the lateral/directional motion as follows: the forces in the Y direction (lateral/directional motion)

$$\dot{v} = Y_v v + Y_p p + Y_r r + Y_\psi \psi + Y_\phi \phi + Y_{\delta A} \delta A + Y_{\delta R} \delta R. \tag{5.62}$$

Rolling moment equation (lateral/directional motion)

$$\dot{p} = L_v v + L_p p + L_r r + L_\psi \psi + L_\phi \phi + L_{\delta A} \delta A + L_{\delta R} \delta R. \tag{5.63}$$

Yawing moment equation (lateral/directional motion)

$$\dot{r} = N_v v + N_p p + N_r r + N_\psi \psi + N_\phi \phi + N_{\delta A} \delta A + N_{\delta R} \delta R. \tag{5.64}$$

Euler (roll) angle rate (lateral/directional motion)

$$\dot{\psi} = r. \tag{5.65}$$

Euler (yaw) angle rate (lateral/directional motion)

$$\dot{\phi} = p. \tag{5.66}$$

Putting these five lateral/directional motion equations into a compact state space form, we get:

$$
\begin{bmatrix} \dot{v} \\ \dot{p} \\ \dot{r} \\ \dot{\psi} \\ \dot{\phi} \end{bmatrix}
=
\begin{bmatrix}
Y_v & Y_p & Y_r & Y_\psi & Y_\phi \\
L_v & L_p & L_r & L_\psi & L_\phi \\
N_v & N_p & N_r & N_\psi & N_\phi \\
0 & 0 & 1 & 0 & 0 \\
0 & 1 & 0 & 0 & 0
\end{bmatrix}
\begin{bmatrix} v \\ p \\ r \\ \psi \\ \phi \end{bmatrix}
+
\begin{bmatrix}
Y_{\delta A} & Y_{\delta R} \\
L_{\delta A} & L_{\delta R} \\
N_{\delta A} & N_{\delta R} \\
0 & 0 \\
0 & 0
\end{bmatrix}
\begin{bmatrix} \delta A \\ \delta R \end{bmatrix}. \tag{5.67}
$$

This is the generic, standard state space representation of the aircraft lateral/directional equations of motion, with v, p, r, ϕ and ψ being the state variables. The aileron deflection, δA, and the rudder deflection, δR are the control variables. It is convention in some texts to replace the side velocity v with the sideslip angle β in these equations. The reader should keep in mind that the variables v and β essentially convey the same information, and can therefore be used interchangeably. Note that, in the linear range, the side velocity v is often used interchangeably with the sideslip angle β because in the linearized model, v is related to β via the expression $\beta = v/U_0$.

Relative Importance of Various Stability Derivatives in Lateral/Directional Motion
As mentioned earlier in the longitudinal case discussion, all stability derivatives appearing in the above A and B matrices are important but some are more important for flight control than others. We now focus on a few stability derivatives of importance in lateral/directional motion. Note that stability derivatives L_β and L_v are related and similarly N_v and N_β are related.

5.5.1 Lateral/Directional Motion Stability Derivatives

5.5.1.1 Motion related
$$Y_v = \frac{\rho S U_0}{2m} C_{y_\beta}.$$

The side force that results from any sideslip motion is usually obtained from the fin of the aircraft, and usually opposes the sideslip motion, i.e. $C_{y_\beta} < 0$. However, for aircraft with a slender fuselage, at large values of the angle of attack the forces can be in an aiding direction. For certain (rare) configurations having a wing of low aspect ratio but required to operate at a large value of angle of attack, this force on the fuselage can counter the resisting force of the fin, which results in the stability derivative C_{y_β} being positive. Such positive values, even if very small, are undesirable because the reversed (or small) side force makes it difficult for a pilot to detect sideslip motion and consequently makes a coordinated turn difficult to achieve. Such values of C_{y_β} also reduce the damping ratio of the Dutch roll mode, whereas C_{y_β} normally makes a large contribution to this damping. In the normal case C_{y_β} is not a derivative which causes great difficulty to AFCS designers.

$$L_\beta = \frac{\rho S U_0^2 b}{2 I_{xx}} C_{l_\beta}.$$

The change in the value of the rolling moment coefficient with sideslip angle C_{l_β} is called the effective dihedral. This derivative is very important in studies concerned with lateral stability and control. It features in the damping of both the Dutch roll and the spiral modes. It also affects the maneuvering capability of an aircraft, particularly when lateral control is being exercised near stall by rudder action only. Usually small negative values of C_{l_β} are wanted, as such values improve the damping of both the Dutch roll and the spiral modes, but such values are rarely obtained without considerable aerodynamic difficulty.

$$N_\beta = \frac{\rho U_0^2}{2} \frac{S b}{I_{xx}} C_{n_\beta}.$$

The change in the yawing moment coefficient with change in sideslip angle C_{n_β} us referred to as the static directional or weathercock stability coefficient. It depends upon the area of the fin and the lever arm. The aerodynamic contribution to C_{n_β} from the fin is positive, but the contribution from the aircraft body is negative. A positive value of C_{n_β} is regarded as static directional stability; a negative value signifies static directional instability. C_{n_β} primarily establishes the natural frequency of the Dutch roll mode and is an important factor in establishing the characteristics of the spiral mode stability. For good handling qualities C_{n_β} should be large, although such values magnify the disturbance effects from side gusts. At supersonic speeds C_{n_β} is adversely affected because the lift curve slope of the fin decreases.

$$L_p = \frac{\rho U_0 S b^2}{4 I_{xx}} C_{l_p}.$$

The change in rolling moment coefficient with change in rolling velocity, C_{l_p}, is referred to as the roll damping derivative. Its value is determined almost entirely by the geometry of the wing. In conjunction with $C_{l_{\delta_A}}$ (q.v.), C_{l_p} establishes the maximum rolling velocity that can be obtained from the aircraft, an important flying quantity. C_{l_p} is always negative, although it may become positive when the win (or parts of it) are stalled.

$$N_p = \frac{\rho U_0 S b^2}{4 I_{xz}} C_{n_p}.$$

The change in rolling moment coefficient with a change in rolling velocity, C_{n_p}, is usually negative, although a positive value is desirable. The more negative C_{n_p} is, the smaller the damping ratio of the Dutch roll mode and the greater the sideslip motion which accompanies entry to, or exit from, a turn.

$$L_r = \frac{\rho U_0 S b^2}{4 I_{xx}} C_{l_r}.$$

The change in rolling moment coefficient with a change in yawing velocity, C_{l_r}, has a considerable effect on the spiral mode, but does not much affect the Dutch roll mode. For good spiral stability, C_{l_r} should be positive but as small as possible. A major contributing factor to C_{l_r} is the lift force from the wing, but if the fin is located either above or below the axis it also makes a substantial contribution to C_{l_r}, being positive or negative dependent upon the fin's geometry.

$$N_r = \frac{\rho U_0 S b^2}{4 I_{zz}} C_{n_r}.$$

The change in yawing moment coefficient with a change in yawing velocity, C_{n_r}, is referred to as the yaw damping derivative. Usually C_{n_r} is negative and is the main contributor to the damping of the Dutch roll mode. It also contributes to the stability of the spiral mode.

5.5.1.2 Control Related

$$Y_\delta = \frac{\rho U_0^2 S}{2m} C_{y_\delta}.$$

The change in side force coefficient with rudder deflection, $C_{y_{\delta R}}$, is unimportant except when considering an AFCS using lateral acceleration as feedback. $C_{y_{\delta A}}$ is nearly always negligible. Because positive rudder deflection produces a positive side force, $C_{y_{\delta R}} < 0$.

$$L_\delta = \frac{\rho U_0^2 S b}{2 I_{xx}} C_{l_\delta}.$$

$C_{l_{\delta R}}$ is the change in rolling moment coefficient that results from rudder deflection. It is usually negligible. Because the rudder is usually located above the axis, positive rudder deflection produces positive rolling motion, i.e. $C_{l_{\delta R}} > 0$.

The change in rolling moment coefficient with a deflection of the ailerons, $C_{l_{\delta A}}$, is referred to as the aileron effectiveness. In lateral dynamics it is the most important control-related stability derivative. It is particularly important for low speed flight where adequate lateral control is needed to counter asymmetric gusts that tend to roll the aircraft.

$$N_\delta = \frac{\rho U_0^2 S b}{2 I_{zz}} C_{n_\delta}.$$

The change in yawing moment coefficient that results from a rudder deflection, $C_{n_{\delta R}}$, is referred to as the rudder effectiveness. When the rudder is deflected to the left (i.e. $\delta_R > 0$) a negative yawing moment is created on the aircraft, i.e. $C_{n_{\delta R}} < 0$.

The change in yawing moment coefficient that results from an aileron deflection, $C_{n_{\delta A}}$, results in adverse yaw if $C_{n_{\delta A}} < 0$, for when a pilot deflects the ailerons to produce a turn, the aircraft will yaw initially in a direction opposite to that expected. When $C_{n_{\delta A}} > 0$, the yaw that is favorable to that turning maneuver, this is referred to as proverse yaw. Whatever sign $C_{n_{\delta A}}$ takes its value ought to be small for good lateral control.

Just as we discussed the approximation of various modes of motion in the longitudinal case, we can do a similar exercise in the lateral/directional case as well. However, there are considerable differences in the nature of these modes of approximation between the longitudinal case and the lateral/directional case. We elaborate on them next.

5.5.2 Lateral/Directional Approximations

In getting these approximations, we simply neglect the $\dot{\psi}$ equation because in the linear range, for small motions, ψ is essentially a constant. Thus we consider only four lateral/directional equations of motion, which are summarized here:

$$
\begin{bmatrix} \dot{v} \\ \dot{p} \\ \dot{r} \\ \dot{\phi} \end{bmatrix} = \begin{bmatrix} Y_v & Y_p & Y_r & Y_\phi \\ L_v & L_p & L_r & L_\phi \\ N_v & N_p & N_r & N_\phi \\ 0 & 1 & 0 & 0 \end{bmatrix} \begin{bmatrix} v \\ p \\ r \\ \phi \end{bmatrix} + \begin{bmatrix} Y_{\delta A} & Y_{\delta R} \\ L_{\delta A} & L_{\delta R} \\ N_{\delta A} & N_{\delta R} \\ 0 & 0 \end{bmatrix} \begin{bmatrix} \delta A \\ \delta R \end{bmatrix} . \tag{5.68}
$$

Noting that the lateral/directional equations of motion consist of again four state variables, which makes finding the eigenvalues of the above A matrix without aid from a computer difficult as in the longitudinal motion case. Again, early aeronautical engineers worked around this by approximating lateral motion as comprised of three separate modes, called roll subsidence, spiral Convergence/divergence, and the Dutch roll.

5.5.3 Roll Subsidence Approximation

Suppose an airplane flying in a steady, level, unaccelerated flight is disturbed by a gust of wind into a roll. Then, the wing that dips down will see an increased angle of attack α, as the air effectively comes up to meet the wing. The other wing, which rolls up, will have the opposite effect (i.e. a decreased effective angle of attack). Obviously, the wing that is dipped down will produce more lift than the wing which is tilted upwards. This lift differential produces a restoring moment that ramps up the roll rate exponentially until the restoring moment balances the disturbing moment, and equilibrium is achieved. From this qualitative discussion, we can say that the roll subsidence mode is a converging maneuver.

Quantitatively, the roll subsidence mode can be approximated with a single degree of freedom, as follows (Figure 5.5)

$$
\dot{p} = -\frac{p}{\tau} \tag{5.69}
$$

where τ is the roll time constant. As it turns out, the eigenvalue corresponding to roll subsidence is negative and purely real. It is also fairly large in magnitude, and thus we can say that roll subsidence is a highly converging maneuver.

5.5.4 Spiral Convergence/Divergence Approximation

Again consider an airplane in steady, level, unaccelerated flight that is disturbed into a roll. This generates a small sideslip velocity v. Then the incoming air hits the vertical stabilizer at an incidence angle β, which generates extra tail lift and thus a yawing

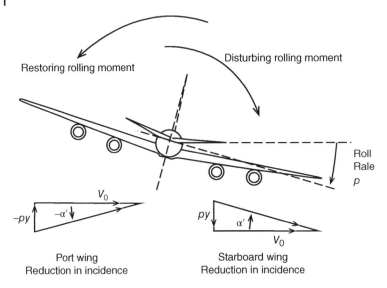

Figure 5.5 Roll subsidence approximation.

moment. This yawing moment actually increases the sideslip in a vicious cycle that, left unchecked, will see the plane slowly diverge in roll, yaw, and altitude. Plainly speaking, the aircraft will spiral into the ground. Therefore it is important that the aeronautical engineer designs the aircraft in such a way as to stabilize the spiral mode (i.e. make it converge instead of diverge).

Mathematically, the spiral approximation is this

$$\begin{bmatrix} 0 \\ \dot{r} \end{bmatrix} = \begin{bmatrix} L_\beta & L_r \\ N_\beta & N_r \end{bmatrix} \begin{bmatrix} \beta \\ r \end{bmatrix}. \tag{5.70}$$

Substituting the algebraic relationship of the first equation into the second differential equation, we get a single scalar differential equation in the state variable r. Thus, it becomes a first order system. This is labeled as spiral mode (Figure 5.6).

As it happens, the eigenvalue associated with the spiral mode is purely real and is given by $\lambda_{spiral} = [L_\beta N_r - L_r N_\beta]/L_\beta$. Stability derivatives L_β (reflecting dihedral effect) and N_r (reflecting yaw rate damping) are usually negative quantities, On the other hand N_β (reflecting directional stability) and L_r (roll moment due to yaw rate) are usually positive quantities. Thus the sign of this real eigenvalue depends on the numerical values for a specific flight condition of a specific aircraft. Therefore the spiral mode becomes stable if this eigenvalue is negative and this is labeled as spiral convergence; unstable if positive. It should be noted that this eigenvalue is smaller in magnitude than that associated with the roll subsidence mode. We can conclude that the spiral mode is characterized by slow, often unstable motion.

5.5.5 Dutch Roll Approximation

Again consider an aircraft in steady, level, unaccelerated flight that is disturbed, this time such that oscillations in yaw ensue. As a wing yaws forward, it will see a higher incoming wind velocity; the wing yawing back sees a slower incoming wind velocity.

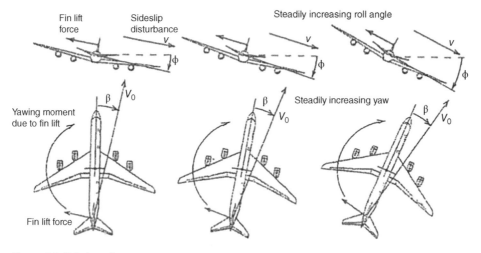

Fin lift force   Sideslip disturbance   Steadily increasing roll angle

Yawing moment due to fin lift

Fin lift force

Steadily increasing yaw

Figure 5.6 Spiral mode.

This produces an oscillatory lift differential. Recall yaw and roll are coupled. Therefore The Dutch roll mode is analogous to the short period. However, while the short period consists of oscillations in angle of attack α, the Dutch roll mode consists of oscillations in yaw that couple into roll (Figure 5.7).

Since the vertical stabilizer is less effective than the horizontal stabilizers, the Dutch roll mode is not damped as well as the short period; the frequency of oscillations between the short and Dutch roll is comparable, however. Quantitatively, the Dutch

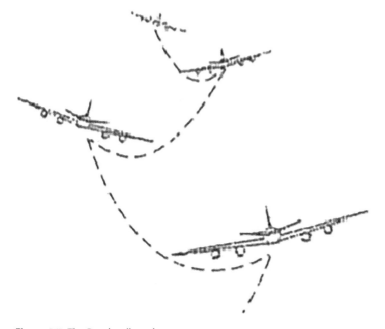

Figure 5.7 The Dutch roll mode.

roll mode is expressed as:

$$\begin{bmatrix} \dot{\beta} \\ \dot{r} \end{bmatrix} = \begin{bmatrix} \dfrac{Y_\beta}{u_o} & \dfrac{Y_r}{u_o} \\ N_\beta & N_r \end{bmatrix} \begin{bmatrix} \beta \\ r \end{bmatrix}. \tag{5.71}$$

5.5.6 Summary of Lateral/Directional Approximation Modes

1. Dutch Roll mode:
   (a) Has lateral oscillations with motion consisting of mainly yawing and rolling together.
   (b) Eigenvalues are always a complex conjugate pair with negative real part.
2. Spiral mode (convergent/divergent):
   (a) Consists mainly of yawing at nearly zero sideslip with some rolling.
   (b) Aerodynamic forces in this mode are very small.
   (c) Flight path: long, smooth return to reference path, if it is of convergent mode.
   (d) The single real eigenvalue is of very low magnitude with it being negative for a convergent situation, positive for a divergent situation.
3. Roll subsidence mode:
   (a) Almost pure rotational motion about the x axis (fuselage axis).
   (b) Significant aerodynamic variables are sideslip angle β and roll rate p.
   (c) The single real eigenvalue is of relatively large magnitude but always negative giving a convergent motion.

Important remark: finally, as mentioned in the Longitudinal motion approximations discussion, the same remark holds for this case as well, namely that the individual element values of the A matrix of all these approximate models do not necessarily coincide with the individual elemental values of the original fourth order matrix that correspond to the state variables in the large order state space system. Hence stability derivative values that appear in the these approximation models are obtained from the aerodynamic data with some approximations.

This completes the discussion of the development the linearized equations of motion for an aircraft. This is a very important step in the overall scheme of things, because we propose to use these linear models to design flight control systems, which constitute the material in Parts II and III of this book.

5.6 Chapter Summary

In this chapter, which specializes the analysis to aircraft dynamics explicitly, we have learnt the concepts related to dynamic stability of the aircraft motion. Recall that static stability deals with a situation that is a snapshot at a given time, which happens to be the equilibrium (or steady state) whereas dynamic stability deals with the actual time histories of the trajectories (for small motions in the linear range) when perturbed from the trim condition. We have also learnt that a typical aircraft, in the dynamic stability situation, exhibits a dynamic behavior that has few specific characteristics and those are called modes of motion. In the longitudinal case, we observed that the aircraft possesses two distinct modes labeled short period mode (motion at constant altitude, but varying angle of attack) and phugoid mode (motion at constant angle of attack but varying altitude). Similarly the roll/yaw motion exhibits modes of motion labeled as the Dutch roll mode, roll subsidence mode, and spiral convergence/divergent mode. For more in depth treatment of any of the concepts presented in this chapter, the reader is encouraged to consult the many excellent textbooks, completely dedicated to the aircraft dynamics and control, [40], [1], [30], [35], [15], [25], [29], [18], [41], [7], [14], [32], [20], [24], [12], [31], [17], [42], [19], [26], [36], [38], [9], [34], [39], [13], [10], [21], [4], [37], [3], [6], [2], [23], [22], [27], [28], [16], [8], [11], [5], [33] referenced at the end of this chapter.

5.7 Exercises

Exercise 5.1. The longitudinal aircraft equations of motion, in linear state space form, are given by

$$
\begin{bmatrix} \dot{u}_{sp} \\ \dot{w} \\ \dot{q} \\ \dot{\theta} \end{bmatrix} = \begin{bmatrix} X_u & X_w & 0 & -g\cos\theta_0 \\ Z_u & Z_w & U_o & -g\sin\theta_0 \\ M_u & M_w & M_q & 0 \\ 0 & 0 & 1 & 0 \end{bmatrix} \begin{bmatrix} u_{sp} \\ w \\ q \\ \theta \end{bmatrix} + \begin{bmatrix} X_{\delta_e} \\ Z_{\delta_e} \\ M_{\delta_e} \\ 0 \end{bmatrix} \delta_e
$$

where the forward speed change u_{sp}, vertical speed change w, pitch rate change q and pitch angle change θ are the state variables; elevator deflection δ_e is the control variable. If the flight path angle γ is defined as

$$
\gamma = \theta - \alpha
$$

$$
= \theta - \frac{w}{U_o}
$$

and the normal acceleration at the center of gravity location is given by

$$
a_{z_{cg}} = \dot{w} - U_o q
$$

where U_o is the constant, nominal forward velocity, obtain the C and D matrices for the above state description if
1. γ is taken as the output variable, and
2. $a_{z_{cg}}$ is taken as the output variable.

Exercise 5.2. The longitudinal aircraft equations of motion, in linear state space form, of a particular aircraft at a particular flight condition, are given by

$$
\begin{bmatrix} \dot{u}_{sp} \\ \dot{w} \\ \dot{q} \\ \dot{\theta} \end{bmatrix} = \begin{bmatrix} -0.0018 & 0.02 & 0.0006 & -0.331 \\ -0.002 & -0.06 & 1 & 0 \\ 0.328 & 11.02 & -0.08 & 0 \\ 0 & 0 & 1 & 0 \end{bmatrix} \begin{bmatrix} u_{sp} \\ w \\ q \\ \theta \end{bmatrix}
$$

where the forward speed change u_{sp}, vertical speed change w, pitch rate change q and pitch angle change θ are the state variables. Investigate the nature of the modes (i.e if it has the two types of modes we discussed before or not and, if they exist, the corresponding damping ratio and undamped and damped natural frequencies for those modes) for this aircraft.

Exercise 5.3. Repeat the exercise of Exercise 5.2 above with the following longitudinal dynamics matrix of another aircraft.

$$
\begin{bmatrix} \dot{u}_{sp} \\ \dot{w} \\ \dot{q} \\ \dot{\theta} \end{bmatrix} = \begin{bmatrix} -0.045 & 0.036 & 0.0006 & -32.2 \\ -0.369 & -2.02 & 168.8 & 0 \\ 0.0019 & -0.039 & -2.948 & 0 \\ 0 & 0 & 1 & 0 \end{bmatrix} \begin{bmatrix} u_{sp} \\ w \\ q \\ \theta \end{bmatrix}.
$$

Exercise 5.4. In this exercise, we consider lateral/directional dynamics matrix of an aircraft.

$$
\begin{bmatrix} \dot{\beta} \\ \dot{p} \\ \dot{r} \\ \dot{\phi} \end{bmatrix} = \begin{bmatrix} -0.254 & 0 & -1.0 & 0.182 \\ -16.02 & -8.40 & 2.19 & 0 \\ 4.488 & -0.350 & -0.760 & 0 \\ 0 & 1 & 0 & 0 \end{bmatrix} \begin{bmatrix} \beta \\ p \\ r \\ \phi \end{bmatrix}.
$$

Identify the lateral/directional modes of motion and their characteristics including the natural frequency and damping ratios for any existing oscillatory modes.

Exercise 5.5. In this exercise, we consider the lateral/directional dynamics matrix of another aircraft, with a different set of values for its stability derivatives. In particular, for this aircraft only the (2,1) element is different from the previous aircraft data.

$$
\begin{bmatrix} \dot{\beta} \\ \dot{p} \\ \dot{r} \\ \dot{\phi} \end{bmatrix} = \begin{bmatrix} -0.254 & 0 & -1.0 & 0.182 \\ -11.02 & -8.40 & 2.19 & 0 \\ 4.488 & -0.350 & -0.760 & 0 \\ 0 & 1 & 0 & 0 \end{bmatrix} \begin{bmatrix} \beta \\ p \\ r \\ \phi \end{bmatrix}.
$$

Identify the lateral/directional modes of motion and their characteristics including the natural frequency and damping ratios for any existing oscillatory modes.

Bibliography

1 M.J. Abzug and E.E Larrabee. *Airplane Stability and Control: A history of the Technologies that made Aviation possible.* Cambridge University Press, Cambridge, U.K, 2 edition, 2002.

2 R.J. Adams. *Multivariable Flight Control.* Springer-Verlag, New York, NY, 1995.

3 D. Allerton. *Principles of flight simulation.* Wiley-Blackwell, Chichester, U.K., 2009.

4 J. Anderson. *Aircraft Performance and Design.* McGraw Hill, New York, 1999.

5 J. Anderson. *Introduction to Flight.* McGraw-Hill, New York, 4 edition, 2000.

6 A. W. Babister. *Aircraft Dynamic Stability and Response.* Pergamon, 1980.

7 J.H. Blakelock. *Automatic Control of Aircraft and Missiles*. Wiley Interscience, New York, 1991.

8 Jean-Luc Boiffier. *The Dynamics of Flight, The Equations*. Wiley, 1998.

9 L.S. Brian and L.L. Frank. *Aircraft control and simulation. John Wiley & Sons, Inc.*, 2003.

10 A. E Bryson. *Control of Spacecraft and Aircraft*. Princeton University Press, Princeton, 1994.

11 Chris Carpenter. *Flightwise: Principles of Aircraft Flight*. Airlife Pub Ltd, 2002.

12 Michael V Cook. *Flight dynamics principles*. Arnold, London, U.K., 1997.

13 Wayne Durham. *Aircraft flight dynamics and control*. John Wiley & Sons, Inc., 2013.

14 B Etkin and L.D. Reid. *Dynamics of flight: stability and control*, volume 3. John Wiley & Sons, New York, 1998.

15 E.T. Falangas. *Performance Evaluation and Design of Flight Vehicle Control Systems*. Wiley and IEEE Press, Hoboken, N.J, 2016.

16 T. Hacker. *Flight Stability and Control*. Elsevier Science Ltd, 1970.

17 G.J. Hancock. *An Introduction to the Flight Dynamics of Rigid Aeroplanes*. Number Section III.5-III.6. Ellis Hornwood, New York, 1995.

18 D.G. Hull. *Fundamentals of Airplane Flight Mechanics*. Springer International, Berlin, Germany, 2007.

19 D. McLean. *Automatic flight control systems. Prentice Hall International Series in System and Control Engineering*, 1990.

20 Duane T McRuer, Dunstan Graham, and Irving Ashkenas. *Aircraft dynamics and automatic control*. Princeton University Press, Princeton,NJ, 1973.

21 A. Miele. *Flight Mechanics: Theory of Flight Paths*. Addison-Wesley, New York, 1962.

22 Ian Moir and Allan Seabridge. *Aicraft Systems: Mechanical, Electrical and Avionics Subsytems Integration*, 3rd Edition. John Wiley & sons, 2008.

23 Ian Moir and Allan Seabridge. *Design and Development of Aircraft Systems*. John Wiley & sons, 2012.

24 Robert C Nelson. *Flight stability and automatic control*. McGraw Hill, New York, 2 edition, 1998.

25 C. Perkins and R. Hage. *Aircraft Performance, Stability and Control*. John Wiley & Sons, London, U.K., 1949.

26 J.M. Rolfe and K.J. Staples. *Flight simulation*. Number 1. Cambridge University Press, Cambridge, U.K., 1988.

27 Jan Roskam. *Airplane Flight Dynamics and Automatic Flight Controls: Part I*. Roskam Aviation and Engineering Corp, 1979.

28 Jan Roskam. *Airplane Flight Dynamics and Automatic Flight Controls: Part II*. Roskam Aviation and Engineering Corp, 1979.

29 J.B. Russell. *Performance and Stability of Aircraft*. Arnold, London, U.K., 1996.

30 D.K. Schmidt. *Modern Flight Dynamics*. McGraw Hill, New York, 2012.

31 Louis V Schmidt. *Introduction to Aircraft Flight Dynamics*. AIAA Education Series, Reston, VA, 1998.

32 D. Seckel. *Stability and control of airplanes and helicopters*. Academic Press, 2014.

33 R. Shevell. *Fundamentals of Flight*. Prentice Hall, Englewood Cliffs, NJ, 2 edition, 1989.

34 Frederick O Smetana. *Computer assisted analysis of aircraft performance, stability, and control*. McGraw-Hill College., New York, 1984.

35 R.E. Stengel. *Flight Dynamics*. Princeton University Press, Princeton, 2004.

36 Brian L Stevens, Frank L Lewis, and Eric N Johnson. *Aircraft control and simulation*. Interscience, New York, 1 edition, 1992.

37 P. Swatton. *Aircraft Performance Theory and Practice for Pilots*. Wiley Publications, 2 edition, 2008.

38 A. Tewari. *Atmospheric and Space Flight Dynamics*. Birkhauser, Boston, 2006.

39 A. Tewari. *Advanced Control of Aircraft,Spacecraft and Rockets*. Wiley, Chichester, UK, 2011.

40 Ranjan Vepa. *Flight Dynamics, Simulation and Control for Rigid and Flexible Aircraft*. CRC Press, New York, 1 edition, 2015.

41 N. Vinh. *Flight Mechanics of High Performance Aircraft*. Cambridge University Press, New York, 1993.

42 P.H. Zipfel. *Modeling and simulation of aerospace vehicle dynamics*. American Institute of Aeronautics and Astronautics Inc., 2000.

6

Spacecraft Passive Stabilization and Control

6.1 Chapter Highlights

In this chapter, we focus our attention on the analysis of passive methods for spacecraft (or satellite) stabilization and control. This chapter on passive methods for stabilization and control parallels the spirit of aircraft static stability and control. With this backdrop, we briefly review various means of passive control methodologies for satellite attitude stabilization and control. Then we obtain the conditions for stabilization via these passive methods.

6.2 Passive Methods for Satellite Attitude Stabilization and Control

The methods available for attitude stabilization and control can be classified into two main categories – passive and active.

Passive systems are those making use of the ambient phenomena and not requiring power or data processing equipment. Active systems use power from within the satellite. The major passive methods of satellite stabilization and control are:

1. Spin stabilization
2. Dual spin
3. Gravity gradient
4. Magnetic
5. Aerodynamic
6. Radiation pressure.

The major active methods are:

1. Momentum exchange devices (like reaction wheels and control moment gyros).
2. Mass expulsion devices (reaction jets).

A more detailed and rigorous classification is as follows.

Flight Dynamics and Control of Aero and Space Vehicles, First Edition. Rama K. Yedavalli.
© 2020 John Wiley & Sons Ltd. Published 2020 by John Wiley & Sons Ltd.

6.2.1 Passive Systems

- Employ environmental/physical torque sources for momentum control.
- Require no on-board power; therefore, no sensors or control logic.
  In this chapter, we focus on these passive methods.
  Major constituents of passive systems and their primary applications include:

- Spin-stabilized systems without a spin speed, reorientation, or spin-axis procession control capability, for inertial orientation of the spin axis with cumulative errors allowed over the mission life.

6.2.1.1 Spin Stabilization

Spin stabilization is an accepted means of maintaining vehicle attitude, since a spinning body has a natural resistance to torques about axes other than the spin axis. The satellite acts like a gyro wheel with high angular momentum. In order that the desired spin axis be maintained, the polar moment of inertia about the spin axis must exceed the lateral moments of inertia. One problem associated with this method is the spin rate decay caused by eddy currents due to interaction with the Earth's magnetic field. Another difficulty is wobbling. Friction dampers are used to damp out any wobbling due to initial thrust misalignment and external torques.

6.2.1.2 Dual Spin Stabilization and Control

The dual spin stabilized satellite consists of two bodies coupled to each other through a bearing. Relative motion between the two bodies is possible about one common axis and usually one of the bodies is spun at a much faster rate with respect to the other. The other body may be totally de-spun or slowly spinning about the common axis.

Normally the spinning part is such that its angular momentum is sufficient to provide the necessary stiffness to the satellite against transverse disturbing torques. The spin axis is nominally kept normal to the orbital plane (i.e. aligned with the pitch axis). While the rotor spin imparts inherent stability to the pitch axis, the pitch attitude control may be accomplished by momentum exchange between the two bodies by suitably accelerating (or decelerating) the spinning part.

6.2.1.3 Gravity Gradient Stabilization and Control

In gravity gradient stabilization the vehicle must have a single axis about which the moment of inertia is a minimum. This axis will then become aligned with the local vertical. A long cylinder or dumbbell configuration are examples of this criterion. This method makes use of the fact that a small difference in gravity force exists between the two extreme tip masses of the booms of a dumbbell shaped satellite. Assuming a satellite with cylindrical symmetry about its Z axis, with moments of inertia I_x and I_z, the gravity-gradient torque is given by:

$$\tau = \frac{3}{2}\omega_0^2(I_x - I_z)\sin 2\theta \quad \text{(dynescm)} \tag{6.1}$$

where ω_0 = the orbit angular rate (rad s^{-1}) and θ = the angle between the satellite's Z axis and the local vertical. Since the satellite has cylindrical mass symmetry, $I_x = I_y$. To develop a substantial torque it is necessary that I_x be very much greater than I_z. We also see that the gravity gradient torque is less effective for satellites at very high altitudes where the orbital period is very great and therefore ω_0^2 is very small.

The natural period of oscillation (libration period) of a gravity stabilized satellite is given by:

$$T|| = \frac{2\pi}{\omega_0 \sqrt{3\left(1 - \frac{I_z}{I_x}\right)}} \quad \text{(s)} \tag{6.2}$$

in the plane of the orbit; and by:

$$T_1 = \frac{\pi}{\omega_0 \sqrt{3\left(1 - \frac{I_z}{I_x}\right)}} \quad \text{(s)} \tag{6.3}$$

in the plane perpendicular to the orbit. For a satellite having an orbital period of 100 min and having $I_x \gg I_z$ we find that $T|| = 57.8$ min and $T_1 = 50.0$ min. These very long libration periods, when combined with the any disturbance torques that are present, make damping of the satellite oscillations a difficult problem. For an Earth satellite to achieve passive gravity gradient stabilization it is necessary to follow certain procedures. These procedures will of course differ somewhat for various satellite missions, but some problems common to all will be discussed herein. It should be presumed that the long extension, or boom, that is required to alter the mass distribution of the satellite will be extended after the satellite is in orbit. The first thing that must be accomplished is to remove virtually all the spin that may have been imparted to the satellite during the launch procedure. A device that rapidly removes the spin energy of a satellite is the so-called "yo-yo" consisting of two weights attached to cables that are wrapped around the satellite. When the weights are released they spin out from the satellite causing a tension in the cables, which results in a retarding torque on the satellite. The next procedure is to align the satellite vertically with the correct side facing downward. This can be accomplished by energizing an electromagnet rigidly attached to the satellite. When the particular, pre-determined face of the satellite is directed downward, the tumbling rate of the satellite at this time will be 1.5 rpo. The satellite is now in a most advantageous condition for capture into gravity gradient attitude stabilization. The boom would then be erected and the electromagnet turned off by radio command from a ground station. The satellite will then have its tumbling angular rate reduced by the ratio of the satellite's moment of inertia. For a typical satellite design, the moment of inertia might be increased by a factor of 100, resulting in a decrease in the satellite's tumbling rate to 0.015 rpo, which is essentially stopped in inertial space. In order to be vertically stabilized the satellite must then achieve a tumbling rate in inertial space of 1.0 rpo. Immediately after the boom is erected the satellite continues in its orbital motion with its Z axis essentially fixed in inertial space. As the satellite moves away from the magnetic pole, a gravity gradient torque continues to act, resulting in planar libration motion of the satellite.

6.2.1.4 Magnetic Control

The interaction of the Earth's magnetic field with magnetic moments fixed or generated within the satellite will produce torques on the vehicle. This torque is given by:

$$\underline{T} = \underline{M} \times \underline{B} \tag{6.4}$$

where $\underline{M}$ = generated magnetic moment and $\underline{B}$ = Earth's magnetic field intensity. The torque developed can then be used for control purposes, provided that a suitable control law for $\underline{M}$ can be chosen.

6.2.1.5 Aerodynamic Control

For a satellite that operates at minimum altitudes, the aerodynamic force acts on the cross-sectional area of the satellite normal to the velocity vector at the center of pressure.

6.2.1.6 Solar Radiation Pressure

In the literature there are many references to the effects of the solar photon pressure field on attitude control and the flight path of the space vehicles. For application purposes, it is necessary only to make certain that the centroid of solar pressure on the craft is "down sun" from the center of mass. The action is analogous to that of a sea anchor that holds the bow of a boat into the wind during a storm.

Attitude control using solar radiation pressure is of much significance only interplanetary vehicles and for near earth satellites, the effect of solar radiation pressure is negligible.

There are, however, some significant operational limitations and design considerations in the utilization of passive control systems:

1. The system will have an extremely low speed of response (on the order of hours) and a limited acceleration capability (small control torques).
2. There is little or no flexibility for changing arbitrarily the nominal spacecraft orientation during mission.
3. They are sensitive to predictable or unpredictable perturbations.
4. Initial stabilization problems.
5. Damping of librational motion is difficult. The advantages are simplicity, high reliability and less power.

6.2.2 Passive/Semi-Active Control

Active attitude control systems provide the greatest design flexibility for the controlled orientation of a spacecraft in the presence of significant perturbations and maneuvers. When passive methods are combined with few features of active control systems we categorize them as passive/semi-active systems. For example, controlling or damping the oscillations of a gravity gradient stabilized satellite or a dual spin stabilized satellite with active nutation dampers can be viewed as passive/semi-active control. We reserve the phrase active control for those cases when independent three axis control is achieved with momentum exchange devises (such as reaction wheels) and mass expulsion devices such as reaction jets.

6.3 Stability Conditions for Linearized Models of Single Spin Stabilized Satellites

Recall the rotational equations of motion in principal axes given by:

$$I_1\dot{\omega}_1 + (I_3 - I_2)\omega_2\omega_3 = M_1 \tag{6.5a}$$

$$I_2\dot{\omega}_2 + (I_1 - I_3)\omega_1\omega_3 = M_2 \tag{6.5b}$$

$$I_3\dot{\omega}_3 + (I_2 - I_1)\omega_1\omega_2 = M_3. \tag{6.5c}$$

Now let us select one of the principal axes, say 3, about which we have the satellite spinning at a constant angular velocity, i.e. $\omega_3 = \Omega = $ constant and we assume that in the ideal or nominal or steady state or equilibrium situation, this spin is pure spin, i.e. the entire angular velocity $\underline{\omega}$ is along only the 3 axis, thus the nominal trajectory is as such:

$$\omega_{30} = \Omega = \text{constant} \tag{6.6}$$

$$M_{10} = M_{20} = M_{30} = 0 \tag{6.7}$$

$$\omega_{10} = \omega_{20} = 0. \tag{6.8}$$

Now let us linearize the above nonlinear set of equations about this nominal situation. So, write:

$$\omega_1 = \omega_{10} + \delta\omega_1 \quad M_1 = M_{10} + \delta M_1 \tag{6.9a}$$

$$\omega_2 = \omega_{20} + \delta\omega_2 \quad M_2 = M_{20} + \delta M_2 \tag{6.9b}$$

$$\omega_3 = \omega_{30} + \delta\omega_3 \quad M_3 = M_{30} + \delta M_3. \tag{6.9c}$$

Substituting these into the above nonlinear equations, and neglecting second and higher order terms and using the above nominal values, we get a set of linearized equations in the perturbation variables as follows:

$$\begin{bmatrix} \dot{\delta\omega_1} \\ \dot{\delta\omega_2} \\ \dot{\delta\omega_3} \end{bmatrix} = \begin{bmatrix} \dot{\omega}_1 \\ \dot{\omega}_2 \\ \dot{\omega}_3 - \Omega \end{bmatrix}$$

$$= \begin{bmatrix} 0 & \left(\dfrac{I_2 - I_3}{I_1}\right)\Omega & 0 \\ \left(\dfrac{I_3 - I_1}{I_2}\right)\Omega & 0 & 0 \\ 0 & 0 & 0 \end{bmatrix} \begin{bmatrix} \delta\omega_1 = \omega_1 \\ \delta\omega_2 = \omega_2 \\ \delta\omega_3 = \omega_3 - \Omega \end{bmatrix}$$

$$+ \begin{bmatrix} \dfrac{1}{I_1} & 0 & 0 \\ 0 & \dfrac{1}{I_2} & 0 \\ 0 & 0 & \dfrac{1}{I_3} \end{bmatrix} \begin{bmatrix} \delta M_1 \\ \delta M_2 \\ \delta M_3 \end{bmatrix}. \tag{6.10}$$

Now let us analyze the stability of this linear model with no input torques (i.e. $\delta M_1 = \delta M_2 = \delta M_3 = 0$), i.e. the uncontrolled motion. Then we have the homogeneous system of linear equations as:

$$\begin{bmatrix} \dot{x}_1(t) \\ \dot{x}_2(t) \\ \dot{x}_3(t) \end{bmatrix} = \begin{bmatrix} 0 & k_1 & 0 \\ k_2 & 0 & 0 \\ 0 & 0 & 0 \end{bmatrix} = \begin{bmatrix} x_1(t) \\ x_2(t) \\ x_3(t) \end{bmatrix} \tag{6.11}$$

where $x_1(t)$, $x_2(t)$ and $x_3(t)$ are the state variables given by

$$x_1(t) = \omega_1(t) \tag{6.12}$$

$$x_2(t) = \omega_2(t) \tag{6.13}$$

$$x_3(t) = \omega_3(t) - \Omega. \tag{6.14}$$

Stability Conditions for Linearized Single Spin Satellite Attitude Motion Clearly the stability of the above linear system is determined by the eigenvalues of the matrix

$$A = \begin{bmatrix} 0 & k_1 & 0 \\ k_2 & 0 & 0 \\ 0 & 0 & 0 \end{bmatrix}. \tag{6.15}$$

We observe that one of the eigenvalues of the matrix is 0, i.e. the equation given by:. $x_3(t) = 0 \rightarrow x_3(t) = $ constant. Obviously the trajectory $\omega_3(t) - \Omega = $ constant. Of course when that constant is non-zero, there is always an error in the angular velocity $\omega_3(t) - \Omega$ and we need a controller to make this error $(\omega_3(t) - \Omega)$ go to zero in the steady state! With regards to the other two angular velocities, ω_1 and ω_2, their stability is determined by the eigenvalues of the matrix

$$A = \begin{bmatrix} 0 & k_1 \\ k_2 & 0 \end{bmatrix} \tag{6.16}$$

which are given by

$$\lambda_{1,2} = \pm\sqrt{k_1 k_2} \tag{6.17}$$

where

$$k_1 k_2 = \left(\frac{I_3 - I_1}{I_2}\right)\left(\frac{I_2 - I_3}{I_1}\right)\Omega^2. \tag{6.18}$$

Clearly, if $k_1 k_2$ is positive (i.e. $k_1 k_2 > 0$), then one eigenvalues is positive real and the other eigenvalues are negative real making the trajectories ω_1 and ω_2 unstable. If $k_1 k_2$ is negative, then we have a pair of pure imaginary roots, making the trajectories pure simple harmonic (sinusoidal motion with zero damping) making them neutrally stable.

Now let us determine the nature of conditions on the moment of inertia for this to happen. The Table 6.1 summarizes the conditions of stability.

Thus spin about the maximum and minimum moments of inertia is neutrally stable and spin about the intermediate moment of inertia axes is unstable. However in practical cases, with some energy dissipation (damping) spin about the minimum moment of inertia is also unstable. So for all practical purposes spin about the maximum moment of inertia is the only desirable situation. However, even in this situation, the system is only neutrally stable, requiring a control system to make it asymptotically stable. The control design aspect is discussed in later chapters.

Table 6.1 Stability conditions for a spin stabilized satellite.

Case	Stability	Physical shape of spacecraft
$I_3 > I_2 \geq I_1$	$k_1 k_2 < 0$, neutrally stable	Spin about maximum M.O.I
$I_3 < I_2 < I_1$	$k_1 k_2 < 0$, Neutrally stable	Spin about minimum M.O.I.
$I_1 > I_3, I_2 < I_3$	$k_1 k_2 > 0$, Unstable	Spin about intermediate M.O.I.

6.4 Stability Conditions for a Dual Spin Stabilized Satellite

Perhaps one of the most innovative ideas in spacecraft dynamics and control is the concept of dual spin stabilization. The major contributors in this area are [13, 17].

The major advantage of the dual spin stabilization is that it overcomes the limitation of a single spin stabilized satellite in that it cannot have the antennas oriented towards a specific point on earth. In a dual spin stabilized satellite, there are two rotating parts, one rotor spinning at a faster rate providing gyroscopic stabilization (like a single spin satellite) and a slowly spinning (one revolution per orbit) platform pointing toward earth. Developing stability conditions for a dual spin stabilized satellite became a challenging task and here we briefly review those stability conditions developed by the above mentioned contributors.

For an axi-symmetric dual spin stabilized spacecraft as depicted in Figure 6.1, the kinetic energy T, magnitude of the angular momentum H, and the nutation frequency ω_n are

$$T = \frac{1}{2}(I_T\omega_T^2 + I_{sP}\omega_P^2 + I_{sR}\omega_R^2) \tag{6.19}$$

$$H^2 = (I_{sP}\omega_P + I_{sR}\omega_R)^2 + (I_T\omega_T)^2 \tag{6.20}$$

$$\omega_n = \frac{I_{sP}\omega_P + I_{sR}\omega_R}{I_T} \tag{6.21}$$

where I_T is the transverse moment of inertia of the spacecraft, I_{sR} is the spin moment of inertia of the rotor, I_{sP} the spin moment of inertia of the platform, ω_T the transverse angular velocity of the spacecraft, ω_P the spin angular velocity of the platform, and ω_R the spin angular velocity of the rotor. We assume no external torque and a frictionless shaft. From the conservation of angular momentum, $\dot{H} = 0$. Differentiating Equation 6.20 with

Figure 6.1 Motion of a dual spin stablilized satellite.

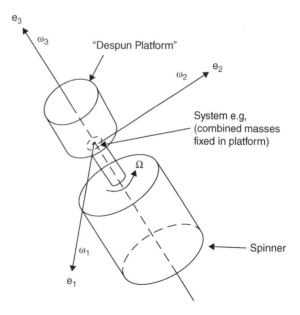

respect to time and applying the conservation of angular momentum, we obtain

$$I_T \omega_T \dot{\omega}_T = -\frac{I_{sP} \omega_P + I_{sR} \omega_R}{I_T} (I_{sP} \dot{\omega}_P + I_{sR} \dot{\omega}_R)$$

$$= -\omega_n (I_{sP} \dot{\omega}_P + I_{sR} \dot{\omega}_R). \tag{6.22}$$

Differentiating Equation 6.19 with respect to time, we arrive at

$$\dot{T} = \dot{T}_P + \dot{T}_R$$

$$= I_T \omega_T \dot{\omega}_T + I_{sP} \omega_P \dot{\omega}_P + I_{sR} \omega_R \dot{\omega}_R. \tag{6.23}$$

Combining Equations 6.22 and 6.23, we have

$$\dot{T} = \dot{T}_P + \dot{T}_R$$

$$= -(\omega_n - \omega_P) I_{sP} \dot{\omega}_p - (\omega_n - \omega_R) I_{sR} \dot{\omega}_R \tag{6.24}$$

$$= -\lambda_P I_{sP} \dot{\omega}_P - \lambda_R I_{sR} \dot{\omega}_R$$

where λ_P and λ_R are the frequencies of oscillating accelerations in the platform and in the rotor, respectively. Because the rotor and the platform are assumed to be uncoupled about the spin axis, the reaction torques that tend to change angular rates are written from Equation 6.24 as

$$I_{sP} \dot{\omega}_P = -\frac{\dot{T}_P}{\lambda_P} \tag{6.25a}$$

$$I_{sR} \dot{\omega}_R = -\frac{\dot{T}_R}{\lambda_R}. \tag{6.25b}$$

Combining Equations 6.22 and 6.25, we obtain

$$I_T \omega_T \dot{\omega}_T = \omega_n \left(\frac{\dot{T}_P}{\lambda_P} + \frac{\dot{T}_R}{\lambda_R} \right). \tag{6.26}$$

Recall that the nutation angle θ and the ω_T are related by

$$\sin \theta = (I_T \omega_T)/H, \tag{6.27}$$

Realizing that θ is no longer constant, differentiating 6.27 with respect to time and using 6.26 we have

$$\dot{\theta} = \frac{2I_T}{\sin 2\theta} \frac{\omega_n}{H^2} \left(\frac{\dot{T}_P}{\lambda_P} + \frac{\dot{T}_R}{\lambda_R} \right) \tag{6.28}$$

Note that single-spin stabilization case is a special case of 6.28, obtained by letting $\lambda_P = \lambda_R =$ and $\dot{T} = \dot{T}_P + \dot{T}_R$.

For a stable spacecraft, the nutation angle must decay with time (i.e. $\dot{\theta} < 0$). Therefore, we arrive at our stability condition:

$$\frac{\dot{T}_P}{\lambda_P} + \frac{\dot{T}}{\lambda_R} < 0. \tag{6.29}$$

Assuming the rotor angular momentum is much larger than the platform angular momentum (i.e. $I_{sR}\omega_R \gg I_{sP}\omega_P$), we can write

$$\lambda_P = \omega_n - \omega_P$$

$$= \frac{I_{sP}\omega_P + I_{sR}\omega_R}{I_T} - \omega_P \tag{6.30a}$$

$$\approx \frac{I_{sR}\omega_R}{I_T}$$

$$\lambda_R = \omega_n - \omega_R$$

$$= \frac{I_{sP}\omega_P + I_{sR}\omega_R}{I_T} - \omega_R \tag{6.30b}$$

$$\approx \left(\frac{I_{sR}}{I_T} - 1\right)\omega_R.$$

Applying Equation 6.30 to inequality 6.29, the stability conditions can be expressed as

$$\frac{\dot{T}_P}{I_{sR}/I_T} + \frac{\dot{T}_R}{(I_{sR}/I_T) - 1} < 0 \tag{6.31}$$

and there are two cases:

1. $I_{sR} > I_T$, in which case the spacecraft is stable if energy dissipation occurs in either the platform or the rotor. Therefore a nutational damper can be placed on either the platform or the rotor.
2. $I_{sR} < I_T$. In this case, the first term in 6.31 is negative (notice that $\dot{T}_P$ and $\dot{T}_R$ are already required to be negative) and the second is positive. The stability condition for this case becomes

$$|\dot{T}_P| > |\dot{T}_R \frac{I_{sR}/I_T}{(I_{sR}/I_T) - 1}| \tag{6.32}$$

As an example, [1], for a dual spin stabilized spacecraft with

$$\frac{I_{sR}}{I_T} = \frac{2}{3} \tag{6.33}$$

the magnitude of the energy dissipation rate of the platform should be at least twice the dissipation rate in the rotor. Hence the damper must be placed on platform.

6.5 Chapter Summary

In this chapter, which specializes in the analysis of spacecraft (satellite) dynamics explicitly, we have briefly reviewed the basic passive and active methods of stabilization and control for spacecraft, with more emphasis on passive methods. We obtained the conditions for stability in terms of the moments of inertia distribution for single spin and dual spin stabilized satellites. For more in depth treatment of any of the concepts presented in this chapter, the reader is encouraged to consult the many excellent textbooks, completely dedicated to spacecraft dynamics and control: [21], [14], [12], [11], [15], [8], [23], [22], [3], [1], [2], [9], [18], [19], [6], [7], [5], [10], [16], [20], [24], [4], [13], [17] referenced at the end of this chapter.

6.6 Exercises

Exercise 6.1. Consider an axi-symmetric dual spin stabilized spacecraft with $I_{SR} = 20$ nm s^{-2} and $I_T = 30$ nm s^{-2} and $\omega_R = 100$ rpm, $I_{SP} = 25$ nm s^{-2} and $\omega_P = 1$ rpm. Find the nutation frequency ω_n in rad s^{-1} and determine the conditions for stability of this satellite.

Exercise 6.2. Consider different ratios of $\dfrac{I_{sR}}{I_T}$, (at least three) and analyze the stability conditions for the dual spin stabilized satellite.

Bibliography

1 B.N. Agrawal. *Design of geosynchronous spacecraft*. Prentice-Hall, Englewood Cliffs, NJ, 1986.
2 R.H. Battin. *An introduction to the mathematics and methods of astrodynamics*. Washington, DC: AIAA, 1990.
3 J.V. Breakwell and R.E. Roberson. *Orbital and Attitude Dynamics*. American Institute of Aeronautics and Astronautics, 1969.
4 A. E Bryson. *Control of Spacecraft and Aircraft*. Princeton University Press, Princeton, 1994.
5 P.R.K Chetty. *Satellite Technology and Its Application*. Tab Books, 1988.
6 Vladimir A. Chobotov. *Spacecraft Attitude Dynamics and Control*. Krieger Publishing Company, 1991.
7 Anton H. de Ruiter, Christopher Damaren, and James R. Forbes. *Spacecraft Attitude Dynamics and Control: An Introduction*. Wiley, 2013.
8 R. Deutsch. *Orbital dynamics of space vehicles*. Prentice-Hall, Englewood Cliffs, NJ, 1963.
9 Pedro Ramon Escobal. *Methods of orbit determination. Krieger*, 1965.
10 Peter Fortescue, Graham Swinerd, and John Stark. *Spacecraft Systems Engineering*, 4th Edition. Wiley Publications, 2011.
11 A.L. Greensite. *Analysis and design of space vehicle flight control systems*. spartan Books, New York, 1970.
12 P.C. Hughes. *Spacecraft attitude dynamics*. Wiley, New York, 1988.
13 A.J. Iorillo. Nutation damping dynamics of axisymmetric rotor stabilized satellites. In *ASME Annual Winter Meeting (November, Chicago)*, New York, 1965. ASME.
14 John L Junkins and James D Turner. *Optimal spacecraft rotational maneuvers*, volume 3. Elsevier, Amsterdam, 1986.
15 M.H. Kaplan. *Modern spacecraft dynamics and control*. Wiley, New York, 1976.
16 Wilfried Ley, Klaus Wittmann, and Wili Hallmann. *Handbook of Space Technology*. Wiley Publications, 2009.
17 P.W. Likins. Attitude stability criteria for dual spin spacecraft. *Journal of Spacecraft and Rockets*, 4(12):1638–1643, 1967.
18 J.J. Pocha. *An Introduction to Mission Design for Geostationary Satellites*. Springer Science & Business Media, 1 edition, 2012.
19 L.W. Pritchard and A.J. Sciulli. *Satellite communication systems engineering*. Prentice-Hall, Englewood Cliffs, NJ, 1986.

20 Hanspeter Schaub and John L. Junkins. *Analytical Mechanics of Space Systems.* AIAA Education Series, 2014.

21 M.J. Sidi. *Spacecraft Dynamics and Control: A Practical Engineering Approach.* Cambridge University Press, New York, N.Y., 1997.

22 W.T. Thomson. *Introduction to space dynamics.* Dover, 1986.

23 J.R. Wertz. *Spacecraft attitude determination and control.* Reidel, Dordrecht, the Netherlands, 1978.

24 Bong Wie. *Space Vehicle Dynamics and Control.* AIAA Education Series, Reston, 2008.

7

Spacecraft Dynamic Stability and Control via Linearized Models

7.1 Chapter Highlights

In this chapter, we focus our attention on the analysis of dynamics specific to spacecraft (or satellites). Since the equations of motion are written in the principal axis frame, the nature of the equations is quite different from the aircraft specific equations. With this backdrop, we obtain the linearized equations of motion for dynamic stability for later use in control system design. We cover the cases of (i) linearized perturbation dynamics for satellites perturbed from the nominal pure spin stabilized state, (ii) the basic translational motion perturbation equations for a satellite perturbed from the nominal circular orbit, and (iii) the basic rotational (attitude) motion perturbation equations for a satellite perturbed from the nominal circular orbit. These linearized models are extremely important for control system design purposes.

7.2 Active Control: Three Axis Stabilization and Control

Active attitude control systems provide the greatest design flexibility for the controlled orientation of a spacecraft in the presence of significant perturbations and maneuvers. These active control systems use sensors and actuators to produce a control signal in a feedback control framework so that the output is made to behave in a desired way. Naturally, active control systems are to be used when the performance specifications of the mission are stringent and the pointing accuracy is of importance. Needless to say, active control system implementation is expensive and is thus used when the mission success warrants it. The typical attitude actuators are momentum exchange devices such as reaction wheels and mass expulsion devices such as reaction jets. A tutorial type summary on attitude sensors and actuators is given in Part D of this book. For now, we briefly mention few features of these actuators.

7.2.1 Momentum Exchange Devices: Reaction Wheels

These utilize the principle of conservation of total angular momentum, whereby, in the absence of externally applied torques, if one part of a closed system increases its momentum by a prescribed amount, the rest of the system loses an equal amount of momentum.

Momentum bias control systems are becoming increasingly popular for the control of spacecraft in both low and synchronous orbits. The major advantage of such systems are their ability to provide three axis control without the need of a yaw sensor. Roll and pitch

Flight Dynamics and Control of Aero and Space Vehicles, First Edition. Rama K. Yedavalli.
© 2020 John Wiley & Sons Ltd. Published 2020 by John Wiley & Sons Ltd.

may be controlled by signals generated from a conventional horizon sensor whereas yaw control is effected through the kinematics of quarter orbit coupling, i.e. a yaw error at one point in the orbit becomes a roll error a quarter of an orbit later.

7.2.2 Momentum Exchange Devices: Gyrotorquers (or CMGs)

A properly designed gyrostabilizer system can function not only as a damper but also as a means of satellite attitude control. A gyrostabilizer is considered to be a rotating wheel gimball in such a way that the spin axis of the wheel can precess about an axis (output axis), normal to the spin axis in response to angular rates about a third axis (input axis) mutually perpendicular to both the spin and output axes. In short, a gyrostabilizer is a gyroscope but used in such a way that it not only senses rates but also supplies torques directly to the vehicle to be stabilized.

7.2.3 Mass Expulsion Devices: Reaction Jets

Conceptually, the simplest means of orienting the satellite is by generating external torques in the opposite direction to that of disturbance torques by mass expulsion. This can be done by placing pans of mass expulsion devices (reaction jets) about each of the control axes. However, since fuel storage is limited, reaction jet control is normally not recommended as a main control device but it is only used in back-up models like:

1. Detumbling
2. Desaturation of momentum exchange devices
3. Orienting the spin vector of the satellite.

One major occasion that warrants the use of reaction jets for control is at the momentum dumping phase. Note that when secular disturbance torques act on the satellite, the momentum wheels (or reaction wheels) need to accelerate continuously to reduce the attitude error generated by these secular disturbance torques, which in turn necessitates the increase in their angular speeds, reaching unsafe levels. So there is a need to dump that momentum (i.e de-saturate the wheel) so that it does not reach that unsafe zone. During that momentum dumping phase, the reaction jets take over to control the attitude error. In most of the situations, it is expedient to employ reaction jets in an on–off manner. In other words, it is prudent to tolerate some error in the attitude until it reaches a threshold level and then activate the reaction jets to reduce the error back to zero and turn them off to save the fuel. During that reaction jet off cycle, the error builds up until the threshold level when again the jets are activated. Deciding the optimal on–off mode of reaction jet control is the most important control system design problem. The resulting attitude motion is then characterized by a limit cycle.

The calculation of the fuel weight necessary to accomplish certain objectives is an important part of the attitude control study involving reaction jets. If for any mode of operation, the thrust history as a function of time is known, then from the relation:

$$T = \frac{\dot{W}}{g_0} V_e = \dot{W} I_{sp} \tag{7.1}$$

T = thrust

$\dot{W}$ is the fuel consumption rate

g_0 is the acceleration due to gravity at the Earth's surface

V_e is the exhaust velocity of the propellant

i_{sp} is the specific impulse of the propellant (fuel).

We may calculate the required fuel weight by:

$$w = \int \frac{T}{I_{sp}} dt. \tag{7.2}$$

The jets and momentum storage devices complement one another, the storage devices counter the cyclical torques on the vehicle without loss of mass, and the jets counter long term secular torques by periodically expelling momentum from the storage system as the storage devices near spin saturation.

In what follows, we focus on the three axis stabilization of a satellite by active control using these attitude sensors and actuators in a feedback control framework for the dynamic stability and control of the satellite. As part of the control design process, we now gather the needed linearized models for the satellite attitude dynamics for use later in the control system design to be discussed in Chapters 8 and 19. The typical axis system for three axis control systems is shown in Figure 7.1.

7.2.4 Linearized Models of Single Spin Stabilized Satellites for Control Design

We have carried out this exercise in the previous chapter when addressing the stability conditions for the single spin stabilization issue. At that time, we focused on

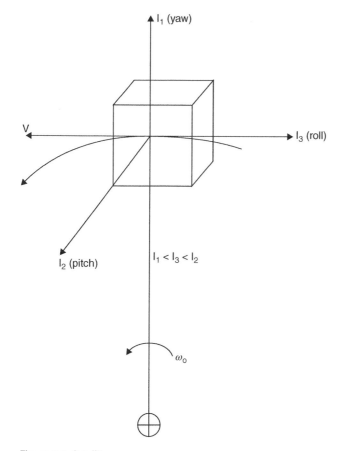

Figure 7.1 Satellite axes.

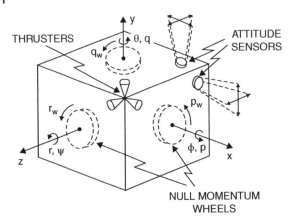

Figure 7.2 Three axis satellite control.

THRUSTERS

ATTITUDE SENSORS

q_w

θ, q

r_w

p_w

r, ψ

ϕ, p

NULL MOMENTUM WHEELS

the uncontrolled (open loop, natural) motion. For completeness, we reproduce that linearized model here.

$$
\begin{bmatrix} \dot{\overline{\omega}}_1 \\ \dot{\overline{\omega}}_2 \\ \overline{\omega}_3 \end{bmatrix} = \begin{bmatrix} \dot{\omega}_1 \\ \dot{\omega}_2 \\ \dot{\omega}_3 - \Omega \end{bmatrix}
$$

$$
= \begin{bmatrix} 0 & \left(\dfrac{I_2 - I_3}{I_1}\right)\Omega & 0 \\ \left(\dfrac{I_3 - I_1}{I_2}\right)\Omega & 0 & 0 \\ 0 & 0 & 0 \end{bmatrix} \begin{bmatrix} \overline{\omega}_1 = \omega_1 \\ \overline{\omega}_2 = \omega_2 \\ \overline{\omega}_3 = \omega_3 - \Omega \end{bmatrix} \tag{7.3}
$$

$$
+ \begin{bmatrix} \dfrac{1}{I_1} & 0 & 0 \\ 0 & \dfrac{1}{I_2} & 0 \\ 0 & 0 & \dfrac{1}{I_3} \end{bmatrix} \begin{bmatrix} \overline{M}_1 \\ \overline{M}_2 \\ \overline{M}_3 \end{bmatrix}.
$$

Active control of this dynamics can be achieved by controlling the motion about all the three axes by active control devices such as momentum/reaction wheels and reaction jets. A pictorial representation of a three axis stabilized satellite is shown in Figure 7.2.

7.3 Linearized Translational Equations of Motion for a Satellite in a Nominal Circular Orbit for Control Design

$$
\underline{\dot{x}} = \underline{f}(\underline{x}, \underline{u}) = \begin{bmatrix} \dot{r} \\ r\dot{\theta}^2\cos^2\phi + r\dot{\phi}^2 - \dfrac{k}{r^2} + \dfrac{u_r}{m} \\ \dot{\theta} \\ -2\dot{r}\dot{\theta}/r + 2\dot{\theta}\dot{\phi}\dfrac{\sin\phi}{\cos\phi} + \dfrac{u_\theta}{mr\cos\phi} \\ \dot{\phi} \\ -\dot{\theta}^2\cos\phi\sin\phi - 2\dot{r}\dot{\phi}/r + u_\phi/mr \end{bmatrix} \tag{7.4}
$$

and

$$y = \underline{C}x = \begin{bmatrix} 1 & 0 & 0 & 0 & 0 & 0 \\ 0 & 0 & 1 & 0 & 0 & 0 \\ 1 & 0 & 0 & 0 & 1 & 0 \end{bmatrix} \underline{x}. \tag{7.5}$$

One solution of these equations (circular, equatorial orbit), is

$$\underline{x}^*(t) = \begin{bmatrix} r^*(t) \\ \dot{r}^*(t) \\ \theta^*(t) \\ \dot{\theta}^*(t) \\ \phi^*(t) \\ \dot{\phi}^*(t) \end{bmatrix} = \begin{bmatrix} r_0 \\ 0 \\ \omega t \\ \omega \\ 0 \\ 0 \end{bmatrix} \tag{7.6}$$

where

$$\underline{u}^*(t) = \underline{0}. \tag{7.7}$$

Linearizing the above nonlinear equations about the nominal circular orbit, we obtain

$$\dot{\hat{x}} = \underline{A}\hat{x}(t) + \underline{B}\hat{u}(t) \tag{7.8}$$

$$\hat{\underline{y}} = \underline{C}\hat{x}(t) \tag{7.9}$$

$$\underline{A} \triangleq \frac{d\underline{f}}{d\underline{x}}\Big|_* = \begin{bmatrix} 0 & 1 & 0 & 0 & 0 & 0 \\ 3\omega^2 & 0 & 0 & 2\omega r_0 & 0 & 0 \\ 0 & 0 & 0 & 1 & 0 & 0 \\ 0 & -\dfrac{2\omega}{r_0} & 0 & 0 & 0 & 0 \\ 0 & 0 & 0 & 0 & 0 & 1 \\ 0 & 0 & 0 & 0 & -\omega^2 & 0 \end{bmatrix} \tag{7.10}$$

$$\underline{B} \triangleq \frac{d\underline{f}}{d\underline{u}}\Big|_* = \begin{bmatrix} 0 & 0 & 0 \\ \dfrac{1}{m} & 0 & 0 \\ 0 & 0 & 0 \\ 0 & \dfrac{1}{mr_0} & 0 \\ 0 & 0 & 0 \\ 0 & 0 & \dfrac{1}{mr_0} \end{bmatrix}. \tag{7.11}$$

We will normalize our units so that $m = r_0 = 1$, and the matrix above will become

$$\dot{\hat{x}}(t) = \begin{bmatrix} 0 & 1 & 0 & 0 & 0 & 0 \\ 3\omega^2 & 0 & 0 & 2\omega & 0 & 0 \\ 0 & 0 & 0 & 1 & 0 & 0 \\ 0 & -2\omega & 0 & 0 & 0 & 0 \\ 0 & 0 & 0 & 0 & 0 & 1 \\ 0 & 0 & 0 & 0 & -\omega^2 & 0 \end{bmatrix} \hat{x}(t) + \begin{bmatrix} 0 & 0 & 0 \\ 1 & 0 & 0 \\ 0 & 0 & 0 \\ 0 & 1 & 0 \\ 0 & 0 & 0 \\ 0 & 0 & 1 \end{bmatrix} \hat{\underline{u}}(t) \tag{7.12}$$

$$\underline{\hat{y}}(t) = \begin{bmatrix} 1 & 0 & 0 & 0 & 0 & 0 \\ 0 & 0 & 1 & 0 & 0 & 0 \\ 1 & 0 & 0 & 0 & 1 & 0 \end{bmatrix} \hat{x}(t). \tag{7.13}$$

The sixth order system may be broken into one fourth order and one second order system as follows. The fourth order system is given by

$$\begin{bmatrix} \dot{\hat{x}}_1 \\ \dot{\hat{x}}_2 \\ \dot{\hat{x}}_3 \\ \dot{\hat{x}}_4 \end{bmatrix} = \begin{bmatrix} 0 & 1 & 0 & 0 \\ 3\omega^2 & 0 & 0 & 2\omega \\ 0 & 0 & 0 & 1 \\ 0 & -2\omega & 0 & 0 \end{bmatrix} \begin{bmatrix} \hat{x}_1 \\ \hat{x}_2 \\ \hat{x}_3 \\ \hat{x}_4 \end{bmatrix} + \begin{bmatrix} 0 & 0 \\ 1 & 0 \\ 0 & 0 \\ 0 & 1 \end{bmatrix} \begin{bmatrix} \hat{u}_1 \\ \hat{u}_2 \end{bmatrix} \tag{7.14}$$

$$\begin{bmatrix} \hat{y}_1 \\ \hat{y}_2 \end{bmatrix} = \begin{bmatrix} 1 & 0 & 0 & 0 \\ 0 & 0 & 1 & 0 \end{bmatrix} \begin{bmatrix} \hat{x}_1 \\ \hat{x}_2 \\ \hat{x}_3 \\ \hat{x}_4 \end{bmatrix} \tag{7.15}$$

and the second order system is given by

$$\begin{bmatrix} \dot{\hat{x}}_5 \\ \dot{\hat{x}}_6 \end{bmatrix} = \begin{bmatrix} 0 & 1 \\ -\omega^2 & 0 \end{bmatrix} \begin{bmatrix} \hat{x}_5 \\ \hat{x}_6 \end{bmatrix} + \begin{bmatrix} 0 \\ 1 \end{bmatrix} \hat{u}_3 \tag{7.16}$$

$$y_3 = \begin{bmatrix} 1 & 0 \end{bmatrix} \begin{bmatrix} \hat{x}_5 \\ \hat{x}_6 \end{bmatrix}. \tag{7.17}$$

In the control system design part of the book, we attempt to design linear feedback controllers based on the above linearized models.

7.4 Linearized Rotational (Attitude) Equations of Motion for a Satellite in a Nominal Circular Orbit for Control Design

Recall that, in Chapter 2, we derived the nonlinear rotational (attitude) equations of motion for a satellite in a nominal circular orbit with a constant orbit angular velocity denoted by ω_o whose value depends on the orbit radius. We now revisit those equations and linearize them assuming small angle approximation and neglecting second and higher order terms. The angular velocity components of the satellite in a circular orbit, expressed in body frame (principal axes frame) are given by

$$\begin{bmatrix} \omega_1 \\ \omega_2 \\ \omega_3 \end{bmatrix} = \begin{bmatrix} 1 & 0 & -\sin\theta \\ 0 & \cos\phi & \cos\theta\sin\phi \\ 0 & -\sin\phi & \cos\phi\cos\theta \end{bmatrix} \begin{bmatrix} \dot{\phi} \\ \dot{\theta} \\ \dot{\psi} \end{bmatrix} - \omega_o \begin{bmatrix} \cos\theta\sin\psi \\ \cos\phi\cos\psi + \sin\phi\sin\theta\sin\psi \\ \cos\phi\sin\theta\sin\psi - \sin\phi\cos\psi \end{bmatrix}. \tag{7.18}$$

Noting that the angular momentum vector of the satellite body given by $\overline{H}$ is $[I_1\omega_1 \quad I_2\omega_2 \quad I_3\omega_3]^T$ and that these components are expressed in the body frame, the resulting moment equations, after linearization, are given by

$$\begin{bmatrix} I_1(\ddot{\phi} - \dot{\psi}\omega_o) + (I_2 - I_3)(\omega_o^2\phi + \omega_o\dot{\psi}) \\ I_2\ddot{\theta} \\ I_3(\ddot{\psi} + \dot{\phi}\omega_o) + (I_2 - I_1)(\omega_o^2\psi - \omega_o\dot{\phi}) \end{bmatrix} = \begin{bmatrix} M_{ext1} \\ M_{ext2} \\ M_{ext3} \end{bmatrix} \tag{7.19}$$

where M_{exti} are the external applied moments along the three principal axes. The above linearization process is left as an exercise for the student/reader (see Exercise 7.1).

Note that in this linearization process, the pitch axis dynamics and control is completely decoupled from the roll/yaw axes dynamics. However, the roll/yaw axis dynamics are coupled even in the linear range. Later in the control system design chapters, we design a control system to stabilize and meet the performance requirements on the satellite to reduce these roll, yaw, and pitch attitude errors.

7.5 Open Loop (Uncontrolled Motion) Behavior of Spacecraft Models

Once we obtain the linearized models for control design purposes, it is important to analyze the open loop system (i.e the uncontrolled system, with no external control torques applied) stability characteristics of the system we wish to control. It turns out that in all the linearized models we considered above, the open loop system happens to be either neutrally stable or even unstable in some cases. This is in quite contrast to the aircraft case where for the majority of commercial aircraft dynamic models the longitudinal modes of motion in open loop such as short period mode or phugoid mode are mostly open loop stable, albeit with insufficient damping ratios. This means that for the majority of spacecraft applications active feedback control is a necessity. This feature clearly highlights the important role to be played by automatic flight vehicle control systems in spacecraft applications.

7.6 External Torque Analysis: Control Torques Versus Disturbance Torques

Since the external applied torques (or moments) have considerable influence on the dynamic behavior of the satellite dynamics, it is important to understand and analyze this external moment scenario in satellite attitude stabilization and control. For a given control objective, some of these external applied moments could be used as control torques for attitude stabilization and control. However, it also happens that some of these external torques present could also become disturbance torques while controlling the satellite using the control torques. So it is critical that the engineer decide which of these torques are used as control torques and which of them are to be treated as disturbance torques for a given mission. While the design of control torques is very much dependent on the performance specifications, it is important to simultaneously model the disturbance torques as well and prepare a budget for the attitude errors generated under the presence of these disturbance torques. This in turn requires an analysis of the nature of these disturbance torques as to whether they are cyclical or secular (constantly acting). Then this knowledge about the disturbance torques needs to be used in the control systems design phase and make sure that the control torques are designed such that the closed loop system not only is stable but deliver satisfactory performance in the sense of meeting the mission design specifications under the presence of the disturbance torques.

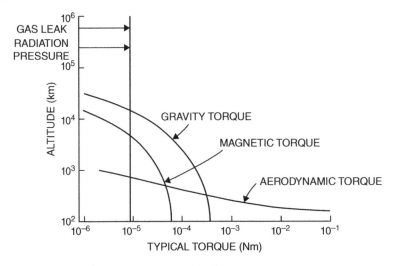

Figure 7.3 Disturbance torque versus altitude.

Typical values of the torques on a small spacecraft as a function of the orbital altitude are shown in Figure 7.3.

7.7 Chapter Summary

In this chapter, which specializes in the analysis of spacecraft (satellite) dynamics explicitly, we have derived the linearized equations of motion for later use in the control system design. The same analysis is used to characterize the various disturbance torques acting on the satellite, tailored to the space/orbit environment. For more in depth treatment of any of the concepts presented in this chapter, the reader is encouraged to consult the many excellent textbooks, completely dedicated to spacecraft dynamics and control: [18], [12], [11], [10], [13], [8], [20], [19], [3], [1], [2], [9], [15], [16], [6], [7], [5], [14], [22], [17], [21], [4], referenced at the end of this chapter.

7.8 Exercises

Exercise 7.1. Carry out the detailed linearization process of the rotational (attitude) equations of motion for a satellite in a circular orbit with constant orbital angular velocity ω_o and verify the accuracy of the model description given in this chapter for that problem.

Exercise 7.2. For the above problem (Exercise 1), obtain the specific A and B matrices for a satellite in geosynchronous equatorial orbit with inertia information taken for few particular satellites given in Appendix A. Then thoroughly analyze the open loop stability of those satellites along with the undamped natural frequency and damping ratio information for any complex conjugate pair eigenvalues.

Bibliography

1 B.N. Agrawal. *Design of geosynchronous spacecraft*. Prentice-Hall, Englewood Cliffs, NJ, 1986.

2 R.H. Battin. *An introduction to the mathematics and methods of astrodynamics*. Washington, DC: AIAA, 1990.

3 J.V. Breakwell and R.E. Roberson. *Orbital and Attitude Dynamics*. American Institute of Aeronautics and Astronautics, 1969.

4 A. E Bryson. *Control of Spacecraft and Aircraft*. Princeton University Press, Princeton, 1994.

5 P.R.K Chetty. *Satellite Technology and Its Application*. Tab Books, 1988.

6 Vladimir A. Chobotov. *Spacecraft Attitude Dynamics and Control*. Krieger Publishing Company, 1991.

7 Anton H. de Ruiter, Christopher Damaren, and James R. Forbes. *Spacecraft Attitude Dynamics and Control: An Introduction*. Wiley, 2013.

8 R. Deutsch. *Orbital dynamics of space vehicles*. Prentice-Hall, Englewood Cliffs, NJ, 1963.

9 Pedro Ramon Escobal. *Methods of orbit determination*. Krieger, 1965.

10 A.L. Greensite. *Analysis and design of space vehicle flight control systems*. spartan Books, New York, 1970.

11 P.C. Hughes. *Spacecraft attitude dynamics*. Wiley, New York, 1988.

12 John L Junkins and James D Turner. *Optimal spacecraft rotational maneuvers*, volume 3. Elsevier, Amsterdam, 1986.

13 M.H. Kaplan. *Modern spacecraft dynamics and control*. Wiley, New York, 1976.

14 John Stark Peter Fortescue, Graham Swinerd. *Spacecraft Systems Engineering, 4th Edition*. Wiley Publications, 2011.

15 J.J. Pocha. *An Introduction to Mission Design for Geostationary Satellites*. Springer Science & Business Media, 1 edition, 2012.

16 L.W. Pritchard and A.J. Sciulli. *Satellite communication systems engineering*. Prentice-Hall, Englewood Cliffs, NJ, 1986.

17 Hanspeter Schaub and John L. Junkins. *Analytical Mechanics of Space Systems*. AIAA, 2014.

18 M.J. Sidi. *Spacecraft Dynamics and Control: A Practical Engineering Approach*. Cambridge University Press, New York, N.Y., 1997.

19 W.T. Thomson. Introduction to space dynamics. Dover, 1986.

20 J.R. Wertz. *Spacecraft attitude determination and control*. Reidel, Dordrecht, the Netherlands, 1978.

21 Bong Wie. *Space Vehicle Dynamics and Control*. AIAA Education Series, Reston, 2008.

22 Wili Hallmann Wilfried Ley, Klaus Wittmann. *Handbook of Space Technology*. Wiley Publications, 2009.

Part II

Fight Vehicle Control via Classical Transfer Function Based Methods

Roadmap to Part II

"The desire to excel is important but what is more important is the desire to prepare"
– Bobby Knight, *Basketball Coach*

Part II covers Fundamentals of Flight Vehicle Control via Classical, Transfer Function Based methods and consists of Chapters 8 through 14. The basics of Laplace transforms and the use of transfer functions in solving ordinary differential equations are presented in Appendix B and form the background needed for the material in this part. Chapter 8 gives an overview of the block diagram based approach in flight control system design, which is common to both aircraft as well as spacecraft situations. The approach taken in this Part II of the book is to introduce the basic control systems concepts in a generic systems level point of view and then illustrate the application of each of those generic control system methods to both aircraft application simultaneously with the spacecraft application, thereby emphasizing the power of a generic systems level approach in control systems analysis and design methods. Thus after introducing few basics of block diagram algebra in Chapter 8, the student is made to learn the art of being able to draw a block diagram for the control problem at hand. Chapter 9 then covers the typical time domain specifications of a control system including the concepts of relative stability and steady state errors. Then a few basic controller structures (proportional, integral and derivative controllers) along with their generalizations, namely lead networks, lag networks, and lead/lag networks, are covered. Then Chapter 10 presents the role of the famous Routh–Hurwitz criterion in assessing the stability of a dynamic system and the illustration of this method to both aircraft and spacecraft situations. Chapter 11 introduces the root locus design methodology. Then Chapter 12 presents the frequency response design methodology via Bode plots. Illustration of application of these design methods in designing autopilot and control augmentation systems for aircraft is given in Chapter 13. A similar exercise is carried out for the satellite attitude control systems design in Chapter 14, focusing a specific satellite as part of a project of interest to the Indian Space Research Organization.

8

Transfer Function Based Linear Control Systems

8.1 Chapter Highlights

In this chapter, we focus on the concept of a transfer function in the Laplace variable domain and learn about the use of this concept in solving linear constant coefficient differential equations. In that connection, we also learn about the many useful features of a transfer function along with the strengths and weaknesses of this concept. Finally, we apply this concept to the application of aircraft linear dynamics as well as of satellite linear dynamics and derive the needed transfer functions for later use in designing control systems using frequency domain techniques.

8.1.1 The Concept of Transfer Function: Single Input, Single Output System

Consider a linear, time invariant system, described by linear state space description (as elaborated in Chapters 1–7), given by

$$\dot{\vec{x}} = A\vec{x} + Bu$$

$$y = C\vec{x} + Du$$

where y and u are of dimension 1 (single input and single output, a SISO system) but $\vec{x}$ is a vector of dimension n. Our intent is to get the time response of output $y(t)$ for a given input function $u(t)$. In other words, we want to know the output as a function of time for a given input function $u(t)$. The relationship between a scalar output for a given scalar input (i.e. a SISO system) in the Laplace (frequency) domain, with the imposition of zero initial conditions, then becomes a simple algebraic ratio of the Laplace transformed output function over the Laplace transformed input function; this is labeled the transfer function between the output and the input.

Thus in general for a SISO system, we can represent the algebraic relationship between the output $Y(s)$ and the input $U(s)$ as shown in the block diagram in Figure 8.1.

$$\frac{Y(s)}{U(s)} = G(s) \tag{8.1}$$

or

$$Y(s) = G(s)U(s).$$

Flight Dynamics and Control of Aero and Space Vehicles, First Edition. Rama K. Yedavalli.
© 2020 John Wiley & Sons Ltd. Published 2020 by John Wiley & Sons Ltd.

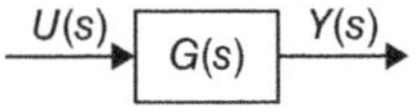

Figure 8.1 A simple block diagram depicting a single input $U(s)$, a single output $Y(s)$ and a transfer function $G(s)$,

Hence, the transfer function between output $Y(s)$ and input $U(s)$ for the above mentioned linear time invariant system is given by

$$
\begin{aligned}
G(s) &= \frac{Y(s)}{U(s)} \\
&= \frac{N(s)}{D(s)}.
\end{aligned}
\tag{8.2}
$$

In the above notation, $N(s)$ denotes the numerator function in the Laplace variable and $D(s)$ denotes the denominator function. Keep in mind that when the ratio of the polynomials $N(s)$ and $D(s)$ is written as above, $N(s) \neq Y(s)$ and $D(s) \neq U(s)$. In other words, $Y(s)$ specifically stands for the Laplace transformed output variable and $U(s)$ stands for the Laplace transformed input variable whereas the numerator $N(s)$ and the denominator $D(s)$ of the transfer function $G(s)$ simply have mathematical polynomial interpretation, not physical variable interpretation.

It is extremely important to keep in mind that the algebraic relationship between the output and input is strictly valid only in the Laplace domain, and not in the time domain. Thus only in the Laplace domain, we have

$$
Y(s) = G(s)U(s).
\tag{8.3}
$$

It is completely **unacceptable and erroneous** to think that

$$
y(t) = g(t)u(t)
\tag{8.4}
$$

and similarly it is erroneous to mix up Laplace domain functions and time domain functions in a single equation or expression.

Also note that for the general single input single output system as shown in 8.1, the transfer function relating the output and the input is also given in terms of the state variable description of the system, as given by

$$
\begin{aligned}
G(s) &= \frac{Y(s)}{U(s)} \\
&= C(s\mathbf{I}_{n \times n} - A)^{-1}B \\
&= \frac{N(s)}{D(s)}.
\end{aligned}
\tag{8.5}
$$

Note that C is a $1 \times n$ matrix and B is a $n \times 1$ matrix and D is a scalar. Then the denominator polynomial is an n th degree polynomial. With A, B, C, D as inputs, MATLAB has a command to obtain the transfer function.

```
[Num, Den] = ss2tf (A,B,C,D,1)
```

Note that the state space to transfer function route is unique. Transfer function to state space, however is not unique.

In summary, to get the transfer function $G(s)$ of a linear time invariant system (i.e. a system described by linear constant coefficient differential equations), we transform those differential equations into Laplace domain functions, and impose zero initial conditions and then take the ratio of the output function $Y(s)$ over the input function $U(s)$. Let us illustrate this procedure with the help of few examples.

8.1.2 An Example for Getting Transfer Functions from State Space Models

Example 8.1 Suppose

$$\begin{bmatrix} \dot{x}_1 \\ \dot{x}_2 \end{bmatrix} = \begin{bmatrix} 1 & 1 \\ 0 & 2 \end{bmatrix} \begin{bmatrix} x_1 \\ x_2 \end{bmatrix} + \begin{bmatrix} 3 \\ 4 \end{bmatrix} u.$$

First, recognize that this is simply a system of simultaneous differential equations

$$\dot{x}_1 = x_1 + x_2 + 3u$$
$$\dot{x}_2 = 2x_2 + 4u.$$

Let the output $y(t)$ be the state variable $x_1(t)$ itself, i.e the C matrix is given by

$$C = [1 \ 0]. \tag{8.6}$$

Note that we are getting the transfer function between an output $Y(s)$ for any any yet unspecified input function $U(s)$ i.e. the transfer function

$$\frac{X_1(s)}{U(s)} = G(s)$$

does not depend on the specific input function. Once we get the transfer function between $Y(s)$ and $U(s)$, then we obtain the output $Y(s)$ for a specific given input $u(t)$ where, in the relationship $Y(s) = G(s)U(s)$, we substitute for $U(s)$, the Laplace transform of the given specific input function $u(t)$. Once we have $Y(s)$ then $y(t)$ is obtained by inverse Laplace transformation.

8.1.3 A Systematic Way of Getting the Transfer Function via the Formula $G(s) = C(sI - A)^{-1}B$

Since the output is $y = x_1$, in the example being considered, we obtain the corresponding C matrix, which in our case is

$$C = [1 \ 0]$$

and we use the formula in Equation 8.5 to obtain the transfer function of a single input single output system as follows:

$$\frac{X_1(s)}{U(s)} = C(sI_{n\times n} - A)^{-1}B$$

$$= [1 \ 0] \left\{ \begin{bmatrix} s & 0 \\ 0 & s \end{bmatrix} - \begin{bmatrix} 1 & 1 \\ 0 & 2 \end{bmatrix} \right\}^{-1} \begin{bmatrix} 3 \\ 4 \end{bmatrix}$$

$$= [1 \ 0] \begin{bmatrix} (s-1) & -1 \\ 0 & (s-2) \end{bmatrix}^{-1} \begin{bmatrix} 3 \\ 4 \end{bmatrix}$$

$$= [1 \ 0] \begin{bmatrix} \dfrac{1}{(s-1)} & \dfrac{1}{(s-1)(s-2)} \\ 0 & \dfrac{1}{(s-2)} \end{bmatrix} \begin{bmatrix} 3 \\ 4 \end{bmatrix}$$

$$= \left[\frac{1}{(s-1)} \quad \frac{1}{(s-1)(s-2)} \right] \begin{bmatrix} 3 \\ 4 \end{bmatrix}$$

$$= \frac{3}{(s-1)} + \frac{4}{(s-1)(s-2)}$$

$$= \frac{3(s-2) + 4}{(s-1)(s-2)}$$

$$= \frac{3s - 2}{(s-1)(s-2)}.$$

Thus

$$\frac{X_1(s)}{U(s)} = \frac{3s - 2}{(s-1)(s-2)}$$

$$= \frac{3s - 2}{s^2 - 3s + 2}.$$

8.1.4 A Brute Force ad hoc Method

We have

$$\dot{x}_1 = x_1 + x_2 + 3u$$
$$\dot{x}_2 = 2x_2 + 4u.$$

Taking Laplace transforms throughout with zero initial conditions, we get

$$sX_1(s) = X_1(s) + X_2(s) + 3U(s)$$
$$sX_2(s) = 2X_2(s) + 4U(s)$$

i.e.

$$(s-1)X_1(s) = X_2(s) + 3U(s)$$
$$(s-2)X_2(s) = 4U(s)$$

i.e.

$$\begin{bmatrix} s-1 & -1 \\ 0 & s-2 \end{bmatrix} \begin{bmatrix} X_1(s) \\ X_2(s) \end{bmatrix} = \begin{bmatrix} 3 \\ 4 \end{bmatrix} U(s).$$

Dividing both sides by $U(s)$,

$$\begin{bmatrix} s-1 & -1 \\ 0 & s-2 \end{bmatrix} \begin{bmatrix} \dfrac{X_1(s)}{U(s)} \\ \dfrac{X_2(s)}{U(s)} \end{bmatrix} = \begin{bmatrix} 3 \\ 4 \end{bmatrix}.$$

Since we want $\frac{X_1(s)}{U(s)}$, apply Cramer's rule.

$$\frac{X_1(s)}{U(s)} = \frac{\begin{vmatrix} 3 & -1 \\ 4 & (s-2) \end{vmatrix}}{\begin{vmatrix} (s-1) & -1 \\ 0 & (s-2) \end{vmatrix}}$$

$$= \frac{3(s-2)+4}{(s-1)(s-2)}$$

$$= \frac{(3s-2)}{(s-1)(s-2)}$$

$$= \frac{3s-2}{s^2-3s+2}$$

which is the same as calculated using the formula given in Equation 8.5.

8.1.5 Use of a Transfer Function in Solving an LTI System of Equations

Once we get the transfer function $G(s)$ between the output and the input for a given LTI system of equations, we can get the output response $y(t)$ for a given specific input function $u(t)$. From Example 8.1 above, we had

$$\frac{X_1(s)}{U(s)} = \frac{3s-2}{(s-1)(s-2)}.$$

Now suppose we want $x_1(t)$ when $u(t) = e^{-t}$, then we first get

$$U(s) = \mathcal{L}[e^{-t}]$$
$$= \frac{1}{s+1}.$$

Then

$$X_1(s) = \left[\frac{(3s-2)}{(s-1)(s-2)}\right] U(s)$$

$$= \left(\frac{(3s-2)}{(s-1)(s-2)}\right)\left(\frac{1}{s+1}\right)$$

$$= \frac{(3s-2)}{(s-1)(s-2)(s+1)}.$$

By partial fraction expansion, we obtain

$$X_1(s) = \frac{A}{(s-1)} + \frac{B}{(s-2)} + \frac{C}{(s+1)}$$

$$= \frac{-\frac{1}{2}}{(s-1)} + \frac{\frac{4}{3}}{(s-2)} + \frac{-\frac{5}{3}}{(s+1)}.$$

Then by inverse Laplace transformation, we get

$$x_1(t) = \mathcal{L}^{-1}[X_1(s)]$$

$$= -\frac{1}{2}e^t + \frac{4}{3}e^{2t} - \frac{5}{3}e^{-t}.$$

It is very important to note that, from Laplace transformation theorems,

$$\mathcal{L}^{-1}(F_1(s)F_2(s)) \neq f_1(t)f_2(t)$$

but

$$\mathcal{L}^{-1}(F_1(s)F_2(s)) = f_1(t) * f_2(t)$$

where $*$ denotes the convolution integral, i.e.

$$f_1(t) * f_2(t) = \int_0^t f_1(\tau)f_2(t - \tau)d\tau$$

$$= \int_0^t f_2(\tau)f_1(t - \tau)d\tau.$$

Similarly

$$\mathcal{L}(f_1(t)f_2(t)) \neq F_1(s)F_2(s)$$

but

$$\mathcal{L}(f_1(t) * f_2(t)) = F_1(s)F_2(s).$$

8.1.6 Impulse Response is the Transfer Function

Now suppose the input $u(t)$ is an impulse of magnitude 1 (i.e. unit impulse), so that $U(s) = 1$. Then

$$Y_{IR}(s) = G(s). \tag{8.7}$$

Therefore

$$y_{IR}(t) = g(t)$$
$$= \mathcal{L}^{-1}[G(s)] \tag{8.8}$$

where the subscript IR indicates that this is the impulse response. Thus, the transfer function can also be defined as the Laplace transformation of the impulse response. Thus once we know the transfer function (or Laplace transform of the impulse response), then we get the response to any other given input. This is a very important concept (that the impulse response in the Laplace domain itself is the transfer function) that plays a major role in identifying transfer function of an LTI system.

8.2 Poles and Zeroes in Transfer Functions and Their Role in the Stability and Time Response of Systems

Consider the transfer function $G(s)$ between a given output $y(t)$ and any input $u(t)$ as described in Equation 8.5, reproduced below for convenience:

$$G(s) = \frac{Y(s)}{U(s)}$$
$$= \frac{N(s)}{D(s)}$$

where $N(s)$ is the numerator polynomial and $D(s)$ is the denominator polynomial. Let us write $G(s)$ in the pole-zero format, i.e.

$$G(s) = \frac{K(s + z_1)(s + z_2)\cdots(s + z_m)}{(s + p_1)(s + p_2)\cdots(s + p_n)}.$$

Thus $s = -z_1$, $s = -z_2,\ldots,s = -z_m$, are the finite zeroes of the transfer function and $s = -p_1$, $s = -p_2,\ldots,s = -p_n$, are the finite poles of the transfer function. Note that, in the general case, there are m finite zeroes and n finite poles. When $m > n$, we label the transfer function as an improper transfer function. When $m = n$, we denote that as a proper transfer function. Finally when $m < n$, we label that as a strictly proper transfer function. Since an improper transfer function can always be split into a proper transfer function plus a strictly proper transfer function (including the possibility that they can be simply constants as well), we focus our attention on these two categories.

Note that poles and zeroes can be real and/or complex conjugate pairs. For example, the following transfer function is a strictly proper transfer function, with real poles and zeroes.

(1)

$$G(s) = \frac{s + 2}{(s + 1)(s + 3)}$$

$s = -2$ is a finite real zero
$s = -1, s = -3$ are two real poles.

(2) The following is also a strictly proper transfer function with real zeroes but with a complex conjugate pair of poles.

$$G(s) = \frac{s + 5}{s^2 + 2s + 4}$$

$$= \frac{s + 5}{(s + 1 + j1.732)(s + 1 - j1.732)}$$

$s = -5$ is a finite real zero
$s_{1,2} = -1 \pm j1.732$ is complex conjugate pair of poles.

Let us now consider the impulse response of a system with the above mentioned general transfer function representation:

$$y(t) = \mathcal{L}^{-1}[G(s)]$$

$$= \mathcal{L}^{-1} \left[\frac{K(s + z_1)(s + z_2) \cdots (s + z_m)}{(s + p_1)(s + p_2) \cdots (s + p_n)} \right]$$

$$= \mathcal{L}^{-1} \left[\frac{A_1}{s + p_1} + \frac{A_2}{s + p_2} + \frac{A_3}{s + p_3} \cdots + \frac{A_n}{s + p_n} \right]$$

$$= A_1 e^{-p_1 t} + A_2 e^{-p_2 t} + A_3 e^{-p_3 t} + \cdots + A_n e^{-p_n t}.$$

Note that if $-p_i$ (i.e. the poles) is real, then the time response is essentially exponential in nature whereas if it is a complex conjugate pair (i.e. terms like $e^{a \pm jb}$) the time response is oscillatory with exponential decay or growth depending on the sign of the real part of the pole. Similarly when they (the poles) are real, if they are negative, it implies exponential decay and if they are positive, it implies exponential growth. Thus whether the poles are real or complex conjugate pairs, the real part of the poles determine decay or growth. If the real parts of the poles are negative, we have a decaying time response as $t \to \infty$, which indicates a stable time response whereas if the real parts of the poles are positive, we have an unbounded growth in the time response as $t \to \infty$ and this amounts to unstable behavior, as shown in Figure 8.2.

So negative real part poles contribute to stable time responses and positive real part poles contribute to unstable behavior in the time response.

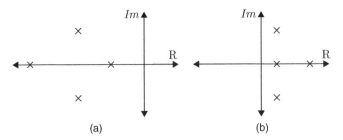

Figure 8.2 (Locations of poles in the complex plane. (a) Stable poles. (b) Unstable poles.

We also observe that the zeroes of the transfer function only affect the amplitudes of the time response.

Thus we can say that the zeroes of the transfer function contribute to the amplitudes of the time response whereas the poles of the transfer function contribute to the stability or instability of the time response with poles in the negative (left) half of the complex plane contributing to stability and poles in the positive (right) half of the complex plane contributing to instability.

8.2.1 Minimum Phase and Non-minimum Phase Transfer Functions

Consider the above transfer function description given by

$$y(t) = \mathcal{L}^{-1}[G(s)]$$
$$= \mathcal{L}^{-1}\left[\frac{K(s+z_1)(s+z_2)\cdots(s+z_m)}{(s+p_1)(s+p_2)\cdots(s+p_n)}\right].$$

If in the above transfer function the poles are all stable, we refer to it as a stable transfer function. If both the zeros as well as poles (all of them) are stable, we refer to it as a minimum phase transfer function. If all the poles are stable but the zeros are all not stable, we refer to it as a non-minimum phase transfer function. This categorization plays an important role later in the control design process. Conceptually, it suffices to realize that it is obviously more difficult (sometimes, not even possible) to design a control system that delivers satisfactory stability and performance characteristics when controlling a non-minimum phase system. Let us now illustrate few of these concepts by another example.

Example 8.2 Given the following LTI differential equation

$$\frac{d^2x}{dt^2} + \frac{dx}{dt} + 8x = \frac{dz}{dt} + 3z$$

obtain the trajectory $x(t)$ when the input trajectory is given by

$$z(t) = e^{-2t}.$$

Initial conditions are zero.

Solution
Taking the Laplace transforms throughout, we have

$$s^2X(s) + sX(s) + 8X(s) = sZ(s) + 3Z(s)$$

where

$$X(s) = \mathcal{L}[x(t)]$$

and

$$Z(s) = \mathcal{L}[z(t)]$$

i.e.

$$(s^2 + s + 8)X(s) = (s + 3)Z(s).$$

Thus the transfer function between $X(s)$ and $Z(s)$ is

$$\frac{X(s)}{Z(s)} = \frac{s + 3}{s^2 + s + 8}.$$

Note that this is a minimum phase transfer function. Now suppose

$$z(t) = e^{-2t}$$

which in the Laplace domain is

$$Z(s) = \frac{1}{(s + 2)}.$$

Then the response of the system to the exponential input above is

$$X(s) = \left(\frac{s + 3}{s^2 + s + 8}\right)\left(\frac{1}{s + 2}\right).$$

The poles of $X(s)$ are

$$s_1 = -2$$

$$s_{2,3} = -\frac{1}{2} \pm j\frac{\sqrt{31}}{2}$$

$$\approx -0.5 \pm j2.7838.$$

To convert from the Laplace space to the time domain,

$$x(t) = \mathcal{L}^{-1}\left[\frac{(s + 3)}{(s + 2)(s^2 + s + 8)}\right]$$

$$= \mathcal{L}^{-1}\left[\frac{A}{(s + 2)} + \frac{Bs + C}{(s^2 + s + 8)}\right].$$

Solving for A, B, C we get $A = 0.1, B = -0.1, C = 1.1$. So

$$X(s) = \frac{0.1}{(s + 2)} + \frac{(-0.1)s + 1.1}{(s + \frac{1}{2})^2 + 7.75}$$

$$= \frac{0.1}{(s + 2)} - \frac{0.1(s + \frac{1}{2})}{(s + \frac{1}{2})^2 + 7.75} + \frac{1.15}{(s + \frac{1}{2})^2 + 7.75}.$$

Looking at the Laplace transform tables, we observe that

$$x(t) = 0.1e^{-2t} - e^{-\frac{1}{2}t}\left[0.1\cos\sqrt{7.75}t - \frac{1.15}{\sqrt{7.75}}\sin\sqrt{7.75}t\right].$$

It is also possible to do the partial fraction expansion in the standard single pole form as follows:

$$X(s) = \frac{(s+3)}{(s+2)(s+0.5+j2.7838)(s+0.5-j2.7838)}$$

$$= \frac{A}{(s+2)} + \frac{B}{(s+0.5+\vec{j}2.7838)} + \frac{C}{(s+0.5-\vec{j}2.7838)}$$

but these time constants B and C turn out to be complex! In fact $C = \bar{B}$ where $\bar{B}$ is the complex conjugate of B. Then

$$x(t) = Ae^{-2t} + Be^{-(0.5+j2.7838)t} + Ce^{-(0.5-j2.7838)}$$

Then using the identities

$$\sin\theta = \frac{e^{j\theta} - e^{-j\theta}}{2j}$$

$$\cos\theta = \frac{e^{j\theta} + e^{-j\theta}}{2}$$

the above can be written as

$$x(t) = Ae^{-2t} + Be^{-(0.5+j2.7838)t} + \bar{B}e^{-(0.5-j2.7838)}$$

$$= Ae^{-2t} + e^{-0.5t}((b_1 + jb_2)e^{-j2.7838t} + (b_1 - jb_2)e^{j2.7838t})$$

$$= Ae^{-2t} + e^{-0.5t}[2b_1 \cos(2.7838t) + 2b_2 \sin(2.7838t)]$$

$$= Ae^{-2t} + B_1 e^{-0.5t} \cos(2.7838t) + B_2 e^{-0.5t} \sin(2.7838t).$$

The main point here is that whenever there is a complex conjugate pair of poles, the corresponding time response is oscillatory. The real part of the complex pole determines the rate of decay or divergence appearing in the exponential term and the imaginary part of the complex pole determines the frequency of the oscillations. If the real part is negative, these oscillations are damped oscillations. If the real part is zero, these are undamped oscillations; finally if the real part is positive, they lead to divergent oscillations. The constants A, B_1, and B_2 simply determine the amplitudes of the oscillations. In the above example, we do have damped oscillations and the overall response is stable and decays with time.

8.2.2 Importance of the Final Value Theorem

Consider again the transfer function $\frac{X(s)}{Z(s)}$ in the previous example. It is given by

$$\frac{X(s)}{Z(s)} = \frac{s+3}{s^2+s+8}. \tag{8.9}$$

Note that the above transfer function has a stable denominator because the three poles are stable. Hence, if we give a unit step as an input to the above system, there exists a steady state value and suppose we are interested in finding that steady state value. This can be easily obtained by the application of the final value theorem given in the table of Laplace transform theorems. Note that, for the final value theorem (FVT) to be valid, the function $sX(s)$ is required to be stable. In the present case, with a step input in $z(t)$, where

$Z(s) = 1/s$, we observe that, indeed $sX(s)$ is stable and so we can get the steady state value of the function $x(t)$, denoted by x_{ss}. Thus, in this case, $x_{ss} = x(\infty) = \lim_{s \to 0} sX(s) = 3/8$.

The utility of the FVT is that once we establish that the final value for $f(t)$ exists (by checking the stability of the function $sF(s)$) the steady state (final) value of the time function f_{ss} is then given by a one-shot application of the FVT without having to compute the entire time history of the function $f(t)$ by the inverse Laplace transform exercise, which offers considerable savings and effort.

Having covered these fundamental concepts on transfer functions, we now shift our attention to the application of these concepts to aircraft and spacecraft dynamics problems.

8.3 Transfer Functions for Aircraft Dynamics Application

Recall that, in Part I of this book, we already obtained the linearized, constant coefficient ordinary differential equations of motion for aircraft and spacecraft problems. We now convert them into input/output transfer function format by specifying one of those motion variables as an input and another one as an output. Let us illustrate this procedure by considering a linear aircraft model.

Example 8.3   A fighter aircraft flying at 200 m s^{-1} at a height of 10000 m has the short period equations of motion

$$\dot{\alpha} = -6\alpha + q$$

$$\dot{q} = -5\alpha - 0.6q - 12\delta_e.$$

It is also known that $\dot{\Theta} = q$, where Θ is the pitch angle, α is the angle of attack, q is the pitch rate, and δ_e is the elevator deflection.

1. Find the transfer function $\frac{q}{\delta_e}(s)$ both by hand and using MATLAB.
2. Also obtain the transfer functions $\frac{\alpha}{\delta_e}(s)$ and $\frac{\Theta}{\delta_e}(s)$

Taking Laplace transforms (with zero initial conditions),

$$s\alpha(s) = -6\alpha(s) + q(s)$$

$$sq(s) = -5\alpha(s) - 0.6q(s) - 12\delta_e(s).$$

This is now an algebraic system of equations. Gathering like terms in the first equation and solve for α as such:

$$\alpha(s) = \frac{q(s)}{s + 6}.$$

Substituting this into the second equation and rearranging the terms, we arrive at:

$$\frac{q(s)}{\delta_e(s)} = \frac{-12(s + 6)}{s^2 + 6.6s + 8.6}.$$

Alternatively we can write the state space model $\dot{x} = Ax + Bu$ as

$$\begin{bmatrix} \dot{\alpha} \\ \dot{q} \end{bmatrix} = \begin{bmatrix} -6 & 1 \\ -5 & -0.6 \end{bmatrix} \begin{bmatrix} \alpha \\ q \end{bmatrix} + \begin{bmatrix} 0 \\ -12 \end{bmatrix} \delta_e.$$

Since q is the output of interest, we can write

$$q = \begin{bmatrix} 0 & 1 \end{bmatrix} \begin{bmatrix} \alpha \\ q \end{bmatrix}$$

which is the C matrix in the state space representation $Y = Cx + Du$. Recognize that the D is simply scalar zero (ie. a zero 1×1 matrix). Therefore

$$\frac{q}{\delta_e}(s) = C(sI - A)^{-1}B$$

$$= \begin{bmatrix} 0 & 1 \end{bmatrix} \begin{bmatrix} s+6 & -1 \\ 5 & s+0.6 \end{bmatrix} \begin{bmatrix} 0 \\ -12 \end{bmatrix}$$

$$\frac{q(s)}{\delta_e(s)} = \frac{-12(s+6)}{s^2 + 6.6s + 8.6}$$

which is same as the answer we obtained above.

Example 8.4 Consider the full set of longitudinal equations of motion of a given aircraft at a given flight condition

$$\dot{q} = -0.65q - 0.2\dot{\alpha} - \alpha - 1.2\delta_e$$
$$\dot{u}_{sp} = 225\delta_{th} + 0.035\alpha - 9.81\Theta - .18u_{sp}$$
$$\dot{\alpha} = q - 0.2u_{sp} - .6\alpha - 0.035\delta_e$$
$$\dot{\Theta} = q.$$

Note the presence of two control variables δ_{th} and δ_e. Also note the presence of the $\dot{\alpha}$ term in the first equation. Then,

1. Find the transfer function $\frac{q}{\delta_e}(s)$ using MATLAB.
2. Also obtain the transfer functions $\frac{\alpha}{\delta_e}(s)$ and $\frac{\Theta}{\delta_e}(s)$. Note that, even though strictly speaking, we need to use different symbols to distinguish Laplace functions from time functions, for better association with the physical variables, we use the same symbols for both in this problem formulation.

Solution
We want to put the given system

$$\dot{q} = -0.65q - 0.2\dot{\alpha} - \alpha - 1.2\delta_e$$
$$\dot{u}_{sp} = 225\delta_{th} + 0.035\alpha - 9.81\Theta - .18u_{sp}$$
$$\dot{\alpha} = q - 0.2u_{sp} - .6\alpha - 0.035\delta_e$$
$$\dot{\Theta} = q$$

into state space format $\dot{x} = Ax + Bu$. We begin to do so by gathering all derivative terms (those indicated by a dot) to the left hand side of the equation. To do so requires eliminating the $\dot{\alpha}$ term from the first equation; we do so by substituting the third equation into the $\dot{\alpha}$ term as follows

$$\dot{q} = -0.65q - 0.2(q - 0.2u_{sp} - 0.6\alpha - 0.035\delta_e) - \alpha - 1.2\delta_e$$
$$= -0.85q + 0.04u_{sp} - 0.88\alpha - 1.193\delta_e.$$

Denoting q, u_{sp}, α, and Θ as the state variables, and δ_e and δ_{th} as the control variables, we arrive at the following state space representation

$$
\begin{bmatrix} \dot{q} \\ \dot{u}_{sp} \\ \dot{\alpha} \\ \dot{\Theta} \end{bmatrix} = \begin{bmatrix} -0.85 & 0.04 & -0.88 & 0 \\ 0 & -0.18 & 0.035 & -9.81 \\ 1 & -0.2 & -0.6 & 0 \\ 1 & 0 & 0 & 0 \end{bmatrix} \begin{bmatrix} q \\ u_{sp} \\ \alpha \\ \Theta \end{bmatrix} + \begin{bmatrix} -1.193 & 0 \\ 0 & 2.25 \\ -0.035 & 0 \\ 0 & 0 \end{bmatrix} \begin{bmatrix} \delta_e \\ \delta_{th} \end{bmatrix}.
$$

1. We desire the transfer function $\frac{q}{\delta_e}(s)$. To obtain this by hand is an exercise in algebra left to the reader. The procedure to obtain the transfer function using MATLAB is shown. We observe that the A matrix that is needed as an argument in the MATLAB command is

$$
A = \begin{bmatrix} -0.85 & 0.04 & -0.88 & 0 \\ 0 & -0.18 & 0.035 & -9.81 \\ 1 & -0.2 & -0.6 & 0 \\ 1 & 0 & 0 & 0 \end{bmatrix}
$$

and because the elevator deflection δ_e is the control variable of interest, the matrix B is taken to be the column matrix

$$
B = \begin{bmatrix} -1.193 \\ 0 \\ -0.035 \\ 0 \end{bmatrix}
$$

because this column matrix is associated with the input δ_e. Since q is the output variable of interest, the C matrix is taken to be

$$
C = \begin{bmatrix} 1 & 0 & 0 & 0 \end{bmatrix}.
$$

The D matrix is simply a zero scalar. We feed this information to MATLAB using its inbuilt function as such

```
[num,den]=ss2tf(A,B,C,D,1)
```

which yields the outputs

```
num = 0 -1.193 -0.8997 -0.1317 0
```

and

```
den = 1 1.63 1.658 0.6472 1.962
```

from which we obtain our transfer function

$$
\frac{q}{\delta_e}(s) = \frac{-1.193s^3 - 0.8997s^2 - 0.1317s}{s^4 + 1.63s^3 + 1.658s^2 + 0.6472s + 1.962}.
$$

2. This time we desire the transfer function $\frac{\Theta}{\delta_e}$. The A matrix will remain the same as in the previous example, because the system we are describing has not changed; we

simply have a different output. The B matrix is still the same as before because the input remains as δ_e. Thus,

$$B = \begin{bmatrix} -1.193 \\ 0 \\ -0.035 \\ 0 \end{bmatrix}.$$

Since, this time, Θ is the output, the corresponding C matrix is taken to be

$$C = \begin{bmatrix} 0 & 0 & 0 & 1 \end{bmatrix}.$$

Using the MATLAB command,

```
[num,den]=ss2tf(A,B,C,D,1)
```

which yields the outputs

```
num = 0  0  -1.193  -0.9041  -0.1325
```

and

```
den = 1  1.63  1.658  0.6472  1.962
```

from we obtain our transfer function

$$\frac{\Theta}{\delta_e}(s) = \frac{-1.193s^2 - 0.9041s - 0.1325}{s^4 + 1.63s^3 + 1.658s^2 + 0.6472s + 1.962}.$$

To obtain the transfer function $\frac{\alpha}{\delta_{th}}$, we now set the B matrix to be the second column of the original B matrix,

$$B = \begin{bmatrix} 0 \\ 2.25 \\ 0 \\ 0 \end{bmatrix}$$

and the C matrix to be

$$\begin{bmatrix} 0 & 0 & 1 & 0 \end{bmatrix}.$$

As previously, the A and D matrices are unchanged. The MATLAB command

```
[num,den]=ss2tf(A,B,C,D,1)
```

then yields the outputs

```
num = 0  0  -0.45  -0.2925  0
```

and

```
den = 1  1.63  1.658  0.6472  1.962
```

noindentfrom which we obtain our transfer function

$$\frac{\alpha}{\delta_{th}}(s) = \frac{-0.45s^2 - 0.2925s}{s^4 + 1.63s^3 + 1.658s^2 + 0.6472s + 1.962}.$$

Next, we switch our attention to the spacecraft dynamics case.

8.4 Transfer Functions for Spacecraft Dynamics Application

Recall that we have already obtained the linearized equations of motion for the attitude dynamics of satellite in a circular orbit in Part I. We can now use those linearized equations of satellite attitude motion, and get the appropriate transfer functions between an output variable and the control (input) variable. For illustrating this procedure, we reproduce the linearized equations of motion of a satellite attitude dynamics, which are given by

$$
\begin{bmatrix} I_1(\ddot{\phi} - \dot{\psi}\omega_o) + (I_2 - I_3)(\omega_o^2\phi + \omega_o\dot{\psi}) \\ I_2\ddot{\theta} \\ I_3(\ddot{\psi} + \dot{\phi}\omega_o) + (I_2 - I_1)(\omega_o^2\psi - \omega_o\dot{\phi}) \end{bmatrix} = \begin{bmatrix} M_{\text{ext1}} \\ M_{\text{ext2}} \\ M_{\text{ext3}} \end{bmatrix} \tag{8.10}
$$

where $M_{\text{ext}i}$ are the external applied moments along the three principal axes.

Note that the pitch attitude angle motion is completely decoupled from the roll/yaw motion. Let us then consider this simple case of getting the transfer function between the pitch angle and the applied external control torque. Noting that

$$
I_2\ddot{\theta} = M_{\text{ext2}}
$$

Let us now separate out the control torque from the disturbance torque and rewrite the forcing function as

$$
I_2\ddot{\theta} = M_c + M_d.
$$

Then the transfer function between output $\theta(s)$ and the input control torque $M_c(s)$ is given by

$$
\frac{\theta}{M_c}(s) = \frac{1}{I_2 s^2}
$$

where I_2 is the principal moment of inertia of the satellite about the pitch axis. Recall that when forming a transfer function between an output and an input, we assume zero initial conditions for the output variable. Also in a SISO system framework, in a transfer function, we can consider only one input and one output at a time. Hence for getting the transfer function between the output pitch angle θ and the control torque M_c, we ignore the disturbance torque as an input. However, we could get the transfer function between the output pitch angle θ and the disturbance torque M_d, as well by ignoring the control torque M_c as an input and obtain,

$$
\frac{\theta}{M_d}(s) = \frac{1}{I_2 s^2}.
$$

Notice that since it is a linear system, superposition holds and thus the total response of the output to both inputs together is simply the sum of the output response to each individual input.

We can carry out a similar exercise for getting the needed transfer functions for the coupled roll/yaw motion as well by utilizing the following state space model of the roll/yaw motion, given by

$$
\begin{bmatrix} I_1(\ddot{\phi} - \dot{\psi}\omega_o) + (I_2 - I_3)(\omega_o^2\phi + \omega_o\dot{\psi}) \\ I_3(\ddot{\psi} + \dot{\phi}\omega_o) + (I_2 - I_1)(\omega_o^2\psi - \omega_o\dot{\phi}) \end{bmatrix} = \begin{bmatrix} M_{\text{ext1}} \\ M_{\text{ext3}} \end{bmatrix}. \tag{8.11}
$$

Denoting ϕ, $\dot{\phi}$, ψ, and $\dot{\psi}$ as the state variables and M_c as the input variable, we can get the state space representation with the appropriate A and B matrices. Similarly based on the specification of the output variable, we can get the appropriate C and D matrices. Getting the needed transfer function between the specified output variable and the control torque M_c, for a satellite in a circular orbit of given radius (and thus a constant ω_0), is left as an exercise for the student.

8.5 Chapter Summary

In this chapter, we have learnt the importance of the concept of the transfer function and its use in the solution of linear constant coefficient ordinary differential equations. We have examined the main properties of transfer functions in the form of poles and zeroes and their role in the natural motion (impulse response) and the stability of the resulting motion. We understood the important difference between a minimum phase and non-minimum phase transfer function. All these concepts form the foundation for more advanced material in in the remainder of this book. Finally, we obtained various needed transfer functions (for use later in the control system design) for both aircraft dynamics application as well as spacecraft dynamics application. Fundamental concepts discussed in this chapter can be found in other textbooks dedicated to control systems such as [1–5].

8.6 Exercises

Exercise 8.1. Using the given information in the Example 8.1 problem, obtain the transfer function

$$G(s) = \frac{X_2}{U}(s).$$

Exercise 8.2. Using the given information in the Example 8.4 problem, obtain the transfer functions

$$G_1(s) = \frac{u_{sp}}{\delta_e}(s)$$

$$G_2(s) = \frac{\alpha}{\delta_e}(s)$$

$$G_3(s) = \frac{\Theta}{\delta_{th}}(s).$$

Exercise 8.3. Using the given information in the Example 8.4 problem, obtain the transfer functions

$$G_1(s) = \frac{\gamma(s)}{\delta_e(s)}$$

$$G_2(s) = \frac{\gamma(s)}{\delta_{th}(s)}$$

where γ is the flight path angle defined as

$$\gamma = \Theta - \alpha.$$

Exercise 8.4. In the satellite attitude dynamics, obtain the following transfer functions

$$G_1(s) = \frac{\phi(s)}{M_c(s)}$$

$$G_2(s) = \frac{\psi(s)}{M_d(s)}$$

for a geosynchronous circular orbit. Use $I_x = I_z = 90$ kg m^2 and $I_y = 60$ kg m^2.

Exercise 8.5. Given

$$Y(s) = \frac{s+3}{s^2 + s + 4}$$

what is the nature of $y(t)$?
(a) Pure exponential
(b) Decaying (damped) sinusoidal
(c) Pure harmonic
(d) Unstable (divergent).
Now suppose the zero associated with the system $Y(s)$ is changed from $s + 3$ to $s - 3$.
What now is the nature of $y(t)$?
(a) Pure exponential
(b) Decaying (damped) exponential
(c) Pure harmonic
(d) Unstable (divergent).
This new $Y(s)$ can best be described as
(a) Minimum phase
(b) Non-minimum phase.

Exercise 8.6. What are the poles of the system

$$H(s) = \frac{20(s + 10)}{s^2 + 110s + 1000}. \tag{8.12}$$

(a) 20 and 100
(b) 20, 110, and 1000
(c) 10, 20, and 100
(d) 10 and 100.

Exercise 8.7. Identify the type of natural response (overdamped, critically damped, or underdamped) associated with each of the following characteristic polynomials:
(a) $s^2 + 8s + 12$
(b) $s^2 + s + 1$
(c) $3s^2 + 9s + 2$
(d) $4^2 + s + 40$
(e) $s^2 + 4s + 4$.

Exercise 8.8. For the Laplace transformed signal $Y(s) = \frac{3s-5}{s^2+4s+2}$, find $y(t)$ for $t \geq 0$.

Exercise 8.9. Use the Laplace transform method to solve $\frac{dy}{dt} + 4y = 6e^{2t}$.

Bibliography

1 J.J. d'Azzo and C.D. Houpis. *Linear control system analysis and design: conventional and modern*. McGraw-Hill, New York, 1988.

2 J.J. d'Azzo and C.H. Houpis. *Linear control system analysis and design: conventional and modern*. McGraw-Hill Higher Education, New York, 1988.

3 C.R. Dorf and R.H. Bishop. *Modern control systems*. Pearson, 12 edition, 2011.

4 G. Hostetter, C. Savant Jr, and R. Stefani. *Design of Feedback Control Systems*. Holt, Rinehart and Winston, 1982.

5 N. Nise. *Control Systems Engineering*. Wiley, 2011.

9

Block Diagram Representation of Control Systems

9.1 Chapter Highlights

In the previous chapter, we presented the methodology for getting transfer functions between a given (single) output and a (single) input in the Laplace (frequency) domain. The corresponding transfer function, denoted by $G(s)$, is represented by a block as in Figure 9.1. In this chapter, we gather the transfer functions of various components of a control system and put them together in a complete block diagram based on the control objective. We then learn how to manipulate the various transfer functions within that block diagram to obtain the transfer function between any desired output and any desired input within that block diagram by following the rules of block diagram algebra. We also introduce various controller structures such as proportional (P), integral (I) and derivative (D), i.e. PID controllers as blocks within the complete block diagram, and demonstrate the role of these controllers in meeting the various design specifications of the control system. Finally we apply them to control problems in aircraft application as well as spacecraft application.

9.2 Standard Block Diagram of a Typical Control System

In this section, we represent a typical control system block diagram in a very simplified, generic fashion with the most critical blocks needed to understand the basic principles of block diagram algebra. The basic idea is to write the block diagram in such a way that we assume the overall input to the control system is the desired behavior of the output and the overall output to the block diagram is the actual behavior of the output. We assume that the actual behavior of the output is always measured by a sensor and this actual behavior is compared to the desired behavior via a comparator, which is represented as a summer junction in the block diagram. Then the error between the desired behavior and the actual behavior is fed to a controller, which then manipulates that error signal in an appropriate fashion and generates the correct signal as an input to the actuator, which supplies the needed input signal to the plant (the system dynamics which is being controlled) until the output of the plant, which is when the actual behavior of the system being controlled comes as close as possible to the desired behavior. Thus the controller plays the most crucial part in reducing the error between the actual behavior and the desired behavior. In other words, we assume all the transfer functions in the block diagram, except that of the controller, are given or known. Then the idea is to design an

Flight Dynamics and Control of Aero and Space Vehicles, First Edition. Rama K. Yedavalli.

Figure 9.1 A unity feedback control system.

appropriate controller transfer function that takes all the other given or known transfer functions into consideration in its design. For a quick approximate analysis and design typically the actuator and sensor are taken as giving instantaneous outputs, which means the transfer functions corresponding to them are constant gains. If that is too unrealistic, we can also incorporate the dynamic models of the actuator and sensor by transfer functions, either as a first order or a second order transfer function. When this is done, we say we are taking the actuator and sensor dynamics into consideration. While this is more realistic, keep in mind that it also complicates the design of the controller. So a typical compromise is to consider the actuator dynamics (which are relatively slow in producing the desired outputs from them) but ignore the dynamics of the sensors which in reality are also very fast acting devices. Also it is possible that the controller can be split into various components and be placed at different places within the block diagram (either in the feedforward path or in the feedback path or in both paths). With this backdrop, in what follows, we assume the simplest form of a block diagram with these different blocks combined into one transfer function in the feedforward path and one transfer function in the feedback path, so that we can easily manipulate and obtain the few fundamental transfer functions needed to analyze the control system behavior. When the entire feedback path is such that the actual behavior of the output is directly fed into the comparator summing junction, we label it as a unity feedback system, as shown in Fig 9.1

Realize that this type of block diagram representation of a control system is not only serving the purpose of a pictorial representation for the control system at hand, but actually contains significant mathematical content. In other words, this type of block diagram is indeed a pictorial representation of mathematical (algebraic) equations. Thus in the Figure 9.1 the signal $E(s)$ is mathematically equivalent to the algebraic equation $E(s) = R(s) - Y(s)$ and similarly

$$G(s) = \frac{Y(s)}{E(s)} \tag{9.1}$$

i.e $Y(s) = G(s)E(s)$. With this understanding, consider the non-unity feedback system in Fig 9.2.

For a system in this form, we define the following:

- The feedforward transfer function $G(s)$ is

$$G(s) = \frac{Y(s)}{E(s)}. \tag{9.2}$$

- The feedback transfer function $H(s)$ is

$$H(s) = \frac{B(s)}{Y(s)}. \tag{9.3}$$

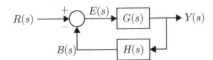

Figure 9.2 A non-unity feedback control system.

- The open loop transfer function $G(s)H(s)$ is

$$G(s)H(s) = \frac{B(s)}{E(s)}. \tag{9.4}$$

It is left as an exercise to the reader to then show that the closed loop transfer function $M(s)$ given by $\frac{Y(s)}{R(s)}$ and the error transfer function given by $\frac{E(s)}{R(s)}$ have the following transfer functions, namely that:

- The closed loop transfer function $M(s)$ is

$$\begin{aligned} M(s) &= \frac{G(s)}{1+G(s)H(s)} \\ &= \frac{Y(s)}{R(s)}. \end{aligned} \tag{9.5}$$

- The error transfer function $\frac{E(s)}{R(s)}$ is

$$\frac{E(s)}{R(s)} = \frac{1}{1 + G(s)H(s)}. \tag{9.6}$$

Sometimes when $r(t)$ and $y(t)$ are of the same units, the error $e(t)$ is defined as $e(t) = r(t) - y(t)$ occasionally, irrespective of whether $H(s)$ is equal to one or not. In that case,

$$\frac{E(s)}{R(s)} = \frac{1 + G(s)[H(s) - 1]}{1 + G(s)H(s)}$$

but the majority of the time, we will use $E(s) = R(s) - B(s)$.

Thus, the process of getting the transfer function between some specified output variable and a specified input variable amounts to some basic block diagram algebra. The following examples convey this procedure.

Example 9.1 Find the closed loop transfer function $\frac{Y(s)}{R(s)}$ in Figure 9.3.

Solution
The diagram in Figure 9.3 is equivalent to the diagram in Figure 9.4.

This conclusion can be arrived at by simply carrying out the algebraic mathematical manipulations that exist within the given block diagram in a systematic way. Start by

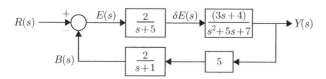

Figure 9.3 Block diagram for Example 9.1.

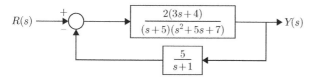

Figure 9.4 Simplification of the Example 9.1 block diagram.

naming every input and output signal of each block within the given block diagram. Then write out down the mathematical (algebraic in nature) relationships between all those input and output signals and then manipulate those mathematical relationships tailored in such a way as to simplify the final block diagram into the standard block diagram. Let us illustrate this procedure for the above block diagram of the given problem.

We observe that

$$B(s) = \frac{5}{s+1} Y(s).$$

We also see that

$$Y(s) = \frac{(3s+4)}{(s^2 + 5s + 7)} \delta E(s)$$

where

$$\delta E(s) = \frac{2}{(s+5)} E(s).$$

Notice that

$$E(s) = R(s) - B(s) = R(s) - \frac{5}{s+1} Y(s).$$

Then, after few more appropriate algebraic manipulations, we get

$$Y(s) \left[1 + \frac{10(3s+4)}{(s+5)(s^2 + 5s + 7)(s+1)} \right] = \left[\frac{2(3s+4)}{(s+5)(s^2 + 5s + 7)} \right] R(s)$$

which then yields the final desired transfer function, namely $Y(s)/R(s)$,

$$\frac{Y(s)}{R(s)} = \frac{\dfrac{2(3s+4)}{(s+5)(s^2 + 5s + 7)}}{1 + \dfrac{10(3s+4)}{(s+5)(s^2 + 5s + 7)(s+1)}}$$

$$= \frac{2(3s+4)}{(s+1)(s+5)(s^2 + 5s + 7) + 30s + 40}$$

$$= \frac{6(s + \frac{4}{3})}{s^4 + 11s^3 + 42s^3 + 97s + 75}.$$

Thus the closed loop transfer function has one zero at $s = -\frac{4}{3}$, and four poles at the roots of the denominator polynomial, which are $s \approx -6.60389$, $s \approx -1.25931$, $s \approx -1.5684 - \vec{j}2.56096$, and $s \approx -1.5684 + \vec{j}2.56096$.

Note that the (open) loop transfer function is given by

$$G(s) = \frac{10(3s+4)}{(s+5)(s^2 + 5s + 7)(s+1)}. \tag{9.7}$$

The feedforward transfer function is

$$G(s) = \frac{2(3s+4)}{(s+5)(s^2 + 5s + 7)}. \tag{9.8}$$

The feedback transfer function is

$$H(s) = \frac{5}{(s+1)}. \tag{9.9}$$

The error transfer function

$$\frac{E(s)}{R(s)} = \frac{1}{1 + \dfrac{10(3s+4)}{(s+5)(s^2+5s+7)(s+1)}}$$

$$= \frac{(s+5)(s^2+5s+7)(s+1)}{(s+5)(s^2+5s+7)(s+1) + 30s + 40}$$

$$= \frac{s^4 + 11s^3 + 42s^2 + 67s + 35}{s^4 + 11s^3 + 42s^2 + 97s + 75}.$$

Example 9.2   Find the closed-loop transfer function for the system depicted in Figure 9.5.

Solution
We simplify the problem according to the steps in Figure 9.6

Figure 9.5 Block diagram for Example 9.2.

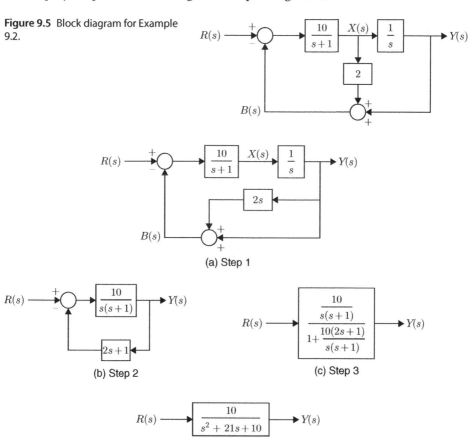

(a) Step 1

(b) Step 2

(c) Step 3

(d) Step 4

Figure 9.6 Simplification of the control block diagram in Example 9.2.

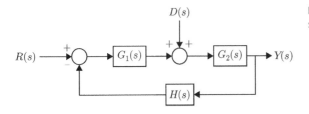

Figure 9.7 Block diagram of a system subjected to disturbance.

After this simplification, we get

$$\frac{Y(s)}{R(s)} = \frac{10}{s^2 + 21s + 10}.$$

9.2.1 A Closed Loop System Subjected to Disturbance

Because all of these systems have to be single input single output (SISO) for the methods we have gone over, we cannot get the transfer function from input and the disturbance at the same time, therefore we have to set the disturbance to zero to get the transfer function between the input and output or set the input to zero to get the transfer function between the disturbance and the output. Consider the system shown in Figure 9.7.

It can be easily shown that

$$\frac{Y(s)}{R(s)}\bigg|_{D(s)=0} = \frac{G_1(s)G_2(s)}{1 + G_1(s)G_2(s)H(s)}$$

$$\frac{Y(s)}{D(s)}\bigg|_{R(s)=0} = \frac{G_2(s)}{1 + G_1(s)G_2(s)H(s)}.$$

At this juncture, it is appropriate to switch the discussion to understanding various time domain response specifications, which are then used as design specifications, so that we attempt to design a controller transfer function to meet those design specifications.

9.3 Time Domain Performance Specifications in Control Systems

Given the transfer function between an output and an input, as seen in Figure 9.8, i.e.

$$\frac{Y(s)}{R(s)} = M(s) = \frac{N(s) = \text{numerator polynomial}}{D(s) = \text{denominator polynomial}}$$

$$Y(s) = M(s)R(s).$$

Our eventual interest is to get $y(t)$ for a given input $r(t)$, see Figure 9.9 and Table 9.1. Typical response or performance characteristics or specifications are in terms of:

(a) Transient response
(b) Steady state response
(c) Relative stability
(d) Frequency response specifications.

Figure 9.8 Simple system.

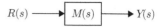

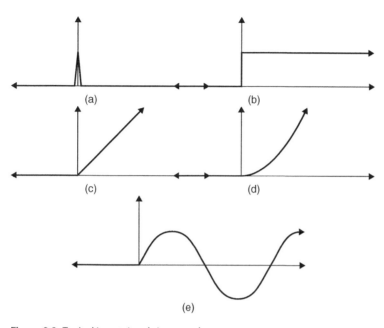

Figure 9.9 Typical input signals in control systems.

Table 9.1 Key to typical input signals in control systems.

Illustration	$r(t), t > 0$	Signal	$R(s)$
9.9a	δ	Unit impulse	1
9.9b	1	Unit step	$\dfrac{1}{s}$
9.9c	t	Unit ramp	$\dfrac{k}{s^2}$
9.9d	t^2	Parabolic	$\dfrac{k}{s^3}$
9.9e	$\sin \omega t$	Pure oscillatory	$\dfrac{\omega}{\omega^2 + s^2}$

The first three are mostly time domain specifications. The time response is composed of two parts:

$$y(t) \;=\; y_t(t) \;+\; y_{ss}(t)$$
$$\qquad\qquad\quad \downarrow \qquad\quad \downarrow$$

That part which
goes to zero as $\;\rightarrow\;$ Transient   Steady state
$$t \rightarrow \infty$$

9.3.1 Typical Time Response Specifications of Control Systems

9.3.1.1 Transient Response: First Order Systems

$$\frac{Y(s)}{R(s)} = M(s) = \frac{N(s)}{(s+a)}$$

i.e. the denominator polynomial is of degree 1.

9.3.1.2 Unit Step Response

$$Y(s) = \frac{N(s)}{s(s+a)} = \frac{k_1}{s} + \frac{k_2}{s+a}$$

$$y(t) = \underset{\text{Steady state}}{k_1} + \underset{\text{Transient response.}}{k_2 e^{-at}}$$

Figure 9.10 shows the first order signals.

$t = \frac{1}{a} = T$ is called the time constant of the system. The smaller the time constant, the faster the system response.

9.3.1.3 Second Order Systems

$$\frac{Y(s)}{R(s)} = M(s) = \frac{N(s)}{s^2 + 2\xi\omega_n s + \omega_n^2}$$

i.e. the denominator polynomial is of degree 2. Suppose

$$\frac{Y(s)}{R(s)} = M(s) = \frac{\omega_n^2}{s^2 + 2\xi\omega_n s + \omega_n^2}.$$

9.3.1.4 Unit Step Response

$$Y(s) = \frac{\omega_n^2}{s(s^2 + 2\xi\omega_n s + \omega_n^2)}$$

$$y(t) = 1 + \frac{e^{-\xi\omega_n t}}{\sqrt{1-\xi^2}} \sin\left[\omega_n\sqrt{1-\xi^2}\,t - \tan^{-1}\frac{\sqrt{1-\xi^2}}{-\xi}\right]$$

Figure 9.11 demonstrates the influence of pole placement on time response. Typical specifications are given in terms of the case $0 < \xi < 1$, illustrated in Figure 9.12.

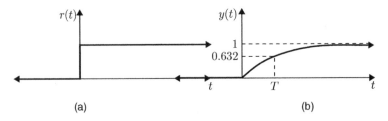

(a) (b)

Figure 9.10 First order signals. (a) Input signal. (b) Output signal.

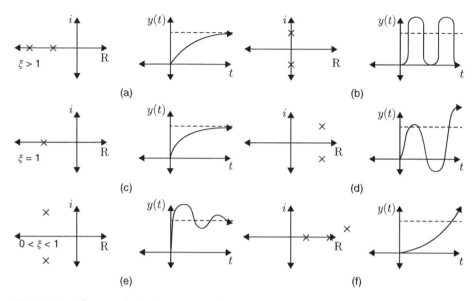

Figure 9.11 Influence of pole placement on time response.

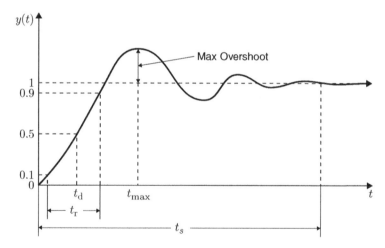

Figure 9.12 Specifications in the time domain.

(a) Maximum overshoot is a measure of relative stability.

$$\%\text{Maximum overshoot} = \frac{\text{Maximum overshoot}}{\text{Final value}} * 100\%.$$

(b) Delay time t_d is the time to reach 50% of the final value.
(c) Rise time t_r is the time taken from 10% to 90% of the final value.
(d) Settling time t_s is the time taken to settle to about $\pm 4\%$ to 5% of the final value.

Typically,

$$t_{\text{max}} = \frac{\pi}{\omega_n \sqrt{1 - \xi^2}}$$

$$\text{Maximum overshoot} = e^{-\frac{\pi\xi}{\sqrt{1-\xi^2}}}$$

$$\%\text{maximum overshoot} = 100e^{-\frac{\pi\xi}{\sqrt{1-\xi^2}}}$$

$$t_d \cong \frac{1 + 0.6\xi + 0.15\xi^2}{\omega_n}$$

$$t_r \cong \frac{1 + 1.1\xi + 1.4\xi^2}{\omega_n}$$

$$t_s \cong \frac{3}{\xi\omega_n}(5\% \text{ tolerance}).$$

It is customary to take $\frac{1}{\xi\omega_n}$ as the time constant of a second order system.

Now that we have gathered these design specifications, the next step is to see what type of controller structures (by which we imply what type of controller transfer functions) would be suitable to meet these design specifications. Naturally, the controller transfer function we inject into the above control system block diagram has significant influence on the closed loop system behavior. So in the next section, we build a few useful controller transfer functions that are likely to meet the standard design specifications.

9.4 Typical Controller Structures in SISO Control Systems

Consider the typical control system block diagram in Figure 9.13. Standard controller structures are:

(1) $G_C(s) = K_P \rightarrow$ proportional controller
(2) $G_C(s) = K_P + K_D s \rightarrow$ proportional derivative (PD) controller
(3) $G_C(s) = K_P + \frac{K_I}{s} \rightarrow$ proportional integral (PI) controller.

However. these typical controllers are ideal cases of more general practical controllers given as follows.

9.4.1 Lead Network or Lead Compensator

Here $|z| < |p|$. See Figure 9.14.

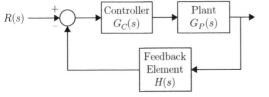

Figure 9.13 Typical control system block diagram.

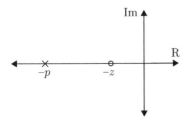

Figure 9.14 Lead compensator poles and zeroes.

Figure 9.15 Lead network.

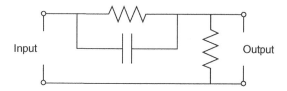

The PD controller is a special case of this.

A lead compensator improves relative stability and is mostly used to satisfy damping ratio and natural frequency requirements. A possible mechanization of this network using RLC (resistor, inductor, and capacitor) elements is shown in Figure 9.15. In electrical engineering notation, this is a high-pass filter.

9.4.2 Lag Network or Lag Compensator

Here $|p| < |z|$. See Figure 9.16.

The PI controller is a special case of this.

A lag compensator reduces steady-state efforts to t^k type inputs (especially for a step input that is like $k = 0, t^0$ input). A possible mechanization of this network using RLC elements is shown in Figure 9.17. In electrical engineering notation, this is a low-pass filter.

Obviously, in the majority of the situations, a combination of these two is warranted, which is termed a lead-lag compensator:

$$G_C(s) = \frac{K(s + z_1)(s + z_2)}{(s + p_1)(s + p_2)}.$$

By carefully adjusting the pole and zero locations as well as the gain K of the compensator, we can design a controller that can satisfy both the relative stability requirements as well as steady-state error requirements.

Let us now illustrate how placing the poles and zeroes of the controller affects the closed loop behavior, which in turn can be used to get guidelines on selecting the gain K.

Figure 9.16 Lag compensator poles and zeroes.

Figure 9.17 Lag network.

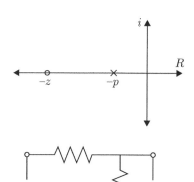

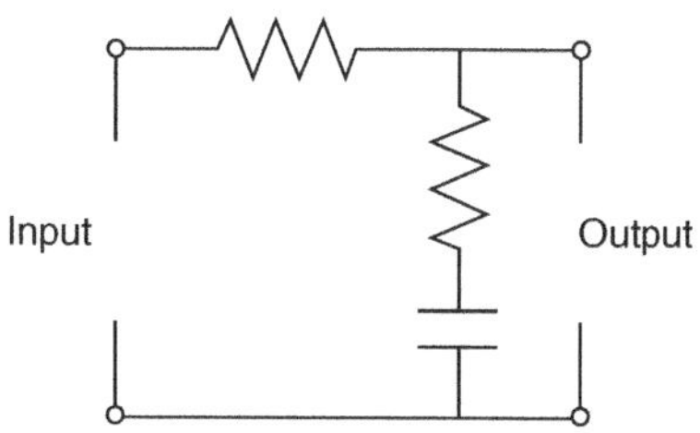

9.4.3 Relative Stability: Need for Derivative Controllers

In general, to improve the relative stability of the closed loop control system, we typically employ derivative (PD) controllers. By relative stability, we mean improving the damping ratio and natural frequency specifications or simply stabilizing an unstable (open loop) system. Let us illustrate this by an example.

Suppose the open loop transfer function of a system is $\frac{K_1}{s^2+4}$. Naturally, this is an undamped second order system and is neutrally stable. If we employ only a proportional controller, we get Figure 9.18; the closed loop transfer function is

$$\frac{Y(s)}{R(s)} = \frac{K_1 K_a}{s^2 + (4 + K_a K_1)}.$$

So whatever gain K you use, you are not changing the neutral stability character of the control system. If instead we use a proportional derivative controller (see Figure 9.19), we have, say, $G_C(s) = K_a(s+5)$, then the closed loop transfer function is

$$\frac{Y(s)}{R(s)} = \frac{K_1 K_a s + 4 + 5 K_a K_1}{(s^2 + K_1 K_a s + 4 + 5 K_a K_1)}.$$

Thus by varying the gains K_a and K_1, we can stabilize the closed loop system.

However, by improving the relative stability characterization, we may have worsened the steady state error response characterization. That is, there is typically a trade-off between relative stability and the steady-state error response! So in general we should be careful as to how much gain we use in the PD controller!

9.4.4 Steady-State Error Response: Need for Integral Controllers

Figure 9.20 shows a closed-loop block diagram. We have seen that

$$\frac{E(s)}{R(s)} = \frac{1}{1 + G_C(s)G_P(s)}.$$

Suppose that

$$G_P(s) = \frac{1}{(s+1)(s+2)} \tag{9.10}$$

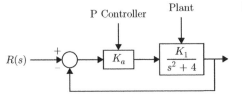

Figure 9.18 System with proportional controller.

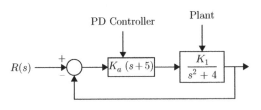

Figure 9.19 System with proportional derivative controller.

Figure 9.20 Closed-loop block diagram.

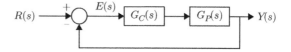

and

$$G_C(s) = K. \tag{9.11}$$

That is, G_C is a proportional controller. Then

$$
\begin{aligned}
\frac{E(s)}{R(s)} &= \frac{1}{1 + \frac{K}{(s+1)(s+2)}} \\
&= \frac{(s+1)(s+2)}{s^2 + 3s + 2 + K}.
\end{aligned} \tag{9.12}
$$

Suppose the input $r(t)$ is a step input, i.e. $R(s) = \frac{1}{s}$. Then the steady-state error due to a step input is

$$
\begin{aligned}
\lim_{s \to 0} sE(s) &= \lim_{s \to 0} \frac{s[(s+1)(s+2)]}{s[s^2 + 3s + 2 + K]} \\
&= \frac{2}{2 + K}.
\end{aligned}
$$

Thus there is a finite steady state error due to a step input. You may want to make this steady state zero. For that we employ integral controllers.

So, in general, we can say that one way to improve the steady state error response (i.e. reducing the steady state errors) is to increase the type of the system's open loop transfer function. Incidentally, the type of an open loop transfer function is simply the number of poles at the origin. Thus if we have no poles at the origin, we call it a type 0 system; if we have one pole at the origin (i.e. an s in the denominator), then it is a type 1 system and so on. This in turn means that we can introduce integration into the controller, i.e. integral control action can be used in general to reduce steady state errors.

The proportional controller in Figure 9.21 gives a finite steady state error for a step input (type 0).

The integral controller in Figure 9.22 gives zero steady state error for a step input (now it is a type 1 system). However, in the integral controller case the gain K has to

Figure 9.21 Proportional controller example.

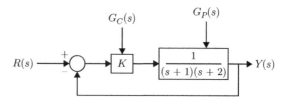

Figure 9.22 Integral controller example.

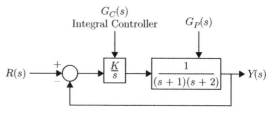

be selected carefully because, while increasing the type of the system, we may make the closed loop system unstable if we do not select the gain K carefully. Compare the closed loop characteristic equations of the systems shown in Figures 9.21 and 9.22 and see, if for the second case, you can find a value of gain K that keeps the closed loop system stable.

9.4.5 Basic Philosophy in Transfer Function Based Control Design Methods

It turns out that basically the control design philosophy in the classical transfer function based methods is to essentially iterate the selection of these gains associated with a PID (or lead-lag) controller transfer functions until a satisfactory response is achieved that meets the design specifications. Thus, philosophically, these design methods are somewhat trial and error based methods. However, the design procedure is conceptually simple and easily understood. It is important to realize that this simple PID control design methodology turns out to be quite effective in many real world applications and is thus deemed quite useful and powerful. Notice that conceptually the PID design methodology embodies the past, present, and future features in the controller, where the proportional (P) gain represents the present, the integral (PI) gain represents the past (looking past) and the derivative (PD) gain represents the future (looking ahead). Thus it is not surprising that they perform very well in achieving the desired closed loop system behavior.

9.5 Chapter Summary

In this chapter, we have presented the main ideas of representing a control system in a block diagram format and learnt a few techniques of block diagram algebra by which we can algebraically manipulate these various transfer functions so that we can obtain a transfer function between any given output variable of interest and a given input variable of interest. We then discussed various controller structures such as a proportional (P), integral(I) and derivative (D) actions, and combinations thereof, labeled PID controllers, and understood in what circumstances these controller structures are used. We then learnt about the corresponding practical versions of them in the form of a lead network (as a more practical transfer function representing PD control action), a lag network (as a more practical transfer function representing PI control action), and a lead/lag network (as a more practical transfer function representing PID control action). We have also presented few important time domain performance specifications such as rise time, settling time and percentage overshoot, etc., so that they can be used as design specifications to be met by the control system so that the actual output behaves as close to the desired output behavior as possible. Since the majority of the practical control systems follow this philosophy, the contents of this chapter are highly important and useful. Fundamental concepts discussed in this chapter can be found in other textbooks dedicated to control systems such as [1–5].

9.6 Exercises

Exercise 9.1. Given that a non-unity feedback system, shown in Figure 9.23

Figure 9.23 Non-unity feedback system.

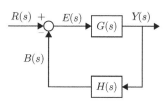

can always be converted to a unity feedback system, as shown in Figure 9.24,

Figure 9.24 Unity feedback system.

get the $G_{eq}(s)$ expression in terms of $G(s)$ and $H(s)$.

Exercise 9.2. Reduce the block diagrams shown in Figures 9.25 and 9.26, finally obtaining the overall system transfer function $Y(s)/R(s)$.

Figure 9.25

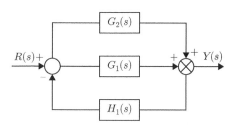

Figure 9.26

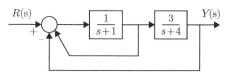

Exercise 9.3. Find the solution of the differential equation:

$$\frac{d^2x}{dt^2} + \frac{dx}{dt} + 8x = \frac{dz}{dt} + 3z.$$

Exercise 9.4. Find the closed loop transfer function $\frac{Y(s)}{R(s)}$ for the block diagram shown in Figure 9.27.

Exercise 9.5. Find the closed loop transfer function $Y(s)/R(s)$ for the block diagram shown in Figure 9.28.

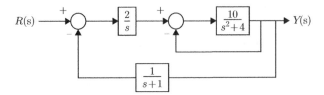

Figure 9.27

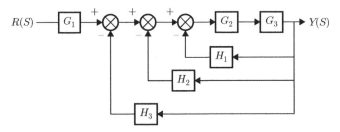

Figure 9.28

Exercise 9.6. Find a controller $G(s)$ such that the overall system of Figure 8.3 is second order and critically damped. A solution to this problem is not unique.

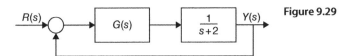

Figure 9.29

Bibliography

1 J.J. d'Azzo and C.D. Houpis. *Linear control system analysis and design: conventional and modern.* McGraw-Hill, New York, 1988.

2 J.J. d'Azzo and C.H. Houpis. *Linear control system analysis and design: conventional and modern.* McGraw-Hill Higher Education, New York, 1988.

3 C.R. Dorf and R.H. Bishop. *Modern control systems.* Pearson, 12 edition, 2011.

4 G. Hostetter, C. Savant Jr, and R. Stefani. *Design of Feedback Control Systems.* Holt, Rinehart and Winston, 1982.

5 N. Nise. *Control Systems Engineering.* Wiley, 2011.

10

Stability Testing of Polynomials

10.1 Chapter Highlights

In this chapter we introduce techniques to determine if a real constant coefficient polynomial has roots with negative real parts or not. As seen from the previous chapter, for a linear time invariant system, the transfer function between a desired output variable and an input variable has a numerator polynomial and a denominator polynomial. For a known given input in the Laplace domain, then the corresponding output variable's time history depends on the pole behavior of that output function, which in turn implies that the output response depends on the roots of the denominator polynomial of that output function (in the Laplace domain). If the poles are stable, meaning that the roots of the denominator polynomial have negative real parts, then the time histories of the output variable converge to a steady state value, making the output have stable behavior. Hence we need a procedure to determine the roots of an arbitrary degree polynomial with real coefficients. In this chapter, we present a popular and important technique known as the Routh–Hurwitz criterion. The attractive feature of this technique is that we can infer the stability of the polynomial without actually solving for all the roots per se. Instead it quickly gives an answer as to whether all the roots have negative real parts or not, just by a few algebraic manipulations with the (real) coefficients of the polynomial. Rigorously speaking we need to make a distinction between a polynomial with real coefficients, which naturally occurs in either the numerator polynomial or a denominator polynomial in a transfer function in the Laplace domain, and another polynomial with real coefficients, which is widely known as the characteristic polynomial of a matrix. The characteristic polynomial of a matrix is obtained through the equation $\text{Det}(\lambda I - A) = 0$, which is an nth degree polynomial for a matrix of order n. Even though this distinction becomes more important in a later part of the book wherein we are interested in the stability of a matrix, i.e. to know whether the eigenvalues of a matrix have negative real parts or not. Note that the eigenvalues of a real matrix are nothing but the roots of the characteristic polynomial of the matrix A. For now, in this chapter, we do not make any distinction between these two types of polynomials and simply consider the problem of deciding whether an nth degree polynomial with real coefficients has negative real part roots or not.

Flight Dynamics and Control of Aero and Space Vehicles, First Edition. Rama K. Yedavalli.
© 2020 John Wiley & Sons Ltd. Published 2020 by John Wiley & Sons Ltd.

10.2 Coefficient Tests for Stability: Routh–Hurwitz Criterion

For first- and second-order polynomials, stability can be easily determined by inspection of the polynomial. First and second order polynomials are stable if all coefficients in the polynomial are non-zero and positive. For example, the polynomials

$$s + 7 = 0$$

and

$$3s^2 + s + 10 = 0$$

are stable polynomials. On the other hand, the polynomial

$$3s^2 + s - 10 = 0$$

is unstable. For higher order polynomials, the positivity of the polynomial coefficients by themselves is only a necessary condition for stability but not sufficient. A polynomial of with all roots in the left side of the complex plane (LHP) has factors of the form $(s + a)$, with $a > 0$. Thus the necessity of the positivity of the coefficients is easy to establish. When multiplied out, the polynomial must have all coefficients of the same algebraic sign (in case, the coefficient of the highest degree is not positive to start with). Also, no coefficient can be zero or missing for stability. For instance, we can tell by inspection the characteristic polynomial $s^2 + bs + c = 0$ will have two (possibly complex-conjugate) LHP roots, so long as $b > 0$ and $c > 0$.

However, for higher order ($n \geq 3$) polynomials, to get a necessary and sufficient condition for the stability of the polynomial we need to resort to a more sophisticated procedure. This test is widely known as the Routh–Hurwitz criterion. We now learn about that simple test.

10.2.1 Stability of Polynomials with Real Coefficients via Polynomial Coefficient Testing: The Routh–Hurwitz Criterion

The Routh–Hurwitz test is a powerful numerical and analytical procedure to determine how many roots of a polynomial are in the RHP and how many are on the imaginary axis. It does not give exact roots of the polynomial, but is usually far quicker than factoring the polynomials. Recall that any given polynomial expression, such as

$$s^5 + s^4 + 3s^3 + 9s^2 + 16s + 10 = 0$$

can be generalized as such

$$a_0 s^n + a_1 s^{n-1} + a_2 s^{n-2} + a_3 s^{n-3} + \cdots + a_{n-1} s + a_n = 0$$

where we assume that the leading coefficient a_0 is positive. A polynomial with the leading coefficient as unity is called a monic polynomial.

As mentioned earlier, a necessary condition for the above polynomial to have negative real part roots is that all the coefficients be positive. Hence, if any of the coefficients are zero or negative, we can easily conclude that the polynomial is unstable.

To start the Routh–Hurwitz test, we then assume that all the coefficients are positive. Then we write the well known Routh–Hurwitz (R–H) table in the following pattern, where we write, in descending powers of s, starting with the highest power of the polynomial, s^n though s^0 to the left of the table. Then we enter the coefficients of the polynomial in the first two rows alternating with the odd degree coefficients in one row and the even in the other.

s^n	a_0	a_2	a_4	a_6	.	.	.	.	.
s^{n-1}	a_1	a_3	a_5	a_7	.	.	.	.	.
s^{n-2}	b_1	b_2	b_3	b_4	.	.	.	.	
s^{n-3}	c_1	c_2	c_3	.	.	.	.		
$\vdots s^2$	f_1	f_2							
s^1	g_1	g_2							
s^0	h_1								

where the entries in the third row and onwards are computed as follows.

s^n	a_0	a_2	a_4	a_6	.	.	.	.	.
s^{n-1}	a_1	a_3	a_5	a_7	.	.	.	.	.
s^{n-2}									
s^2									
s^1									
s^0									

The Routh–Hurwitz test involves filling in the rest of the table in a certain way. The number that will be in the circle above the formula in the third row is denoted by b_1, which is computed as follows:

$$\frac{-\begin{vmatrix} a_0 & a_2 \\ a_1 & a_3 \end{vmatrix}}{a_1} = b_1.$$

The rest of the entries in the third row are denoted by b_i ($i = 2, 3, \ldots$) and are computed as follows:

$$\frac{-\begin{vmatrix} a_0 & a_4 \\ a_1 & a_5 \end{vmatrix}}{a_1} = b_2.$$

The next entries in the fourth row, denoted by c_i are computed as follows:

$$\frac{-\begin{vmatrix} a_1 & a_3 \\ b_1 & b_2 \end{vmatrix}}{b_1} = c_1.$$

The rest of the entries in the fourth row are denoted by c_i ($i = 2, 3, \ldots$) and are computed as follows:

$$\frac{-\begin{vmatrix} a_1 & a_5 \\ b_1 & b_3 \end{vmatrix}}{b_1} = c_2.$$

This procedure is continued until all the rows until s^0 are filled. Then the necessary and sufficient condition for the roots of the characteristic polynomial to have negative real parts and thus be Hurwitz stable is that all the entries of the first column of the R–H table be all positive.

Example 10.1   Let us now illustrate this procedure with a simple example. Consider the same characteristic polynomial mentioned before, namely,

$$s^5 + s^4 + 3s^3 + 9s^2 + 16s + 10 = 0.$$

Let us form the R–H table for this polynomial.

s^5	1	3	16
s^4	1	9	10
s^3	-6	6	
s^2			
s^1			
s^0			

$$b_1 = \dfrac{-\begin{vmatrix} 1 & 3 \\ 1 & 9 \end{vmatrix}}{1} = -6.$$

Continuing the process

$$b_2 = \dfrac{-\begin{vmatrix} 1 & 16 \\ 1 & 10 \end{vmatrix}}{1} = \dfrac{-\begin{vmatrix} 1 & 16 \\ 1 & 10 \end{vmatrix}}{1} = 6.$$

The next entries in the s^2 row are

$$c_1 = \dfrac{-\begin{vmatrix} 1 & 9 \\ -6 & 6 \end{vmatrix}}{-6} = 9$$

and further entries would be zeroes.

s^5	1	3	16
s^4	1	9	10
s^3	-6	6	
s^2	9	10	
s^1			
s^0			

Continuing this process would finally complete the R–H table, which now looks like:

s^5	1	3	16
s^4	1	9	10
s^3	-6	6	
s^2	9	10	
s^1	114/9		
s^0	10		

The number of RHP roots of the polynomial is equal to the number of algebraic sign changes in the first (left most) column of the table, starting from top to bottom. This example has two RHP roots because the sign changes twice, once from positive one to negative six and then back to positive nine. Hence this polynomial (and the matrix that gives rise to this characteristic polynomial) is unstable. We could have concluded that the polynomial is unstable at the time we obtained the b_1 coefficient as −6, but in this example we continued further to complete the entire R–H table to simply know how many RHP roots are there in the poynomial because that is actually what the objective of the RH criterion is.

Example 10.2 Let us now illustrate this procedure with another simple example. Consider the characteristic polynomial mentioned before, namely,

$$s^5 + s^4 + 3s^3 + 9s^2 + 16s + 10 = 0.$$

Let us form the R–H table for this polynomial.

s^5	1	12	16
s^4	1	9	10
s^3	3	6	
s^2			
s^1			
s^0			

$$b_1 = \frac{-\begin{vmatrix} 1 & 12 \\ 1 & 9 \end{vmatrix}}{1} = 3.$$

Continuing the process

$$b_2 = \frac{-\begin{vmatrix} 1 & 16 \\ 1 & 10 \end{vmatrix}}{1} = \frac{-\begin{vmatrix} 1 & 16 \\ 1 & 10 \end{vmatrix}}{1} = 6.$$

The next entries in the s^2 row are

$$c_1 = \frac{-\begin{vmatrix} 1 & 9 \\ 3 & 6 \end{vmatrix}}{3} = 7$$

and further entries would be zeroes. Completing the R–H table would result in

s^5	1	3	16
s^4	1	9	10
s^3	3	6	
s^2	7	10	
s^1			
s^0			

Continuing this process would give you:

s^5	1	12	16
s^4	1	9	10
s^3	3	6	
s^2	7	10	
s^1	12/7		
s^0	10		

In this example, the first column entries of the R–H table are all positive (no sign changes). Hence this polynomial is Hurwitz stable.

Notice that the stability of polynomial is quite sensitive to the numerical values of the coefficients of the polynomial.

Sometimes, based on the numerical values of the coefficients, we may run into some unusual situations that may prevent us from completing the R–H table. We now discuss few of those situations.

10.3 Left Column Zeros of the Array

Consider the polynomial:

$$s^4 + s^3 + 2s^2 + 2s + 3.$$

A snag develops in the Routh–Hurwitz test:

$$
\begin{array}{c|ccc}
s^4 & 1 & 2 & 3 \\
s^3 & 1 & 2 \\
s^2 & 0 & 3 \\
s^1 & & \\
s^0 & &
\end{array}
$$

The first entry in the row corresponding to s^1 will be

$$
b_1 = \frac{-\begin{vmatrix} 1 & 2 \\ 0 & 3 \end{vmatrix}}{0}
$$

which involves division by zero. This situation can be resolved by replacing the left zero by a small positive number, denoted by ϵ and continuing the process:

$$
\begin{array}{c|ccc}
s^4 & 1 & 2 & 3 \\
s^3 & 1 & 2 \\
s^2 & \epsilon & 3 \\
s^1 & \dfrac{2\epsilon - 3}{\epsilon} & \\
s^0 & 3 &
\end{array}
$$

Since,

$$
\lim_{\epsilon \to 0} 2 - \frac{3}{\epsilon} = -\infty
$$

$$
\begin{array}{c|c}
s^4 & 1 \\
s^3 & 1 \\
s^2 & 0^+ \\
s^1 & -\infty \\
s^0 & 3
\end{array}
$$

Since there are sign changes in the first column, this polynomial is unstable.

An alternative method for circumventing left column zeros is to introduce additional known roots to the polynomial, thereby increasing its order, and that would allow us to change the coefficients so that the left column zero does not occur. Another method is to substitute $\left(\frac{1}{p}\right)$ for s and get a new polynomial in p and then complete the test with that new polynomial. The exercises at the end of the chapter will help the student to illustrate these approaches.

10.4 Imaginary Axis Roots

There are three basic types of factors possible in an even degree polynomial.

$$(s + ja)(s - ja) = (s^2 + a^2)$$
$$(s + a)(s - a) = (s^2 - a^2)$$
$$(s + a + jb)(s + a - jb)(s - a + jb)(s - a - jb) = s^4 + 2(b^2 - a^2)s^2 + (a^2 + b^2)^2.$$

For the polynomial

$$s^4 + s^3 + 5s^2 + 3s + 6$$

s^4	1	5	6
s^3	1	3	
s^2	2	6	
s^1	0	0	

When the entire row has zeroes, it indicates the possible presence of pure imaginary axis (i.e zero real part) roots. We then form the so-called auxiliary polynomial (or divisor polynomial) corresponding to the row above the row with all zeroes. In the above, the auxiliary (or divisor) polynomial is given by

$$2s^2 + 6 \quad \text{(divisor of the original polynomial)}.$$

We then replace the row of zeroes by the coefficients of the derivatives of the divisor polynomial,

$$\frac{d}{ds}(2s^2 + 6) = 4s$$

and then complete the R–H table. For this example, the R–H table is then:

s^4	1	5	6
s^3	1	3	
s^2	2	6	
s^1	4	8	
s^0	6		

Since there are no sign changes, there are no RHP roots. However, there are two roots on the imaginary axis (roots with zero real parts). This tells us that the above polynomial is unstable due to the presence of roots on the imaginary axis, but with no RHP roots.

We now present an application in control systems where the R–H criterion can play significant role in the design of controllers. The following example illustrates this application.

10.5 Adjustable Systems

We can also use the R–H criterion to determine the range of values for stability for an adjustable parameter within a given characteristic equation. Let us illustrate this procedure with a simple example.

Example 10.3 Find the range or ranges of an adjustable parameter K for which the closed loop of a unity feedback system whose open loop transfer function $T(s)$ given by

$$T(s) = \frac{2s + K}{s^2(s^2 + 2s + 4)}$$

is stable.

Solution
Forming the closed loop transfer function $(1 + T(s)) = 0$, we obtain the closed loop characteristic polynomial as

$$s^4 + 2s^3 + 4s^2 + 2s + K = 0.$$

From inspection of the closed loop characteristic polynomial, we can already observe that $K > 0$ is a necessary condition for stability. Completing the Routh–Hurwitz test in terms of K will help us to get the necessary and sufficient condition for stability. So we form the R–H table as such

$$
\begin{array}{c|ccc}
s^4 & 1 & 4 & K \\
s^3 & 2 & 2 & \\
s^2 & 3 & K & \\
s^1 & \dfrac{6 - 2K}{3} & & \\
s^0 & K & &
\end{array}
$$

We know there must be no sign changes to ensure zero RHP roots. Therefore, from the last two lines of the R–H table, we have our upper and lower bounds for K

- $K > 0$ and
- $\frac{6-2K}{3} > 0$.

Solving the second inequality, we obtain the the the upper bound $K < 3$. Therefore, we see $0 < K < 3$ is the acceptable range of the adjustable parameter to ensure stability of the closed loop system.

Thus, it can be seen that the R–H criterion can be effectively used to make some useful design decisions in control systems.

10.6 Chapter Summary

In this chapter, we presented a highly popular and important technique, known as the Routh–Hurwitz criterion, for determining whether a given a polynomial of arbitrary degree with real coefficients has all its roots with negative real parts or not, without actually solving for the roots of that polynomial. This simple coefficient test is a quick means to assess whether the given transfer function with its denominator polynomial known is stable or not. The stability issue for a linear control system in Laplace domain is a very important first step in designing control systems in later chapters for achieving or meeting various other performance specifications. Naturally, the foremost concern in designing a control system would be the stability of the closed loop system before we embark on the task of improving its performance to meet all the other design specifications.As such the content of this chapter is very important and useful. Fundamental concepts discussed in this chapter can be found in other textbooks dedicated to control systems such as [1–5].

10.7 Exercises

Exercise 10.1. Examine the stability of the polynomial

$$s^3 + 6s^2 + 11s + 6 = 0.$$

Exercise 10.2. Investigate the stability of the polynomial

$$s^4 + 6s^3 + 26s^2 + 56s + 80 = 0.$$

Exercise 10.3. Investigate the stability of the polynomial

$$s^4 + 10s^3 + 35s^2 + 50s + 24 = 0.$$

Exercise 10.4. Investigate the stability of the polynomial

$$s^5 + 2s^4 + 3s^3 + 6s^2 + 2s + 1 = 0.$$

Exercise 10.5. The open loop transfer function of a unity feedback system is given by

$$G(s) = \frac{K}{s(1 + sT_1)(1 + sT_2)}.$$

Derive an expression for gain K in terms of T_1 and T_2 for the closed loop system to be stable.

Exercise 10.6. Find the range(s) of the adjustable parameter $K > 0$ for which the systems of Figure 10.1 and 10.2 are stable.

Figure 10.1

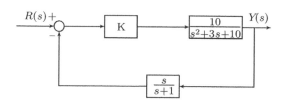

Figure 10.2

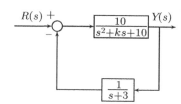

Bibliography

1 J.J. d'Azzo and C.D. Houpis. *Linear control system analysis and design: conventional and modern.* McGraw-Hill, New York, 1988.

2 J.J. d'Azzo and C.H. Houpis. *Linear control system analysis and design: conventional and modern.* McGraw-Hill Higher Education, New York, 1988.

3 C.R. Dorf and R.H. Bishop. *Modern control systems.* Pearson, 12 edition, 2011.

4 G. Hostetter, C. Savant Jr, and R. Stefani. *Design of Feedback Control Systems.* Holt, Rinehart and Winston, 1982.

5 N. Nise. *Control Systems Engineering.* Wiley, 2011.

11

Root Locus Technique for Control Systems Analysis and Design

11.1 Chapter Highlights

In this chapter, we present an elegant, graphical technique called the root locus technique, so that we can visualize the closed loop poles as the gain K is varied using only the open loop transfer function information. Thus, the root locus is the locus of the roots of the closed loop characteristic polynomial as the gain K is varied but without actually solving for the roots of the closed loop characteristic polynomial. This graphical technique basically shows the locus of the closed loop poles as a function of the gain K by using the pole/zero information of the open loop transfer function. So by drawing the root locus based on the open loop transfer function, we can visualize what the closed loop poles are as the gain K is varied. Root locus is a powerful control systems analysis and design method that is widely used in many engineering system applications. In this chapter, we present the rules by which we can quickly and approximately construct the root locus and then use the drawn root locus to design an appropriate controller transfer function along with the gain K to meet the various closed loop design specifications in terms of the desired pole locations.

11.2 Introduction

Consider the open loop transfer function, defined as $KG(s)H(s)$, where we can usually easily determine the poles. Note that they do not vary with the gain K. However, determining the poles of the closed loop transfer function given by

$$T(s) = \frac{KG(s)}{1 + KG(s)H(s)} \tag{11.1}$$

can be difficult because we would have to factor the denominator, which becomes a function of the gain K. Thus as the gain K changes the roots of the closed loop characteristic polynomial (namely $1 + KG(s)H(s) = 0$, which in turn are the poles of the closed loop transfer function $T(s)$) keeps changing. Clearly, it would be extremely helpful if we graphically visualize where those closed loop poles reside in the complex plane as gain K is varied. This is what the root locus is, namely the locus of the closed loop poles in the complex plane as the gain K is varied. The most attractive feature of root locus technique is that this locus can be drawn by simply using all the known information about the open loop transfer function without actually being concerned about building the actual closed

Flight Dynamics and Control of Aero and Space Vehicles, First Edition. Rama K. Yedavalli.
© 2020 John Wiley & Sons Ltd. Published 2020 by John Wiley & Sons Ltd.

loop transfer function itself. Thus one of the objectives of this chapter is to learn how to draw the root locus quickly, albeit approximately, so that we can make quick conclusions and decisions about what gain K needs to be used in the control design so that the closed loop system meets the design specifications. Naturally, the root locus technique is ideally suited for cases where the design specifications are in terms of desired closed loop pole locations. To get deeper insight into the behavior of closed loop poles as the gain K changes, let us write the above closed loop transfer function in a more detailed way as follows:

$$G(s) = \frac{N_G(s)}{D_G(s)}$$

$$H(s) = \frac{N_H(s)}{D_H(s)}.$$

Then the closed-loop transfer function becomes

$$T(s) = \frac{KN_G(s)D_H(s)}{D_G(s)D_H(s) + KN_G(s)N_H(s)}. \tag{11.2}$$

Notice that when the gain K is zero, the closed loop poles are nothing but the poles of the open loop transfer function. On the other extreme, as gain K tends to ∞, the closed loop poles reach towards the open loop zeroes. So we already obtain one rule for drawing the root locus approximately, namely that the root locus starts from the poles of the open loop transfer function and ends at the zeroes of the open loop transfer function. Note that, any complex number, $\sigma + jw$ described in Cartesian coordinates can be graphically represented by a vector. The function defines the complex arithmetic to be performed in order to evaluate $F(s)$ at any point, s. Thus it can be seen that

$$M = \frac{\prod \text{zero length}}{\prod \text{pole lengths}} = \frac{\prod_{i=1}^{m} |(s + z_i)|}{\prod_{j=1}^{n} |(s + p_j)|}. \tag{11.3}$$

The angle, θ, at any point, s, in the complex plane, is

$$\theta = \sum \text{zero angles} - \sum \text{pole angles} \tag{11.4}$$

$$= \sum_{i=1}^{m} \angle(s + z_i) - \sum_{j=1}^{n} \angle(s + p_j). \tag{11.5}$$

We now use the above conceptual observations to help us determine the conditions that need to be satisfied for any point s in the complex plane to be qualified as a point on the root locus. Thus there are few rules to be followed for us to build the root locus. In what follows, we elaborate on these properties to be obeyed by the points on the root locus in an orderly fashion.

11.3 Properties of the Root Locus

The closed loop transfer function is

$$T(s) = \frac{KG(s)}{1 + KG(s)H(s)}. \tag{11.6}$$

A closed loop pole is that point s in the complex plane that satisfies the following constraint, namely that at that point the closed loop characteristic polynomial (i.e. the polynomial in the denominator) becomes zero, or

$$KG(s)H(s) = -1 = 1\angle(2k+1)180° \quad k = 0, \pm 1, \pm 2, \pm 3, \dots. \tag{11.7}$$

A value of s is a closed-loop pole if

$$|KG(s)H(s)| = 1 \tag{11.8}$$

and

$$\angle KG(s)H(s) = (2k+1)180° \tag{11.9}$$

where K is

$$K = \frac{1}{|G(s)H(s)|}. \tag{11.10}$$

The value of gain is evaluated using

$$K = \frac{1}{|G(s)H(s)|} = \frac{1}{M} = \frac{\prod \text{pole lengths}}{\prod \text{zero length}}. \tag{11.11}$$

Let us demonstrate this relationship for the second order system.

Example 11.1   If we have an open-loop transfer function of $KG(s)H(s) = \frac{K}{s(s+20)}$, determine the nature of the closed loop poles.

Solution
Because this is a second order system factoring the denominator is not that difficult. So to convey the main concept of root locus drawing, let us use simple standard way of getting the actual closed loop poles. Let us use Equation 11.2 to find the closed-loop poles.

$$
\begin{aligned}
T(s) &= \frac{KN_G(s)D_H(s)}{D_G(s)D_H(s) + KN_G(s)N_H(s)} \\
&= \frac{K}{s(s+20) + K} \tag{11.12} \\
&= \frac{K}{s^2 + 20s + K}.
\end{aligned}
$$

Looking at the denominator we can see that the closed-loop poles depend on the gain K, so as K is varied, so do the locations of the closed loop poles. Varying the gain K we can generate Table 11.1. If we graph the table, which can be seen in Figure 11.2, we can see a pattern emerging. When the gain is less than 100, the system is overdamped. At 100, the system is critically damped and above that it is underdamped.

The thought process outlined in the above example is the basis for the root locus. If we drew a line connecting the different poles for each possible gain K, we would have a continuous locus for all the closed loop poles as a function of the gain K. That is exactly what we mean by root locus. For a system with the open loop transfer function given by

$$KG(s)H(s) = \frac{K}{s(s+20)} \tag{11.13}$$

the root locus is finally shown in Figure 11.1.

Table 11.1 Closed-loop pole location as a function of the gain K.

Gain K	Pole 1	Pole 2
0	0	−20
20	−18.94	−1.05
40	−17.74	−2.25
60	−16.32	−3.67
80	−14.47	−5.52
100	−10	−10
120	−10 + j 4.47	−10 − j 4.47
140	−10 + j 6.32	−10 − j 6.32
160	−10 + j 7.74	−10 − j 7.74
180	−10 + j 8.94	−10 − j 8.94
200	−10 + j 10	−10 − j 10

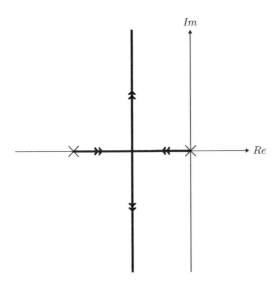

Figure 11.1 Root locus for the transfer function given in Equation 11.13.

Figure 11.2 depicts the gain at each point on the root locus. Notice that getting the closed loop pole locations at every gain K was relatively easy because this was only a second order system. However, for higher order systems this would become increasingly difficult and impractical. Which is why determining the root locus for a more general higher order open loop transfer function requires a far more sophisticated procedure, which was developed by Evans using some fundamental simple logical

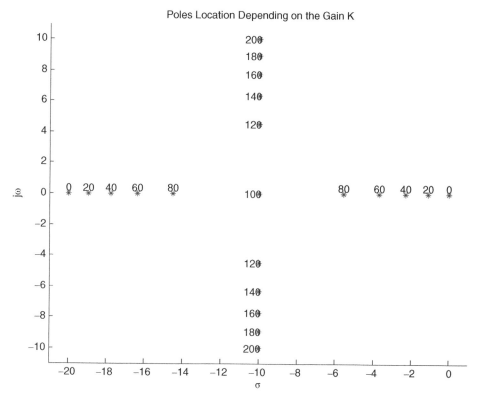

Figure 11.2 MATLAB poles graphed on the complex plane for the transfer function given in Equation 11.13.

arguments. This attractive graphical technique of drawing the root locus quickly and approximately (to serve the design purpose) is possible by following some rules of root locus construction devised by Evans. Thus Evans can be regarded as the father of the root locus. In what follows, we briefly summarize these rules of root locus construction laid out by Evans [4, 5]. Interestingly, Evans developed this root locus technique while working in the aircraft industry. In this new millennium with advanced use of computer technology, it is true that we can get the actual and exact root locus in a relatively easier way compared to the time when Evans initiated this process. What is important to realize is that the ability to draw the root locus quickly and approximately the Evans way can never be underestimated because this ability to draw root locus following these rules can still be used to verify and confirm the correctness of the exact root locus because, after all, in this computer age, it is still possible to enter the data in an erroneous way, and not know about it. Thus the danger of garbage in, garbage out applies very well to any computer generated code or procedure. Thus learning about the Evans root locus technique is still very important and beneficial for a student mastering control systems even in this computer age. With that understood, we now briefly summarize the rules.

11.4 Sketching the Root Locus

The following five rules are used in sketching the root locus. The rules yield a sketch that gives intuitive insight into the behavior of a control system with minimal calculations.

1. Number of branches. The number of branches of the root locus equals the number of closed-loop poles.
2. Symmetry. The root locus is symmetrical about the real axis.
3. Real axis segments. On the real axis, for $K > 0$ the root locus exists to the left of an odd number of real axis, finite open-loop poles and/or finite open-loop zeroes.
4. Starting and ending points. The root locus begins at the finite poles of $G(s)H(s)$ and ends at the finite and infinite zeros of $G(s)H(s)$.
5. Behavior at infinity. The root locus approaches straight line asymptotes as the locus approaches infinity. Further, the equation of the asymptotes is given by the real axis intercept, σ_a, and angle, θ_a, as follows:

$$\sigma_a = \frac{\sum \text{finite poles} - \sum \text{finite zeros}}{\#\text{finite poles} - \#\text{finite zeros}} \tag{11.14}$$

$$\theta_a = \frac{(2k+1)\pi}{\#\text{finite poles} - \#\text{finite zeros}} \tag{11.15}$$

where $k = 0, \pm1, \pm2, \pm3$ and the angle is given in radians with respect to the positive extension of the real axis.

Example 11.2 Sketch the root locus for the transfer function

$$T(s) = \frac{s+4}{(s+2)(s+6)(s+8)}.$$

Solution
First we will look at the number of branches. The transfer function has three finite poles, so it will have three branches. Next the starting points are the finite poles, so they will be at $(-2, 0), (-6, 0), (-8, 0)$. The end points are at the finite and infinite zeros, so there will be one finite ending point at $(-4, 0)$ and two end points at infinity. Now looking at the real axis segments, we know that the root locus exists left of an odd number of real axis, finite poles and/or finite zeros, so the root locus will be $(-2, 0)$ to $(-4, 0)$ and $(-6, 0)$ to $(-8, 0)$ on the real axis. Now that we know what it looks like on the real-axis we will look at its behavior at infinity. So we need to find the asymptote intercepts and angles.

$$\sigma_a = \frac{\sum \text{finite poles} - \sum \text{finite zeros}}{\#\text{finite poles} - \#\text{finite zeros}}$$

$$= \frac{[(-2) + (-6) + (-8)] - [-4]}{3 - 1}$$

$$= \frac{-12}{2}$$

$$= -6$$

$$\theta_a = \frac{(2k+1)\pi}{\#\text{finite poles} - \#\text{finite zeros}}$$

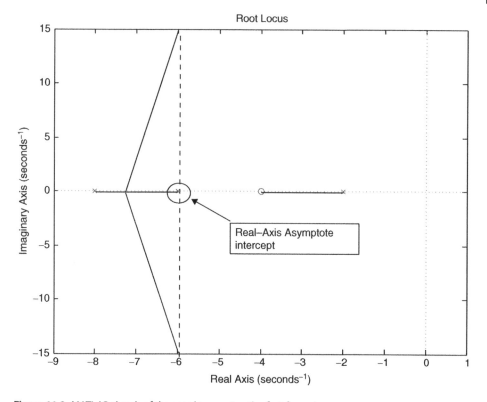

Figure 11.3 MATLAB sketch of the root locus using the first five rules.

$$= \frac{(2k + 1)\pi}{3 - 1}$$

$$= \frac{(2k + 1)\pi}{2}$$

$$= \frac{\pi}{2} \quad (\text{for} \quad k = 0) \quad \text{and} \quad \frac{3\pi}{2} \quad (\text{for} \quad k = 1).$$

Finally we know it will be symmetric about the real axis, so using everything we have calculated up to this point, Figure 11.3 shows what our sketch should look like.

The actual root locus can be see in Figure 11.4. With the rules in the next section we will be able to get our sketch a lot closer to the actual root locus.

11.5 Refining the Sketch

The rules in the previous section help generate the root locus sketch rapidly. Now if we want more details for a more accurate sketch we can further refine the sketch by few important points of interest and the gains associated with them. With that in mind, we now attempt to calculate the real axis breakaway and break-in points, the $j\omega$ axis crossing, angles of departure from complex poles, and angles of arrival to complex zeros.

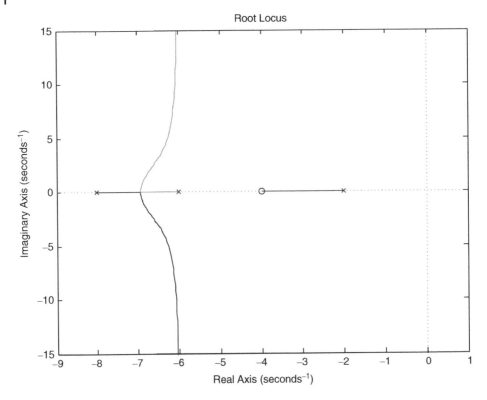

Figure 11.4 MATLAB root locus example.

11.5.1 Real Axis Breakaway and Break-In Points

The point where the root locus leaves the real axis, $-\sigma_1$ is called the breakaway point, and the point where the locus returns to the real axis, σ_2 is called the break-in point.

At the breakaway or break-in point, the branches of the root locus form an angle of $180°/n$ with the real axis, where n is the number of closed-loop poles arriving at or departing from the single breakaway or break-in point on the real axis One of the methods to find the points is to use differential calculus to find the maximum and minimum of the gain K.

$$K = -\frac{1}{G(s)H(s)}. \tag{11.16}$$

For points along the real axis where breakaway and break-in points could exist $s = \sigma$. Therefore the equation becomes

$$K = -\frac{1}{G(\sigma)H(\sigma)}. \tag{11.17}$$

So if we take the derivative of the equation and set it to zero we will find the maximum and minimum, which will give us the breakaway and break-in points.

A second way to find the breakaway and break-in points is a variation of the first method and is called the transition method. It eliminates the differentiation, giving us this relationship

$$\sum_{1}^{m} \frac{1}{\sigma + z_i} = \sum_{1}^{n} \frac{1}{\sigma + p_i}. \tag{11.18}$$

11.5.2 The $j\omega$ Axis Crossings

The $j\omega$ axis crossing is a point on the root locus that separates the stable operation of the system from the unstable operation. The value of ω at the axis crossing yields the frequency of oscillation, while the gain at the $j\omega$ axis crossing usually yields the maximum positive gain for the system to be stable.

To find the $j\omega$ axis crossing, it is possible to use the Routh–Hurwitz criterion in conjunction with the root locus procedure. Recall that we can infer that a row of zeroes in the R–H table indicates an imaginary axis crossing. Thus it is relatively easy to find the gain at which the $j\omega$ axis crossing occurs. Then going up one row and solving for the roots will yield the frequency at the $j\omega$ axis crossing.

11.5.3 Angles of Departure and Arrival

The root locus departs from complex, open-loop poles and arrives at complex, open-loop zeros. If we assume a point on the root locus ϵ close to a complex pole, the sum of angles drawn from all finite poles and zeros to this point will be an odd multiple of $180°$. Except for the pole ϵ that is close to the point under consideration, we assume all angles of all other poles and zeros are drawn directly to the pole near the point. Thus the only unknown angle is the one sum of the angle drawn from the pole close to ϵ. We can then solve for the unknown angle, which is also the angle of departure from the complex pole.

$$-\theta_1 + \theta_2 + \theta_3 - \theta_4 - \theta_5 - \theta_6 = (2k + 1)180° \tag{11.19}$$

or

$$\theta_1 = \theta_2 + \theta_3 - \theta_4 - \theta_5 - \theta_6 - (2k + 1)180°. \tag{11.20}$$

Using a similar process we can find the angle of arrival to a complex zero

$$\theta_2 = \theta_1 - \theta_3 + \theta_4 + \theta_5 - \theta_6 + (2k + 1)180°. \tag{11.21}$$

Example 11.3   Draw the root locus for the system with the open-loop transfer function

$$KG(s)H(s) = \frac{(s - 1)(s - 4)}{(s + 3)(3 + 6)}.$$

Solution

We will start again using the first five rules and the add the new rules we have just gone over. First we can see that there will be two branches because we have two closed-loop poles. Next looking at the starting and ending points we see that the two starting points are $(-3, 0)$ and $(-6, 0)$ and the ending points are $(1, 0)$ and $(4, 0)$. Now the real axis segments exist from $(-3, 0)$ to $(-6, 0)$ and from $(1, 0)$ to $(4, 0)$. Because all the zeros are finite, there is no need to calculate the behavior at infinity. Now we need to find out what the

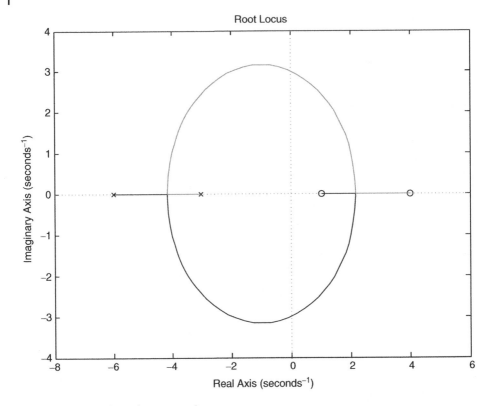

Figure 11.5 MATLAB root locus example.

root locus does between the two real axis segments, first we will calculate where they breakaway and break-in to the axis.

$$\sum_{1}^{m} \frac{1}{\sigma + z_i} = \sum_{1}^{n} \frac{1}{\sigma + p_i}$$

$$= \frac{1}{\sigma - 1} + \frac{1}{\sigma - 4}$$

$$= \frac{1}{\sigma + 3} + \frac{1}{\sigma + 6}.$$

Simplifying,

$$\sigma^2 + 2\sigma - 9 = 0.$$

Then the breakaway point is $\sigma_1 = -1 - \sqrt{10} = -4.16$ and the break-in point $\sigma_2 = -1 + \sqrt{10} = 2.16$. Now we have all the information we need to draw a good sketch of the root locus which can be found in Figure 11.5.

Now that we know how to draw the root locus, it is important to note that the root locus is very sensitive to the open loop pole and zero locations. Sometimes even the slightest changes in the open loop pole and zero locations can have a drastic impact on the look of the root locus, making it look very different. See these root loci in Figure 11.6.

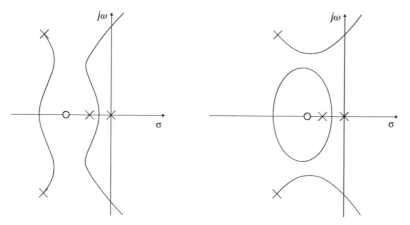

Figure 11.6 Differences in root locus.

Until now, we have simply learnt the importance of the root locus and the way to draw it quickly. However the major contribution of the root locus technique lies in its utility in control system design. Its powerful and attractive role can be realized when we use it for design purposes. Let us now elaborate on the use of root locus technique in a control design scenario.

11.6   Control Design using the Root Locus Technique

Suppose for a system the open loop transfer function is

$$KG(s)H(s) = \frac{K}{(s+1)(s+2)}.$$

The root locus for this transfer function is shown in Figure 11.7, which shows that the closed loop system is stable for all positive values of gain K. However, the system is type 0 and thus there is a steady state error due to a step input. Suppose you want to make the steady state error due to a step input zero. Then you would insert an integrator in the loop to make it a type 1 system. Then the new open loop transfer function is

$$KG(s)H(s) = \frac{K}{s(s+1)(s+2)}$$

i.e. we used an integral controller.

Figure 11.7 Root Locus plot for the above given open loop transfer function.

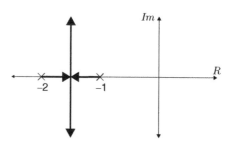

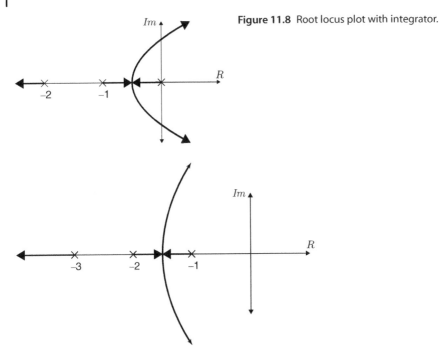

Figure 11.8 Root locus plot with integrator.

Figure 11.9 Root locus plot with a pole at $s = -3$.

If you now draw the root locus for this new system it looks like Figure 11.8. This tells us that with the integrator in there, the closed loop system is no longer stable for all values of K but is stable only for some specific range of K. Instead if we added a pole at, say, $s = -3$, i.e. $(s + 3)$ then the open loop transfer function would be

$$KG(s)H(s) = \frac{K}{(s + 1)(s + 2)(s + 3)}$$

and the root locus would be Figure 11.9. Thus the gain at which instability occurs would be much higher, i.e. we could have more gain range to use before instability occurs. Obviously the locations of the open loop poles and zeroes have a drastic influence on the shape of the root locus.

Suppose we added a zero to the open loop transfer function (i.e. used a proportional derivative controller). Then the open loop transfer function is

$$KG(s)H(s) = \frac{K(s + 3)}{(s + 1)(s + 2)}$$

and the root locus would be as in Figure 11.10. Now we see that the closed loop system is stable for all values of gain K. That is why we say derivative control (with appropriate zero location) improves relative stability, but does not do any improvement for steady state errors but integral control improves steady state error performance but deteriorates the relative stability. So from this discussion, it is clear that the gain, K, and the locations of the poles and zeroes of the controller have a direct influence on the root locus, which helps us to decide whether these controller parameters can be used for a satisfactory design. Note that the root locus is very sensitive to the numerical values

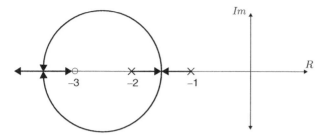

Figure 11.10 Root locus plot with proportional derivative.

for these controller parameters (namely the gain, pole, and zero locations) and thus one has to be cautious about the numerical values being used in the analysis and design procedures.

In the exercises at the end of this chapter, the reader will get a chance to design a controller that is required to satisfy different specifications and the root locus can be used to achieve these objectives.

11.7 Using MATLAB to Draw the Root Locus

Getting a very accurate, well defined root locus can be hard to draw by hand and can involve complex calculations. MATLAB is not only a great way to draw the root locus, but to also find the gain K at any point. There are only a few commands that one needs to know to draw the root locus.

1. To input the transfer function in MATLAB there are a few different commands.
   (a) The first way is if you have the equation simplified and you know all the zeros and poles of the open-loop transfer function, then using the command `sys = zpk(z,p,k)` will create the system in MATLAB. In the commands z and p are the vectors of real or complex poles and zeros, and k is the gain.
   (b) The second way is when you have the transfer function in a polynomial form `sys = tf(num,den)`, where num is the numerator vector and den is the denominator vector.
2. Now to get the root locus you can use the command `rlocus(sys)`.

Example 11.4 Draw the root locus of the system

$$KG(s)H(s) = \frac{(s+2)^2}{s^2(s+10)(s^2+6s+25)}.$$

Solution
First we will get it in a polynomial form so we can use both methods on it. Doing so, we get

$$KG(s)H(s) = \frac{s^2+4s+4}{s^5+16s^4+85s^3+250s^2}.$$

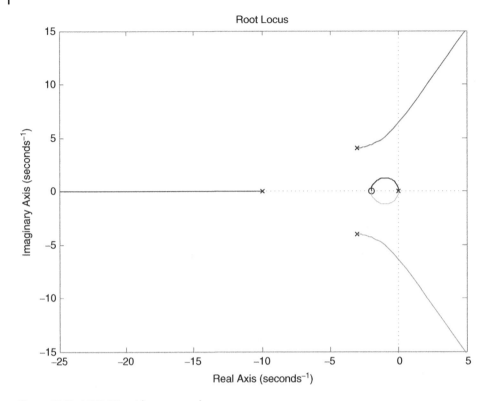

Figure 11.11 MATLAB root locus example.

As detailed above, there are two approaches to define a system in MATLAB:

- `sys = zpk(z,p,k)`, where `z=[-2 -2]`, `p=[ 0 0 -10 -3-4*i -3+4*i]`, and `k=[1]`
- `sys = tf(num,den)`, where `num=[1 4 4]` and `den=[1 16 85 250 0 0]`. Notice the two zeros in the `den` input field are the coefficients of the s^1 and s^0 terms.

Having now defined `sys` in MATLAB, we use the `rlocus(sys)` command, which outputs the root locus for the specified transfer function. The output is presented in Figure 11.11.

11.8 Chapter Summary

In this chapter, we have learnt the concept of the root locus and the rules that help us draw the root locus using the open loop transfer function information. We then applied this root locus technique in a design setting to synthesize control systems to meet various closed loop system performance specifications, mostly given in the form of desired closed loop pole locations. Fundamental concepts discussed in this chapter can be found in other textbooks dedicated to control systems such as [1–3, 6, 7].

11.9 Exercises

Exercise 11.1. Sketch the general shape of the root locus for each of the open-loop pole-zero plots shown Figure 11.12.

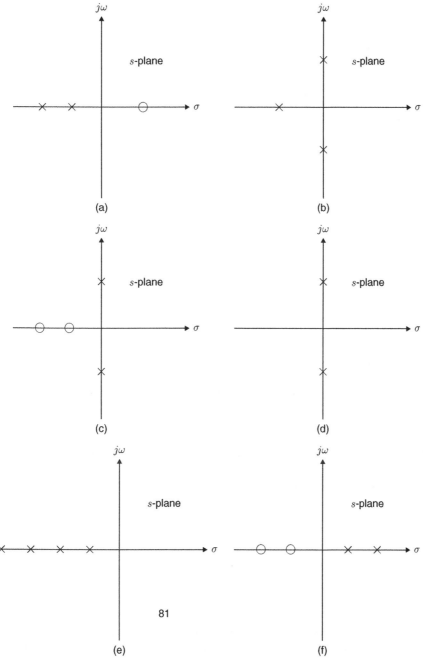

Figure 11.12 Figures for Exercise 11.1.

Exercise 11.2. Sketch the root locus for the unity feedback systems for the following transfer functions.

(a)

$$G(s) = \frac{K(s+2)(s+5)}{(s^2+9s+25)}$$

(b)

$$G(s) = \frac{K}{(s+1)^3(s+5)}.$$

Exercise 11.3. For the open-loop pole zero plot shown in Figure 11.13, sketch the root locus and find the break-in point.

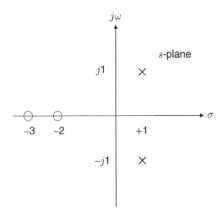

Figure 11.13

Exercise 11.4. For the open-loop transfer function

$$KG(s)H(s) = \frac{10}{(s+2)(s+p_1)} \tag{11.22}$$

obtain an equivalent transfer function of a control system whose equivalent open loop transfer function can be written as

$$KG(s)H(s) = p_1 G_{eq}(s)$$

so that p_1 takes the role of the gain K, allowing us to plot the root locus as a function of p_1.

Exercise 11.5: Take a unity feedback system

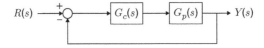

and let

$$G_p(s) = \frac{1}{s(s^2+4s+5)}. \tag{11.23}$$

(a) Design a cascade lag compensator to reduce the steady state error for a ramp input by tenfold over the uncompensated system. Analyze the effect on relative stability. Note the uncompensated system is the one with $G_c(s) = 1$.

(b) For the same $G_p(s)$ as above, design a cascade lead compensator such that there is a reasonable trade off between relative stability and steady state error. Use

$$G_c(s) = \frac{K(s+a)}{s+b}.$$ (11.24)

Bibliography

1 J.J. d'Azzo and C.D. Houpis. *Linear control system analysis and design: conventional and modern.* McGraw-Hill, New York, 1988.

2 J.J. d'Azzo and C.H. Houpis. *Linear control system analysis and design: conventional and modern.* McGraw-Hill Higher Education, New York, 1988.

3 C.R. Dorf and R.H. Bishop. *Modern control systems.* Pearson, 12 edition, 2011.

4 W.R. Evans. Control system synthesis by root locus method. volume 69, pages 66–69.

5 W.R. Evans. Graphical analysis of control systems. volume 67, pages 547–551.

6 G. Hostetter, C. Savant Jr, and R. Stefani. *Design of Feedback Control Systems.* Holt, Rinehart and Winston, 1982.

7 N. Nise. *Control Systems Engineering.* Wiley, 2011.

12

Frequency Response Analysis and Design

12.1 Chapter Highlights

This chapter presents a very important and useful material that is very widely used in the analysis and design of control systems. It essentially deals with the study of the steady state response of system output for a given periodic oscillatory input function such as a trigonometric sine or a cosine function. Hence the label, frequency response. The motivation for thoroughly studying the response to a sine/cosine function comes from the fact that any analytic function $f(t)$ can be expressed as the summation of sine and cosine functions in a Fourier series and thus if we understand the behavior of the output to a sine/cosine function, then for a linear system, since superposition holds, one can get the response to any arbitrary time function as a sum of the responses of the frequency responses at various frequencies. In this chapter we present another graphical technique labeled the Bode plot by which one can quickly draw the amplitude and phase response of the output function (in the Laplace domain) as a function of the frequency ω. The Bode plot gives very useful information about the speed of response of the output in terms of measures such as bandwidth, peak resonance, etc. In addition, measures of stability margins called gain margin and phase margin can easily be determined from the Bode plots [1].

12.2 Introduction

The output response of a linear system represented by a transfer function to sinusoidal inputs is called the system's frequency response. It can be obtained from knowledge of its pole and zero locations.

Let us consider a system described by

$$\frac{Y(s)}{U(s)} = G(s) \tag{12.1}$$

where the input $u(t) = A \sin(\omega_0 t)1(t)$. The input sinusoidal function has a Laplace transform

$$U(s) = \frac{A\omega_0}{s^2 + \omega_0^2}. \tag{12.2}$$

Flight Dynamics and Control of Aero and Space Vehicles, First Edition. Rama K. Yedavalli.
© 2020 John Wiley & Sons Ltd. Published 2020 by John Wiley & Sons Ltd.

With zero initial conditions, the Laplace transform of the output is

$$Y(s) = G(s)\frac{A\omega_0}{s^2 + \omega_0^2}. \tag{12.3}$$

A partial fraction expansion, given by

$$Y(s) = \frac{\alpha_1}{s - p_1} + \frac{\alpha_2}{s - p_2} + \cdots + \frac{\alpha_n}{s - p_n} + \frac{\alpha_0}{s + j\omega_0} + \frac{\alpha_0^*}{s + j\omega_0} \tag{12.4}$$

where $p_1, p_2, \ldots, p_n$ are the poles of $G(s)$. In the above, the residual α_0^* is the complex conjugate of α_0. Let us assume the transfer function $G(s)$ is a stable transfer function. Then the steady state output response would consist of only the response pertaining to the last two partial fractions containing last two residuals α_0 α_0^*. Thus the output response would be another sinusoidal function (just like the input) except that its amplitude is different from the input amplitude and in addition would have a phase angle ϕ. The phase angle ϕ and the ratio of output amplitude over input amplitude, denoted as M, of this output response in the time domain are then given by

$$\phi = \tan^{-1}\left[\frac{\text{Im}(\alpha_0)}{\text{Re}(\alpha_0)}\right] \tag{12.5}$$

$$M = |G(j\omega_0)| \tag{12.6}$$

$$\phi = \tan^{-1}\left[\frac{\text{Im}[G(j\omega_0)]}{\text{Re}[G(j\omega_0)]}\right] \tag{12.7}$$
$$= \angle G(j\omega_0).$$

In polar form,

$$G(j\omega_0) = Me^{j\phi}. \tag{12.8}$$

The magnitude M is given by $|G(j\omega)|$, and the phase ϕ is given by $\angle[G(j\omega)]$; that is, the magnitude and the angle of the complex quantity $G(s)$ are evaluated with s taking on the values along the imaginary axis ($s = j\omega$). In other words, the steady state output response is completely determined by the angle of the transfer function through which the input is passing, and by the magnitude of the transfer function through which the input is passing. This information about the phase and magnitude of the transfer function constitutes the frequency response of that transfer function or of the linear system represented by that transfer function. The frequency response of a linear system sheds considerable insight into the various speed of response characteristics of that linear system such as how fast the output is produced for a given input and how rich the output information is, i.e. how much of the input information is contained in that output and so on. The metrics by which this speed of response is characterized are called frequency response specifications, which include labels such as bandwidth, cut off (break) frequencies, peak resonant frequency, gain margin and phase margin among others. We now elaborate on these frequency response specifications.

12.3 Frequency Response Specifications

If we consider any control system with an output and an input with the understanding that the output follow the input, in the ideal case when the output exactly duplicates the

Figure 12.1 Ideal frequency response.

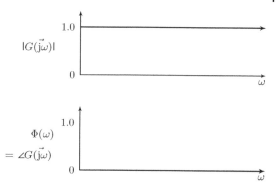

Figure 12.2 Frequency response specifications.

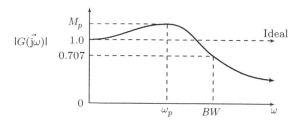

input, we observe that the ideal frequency response would be as shown in Figure 12.1. However, obviously, in practice, the frequency responses are never like these. So by looking at the actual frequency responses and their closeness to the ideal response, we can get a feel for the goodness of the control system. So these measures of goodness are the frequency response specifications, which are as follows (shown in Figure 12.2).

i) M_p is the peak resonance, the maximum value of $|G(\vec{j\omega})|$. This is related to relative stability because a large peak resonance leads to a large overshoot in the time response.

ii) ω_p is the resonant frequency, the frequency at which peak resonance occurs.

iii) BW is the bandwidth, the most important frequency response specification. It is the frequency at which the magnitude drops to 70.7% of its zero-frequency level. The bandwidth is a measure of the speed of response.

Clearly a large bandwidth is desirable because it means a fast acting system. A low bandwidth means the system is slow and sluggish. In other words, an instantaneous response in the time domain is equivalent to an infinite bandwidth in the frequency domain.

Two other important specifications in the frequency domain are the gain margin and the phase margin. These two concepts are related to determining the stability of the closed-loop system, again just by looking at the frequency response of the open loop transfer function. Thus if we plot the frequency response of the open loop transfer function for a given gain K, then the gain margin is simply the number of times the gain can be increased before it becomes unstable (assuming the instability occurs when we increase the gain, which is the most common situation). So the gain margin, as the name implies, is the cushion the system has before instability arises. So in a way, it is a measure of the degree of stability.

12.3.1 Frequency Response Determination

Example 12.1 Given a transfer function

$$G(s) = \frac{1}{(s+1)}$$

determine its frequency response.

Solution
We need to calculate $|G(\vec{j\omega})|$ and $\angle G(\vec{j\omega})$. The magnitude is given by

$$M(\omega) = |G(\vec{j\omega})|$$

$$= \frac{1}{|\vec{j\omega}+1|}$$

$$= \frac{1}{\sqrt{1+\omega^2}}$$

and the phase is given by

$$\Phi(\omega) = \angle G(\vec{j\omega}) = \angle 1 - \angle(1 + \vec{j\omega})$$

$$= 0 - \tan^{-1}\frac{\omega}{1}$$

$$= -\tan^{-1}\omega.$$

If we plot $M(\omega)$ and $\Phi(\omega)$ as a function of frequency, we get the plots in Figure 12.3. These are called Bode plots, named after their inventor Bode, who happened to be an electrical engineer working in the radio communications field.

The MATLAB routine [Mag, Phase, ω] = bode [Num, Den] or bode [Num, Den, ω] gives the Bode plots for more complicated transfer functions.

As with the root locus in the previous chapter, here it is also advantageous to learn the capability of drawing Bode plots quickly and approximately, even though using MATLAB we can draw the exact Bode plot. So in what follows, we again try to learn few rules to draw the Bode plots quickly in what we label as straight line approximations of Bode plots.

Figure 12.3 Bode plots.

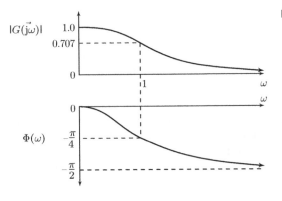

12.4 Advantages of Working with the Frequency Response in Terms of Bode Plots

1. Dynamic compensator designs can be based entirely on Bode plots.
2. Bode plots can be determined experimentally.
3. Bode plots of systems in series simply add, which is quite convenient.
4. The use of log scale permits a much wider range of frequencies to be displayed on a single plot than is possible with linear scales.

12.4.1 Straight Line Approximation of Bode Plots

In working with frequency response, it is more convenient to replace s with $j\omega$ and to write the transfer function in the Bode form

$$KG(j\omega) = K_0 \frac{(j\omega\tau_1 + 1)(j\omega\tau_2 + 1)\dots}{(j\omega\tau_a + 1)(j\omega\tau_b + 1)\dots}. \tag{12.9}$$

The transfer functions can also be rewritten as

$$KG(j\omega) = K_0 \frac{j\omega\tau_1 + 1}{j\omega\tau_a + 1}. \tag{12.10}$$

Then

$$\angle KG(j\omega) = \angle K_0 + \angle(j\omega\tau_1 + 1) - \angle(j\omega)^2 - \angle(j\omega\tau_1 + 1) \tag{12.11}$$

$$\log |KG(j\omega)| = \log |K_0| + \log |j\omega\tau_1 + 1| - \log |(j\omega)^2| - \log |j\omega\tau_a + 1|. \tag{12.12}$$

In decibels, the equation becomes

$$|KG(j\omega)|_{db} = 20 \log |K_0| + 20 \log |j\omega\tau_1 + 1|$$
$$- 20 \log |(j\omega)^2| - 20 \log |j\omega\tau_a + 1|. \tag{12.13}$$

To sketch the complete Bode plot of a given transfer function in a quick and approximate way, we employ the concept of straight line approximation. In other words, we approximate the actual curves we get as a function of frequency into straight lines in certain frequency ranges. These straight lines are of different slopes based on the pole zero nature of the given transfer function. To be able to estimate those slopes, we construct a preliminary straight line approximation plot for each pole and zero present in the transfer function and then algebraically add or subtract them based on the nature of the poles and zeroes present in the transfer function. For this, let us first consider the possible scenarios of pole zero terms we encounter in a typical transfer function.

All of the transfer functions we have dealt with so far have been composed of one of these three classes of terms:

1. $k_0(j\omega)^n$
2. $(j\omega\tau + 1)^{\pm 1}$
3. $[(\frac{j\omega}{\omega_n})^2 + 2\xi \frac{j\omega}{\omega_n} + 1]^{\pm 1}$.

1. The first term $k_0(j\omega)^n$

$$\log k_0 |(j\omega)^n| = \log k_0 + n \log |(j\omega)|. \tag{12.14}$$

The magnitude plot of this term is a straight line with a slope of $n \times (20$ db/ decade). This term is the only class of term that affects the slope at the lowest frequencies because all other terms are constant in the region. The phase of $k_0(j\omega)^n$ is $n \times 90°$.

2. $(j\omega\tau + 1)^{\pm 1}$. Because of the nature of this term it approaches one asymptote at very low frequencies and another one at very high frequencies.

   (a) For $j\omega\tau \ll 1$, $(j\omega\tau + 1) \cong 1$

   (b) For $j\omega\tau \gg 1$, $(j\omega\tau + 1) \cong j\omega\tau$.

   For our Bode plot sketch we will define a break point $\omega = \frac{1}{\tau}$, where before the break point the slope will be equal to zero and after the break point it will behave like $k_0(j\omega)^n$ and have a slope equal to $n \times (20$ db/decade). The phase curve can be drawn in a similar fashion using high and low frequency asymptotes.

   (a) For $j\omega\tau \ll 1$, $\angle 1 = 0°$

   (b) For $j\omega\tau \cong 1$, $\angle j\omega\tau + 1 \cong 45°$

   (c) For $j\omega\tau \gg 1$, $\angle j\omega\tau = 90°$.

3. $[(\frac{j\omega}{\omega_n})^2 + 2\xi\frac{j\omega}{\omega_n} + 1]^{\pm 1}$ This term is very similar to the last but with a few differences. First the break point is now $\omega = \omega_n$. Also the magnitude change of the slope is $n \times (40$ db/decade), which is twice that of the previous term. Finally the phase curve is $n \times 180°$ and the transition though the break point varies with the damping ratio ξ. Because of the dependence on the damping ratio a rough sketch can be drawn using the following formula

$$|G(j\omega)| = \frac{1}{2\xi} \quad \text{at} \quad \omega = \omega_n. \tag{12.15}$$

Because of this we usually drawn a small peak at the break point to show the resonant peak.

12.4.2 Summary of Bode Plot Rules

1. Manipulate the transfer function into the Bode form given.
2. Determine the value of n for the $K_0(j\omega)^n$ term (class 1). Plot the low frequency magnitude asymptote through the point K_0 at $\omega = 1$ with a slope of n(or $n \times 20$ db/decade).
3. Complete the composite magnitude asymptotes: extend the low frequency asymptote until the first frequency break point. Then step the slope by ± 20 db or ± 40 db, depending on whether the break point is from a first or second order term in the numerator or denominator. Continue through all break points in ascending order.
4. Sketch in the approximate magnitude curve: increase the asymptote value by a factor of 1.4 ($+3$ db) at first order numerator break points, and decrease it by a factor of 0.707 (-3 db) at first order denominator break points. At second order break points, sketch in the resonant peak (or valley).
5. Plot the low-frequency asymptote of the phase curve, $\phi = n \times 90°$.
6. As a guide, sketch in the approximate phase curve by changing the phase by $\pm 90°$ or $\pm 180°$ at each break point in ascending order. For first order terms in the numerator, the change of phase is $+90°$; for those in the denominator the change is $-90°$. For second order terms, the change is $\pm 180°$.
7. Locate the asymptotes for each individual phase curve so that their phase change corresponds to the steps in the phase toward or away from the approximate curve indicated by Step 6.

8. Graphically add each phase curve. Use grids if an accuracy of about $\pm 5°$ is desired. If less accuracy is acceptable, the composite curve can be done by eye. Keep in mind that the curve will start at the lowest frequency asymptote and end on the highest frequency asymptote and will approach the intermediate asymptotes to an extent that is determined by how close the break points are to each other.

Let us now illustrate the procedure with an example.

Example 12.2   Draw the Bode plot for the transfer function

$$KG(s)H(s) = \frac{s+3}{s(s^2 + s + 25)}.$$

Solution

1. Rearranging the transfer function into the Bode form gives

$$KG(s)H(s) = \frac{3(\frac{s}{3} + 1)}{25s(\frac{s^2}{25} + \frac{s}{25} + 1)}$$

$$= \frac{\frac{3}{25}(\frac{s}{3} + 1)}{s(\frac{s^2}{5} + \frac{s}{25} + 1)}.$$

2. The low frequency asymptote is -20 db/decade because of the s in the denominator.
3. Looking at the transfer function we can see that there will be two break points, one at 3 from the zero and one at 5 from the natural frequency ω_n of the complex pole. Because the break point at 3 is from a zero, the slope will increase by 20 db/decade,

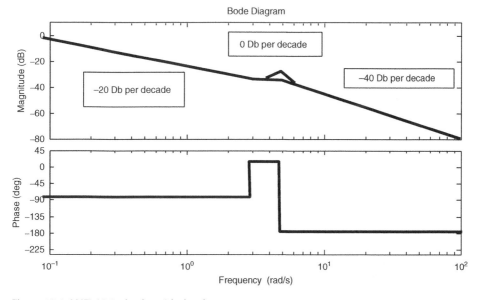

Figure 12.4 MATLAB Bode plot with sketch overtop.

which will change it from -20 db/decade to 0 db/decade. Then the complex pole at 5 will change the slope by -40 db/decade.

4. See the sketch in Figure 12.4. Notice the resonant peak at $\omega = 5$.
5. The low frequency phase will be $\phi = -90°$.
6. The zero at 3 will raise the phase by $90°$ and the complex pole at 5 will lower the phase by $180°$.
7. The phase curve is sketched in the bottom part of Figure 12.4.

12.5 Examples on Frequency Response

Example 12.3

1. (a) $G(s) = \frac{1}{s(s+1)}$. To get the frequency response, we need to compute the magnitude of $G(s)$ along $s = \vec{j}\omega$ and the angle of $G(s)$ along $s = \vec{j}\omega$.

$$|G(\vec{j}\omega)| = \frac{1}{|\vec{j}\omega(\vec{j}\omega + 1)|} = \frac{1}{|(\vec{j}\omega)^2 + \vec{j}\omega|} = \frac{1}{|-\omega^2 + \vec{j}\omega|}$$

$$= \frac{1}{\sqrt{(-\omega^2)^2 + (\omega)^2}} = \frac{1}{\sqrt{\omega^4 + \omega^2}} = \frac{1}{\omega\sqrt{(\omega^2 + 1)}}.$$

Similarly

$$\angle G(\vec{j}\omega) = 0 - \angle(-\omega^2 + \vec{j}\omega) = -\tan^{-1}\frac{\text{Imaginary part}}{\text{Real part}}$$

$$= -\tan^{-1}\frac{\omega}{-\omega^2} = -\tan^{-1}\frac{1}{-\omega} \text{ (Second quadrant).}$$

Plotting these two quantities as a function of ω, we get Figure 12.5.
At $\omega = 1$, $\angle G(\vec{j}\omega) = -135°$. At $\omega = \infty$, $\angle G(\vec{j}\omega) = -180°$.

(b) $G(s) = \frac{1}{1+5s}$, so $G(\vec{j}\omega) = \frac{1}{1+5\vec{j}\omega}$.

$$|G(\vec{j}\omega)| = \frac{1}{\sqrt{1 + 25\omega^2}}$$

$$\angle G(\vec{j}\omega) = 0 - \tan^{-1}\frac{5\omega}{1} \text{ (First quadrant)}$$

Figure 12.5 Bode plot for Example 12.1(a).

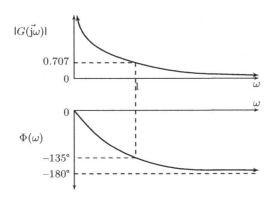

Figure 12.6 Bode plot for Example 12.1(b).

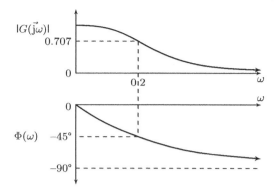

At $\omega = 0.2$, $\angle G(\vec{j\omega}) = -45°$. At $\omega = \infty$, $\angle G(\vec{j\omega}) = -90°$.

See the resulting Bode Plots in Figure 12.6

12.5.1 Bode's Gain Phase Relationship

An important Bode plot theorem is:

For any stable minimum-phase system (that is, one with no RHP zeros or poles), the phase of G(jω) is uniquely related to the magnitude of G(jω).

which can be simplified with an approximation

$$\angle G(j\omega) \cong n \times 90°$$ (12.16)

where n is the slope of $G|(j\omega)|$ in units of decades of amplitude per decades of frequency.

Adjust the slope of the magnitude curve $|KG(j\omega)|$ so that it crosses over magnitude $1(0 \text{ db})$ with a slope of -1 (-20 db) for a decade around ω_e.

12.5.2 Non-minimum Phase Systems

A transfer function that has all poles and zeros in the LHP is a minimum phase transfer function, which has a lot of special properties. One is that they are stable for all gains k and that you can use Bode plot rules to approximate the phase change. Consider the transfer functions

$$G_1(s) = 10\frac{s+1}{s+10}$$ (12.17)

$$G_2(s) = 10\frac{s-1}{s+10}.$$ (12.18)

Both transfer functions have the same magnitude for all frequencies of

$$|G_1(j\omega)| = |G_2(j\omega)|.$$ (12.19)

As you can see in Figure 12.7 the magnitude of both transfer functions is the same but the phase is different. The phase for the minimum phase transfer function follows the rules that were laid out above, but the non-minimum phase transfer function does not, and cannot be modeled by the rules above. Because of this, this chapter will only deal with minimum phase transfer functions.

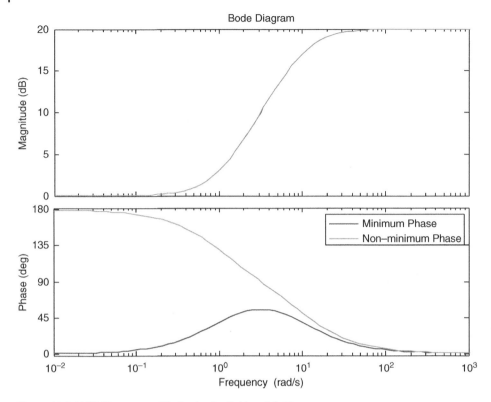

Figure 12.7 MATLAB generated Bode plot for $G_1(s)$ and $G_2(s)$.

12.6 Stability: Gain and Phase Margins

A large number of control systems behave in a pattern similar to the ones we have seen, where they are stable for small gain K values but as K gets bigger the system becomes unstable. To measure the stability of these types of systems we can employ two metrics: gain margin and phase margin.

The gain margin (GM) is the factor by which the gain can be raised before instability results. The GM can also be determined from a root locus with respect to K by noting two values of K: (1) at the point where the root locus crosses the $j\omega$ axis, and (2) at the nominal closed-loop poles. The GM is the ratio of these two values. The gain margin can also be directly read off of a Bode plot my measuring the vertical line between the $|KG(j\omega)|$ curve and the $|KG(j\omega)| = 1$ curve at the frequency where $\angle KG(j\omega) = 180°$. Analytically, the gain margin is the reciprocal of the magnitude $|G(j\omega)|$ at the frequency where the phase angle is $-180°$. Defining the phase crossover frequency ω_c to be the frequency at which the phase angle of the open-loop transfer function equals $-180°$ gives the gain margin K_g:

$$K_g = \frac{1}{|G(j\omega_1)|}. \tag{12.20}$$

In terms of decibels,

$$K_g \text{ db} = 20 \log K_g = -20 \log |G(j\omega_1)|. \tag{12.21}$$

The phase margin(PM) is the amount by which the phase of $G(j\omega)$ exceeds $-180°$ when $|KG(j\omega)| = 1$. The term crossover frequency ω_c is used to denote the frequency at which the gain is 1 or 0 db. The phase margin is more commonly used to specify control system performance because it is most closely related to the damping ratio of the system.

$$PM = \tan^{-1} \left[\frac{2\zeta}{\sqrt{\sqrt{1 + 4\zeta^4} - 2\zeta^2}} \right]. \tag{12.22}$$

A good approximation that can be used up to phase margins of $70°$ is

$$\zeta \cong \frac{PM}{100}. \tag{12.23}$$

The phase margin is also that amount of additional phase lag at the gain crossover frequency required to bring the system to the verge of instability. The gain crossover frequency is the frequency at which $|G(j\omega)|$, the magnitude of the open-loop transfer function, is unity. The phase margin γ is $180°$ plus the phase angle ϕ of the open-loop transfer function at the gain crossover frequency, or

$$\gamma = 180° + \phi. \tag{12.24}$$

The gain margin for the second-order system is infinite $GM = \infty$, because the phase curve does not cross $-180°$ as the frequency increases. It would also be true for any first or second order system.

$PM = 30°$ is often judged to be the lowest adequate value.

12.6.1 Gain and Phase Margins Determined Analytically

Example 12.4 Consider a system with open loop transfer function

$$KGH = \frac{10}{s(s + 1)(s + 5)}. \tag{12.25}$$

Let us say we need the gain and phase margins for this system.

Gain Margin. This needs the phase crossover frequency ω_c. It is that frequency at which the phase is $-180°$.

$$\angle KGH = \frac{\angle 10}{\angle (s(s + 1)(s + 5))} \Big|_{s=j\omega}$$

$$= \angle 10 - \angle (\vec{j}\omega(\vec{j}\omega + 1)(\vec{j}\omega + 5))$$

$$= \angle 10 - \angle (-6\omega^2 + \vec{j}(5\omega - \omega^3))$$

$$= 0 - \angle [\vec{j}(5\omega - \omega^3) - 6\omega^2]$$

$$= -\tan^{-1} \frac{5\omega - \omega^3}{-6\omega^2}.$$

Equating the above to $-180°$,

$$\therefore \tan^{-1} \frac{5\omega - \omega^3}{-6\omega^2} = 180°.$$

Solving, we see the crossover frequency $\omega_c = 2.24 \frac{\text{rad}}{\text{s}}$. Now at this phase crossover frequency, the magnitude of KGH is to be found.

$$\left| \frac{10}{-6\omega^2 + \vec{j}(5\omega - \omega^3)} \right|_{\omega=\omega_c} = \frac{10}{\sqrt{36\omega_c^4 + (5\omega_c - \omega_c^3)^2}} = \frac{10}{30} = \frac{1}{3}$$

$$\therefore GM = \frac{1}{|KGH|_{\omega=\omega_c}} = \frac{1}{\frac{1}{3}} = 3$$

In dB GM $= 20 \log_{10} 3 = \boxed{9.542 \text{ dB}}$.

Phase Margin. We need the gain crossover frequency ω_g. It is that frequency at which the magnitude of KGH is equal to 1, i.e. $|KGH| = 1$. Now

$$\left| \frac{10}{\vec{j}\omega(\vec{j}\omega + 1)(\vec{j}\omega + 5)} \right| = \frac{10}{\sqrt{(5\omega - \omega^3)^2 + (-6\omega^2)^2}} \bigg|_{\omega=\omega_g} = 1$$

$$\frac{100}{36\omega_g^4 + (5\omega_g - \omega_g^3)^2} = 1 \Rightarrow \omega_g^6 + 26\omega_g^4 + 25\omega_g^2 - 100 = 0$$

or $\beta^3 + 26\beta^2 + 25\beta - 100 = 0$ where $\beta = \omega_g^2$.

Solving for β, we get

$$\beta = \begin{cases} -24.8310 \\ -2.6747 \\ 1.5057 \end{cases}$$

So $\omega_g = \sqrt{1.5057} = .227 \approx 1.23 \frac{\text{rad}}{\text{s}}$.

Now

$$\Phi = \angle KGH|_{\omega=\omega_g} = -\tan^{-1} \frac{(5\omega_g - \omega_g^3)}{-6\omega_g^2} = -\tan^{-1} \frac{4.25}{-9.0777}$$

$$= -154.7°$$

$$\gamma = PM = 180° + \Phi = 180° - 154.7° = \boxed{25.3°}.$$

Example 12.5 Now, consider another example with the transfer function given by:

$$G(s) = \frac{2000}{(s+2)(s+7)(s+16)} = \frac{2000}{s^3 + 25s^2 + 158s + 224}$$

$$G(\vec{j}\omega) = \frac{2000}{(\vec{j}\omega + 2)(\vec{j}\omega + 7)(\vec{j}\omega + 16)} = \frac{2000}{(\vec{j}\omega)^3 + 25(\vec{j}\omega)^2 + 158(\vec{j}\omega) + 224}$$

$$= \frac{2000}{(224 - 25\omega^2) + \vec{j}(158\omega - \omega^3)}.$$

Now to find the gain margin, we need to find the phase crossover frequency ω_c. It is that frequency at which the $\angle G(j\omega) = -180°$. So

$$\angle G(j\omega) = 0 - \tan^{-1}\frac{158\omega - \omega^3}{224 - 25\omega^2} = -180°$$

$$\tan 180° = 0 = \frac{158\omega - \omega^3}{224 - 25\omega^2}$$

$$= 158\omega - \omega^3.$$

Solving for the positive root, we find $\omega = \omega_c = 12.6$. Now the magnitude $|G(j\omega)|$ at this frequency ω_c is $|G(j\omega_c)|$ and is given by

$$|G(j\omega_c)| = \frac{2000}{\sqrt{(224 - 25\omega_c^2)^2 + (158\omega_c - \omega_c^3)^2}}$$

$$= 0.5367.$$

Therefore the gain margin K_g is:

$$K_g = \frac{1}{|G(j\omega_c)|}$$

$$= \frac{1}{0.5367}$$

$$= 1.863.$$

And in decibels the gain margin is

$$10 \log K_g = 10 \log_{10}(1.8637)$$

$$= 5.404 \text{ dB}.$$

Now to find the phase margin, we need to calculate the gain crossover frequency ω_g, which is that frequency at which the $|G(j\omega_g)| = 1$.

$$|G(j\omega_g)| = \frac{2000}{\sqrt{(224 - 25\omega_g^2)^2 + (158\omega_g - \omega_g^3)^2}} = 1.$$

Squaring both numerator and denominator, we get

$$\frac{(2000)^2}{(224 - 25\omega_g^2)^2 + (158\omega_g - \omega_g^3)^2} = 1.$$

Expanding, we get

$$\omega_g^6 + 307\omega_g^4 + 14081\omega_g^2 - 3949824 = 0.$$

Defining $\omega_g^2 = \beta$, we get

$$\beta^3 + 307\beta^2 + 14081\beta - 3949824 = 0.$$

Getting the roots of this polynomial by MATLAB, we have

$$\beta = (-1.9555 \pm \vec{j}0.9344) \times 10^2$$

$$= 0.8409 \times 10^2 = 84.09.$$

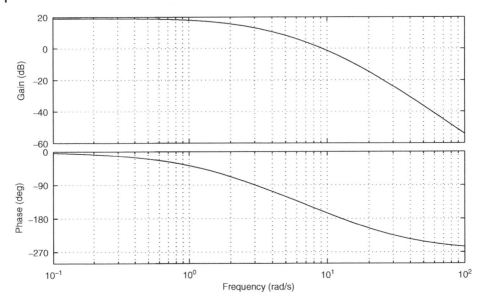

Figure 12.8 MATLAB generated Bode plot for Example 12.2.

Therefore

$$\omega_g = \sqrt{\beta}$$
$$= \sqrt{84.09}$$
$$= 9.17 \frac{rad}{s}$$

and

$$\angle G(j\vec{\omega}_g) = -\tan^{-1}\frac{158\omega_g - \omega_g^3}{224 - 25\omega_g^2} \text{ where } \omega_g = 9.17 \frac{rad}{s}$$

$$\cong -160°$$

$$\therefore \text{Phase margin, PM} = 180 + \angle G(j\vec{\omega}_g) = 180° - 160°$$

$$= 20°$$

$$\boxed{GM = 5.4 \text{ dB and PM} = 20°.}$$

The gain margin and phase margin can be easily read off the Bode plot in Figure 12.8.

12.6.2 Steady State Errors

To find the steady state error of a system you have to start out with the error transfer function

$$\frac{E(s)}{R(s)} = \frac{1}{1 + G(s)H(s)}. \tag{12.26}$$

Then using the final value theorem we get

$$e_{ss} = \frac{R(s)}{(1 + K_p)s}.$$ (12.27)

So depending on the system type and input type, only one of the static error constants are finite and significant. The static position coefficient is

$$e_{ss} = \frac{1}{1 + K_p}$$ (12.28)

and for a step input K_p is constant and the error is $\frac{1}{1+K_p}$. For a unity feedback system with a unit ramp input, the steady state error is

$$e_{ss} = \frac{1}{K_v}$$ (12.29)

which is called the velocity error. For a second order system with a parabolic input we get an acceleration error which is

$$e_{ss} = \frac{1}{K_a}.$$ (12.30)

For steady state errors see Table 12.1.

12.6.3 Closed-Loop Frequency Response

Consider a system in which $|KG(j\omega)|$ shows the typical behavior

$$|KG(j\omega)| \gg 1$$ (12.31)

$$|KG(j\omega)| \ll 1.$$ (12.32)

The closed-loop frequency response magnitude is approximated by

$$|\tau(j\omega)| = \left| \frac{KG(j\omega)}{1 + KG(j\omega)} \right|$$

$$\cong \begin{cases} 1 & \omega \ll \omega_c \\ |KG| & \omega \gg \omega_c \end{cases}.$$ (12.33)

$|\tau(j\omega)|$ depends greatly on the PM in the vicinity of ω_c, and because of this, the following approximation was generated

$$\omega_c \leq \omega_{BW} \leq 2\omega_c$$ (12.34)

Table 12.1 Steady State Errors.

Input	Steady state error formula	Type 0 Static error constant	Error	Type 1 Static error constant	Error	Type 2 Static error constant	Error
Step, $u(t)$	$\dfrac{1}{1 + K_p}$	K_p =constant	$\dfrac{1}{1 + K_p}$	K_p = inf	0	K_p = inf	0
Ramp, $tu(t)$	$\dfrac{1}{K_v}$	$K_v = 0$	inf	K_v =constant	$\dfrac{1}{K_v}$	K_v = inf	0
Parabola, $\dfrac{1}{2}2u(t)$	$\dfrac{1}{K_a}$	$K_a = 0$	inf	$K_a = 0$	inf	K_a =constant	$\dfrac{1}{K_a}$

12.7 Notes on Lead and Lag Compensation via Bode Plots

12.7.1 Properties of the Lead Compensator

- The lead compensator is used to improve the transient response or relative stability.
- The lead compensator is basically a high pass filter (i.e. the high frequencies are passed but low frequencies are attenuated).
- The lead compensator increases the gain crossover frequency (i.e. it shifts the compensated new gain crossover frequency to the right of the old uncompensated gain crossover frequency).
- A proportional derivative controller is a special case of the lead compensator.
  The transfer function of the lead compensator has the following form:

$$D(s) = \frac{(1 + Ts)}{(1 + \alpha Ts)} \quad \text{where } \alpha < 1$$
$$= \frac{K_C(s + z)}{(s + p)}$$

(i.e. the zero of the compensator is in front of the pole in the left half of the plane). The magnitude and phase characteristics of the lead compensator are shown in Figure 12.9.

12.7.2 Properties of the Lag Compensator

The lag compensator works in the opposite way. It has the following characteristics (see Figure 12.10.

- $D(s) = \dfrac{1 + Ts}{1 + \alpha Ts} = K_{\text{lag}} \dfrac{(s + z)}{(s + p)}, \alpha > 1$
- We normally place the pole and zero of a lag network very close to each other.
- The lag compensator is basically a low pass filter (i.e. low frequencies are passed but high frequencies are attenuated).
- The lag compensator decreases the gain crossover frequency (i.e. it shifts the new compensated gain crossover frequency to the left of the old uncompensated gain crossover frequency).
- A proportional integral controller is a special case of the lag compensator.
- The lag compensator is thus used to improve steady state error behavior.

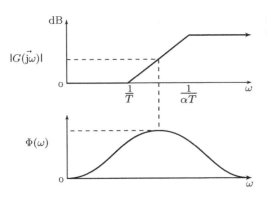

Figure 12.9 Bode plot for lead compensation.

Figure 12.10 Bode plot for lag compensation.

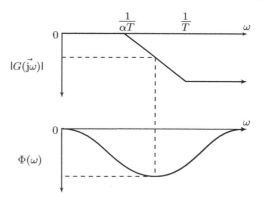

12.7.3 Steps in the Design of Lead Compensators Using the Bode Plot Approach

1. Determine the open loop gain K to satisfy the requirement on the steady state error coefficients.
2. Using the gain K thus determined, evaluate the phase margin of the uncompensated system.
3. Determine the necessary phase lead Φ_a to be added to the system.
4. Determine the attenuation factor α by using $\alpha = \dfrac{1 - \sin \Phi_a}{1 + \sin \Phi_a}$.
5. Select $\dfrac{1}{T}$ to be that frequency where the magnitude of the uncompensated system is equal to $-20 \log \left(\dfrac{1}{\sqrt{\alpha}} \right)$ or, very crudely, select $\dfrac{1}{T}$ to be a little left of the gain crossover frequency of the uncompensated system.
6. Once you have $\dfrac{1}{T}$ and $\dfrac{1}{\alpha T}$, determine K_C, z, and p.
7. Now you have the complete transfer function of the compensator. Redraw the magnitude and phase of the compensated open loop transfer function $D(s)$, $G_p(s)$ and make sure that the phase margin and gain margin and the steady state error requirements are met.

12.7.4 Steps in the Design of Lag Compensators Using Bode Plot Approach

1. Determine the open loop gain K such that the requirement on the particular error coefficient is satisfied.
2. Using the gain K thus determined, draw the Bode plots of the uncompensated system and determine the phase and gain margins of the uncompensated system.
3. If the specifications on the phase and gain margins are not satisfied, then find the frequency point where the phase angle of the open loop transfer function is equal to $-180°$ plus the required phase margin. The required phase margin is the specified phase margin plus 5–$10°$. Choose this frequency as the new gain crossover frequency.
4. Choose the corner frequency $\omega = 1/T$ (corresponding to the zero of the lag network) one decade below the new gain crossover frequency.
5. Determine the attenuation necessary to bring the magnitude curve down to 0 db at the new gain crossover frequency. Noting that this attenuation is $-20 \log \beta$,

determine the value of β. Then the other corner frequency (corresponding to the pole of the lag network) is determined from $\omega = 1/(\beta T)$.

6. Now you have the complete transfer function of the compensator. Redraw the magnitude and phase of the compensated open loop transfer function $D(s)$, $G_p(s)$ and make sure that the phase margin and gain margin and the steady state error requirements are all met.

12.8 Chapter Summary

In this chapter we have understood the meaning and importance of frequency response in the analysis and design of control systems. We have seen how the ability to draw the Bode plots via straight line approximations helps us to quickly get an idea of the speed of response of the output for a sinusoid input, using the information about various break frequencies leading to concepts such as low pass filter, high pass filter, and band pass filter, and their usefulness in shaping the controller transfer functions. We also have observed the complementary relationship between frequency response and time response in the sense of the larger the bandwidth, the faster the time response.

Fundamental concepts discussed in this chapter can be found in other textbooks dedicated to control systems such as [2–6].

12.9 Exercises

Exercise 12.1.

(a) Calculate the magnitude and phase of

$$G(s) = \frac{1}{s + 20}$$

by hand for ω=1, 2, 5, 10, 20, 50, and 100 rad s^{-1}.

(b) Sketch the asymptotes for $G(s)$ according to the Bode plot rules, and compare these with your computed results from part (a).

Exercise 12.2. Sketch the asymptotes of the Bode plot magnitude and phase for each of the following open-loop transfer functions. After completing the hand sketches, verify tour results using MATLAB.

(a)

$$G(s) = \frac{1}{(s + 1)(s^2 + 4s + 2)}$$

(b)

$$G(s) = \frac{s}{(s + 1)(s + 30)(s^2 + 2s + 1000)}$$

(c)

$$G(s) = \frac{s + 2}{s(s + 1)(s + 10)(s + 30)}$$

(d)
$$G(s) = \frac{(s+4)}{s(s+10)(s+2s+2)}$$

(e)
$$G(s) = \frac{(s+4)}{s^2(s+10)(s^2+6s+25)}.$$

Exercise 12.3. The block diagram of a control system as shown in Figure below

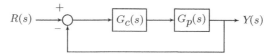

Treating the uncompensated system as
$$G_p(s) = \frac{2500}{s(s+25)}$$

analyze the open loop transfer function properties, in particular
(a) the phase margin
(b) the gain crossover frequency
(c) the gain margin.
Then apply the classical control theory we have learnt in this chapter to
(d) design a lead network to increase the phase margin to 45°.
Use MATLAB to verify all your calculations, along with the analytical calculations of phase margin for the uncompensated system. Repeat the exercise by varying the values of α and τ and observe the trend of the phase margin. In particular,
(e) obtain the results for a new phase margin of 60°.

Exercise 12.4. Take a unity feedback system as shown in Figure below

and let
$$G_p(s) = \frac{1}{s(s^2+4s+5)}. \tag{12.35}$$

(a) Design a cascade lag compensator to reduce the steady state error for a ramp input by tenfold over the uncompensated system. Analyze the effect on relative stability. Note the uncompensated system is the one with $G_c(s) = 1$.
(b) For the same $G_p(s)$ as above, design a cascade lead compensator such that there is a reasonable trade off between relative stability and steady state error. Use

$$G_c(s) = \frac{K(s+a)}{s+b}. \tag{12.36}$$

Bibliography

1 H.W. Bode. *Network Analysis and Feedback Amplifier Design.* Van Nostrand, Princeton, NJ, 1945.

2 J.J. d'Azzo and C.D. Houpis. *Linear control system analysis and design: conventional and modern.* McGraw-Hill, New York, 1988.

3 J.J. d'Azzo and C.H. Houpis. *Linear control system analysis and design: conventional and modern.* McGraw-Hill Higher Education, New York, 1988.

4 C.R. Dorf and R.H. Bishop. *Modern control systems.* Pearson, 12 edition, 2011.

5 G. Hostetter, C. Savant Jr, and R. Stefani. *Design of Feedback Control Systems.* Holt, Rinehart and Winston, 1982.

6 N. Nise. *Control Systems Engineering.* Wiley, 2011.

13

Applications of Classical Control Methods to Aircraft Control

13.1 Chapter Highlights

This chapter's objective is to apply the various frequency domain and classical control design techniques we have learnt in the previous chapters to the specific field of aircraft flight control systems. In that connection, we recall the aircraft dynamics and from that identify the output variables, input variables and the measurement variables and represent the control objective in a block diagram format and get the needed transfer functions from the equations of motion and then apply the theoretical tools we have learnt up to now for the specific application of aircraft flight control systems. It so happens that the aircraft flight control literature naturally always follows the theoretical developments in control systems theory. Thus, it is no wonder that we cover the subject of aircraft flight control systems after we learn the theoretical techniques such as the R–H criterion, root locus and Bode plots.

Since this book's objective is to treat both aircraft as well as spacecraft flight control systems in a unified framework, the application of classical control theory to aircraft flight control problems cannot be as exhaustive as those in books specifically dedicated to aircraft flight control systems. For that reason, in this chapter, we basically cover the essential material related to aircraft flight control systems, the literature on which is quite voluminous. Hence we approach this subject in a more conceptual fashion. In the spirit of many excellent books that are written on the specific subject of aircraft control systems, [40], [1], [30], [35], [15], [25], [29], [18], [41], [7], [14], [32], [20], [24], [12], [31], [17], [42], [19], [26], [36], [38], [9], [34], [39], [13], [10], [21], [4], [37], [3], [6], [2], [23], [22], [27], [28], [16], [8], [11], [5], [33], we briefly divide the coverage in the form of longitudinal autopilots along with a brief discussion on automatic landing control systems. Then we briefly review a few specific lateral/directional aircraft flight control systems as well. Note that, conceptually, the same control design philosophy is applied to both of these types of control systems with the understanding that we simply use the aircraft longitudinal equations of motion for building the longitudinal dynamics transfer functions whereas we use the lateral/directional equations of motion when building the lateral/directional transfer functions.

Flight Dynamics and Control of Aero and Space Vehicles, First Edition. Rama K. Yedavalli.
© 2020 John Wiley & Sons Ltd. Published 2020 by John Wiley & Sons Ltd.

13.2 Aircraft Flight Control Systems (AFCS)

Recall from Part I of the book (on Flight Vehicle Dynamics) that, in the linearized equations of motion for aircraft flight dynamics, the longitudinal equations of motion were decoupled from the lateral/directional equations of motion. Thus, in the control systems design exercise, it is customary to treat the subject of AFCS also in two phases, one for controlling longitudinal motion variables (i.e. pitch motion) and the other for controlling lateral/directional variables (i.e roll/yaw motion). Recall that the state variables in the longitudinal motion in the linear regime are the (i) forward speed change (i.e. change from the equilibrium state) (u_{sp}), (ii) angle of attack (α) or the vertical velocity (w) change, (iii) the pitch angle change, θ, and the pitch rate change, q. Typically in longitudinal motion, the control variables are the elevator angle δ_e (and any other aerodynamic control surface deflections). The control variable related to the engine (propulsion system) is the throttle angle δ_{th}. In industry, it is also customary to categorize AFCS in accordance with the control objective of the control system. For example, for an aircraft with a stable phugoid or short period mode with an inadequate open loop damping ratio of the mode, if the control objective is to increase the damping ratio of the mode in the closed loop, that control system is termed as a stability augmentation system (SAS). Thus roll dampers, pitch dampers and yaw dampers can be classified as SASs. On the other hand, if the control objective is to achieve a particular type of response for a given motion variable, then that control system is categorized as a control augmentation system (CAS). A steady coordinated turn control system control, a systems to achieve a desired roll rate or pitch rate or normal acceleration, etc., can be classified as a CAS. Finally control systems that regulate a given motion variable (i.e hold its value constant), thereby giving pilot relief are called autopilots. For example Mach hold (holding speed constant, like in cruise control), altitude hold, pitch attitude hold, bank angle hold, heading hold/VOR hold, etc. could be classified as autopilots. However, there is no obligation to strictly follow this nomenclature. It is more a matter of familiarity with industry practice. At the conceptual level, all of these are simply control systems designed to achieve a particular control objective. With this background, we now briefly discuss the specifics of a few of these control systems, mostly borrowing from the existing literature, referenced at the end of this chapter.

13.3 Longitudinal Control Systems

In this section, we focus on all those control systems that attempt to control longitudinal motion variables, such as speed u_{sp}, angle of attack α, pitch angle θ, and pitch rate q. For brevity, we focus on only a few of these control objectives, and illustrate the basic steps of design through block diagram representation of that particular control objective. The understanding is that a similar conceptual procedure can be applied to any other control objective. With that in mind, we present a brief account of a control system whose objective is to regulate the pitch angle change to zero. Thus the nomenclature introduced before this is a pitch displacement autopilot.

13.3.1 Pitch Displacement Autopilot

The simplest form of autopilot, which is the type that first appeared in aircraft and is still being used in some of the older transport aircraft, is the pitch displacement autopilot, namely keeping the perturbed pitch angle θ to remain constant at a given value, which could be zero. This autopilot was designed to hold the aircraft in straight and level flight with little or no maneuvering capability.

For this type of autopilot the aircraft is initially trimmed to straight and level flight, the reference aligned, and then the autopilot engaged. If the pitch altitude varies from the reference, a voltage e_g, is produced by the signal generator on the vertical gyro. This voltage is then amplified and fed to the elevator servo. The elevator servo can be electromechanical or hydraulic with an electrically operated valve. The servo then positions the elevator, causing the aircraft to pitch about the Y axis and so returning it to the desired pitch attitude. The elevator servo is, in general, at least a second order system, but if properly designed, its natural frequency is higher than that of the aircraft. If the damping ratio is high enough, the elevator servo can be represented by a sensitivity (gain) multiplied by a first order time lag. Representative characteristic times vary from 0.1 to 0.03 s. The longitudinal transfer function is generated from the dynamic model of the short period approximation.

A typical block diagram for this control system is presented in Figure 13.1.

It needs to be kept in mind that the specific controller we design is very much dependent on the specifics of the dynamics of that particular aircraft. For example, the short period dynamics transfer function of a particular aircraft, such as a conventional transport flying at 150 mph at sea level could have a transfer function given by

$$\frac{\theta(s)}{\delta_e(s)} = \frac{(s + 3.1)}{s(s^2 + 2.8s + 3.24)}.$$ (13.1)

The representative block diagram and the corresponding root locus are shown in Figures 13.2 and 13.3.

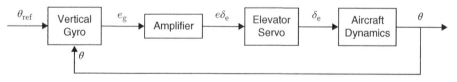

Figure 13.1 Displacement autopilot.

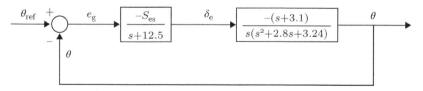

Figure 13.2 Block diagram for the conventional transport and autopilot.

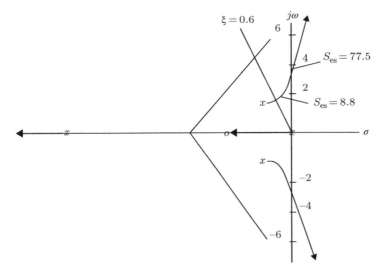

Figure 13.3 Root locus for conventional transport and autopilot.

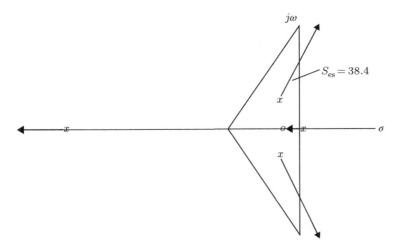

Figure 13.4 Root locus for jet transport and autopilot.

Suppose now a second aircraft, which is a jet transport flying at 600 ft s⁻¹ at 40,000 ft has a transfer function, given by

$$\frac{\theta(s)}{\delta_e(s)} = \frac{(s + 0.306)}{s(s^2 + 0.805s + 1.325)} \tag{13.2}$$

Then for this aircraft's specific transfer function, the root locus turns out to be as shown in Figure 13.4.

Naturally, if we decide to design the controller gain by the root locus method, care needs to be taken to observe that the gain range for stability is quite different for one aircraft compared to the other aircraft. Carefully compare the above two root locus diagrams (shown in Figures 13.3 and 13.4) to realize that the gain determination as well as

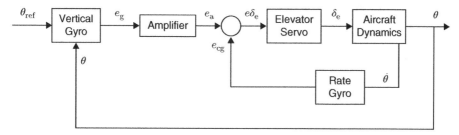

Figure 13.5 Displacement autopilot with pitch rate feedback for damping.

the ranges of these gains allowed to keep the closed loop system stable could be quite different for these two distinct aircraft with distinct transfer functions.

13.3.2 Pitch Displacement Autopilot Augmented by Pitch Rate Feedback

Since the gain range for stability is limited in the above designs, one way to improve on these designs is to augment the system by adding pitch rate feedback, which amounts to PD control action because we are now feeding the derivative of pitch angle, namely the pitch rate q. The corresponding block diagrams for this are given in Figures 13.5 and 13.6.

It is interesting as well as educational to observe that, with pitch rate feedback added, the corresponding root locus becomes as shown Figure 13.7.

The root locus shown in Figure 13.7 clearly shows the improvement in the control gain range because this time the closed loop system is stable for all positive gains in the rate gyro/amplifier gain combination.

Although most aircraft are designed to be statically stable (C_{ma} is negative), certain flight conditions can result in large changes in the longitudinal stability, see Figure 13.8. Such changes occur in some high performance aircraft as they enter the transonic region. However, an even more severe shift in the longitudinal stability results in some high performance aircraft at high angles of attack, a phenomenon referred to as pitch-up.

Pitch-up is most likely to occur in aircraft that have the horizontal stabilizer mounted well above the wing of the aircraft; a common place is the top of the vertical stabilizer. This is sometimes done to obtain the end-plate effect on the vertical stabilizer and thus increase the effectiveness of the vertical stabilizer. Another factor that contributes to

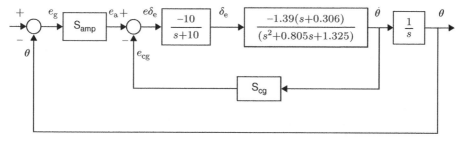

Figure 13.6 Block diagram for the jet transport and displacement autopilot with pitch rate feedback added for damping.

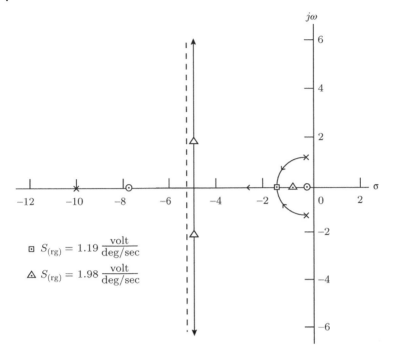

$$\square\ S_{(rg)} = 1.19 \frac{\text{volt}}{\text{deg/sec}}$$

$$\triangle\ S_{(rg)} = 1.98 \frac{\text{volt}}{\text{deg/sec}}$$

Figure 13.7 Root locus for the inner loop of the jet transport and autopilot with pitch rate feedback.

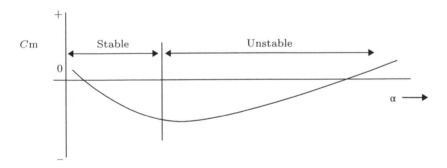

Figure 13.8 *Cm* versus α for an aircraft subject to pitch-up.

this unstable flight condition is a wing with a low aspect ratio. Such a wing has a large downwash velocity that increases rapidly as the angle of attack of the wing is increased.

As the high horizontal tail moves down into this wing wake, pitch-up occurs if the downwash velocity becomes high enough. Pitch-up may occur in straight-wing as well as swept-wing aircraft. For the swept-wing aircraft the forward shift of the center of pressure of the wing at high angles of attack is also a contributing factor. A practical solution of the pitch-up problem is to limit that aircraft to angles of attack below the critical angle of attack; however, this also limits the performance of the aircraft. An aircraft that is subject to pitch-up will generally fly at these higher angles of attack; thus an automatic control system that makes the aircraft flyable at angles of attack greater than the critical

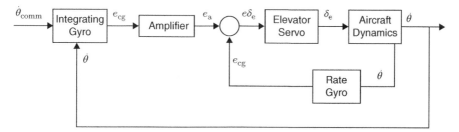

Figure 13.9 Pitch orientational control system.

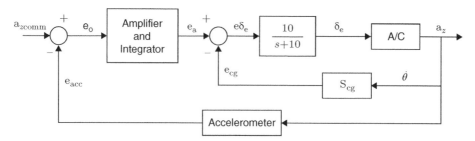

Figure 13.10 Block diagram of an acceleration control system. $S_{acc} = 1 \text{ V g}^{-1}$.

angle of attack increases the performance capabilities of the aircraft. The pitch orientational control system, shown in Figure 13.9, if properly designed, provides this control.

13.3.3 Acceleration Control System

Acceleration control systems are generally needed for fighter type aircraft. Figure 13.10 is a block diagram for an acceleration control system.

The required transfer function can be derived from the Z force equation given below.

$$a_z = \dot{w} - U_0 \dot{\theta} \tag{13.3}$$

Taking a_z as the output variable, we can accordingly get the corresponding C and D matrices and get the appropriate transfer function between a_z and δ_e. We could either use simply the short period approximation equations or the entire longitudinal linear model equations to get the needed transfer function.

For a typical fighter-type high performance aircraft, a representative transfer function for a specific aircraft at a flight condition with a relatively high angle of attack looks as follows:

$$\frac{a_z(s)}{\delta_e(s)} = \frac{-77.7(s^2 - 60)}{s^2 + 0.9s + 8}. \tag{13.4}$$

Factoring,

$$\frac{a_z(s)}{\delta_e(s)} = \frac{-77.7(s + 7.75)(s - 7.75)}{s^2 + 0.9s + 8} \frac{\text{ft s}^{-2}}{\text{rad}}. \tag{13.5}$$

However, the units for the transfer function of the elevator servo are degree/volt, and for the accelerometer they are volt/gram. The units of Equation 13.5 can be changed to gram/degree by dividing by 32.2 (ft s^{-2}) g^{-1} and 57.3 deg rad^{-1}. Thus Equation 13.5 becomes

$$\frac{a_z(s)}{\delta_e(s)} = \frac{-0.042(s + 7.75)(s - 7.75)}{s^2 + 0.9s + 8} \, \frac{g}{\text{deg}}. \tag{13.6}$$

An examination of Equation 13.6 indicates that there is a zero in the right half plane, thus indicating a non-minimum phase transfer function. This type of occurrence of a non-minimum phase transfer function is common for a high angle of attack flight condition. This means that for a positive step input of δ_e the steady state sign of a_z will be positive, which is consistent with the sign convention already established. Thus the sign at the summer for the acceleration feedback must be negative for negative feedback. The sign for the elevator servo remains positive, so that a positive $a_{z(comm)}$ yields a positive a_z output.

The closed-loop transfer function for the inner loop of the acceleration control system (see Figure 13.11) for $S_{(cg)} = 0.23$ is

$$\frac{\dot{\theta}(s)}{e_a(s)} = \frac{-150(s + 0.4)}{(s + 3)(s^2 + 7s + 24)} \, \frac{\text{deg s}^{-1}}{V}. \tag{13.7}$$

The block diagram for the outer loop is shown in Figure 13.12 for $S_{(cg)} = 0.23$.

The $\frac{a_z}{\dot{\theta}}$ block is required to change the output of the inner loop $\dot{\theta}$ to the required output for the outer loop a_z. The transfer function for this block can be obtained by taking the ratio of $\frac{a_z(s)}{\delta_e(s)}$ and $\frac{\dot{\theta}(s)}{\delta_e(s)}$ transfer functions; thus

$$\frac{a_z(s)}{\dot{\theta}(s)} = \frac{-0.042(s + 7.75)(s - 7.75)}{-15(s + 0.4)} \, \frac{g}{\text{deg s}^{-1}}. \tag{13.8}$$

Then the $a_z(s)/e_a(s)$ transfer function is the product of Equations 13.7 and 13.8

$$\frac{a_z(s)}{e_a(s)} = \frac{-0.42(s + 7.75)(s - 7.75)}{(s + 3)(s^2 + 7s + 24)} \, \frac{g}{V}. \tag{13.9}$$

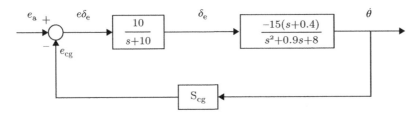

Figure 13.11 Block diagram for the inner loop of the acceleration control system.

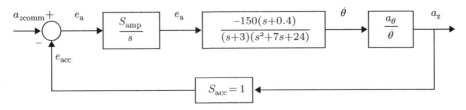

Figure 13.12 Block diagram for the outer loop for the acceleration control system.

This operation changes only the numerator of the forward transfer function by replacing the zero of the $\frac{\theta(s)}{\delta_e(s)}$ transfer function with the zeros of the $\frac{a_z(s)}{\delta_e(s)}$ transfer function. The denominator remains the same. The closed loop transfer function is

$$\frac{a_z(s)}{a_{z(\text{comm})}(s)} = \frac{-0.93(s + 7.75)(s - 7.75)}{(s^2 + 2.2s + 2.4)(s^2 + 7.8s + 24)}. \tag{13.10}$$

Although the accelerometer control system provides good operation, there are some practical problems. One of these is that the accelerometer cannot distinguish between the acceleration due to gravity and accelerations caused by aircraft motion. The acceleration of gravity can be balanced out so that in straight and level flight at normal cruise air speed and altitude the output of the accelerometer is zero. However, at different angles of attack the accelerometer output is not zero. For example, if the angle of attack changed by 10∘ from the value at which the accelerometer was nulled, the output would correspond to +/−0.5 ft/sec². The accelerometer can be adjusted so that it is insensitive to accelerations that are less than 1 ft s⁻², thus eliminating this problem. Another problem that would probably be harder to overcome is the unwanted accelerations arising from turbulence. This shows up as noise and has to be filtered out. As a result of these problems, and because there are not many requirements that call for an aircraft maneuvering at constant acceleration, the acceleration autopilot is not often employed. However, there are some requirements that make the acceleration autopilot ideal. An example is the necessity to perform a maximum performance pull-up in connection with a particular tactical maneuver.

13.4 Control Theory Application to Automatic Landing Control System Design

One of the most interesting and useful applications of control theory in the aircraft flight control field belongs to the design of an automatic landing control system. The goal of the automatic landing control system is to be able to land the aircraft in all weather conditions. To achieve this, we need a way to land the aircraft without any visual reference to the runway. This in turn can be accomplished by an automatic landing control system, which would guide the aircraft down a predetermined glide slope and then at a preselected altitude to reduce the rate of descent and cause the aircraft to flare out and touch down with an acceptably low rate of descent. Thus there are four important phases, (i) glide path stabilization and simultaneous (ii) speed control, then (iii) altitude hold and finally (iv) the flare control.

A pictorial representation of the geometry of the landing phase of an aircraft is given in Figure 13.13.

13.4.1 Glide Path Coupling Phase: Glide Path Stabilization by elevator and Speed Control by Engine Throttle

In the glide path coupling phase, the aircraft's pitch attitude angle θ and simultaneously the speed u_{sp} (among the longitudinal state variables) need to be controlled. That means we need one control system to achieve the glide slope stabilization and another control

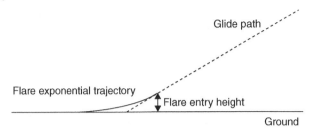

Figure 13.13 Automatic landing geometry.

system to simultaneously control the landing speed. Thus there are two control variables needed during this phase. The pitch angle control for guide slope coupling is done with the elevator angle as the control variable while the speed is controlled by the engine throttle as the control variable. For controlling the pitch angle, as described before, we use the elevator angle δ_e as the control variable with θ as the output variable. Thus in this part of the control system, we essentially work with the transfer function between pitch angle and the elevator angle. As usual, the elevator servo is modeled by a first order transfer function with a 0.1 s lag. Typically we assume that sensors are fast enough that their lags could be neglected to make the control system design more tractable. The control system design of this pitch attitude control would then be in the same lines as the pitch attitude control system discussed before.

So let us now discuss the speed control system aspect. For this, we obtain the transfer function between the forward speed change u_{sp} and the throttle servo angle δth. Typically the throttle servo and engine response transfer function is modeled by first order transfer function with a single 5 s lag. Depending on the complexity of the dynamics, a higher order transfer function may be warranted for some aircraft.

After making sure that there are no close pole zero pairs (so as to avoid any pole zero cancellation), a typical, representative transfer function between speed and throttle servo could be given by

$$\frac{u_{sp}}{\delta_{th}} \approx \frac{K(s + 0.2736 \pm j0.1116)(s + 0.001484)}{(s + 0.2674 \pm j0.1552)(s + 0.0002005)(s + 0.06449)}.$$

Note that the oscillatory poles are shown as simple poles with complex conjugate roots in the above denominator. This transfer function can be used to perform frequency domain design of a speed loop compensator, the details of which are omitted. Please see books such as [7] and [36] for more details on the control system design steps.

13.4.2 Glide Slope Coupling Phase

The geometry associated with the glide slope coupling problem is shown in Figures 13.14 and 13.15.

Let U be the constant steady state forward speed, and d be the perpendicular distance between the glide slope line and the aircraft's cg location (where we now assume the aircraft to be a point mass) and γ be the angle between the forward velocity direction and the horizontal (flight path angle). We also assume the constant angle between ground (runway horizontal) and the glide slope line, a typical value of 2.5°. Then the rate of

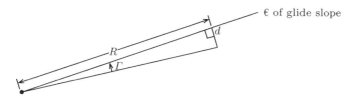

Figure 13.14 Effect of beam narrowing.

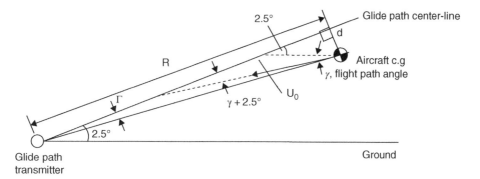

Figure 13.15 Geometry of the glide slope problem. Note: d and γ are negative.

change of distance d is given, from the geometry of the problem, as

$$\dot{d} = U \sin (\gamma + 2.5)° \tag{13.11}$$

which can be approximated as

$$\dot{d} = U/57.3(\gamma + 2.5). \tag{13.12}$$

Then the glide slope error angle, Γ, in degrees (from the figure), can be approximately written as

$$\Gamma = 57.3d/R \tag{13.13}$$

where R is the range distance as shown in the diagram. Then the objective behind a control system (to be designed) is to reduce or bring the angle Γ to zero. Thus an appropriate block diagram for this glide slope coupling control system can be built as shown in Figure 13.16, with Γ as the output and $\Gamma_{ref} = 0$ as the reference input. The coupler controller block's transfer function could be taken as a lead/lag compensator of the type $K(1 + 0.1/s)(s + 0.5)/(s + 5)$.

A representative overall open loop transfer function of the glide slope coupling control system can be seen to be

$$\frac{\Gamma(s)}{e_\Gamma(s)} = \frac{K_T(s + 0.1)(s + 4.85)(s - 4.35)}{(s^2(s + 5)(s + 5.5)(s^2 + 5.4s + 11.4)}$$

where the total gain K_T of the above open loop transfer function is taken as $K_T = K/R$. The block diagram for the above control system is given in Figure 13.17.

The important point to note is that the physical range R distance is serving as a variable control gain within the control system. Thus the idea is to use the root locus technique

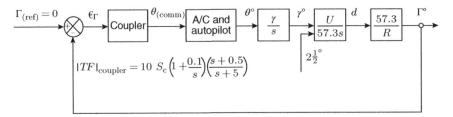

Figure 13.16 Block diagram of an automatic glide slope control system.

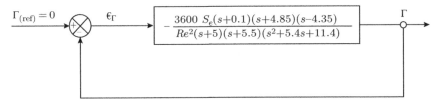

Figure 13.17 Simplified block diagram of automatic glide slope control system.

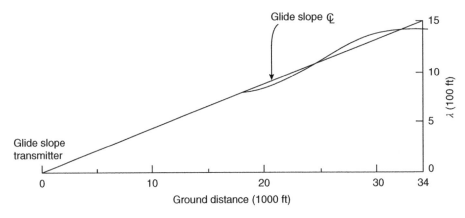

Figure 13.18 Response of a glide slope.

to design a gain K_T for stable operation of the closed loop system and then infer the range R from this control system gain. See [7] for details of a particular design exercise. A simulation of this particular design is shown in Figure 13.18.

From this simulation it is seen that typically a range of about 1500 ft provides a stable operation for a glide slope coupling control system.

Once the glide slope coupling control system is designed, then the next step is to design the flare controller.

13.4.3 Flare Control

At an altitude between 20 and 40 ft above the end of the runway, an automatic flare control system needs to be engaged. This height at which the flare maneuver starts is called the decision height. This trajectory is called the landing flare. Its geometry is as shown in Figure 13.19.

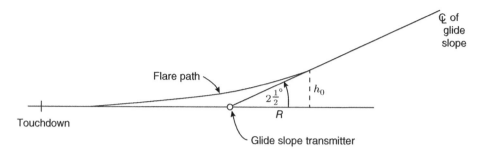

Figure 13.19 Simplified geometry for the flare path.

The principal sensor during this phase is a radar altimeter, and satisfactory performance requires tight control of the aircraft altitude and its rate of descent.

13.4.4 Determination Flare Control Parameters

During the flare maneuver, the glide path angle has to be changed from roughly $-3°$ to a positive value for a smooth, belly-first touchdown. The automated flare control system must control the rate of descent of the aircraft's center of gravity to match the idealized exponential path described by

$$h = h_o e^{-t/\tau} \tag{13.14}$$

while at the same time managing the pitch of the aircraft in preparation for touchdown.

Modern digital computer based flight control systems allow functions such as throttle to be programmed as a function of altitude.

The flare requires a very tight control loop and this design has considerable influence on the nature of landing such as hard or soft landing. Thus selecting appropriate parameters for this flare controller is of extreme importance. In what follows, we present a brief account of the guidelines for determining these parameters.

The equations and constraints for the exponential model that generates the flare command are as follows:

$$h(t) = h_0 e^{\frac{-t}{\tau}}; \text{ therefore, } \dot{h} = -\frac{h}{\tau}$$

where $h_0 = h(0)$, the decision height.

$$h(0) = -\tau \dot{h}(0) = \tau V_T \sin \gamma \text{ (constraint)}.$$

Here V_T is the constant velocity of the aircraft cg during this phase and *gamma* is the flight path angle, which is assumed to be small.

Since γ is very shallow throughout the flare, so the horizontal component of the aircraft's velocity, $V_T \cos \gamma$ is approximately equal to V_T and also will be assumed constant.

$$4\tau V_T = (h_0 / \tan \gamma) + D = \tau V_T \cos \gamma + D$$

where D is the total horizontal distance of the runway to be used until the aircraft comes to a complete stop. Thus this distance includes the horizontal distance R at the touchdown point from the start of the flare maneuver plus the additional distance on the runway to come to a complete stop after the touch down. Thus the total distance D

Figure 13.20 Late touchdown forced airbrake to be used. (Illustration: Wikimedia).

depends on the specific situation with the runway conditions. Since γ is a small angle, $\cos \gamma \approx 1.0$, this equation leads to

$$\tau = D/3V_{\mathrm{T}}.$$

The interesting point here is that the time constant parameter *tau* is related to the runway distances, making both of them design variables, which in turn affects the flare maneuver parameters. These parameters then decide the nature of the landing (hard or soft landing); an example is shown in Figure 13.20.

As an example, with $D = 2000$ ft and $V_{\mathrm{T}} = 250$ ft s^{-1}, they become $\tau = 2.667$ s and $h_0 = 29.1$ ft.

A control system block diagram then has the rate of descent $\dot{h}$ as the output variable with the desired rate of descent $\dot{h}_{\mathrm{r}}$ as the reference signal. Note that this reference signal, which is a function of the time constant parameter τ, comes as a guidance command from the guidance system.

13.4.5 Altitude Hold and Mach Hold Autopilots

With the knowledge we gained from the above examples, it is relatively straightforward to write the appropriate block diagrams and design a control logic for Mach hold and altitude hold autopilots. For an altitude hold autopilot, we need to get the transfer function between the longitudinal state variable, height h and whatever control variable is assumed, typically a control surface deflection such as the elevator angle. Similarly for the Mach hold autopilot, we need to get the transfer function between the longitudinal state variable, the forward speed change u_{sp} and whatever control variable is assumed, typically a control surface deflection such as the elevator angle.

13.4.6 Conceptual Control System Design Steps

To summarize, the major task in the control system design is to clearly understand and identify what the output (controlled) variable is, what the control (input) variable is, and then write an appropriate block diagram with all rest of the actuator, sensor, summer junction, and controller components, and then gather all the transfer functions needed in each of those component blocks and start analyzing the open loop system characteristics. Finally the controller transfer function parameters can be designed using the root locus and Bode plot based design algorithms so that the finalized closed loop transfer function meets all the design specifications.

13.5 Lateral/Directional Autopilots

As mentioned earlier, keeping the objective of this book in mind, we limit our discussion on this topic to the very essential conceptual level. This is justified because as far as the application of classical control theory to the lateral/directional aircraft flight control problems is concerned, what is needed is to essentially replace the longitudinal dynamics transfer functions we considered before to lateral/directional dynamics transfer functions by appropriate labeling of the outputs and inputs involved in the lateral/directional aircraft dynamics. One of the standard autopilots in this topic is the washout circuit designed in the yaw damping control system. Another interesting application pertains to the steady coordinated turn problem. This control system is heavily dependent on the specificity of the lateral/directional dynamics and is worth looking into and getting familiar with by reading dedicated books on aircraft flight control systems, the references for which are given at the end of this chapter. We cover this topic at a conceptual level, without any detailed discussion of the actual control system design steps, for which the reader is encouraged to consult other textbooks dedicated to aircraft control.

13.5.1 Steady Coordinated Turn Control System

There are four design philosophies used in attempting to achieve steady coordinated turn, which are very much based on the nature of lateral/directional motion equations that bring out the features of a steady coordinated turn. For this recall that the lateral/directional state variables are (i) side slip angle β, (ii) roll rate p, (iii) yaw rate r, and finally (iv) bank angle ϕ. The heading angle ψ is typically omitted as a controlled variable, because it is mostly a constant. The control variables are (i) the aileron deflection δ_A and (ii) the rudder deflection δ_R. So the problem of achieving a steady coordinated turn is predicated by the definition of the output variable we wish to select. We now list the four design philosophies.

Design philosophy I. Here we attempt to achieve steady coordinated turn by regulating the side slip angle β to zero. Thus the needed transfer functions are the $\beta(s)/\delta_A(s)$ and $\beta(s)/\delta_R(s)$.

To improve performance, we can even employ an inner loop controlling the rate of change of β, i.e. $\dot{\beta}$, as well.

Design philosophy II. Here we attempt to achieve steady coordinated turn by regulating the lateral acceleration a_y to zero. Thus the needed transfer functions are $a_y(s)/\delta_A(s)$ and $a_y(s)/\delta_R(s)$.

To improve the performance, we can even employ an inner loop controlling the yaw rate, i.e. r, as well.

Design philosophy III. Here we attempt to achieve steady coordinated turn by controlling the yaw rate r to a very specific desired value, i.e r_{command}. This is labeled as the computed yaw rate method. Thus the required transfer functions are $r(s)/\delta_A(s)$ and $r(s)/\delta_R(s)$. The commanded (desired and computed) yaw rate r_{command} is given by the equation

$$r_{\text{command}} = (g/V_T)\sin\phi \tag{13.15}$$

where g is the acceleration due to gravity, V_T is the true airspeed along the lateral y axis, and ϕ is the bank angle. We assume that the right hand side information is known and is available through measurements of those variables.

Design philosophy IV. This method is labeled as the rudder coordination computer in which the rudder angle required is computed for a given amount of aileron angle to achieve a steady coordinated turn.

For this method to work, we need to assume that the transfer functions for aileron and rudder servos are the same and thus equal to each other. This may pose some problems in some practical situations but is sufficiently robust that it can also be employed without that much concern in the majority of situations where the above assumption is satisfied.

13.5.2 Inertial Cross Coupling

This is an important phenomenon specific to aircraft linear dynamics. As we have seen from the discussion in previous sections, the linear control design methods were applied by invoking the decoupling between longitudinal dynamics and the lateral/directional dynamics of the aircraft dynamics for small motions in the linear range. Thus in some sense, the control systems were designed in a modular way, separately for each control objective at hand. However, as the aircraft geometries evolved over time based on aerodynamic considerations over the large flight envelope, slowly it turned out that for some high performance aircraft with slender wing bodies, the weight distribution started to change in such a way that more weight became concentrated in the fuselage as the aircraft's wings became thinner and shorter. This in turn changed the moments of inertia distribution along the 3 axis, wherein the moment of inertia about the longitudinal x axis decreased while the those along the roll/yaw axes started increasing. This uneven moment of inertia distribution no longer allowed the assumption of decoupling of dynamics between the longitudinal motion and the roll/yaw motion. This coupling between the longitudinal and lateral/directional motion of these types of aircraft is labeled as the inertial cross coupling phenomenon. This phenomenon results in the need to make modifications to the traditional modular design practice. The intent was then to see how far we can stretch these modular linear control systems (i.e linear control systems designed separately for longitudinal dynamics and for lateral/directional dynamics) to handle this cross coupling between these two types of dynamics in an integrated way. Accordingly, an integrated control system that accounts for this inertial cross coupling was designed.

13.6   Chapter Summary

In this chapter, we have covered the application of the classical control theory tools to the specific area of aircraft flight control systems. In line with the objective of this book, our emphasis was to first master the theoretical techniques of classical control theory and then apply them to flight vehicle control systems field. Again, keeping the overall (deliberate) scope of the book, we limited our coverage to a reasonable level on aircraft longitudinal control systems and then treating the lateral/directional control systems in a peripheral way as the essential skill remains the same whether it is a longitudinal dynamics related transfer function or lateral/directional related transfer function. This philosophy of conceptual coverage of this application area simply means that the student who is interested in pursuing deeper knowledge of aircraft control systems is expected to consult many praiseworthy books that deal specifically with aircraft flight dynamics and control, which are listed at the end of this chapter in the reference section.

With this understanding, it is now time to move on to learn few basics of spacecraft flight control systems in the next chapter.

Bibliography

1 M.J. Abzug and E.E Larrabee. *Airplane Stability and Control: A history of the Technologies that made Aviation possible*. Cambridge University Press, Cambridge, U.K, 2 edition, 2002.

2 R.J. Adams. *Multivariable Flight Control*. Springer-Verlag, New York, NY, 1995.

3 D. Allerton. *Principles of flight simulation*. Wiley-Blackwell, Chichester, U.K., 2009.

4 J. Anderson. *Aircraft Performance and Design*. McGraw Hill, New York, 1999.

5 J. Anderson. *Introduction to Flight*. McGraw-Hill, New York, 4 edition, 2000.

6 A. W. Babister. *Aircraft Dynamic Stability and Response*. Pergamon, 1980.

7 J.H. Blakelock. *Automatic Control of Aircraft and Missiles*. Wiley Interscience, New York, 1991.

8 Jean-Luc Boiffier. *The Dynamics of Flight, The Equations*. Wiley, 1998.

9 L.S. Brian and L.L. Frank. *Aircraft control and simulation*. John Wiley & Sons, Inc., 2003.

10 A. E Bryson. *Control of Spacecraft and Aircraft*. Princeton University Press, Princeton, 1994.

11 Chris Carpenter. *Flightwise: Principles of Aircraft Flight*. Airlife Pub Ltd, 2002.

12 Michael V Cook. *Flight dynamics principles*. Arnold, London, U.K., 1997.

13 Wayne Durham. *Aircraft flight dynamics and control*. John Wiley & Sons, Inc., 2013.

14 B Etkin and L.D. Reid. *Dynamics of flight: stability and control*, volume 3. John Wiley & Sons, New York, 1998.

15 E.T. Falangas. *Performance Evaluation and Design of Flight Vehicle Control Systems*. Wiley and IEEE Press, Hoboken, N.J, 2016.

16 T. Hacker. *Flight Stability and Control*. Elsevier Science Ltd, 1970.

17 G.J. Hancock. *An Introduction to the Flight Dynamics of Rigid Aeroplanes.* Number Section III.5-III.6. Ellis Hornwood, New York, 1995.

18 D.G. Hull. *Fundamentals of Airplane Flight Mechanics.* Springer International, Berlin, Germany, 2007.

19 D. McLean. Automatic flight control systems. *Prentice Hall International Series in System and Control Engineering*, 1990.

20 Duane T McRuer, Dunstan Graham, and Irving Ashkenas. *Aircraft dynamics and automatic control.* Princeton University Press, Princeton, NJ, 1973.

21 A. Miele. *Flight Mechanics: Theory of Flight Paths.* Addison-Wesley, New York, 1962.

22 Ian Moir and Allan Seabridge. *Aicraft Systems: Mechanical, Electrical and Avionics Subsytems Integration*, 3rd Edition. John Wiley & sons, 2008.

23 Ian Moir and Allan Seabridge. *Design and Development of Aircraft Systems.* John Wiley & sons, 2012.

24 Robert C Nelson. *Flight stability and automatic control.* McGraw Hill, New York, 2 edition, 1998.

25 C. Perkins and R. Hage. *Aircraft Performance, Stability and Control.* John Wiley & Sons, London, U.K., 1949.

26 J.M. Rolfe and K.J. Staples. *Flight simulation.* Number 1. Cambridge University Press, Cambridge, U.K., 1988.

27 Jan Roskam. *Airplane Flight Dynamics and Automatic Flight Controls: Part I.* Roskam Aviation and Engineering Corp, 1979.

28 Jan Roskam. *Airplane Flight Dynamics and Automatic Flight Controls: Part II.* Roskam Aviation and Engineering Corp, 1979.

29 J.B. Russell. *Performance and Stability of Aircraft.* Arnold, London, U.K., 1996.

30 D.K. Schmidt. *Modern Flight Dynamics.* McGraw Hill, New York, 2012.

31 Louis V Schmidt. *Introduction to Aircraft Flight Dynamics.* AIAA Education Series, Reston, VA, 1998.

32 D. Seckel. *Stability and control of airplanes and helicopters.* Academic Press, 2014.

33 R. Shevell. *Fundamentals of Flight.* Prentice Hall, Englewood Cliffs, NJ, 2 edition, 1989.

34 Frederick O Smetana. *Computer assisted analysis of aircraft performance, stability, and control.* McGraw-Hill College., New York, 1984.

35 R.E. Stengel. *Flight Dynamics.* Princeton University Press, Princeton, 2004.

36 Brian L Stevens, Frank L Lewis, and Eric N Johnson. *Aircraft control and simulation.* Interscience, New York, 1 edition, 1992.

37 P. Swatton. *Aircraft Performance Theory and Practice for Pilots.* Wiley Publications, 2 edition, 2008.

38 A. Tewari. *Atmospheric and Space Flight Dynamics.* Birkhauser, Boston, 2006.

39 A. Tewari. *Advanced Control of Aircraft,Spacecraft and Rockets.* Wiley, Chichester, UK, 2011.

40 Ranjan Vepa. *Flight Dynamics, Simulation and Control for Rigid and Flexible Aircraft.* CRC Press, New York, 1 edition, 2015.

41 N. Vinh. *Flight Mechanics of High Performance Aircraft.* Cambridge University Press, New York, 1993.

42 P.H. Zipfel. *Modeling and simulation of aerospace vehicle dynamics.* American Institute of Aeronautics and Astronautics Inc., 2000.

14

Application of Classical Control Methods to Spacecraft Control

14.1 Chapter Highlights

In this chapter, we focus on the application of classical control theory to spacecraft/satellite attitude control systems. There are excellent books that deal specifically with spacecraft/satellite control systems, which are given in the references section of this chapter. With the intention of not repeating that coverage here in this book, in this chapter, we cover the control design exercise that this author was personally involved in his early higher education phase. The specific satellite control problem this author was involved in was the attitude control system for a satellite with near resemblance to the first satellite the Indian Space Research Organization launched, named the Aryabhata satellite. We first cover the pitch axis attitude control system and then cover the roll/yaw coupled control system. The emphasis is on the conceptual formulation of the problem in the classical control theory framework. The essential material is taken from [6].

14.2 Control of an Earth Observation Satellite Using a Momentum Wheel and Offset Thrusters: Case Study

14.2.1 Overview

Satellites for remote sensing of Earth resources need to be Earth oriented. The presence of various disturbances, both internal and environmental, necessitates an effective attitude control system. For attitude stabilization and control a variety of systems have evolved in the past few decades. Among them is the use of a fixed momentum biased reaction wheel and gas jets. The wheel provides roll/yaw stiffness and controls pitch axis motion [1]. For removing undesirable momentum (for momentum dumping) a pair of gas jets are employed. A second pair of thrusters offset from the yaw axis are used for roll/yaw control and active nutation damping [4]. The geostationary Canadian Telecommunication System CTS used such a system with success [5]. A similar system has also been suggested for use by India for a geostationary communication satellite [7]. This section extends the applicability of the above concept to a near Earth, Sun synchronous satellite that is subjected to larger environmental torques and needs continuous control of the satellite axis to compensate for orbital precession [2]. The system is attractive as it avoids yaw sensing and requires a lower number of moving parts for high accuracy three axis control. Some assumed parameters:

Flight Dynamics and Control of Aero and Space Vehicles, First Edition. Rama K. Yedavalli.
© 2020 John Wiley & Sons Ltd. Published 2020 by John Wiley & Sons Ltd.

- Weight of satellite = 300–400 kg
- Power = 40–60 W
- Lifetime = 1 yr
- Nearly spherical, with radius = 0.5 m
- Moment of inertia = 100, 80, 80 kg m^2
- Distance between center of pressure and center of mass = 10 cm
- Pointing accuracy = 0.5°
- Power limitation (continuous) on control system = 2–4 W
- Weight limitation = 20–30 kg
- Horizon sensor accuracy = 0.1°.

14.2.2 Formulations of Equations

Two body-centered orthogonal coordinate systems are used for modeling the problem in Figure 14.1: (a) a body-centered orbital coordinate frame $x_0 y_0 z_0$ with x_0 along the local vertical and z_0 normal to the orbital plane, and (b) the satellite's principal coordinate frame xyz, which is related to the former by three Eulerian rotations denoting the pitch (ψ), roll (ϕ), and yaw (λ) sequence. For a satellite with a rotor about the pitch axis the

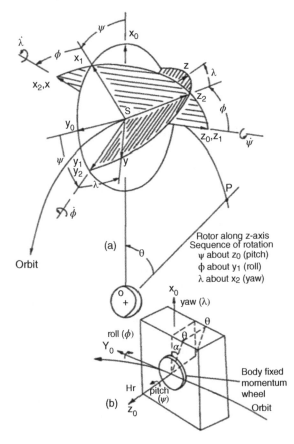

Figure 14.1 (a) Geometry of flight path; (b) nominal on-orbit attitude of satellite.

equations of motion are:

$$I_x \dot{\omega}_x - (I_y - I_z)\omega_y\omega_z + \omega_y H_r = T_x$$
$$I_y \dot{\omega}_y - (I_z - I_x)\omega_x\omega_z + \omega_x H_r = T_y$$
$$I_z \dot{\omega}_z - (I_y - I_x)\omega_x\omega_y + H_r = T_z.$$

The body rates can be expressed in terms of orbital and Eulerian angle rates (ignoring orbital perturbations) as

$$\omega_x = \dot{\lambda} - (\dot{\psi} + \dot{\theta})\sin\phi$$
$$\omega_y = \dot{\phi}\cos\lambda + (\dot{\psi} + \dot{\theta})\sin\lambda\cos\phi$$
$$\omega_z = (\dot{\psi} + \dot{\theta})\cos\lambda\cos\phi - \dot{\phi}\sin\lambda.$$

On linearization the equations of motion for an axi-symmetric satellite ($I_x = I_y = I$) transform into

$$I\ddot{\lambda} + \dot{\theta}\lambda[H_r - \dot{\theta}(I - I_z)] + \dot{\phi}[H_r - \dot{\theta}(2I - I_z)] = T_x$$
$$I\ddot{\phi} + \dot{\theta}\phi[H_r - \dot{\theta}(I - I_z)] - \dot{\lambda}[H_r - \dot{\theta}(2I - I_z)] = T_y$$
$$I_z\ddot{\psi} + H_r = T_z.$$

For small perturbations, therefore, pitch dynamics is decoupled from roll/yaw dynamics. The environmental torques disturbing the satellite attitude arise mainly from the Earth's magnetic field and atmosphere [2]. Table 14.1 gives their magnitude and nature.

Table 14.1 Disturbance torque on the satellite.

Axis	Source	Magnitude	Nature
Principal axes:			
S_x	Magnetic	0.735×10^{-5}	Periodic at
S_y	Magnetic	1.47×10^{-5}	orbital frequency
S_z	Magnetic	1.65×10^{-5}	
z	Atmospheric	1×10^{-5}	Secular
All	Others	$< 1 \times 10^{-6}$	
Nodal frames:			
OX_n	Magnetic	0.35×10^{-5}	Secular
		1.1×10^{-5}	Periodic at
OY_n	Magnetic	1.1×10^{-5}	twice the orbital frequency
OZ_n	Magnetic	1.65×10^{-5}	Periodic at orbital frequency
	Atmospheric	1×10^{-5}	Secular
All	Others	1×10^{-6}	

14.2.3 Design of Attitude Controllers

Equipment proportioning

A preliminary choice of the values for some of the design parameters like rotor momentum, jet thrust level, gas requirements, etc. can be made from static considerations as follows:

1. A cyclic torque of 1.1×10^{-5} N m having twice the orbital frequency and a secular torque of 0.35×10^{-5} N m acts on the satellite perpendicular to the rotor spin axis. Under the action of cyclic disturbance the spin axis oscillates about a mean position, while a secular disturbance causes a constant drift of the spin axis from the orbit normal. The drift Δ at any instant can be estimated as

$$\Delta = \frac{\int_0^t T \, dt}{H_r}, \quad q = x \text{ or } y. \tag{14.1}$$

From the mission roll/yaw pointing accuracy requirements, this drift should not be more than $\pm 0.5°$. Accounting for the sensor inaccuracies of $0.1°$, only $0.4°$ may be permitted. From the considerations of efficient fuel utilization, the availability of ground command (in case of any failure of onboard logic), and orbital precession ($0.985°/\text{day}$) a total drift of more than $0.8°$ ($\pm$ side to $\mp$ side) may not be permitted in 12 h. Then, the necessary angular momentum to be imparted to the wheel to overcome the secular disturbances is 10 N m s. With this angular momentum, the drift/orbit due to secular disturbances is estimated to be $0.115°$. The amplitude of the oscillations due to cyclic disturbances will be approximately $0.1°$.

2. Under the action of disturbances the reaction wheel acts as a momentum storage device. The angular momentum, acquired due to secular torques, has to be damped occasionally using reaction jets. If the change in angular momentum of the wheel from the base value is denoted by ΔH and this is damped by the operation of jets for time t_c, then the jet torque required for momentum dumping is given by

$$T_j = \frac{\Delta H}{t_c} = \frac{\int_0^t T_z \, dt}{t_c} \tag{14.2}$$

where t_d is the time between two successive momentum dumping operations.

3. Under ideal conditions, the output torque from the pitch control motor should equal the jet torque. Hence the output power required of the motor to cater to this jet torque is given by

$$P_{out} = T_j \omega_r \tag{14.3}$$

4. The change in speed $\Delta \omega_r$ (due to change in angular momentum ΔH) in t_d is generally limited to ± 3 to 10% of the base speed. Choosing a jet torque of 0.02 N m, t_d/t_c is found to equal 2000. The corresponding values of other parameters found using the above relations are given in Table 14.2.

5. The minimum weight of the gas required for dumping operation over the specified useful lifetime of the satellite is given by

$$W_{gas} = \frac{\Delta H^*}{g_0 r I_{sp}}$$

$$= \frac{\int_0^{t^*} T_z \, dt}{g_0 r I_{sp}} \tag{14.4}$$

Table 14.2 Choice of design parameters.

P_{out}	Base speed	J_1	$\Delta\omega_r = \pm3\%$		$\Delta\omega_r = \pm5\%$		$\Delta\omega_r = \pm10\%$	
W	N rpm	kg m^2	t_c	t_d	t_c	t_d	t_c	t_d
1	500	0.192						
2	1000	0.095	14 s	8.4 h	25 s	14 h	50 s	28 h
3	1500	0.064						

where t^* is the mission lifetime. Using dry compressed air ($I_{sp} = 70$ s) and moment arm of 0.5 m, the yearly gas requirement is found to be 0.9 kg for momentum dumping and 0.32 kg for axis control. In addition, 0.224 kg of gas per year is required to overcome changes due to orbital precession. Thus the total minimum requirement is 1.5 kg. The actual amount, of course, depends upon the control logic, efficiency of the total system, leakage, etc.

Pitch control logic

Laplace transformation of pitch dynamics gives

$$s^2 I_z \psi(s) = T_z(s) + L(s) \tag{14.5}$$

where the control torque $L = -\dot{H}_r$. The control block diagram of the pitch control loop is shown in Figure 14.2. Proportional plus integral cascade compensation is chosen for simplicity and accuracy. For the rotor motor unit, L depends upon the input voltage E_m to its control winding as in [3]

$$\frac{L_s(s)}{E_m(s)} = \frac{K_m J_1 s}{\tau_m(s + 1/\tau_m)}. \tag{14.6}$$

With this, the transfer functions of the system are

$$\frac{\psi(s)}{\psi_e(s)} = \frac{K(s + K_t)}{s^2(s + 1/\tau_m)} \quad \text{where} \quad K = \frac{K_a K_m J_1}{\tau_m I_z}$$

$$= \frac{K(s + K_1)}{s^2(s + 1/\tau_m) + K(s + K_1)}$$

$$= \frac{K(s + K_1)}{(s^2 + 2\xi\omega_n s + \omega_n^2)(s + \delta)}$$

$$= -\frac{L(s)}{T(s)} \tag{14.7}$$

and

$$\frac{\psi_e(s)}{T(s)} = \frac{-(s + 1/\tau_m)}{I_z(s^2 + 2\xi\omega_n s + \omega_n^2)(s + \delta)}. \tag{14.8}$$

From these relations,

$$\omega_n^2 = \frac{2\xi\delta K_1}{\delta - K_1} \tag{14.9}$$

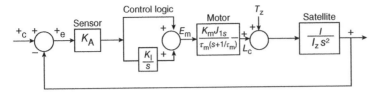

Figure 14.2 Block diagram for pitch control.

$$\frac{1}{\tau_m} = \delta + 2\xi\omega_n \tag{14.10}$$

$$\omega_n^2\delta = KK_1 \tag{14.11}$$

Also from the Routh–Hurwitz criterion, $K_1 < 1/\tau_m$. Here we choose $K_1 = 0.1/\tau_m$. Since δ is the dominant pole in the error-torque transfer, a reasonable value of $\xi = 0.5$ is taken. With these choices, all the parameters of the transfer functions are given in terms of the motor unit time constant only. This single parameter can be obtained from the overall error budget and motor power limitations.

In the steady state (nominal operation mode), the satellite is subjected to small secular and periodic environmental torques (Table 14.1). Taking into account the sensor inaccuracy and giving allowance to modeling errors one may aim at a steady-state error limitation of 0.3°. For this, from Equation 14.8, the motor time constant $\tau_m \approx 190$ s.

During momentum dumping operation a jet torque of 0.02 N m acts for t_c. Though the steady state error due to this pulse is zero, the transient error builds up to too large a value to be tolerated for the motor rotor unit time constant found above. The time constant should, therefore, be selected on the basis of the transient error during the momentum dumping mode. For finding the time t when the transient error reaches the maximum and for obtaining τ_m to limit it to a desired value, one needs to solve a rather involved transcendental equation [8]. What is attempted, alternatively, is to complete the design by root locus, determine $\psi_{e,max}$, and tabulate the results for different values of time constant τ_m. A typical root locus plot for a particular time constant ($\tau_m = 7.7$ s) is shown in Figure 14.3. The results obtained for $t_c = 14$s and $T_j = 0.02$Nm are tabulated in Table 14.3.

To have less error, a fast acting motor should be chosen. However, there is limitation on the available power. The former requirement is in conflict with the latter. From power

Table 14.3 Salient error response features for different motor-rotor unit time constants.

τ_m s	Total steady state error	Maximum error during dumping mode	Time taken to reach maximum errors	Time elapsed from the start to converge to 0.5°, s
190	0.3°	Very large	-	-
18	0.023°	4°	50	250
7.7	0.00374°	1.6°	26	91
4.6	0.00147°	0.91°	20	34

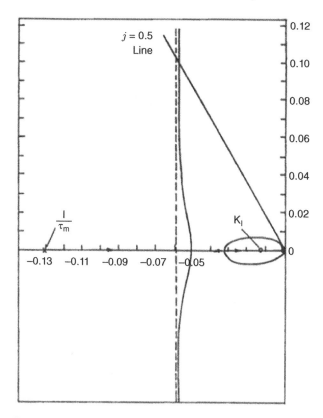

Figure 14.3 Root locus for the case $\tau_m = 7.7$s.

consideration it is advisable to use a wheel with low base speed. From Table 14.2, a wheel with base speed of 500 rpm and inertia of 0.192 kg m² is selected. For a rated no load speed $\omega_m = 550$ rpm, the power ratings for different time constants are tabulated in Table 14.4. In this case, the actual output power for dumping is 1 W and assuming (to be conservative) the efficiency of the motor to be only about 16–20% during the dumping operation, the input power required is 6 W. It may be noted from Tables 14.3 and 14.4 that in meeting the power limitation, the error during dumping mode is to be allowed to a large value.

A tolerable error, meeting the power limitation, can be attained if the dumping is done by intermittent torque pulses, instead of a single torque pulse of duration t_c. The time between two consecutive torque pulses should be so chosen as to minimize the error as well as the total dumping operation time. A null-point firing scheme can be applied to advantage [7].

Pitch Response

Typical response plots of the system are obtained for $\tau_m = 7.7$ s. For other time constants, the responses are qualitatively similar. Figure 14.4(a) shows the error response to a secular (step) disturbance. The near exponential build up of the error to its steady state value of around 0.004° may be noted. The response of the system to a step input command in pitch attitude is shown in Figure 14.4(b). The same plot represents the response

Table 14.4 Power rating requirements for different pitch controller parameters.

[Base speed = 500 rpm; no load speed at 21V (dc) = 550 rpm; J_1 = 0.192 kgm²]

τ_m s	L_s^* rated stall torque, Nm	P_{out}^* rating W	P_{in}^* rating W	k_n V rad^{-1}	k_1 s^{-1}	k_m rad, s^{-1} V^{-1}
18.0	0.61	6.6	20	8.55	0.0056	2.74
7.7	1.43	15.5	46	22.2	0.013	2.74
4.6	2.4	26	78	35.2	0.0216	2.74

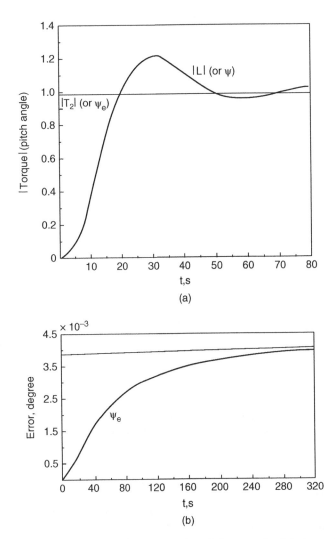

Figure 14.4 Response during normal mode for a secular disturbance torque (or step pitch angle command).

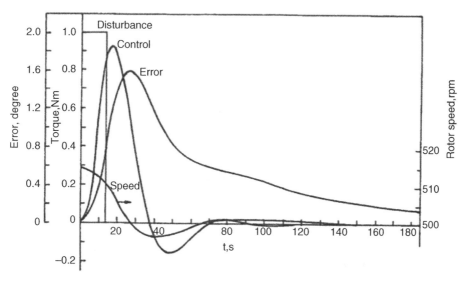

Figure 14.5 Control torque, error angle, and motor speed during dumping mode.

of the control torque for a step change in disturbance torque. The rise time is about 15 s and the settling time is approximately 54 s, with a percentage overshoot of 22%. Figure 14.5 shows the jet torque, motor control torque error response, and the motor speed response during the dumping phase (with $t_c = 14$ s). The maximum pitch error developed is 1.6° and occurs 26 s after the initiation of dumping. The error reduces to 0.5° and the speed becomes nominal in about 90 s. The peak error and power demand during the dumping mode exceeds the specified limit. However, the overall dumping phase lasts only for minutes and only two or three times per day; this may not pose a serious problem.

Spin Axis Control and Nutation Damping

The spin axis correction involves changing the direction of total angular momentum of the satellite and then removal of any undesired motion of the spin axis. The change in the direction of angular momentum can be brought about only by external torquing either using magnetic coils or reaction jets. In the present case reaction jets have been considered for axis correction.

Nutation induced by jet torquing can be removed either by passive nutation damping or by active devices. For the given inertia ratio and estimated momentum the nutation frequency ($\approx H_r/I$) is found to be very small (≈ 0.125 rad s^{-2}). As such, conventional passive nutation dampers may not be effective. Active nutation attenuation using reaction jets is considered here [4]. In this system, the axis correction jets offset from the roll axis (Figure 14.1) are actuated by the roll sensors. The small yaw torque component generated due to the offset provides yaw damping. Figure 14.6, extracted from reference [4] shows the block diagram of the roll/yaw controller. The derived rate modulator, which sets the on time of the jets, employs two one-shot multi-vibrators set to the nominal on-time so as to reduce the effects of modulator hysteresis variations on the size of the minimum impulse bit. The dashed box contains an active nutation scheme. The scheme

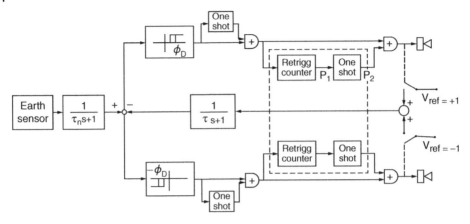

Figure 14.6 Block diagram of spin-axis controller.

prevents the system from getting into a hard two sided limit cycle. As detailed in [4], the system works as follows:

After a dead band actuated control pulse, a dead beat pulse of the same polarity is fired, approximately $\frac{5}{8}T_N$ later (where T_N is the nutation period), provided no additional dead band actuated pulse of the same polarity has been fired in this interval. The various control parameters of the roll/yaw controller, offset angle and the size of the minimum impulse bit are found using the following relations:

1. In order to avoid unnecessary offset thruster firings and to forestall noise induced limit cycling across the roll dead zone, the noise filter time constant τ_n is selected such that maximum noise attenuation is obtained without affecting the control dynamics adversely. It is suggested to select the filter corner frequency as $1/\tau_n = 10\omega_p$. To adequately damp the dynamic mode corresponding to the orbital frequency, one can choose the thruster offset angle as

$$\alpha = \tan^{-1}\left(2\frac{\sqrt{\theta I_z}}{H_r}\right) \tag{14.12}$$

2. The modulator parameters K_f and τ are selected such that the closed loop nutation pole pair, which are the dominant roots, lie within a suitable region in the s plane. Let the closed loop nutation pole pair be given by

$$\delta_1 = -a \pm jb. \tag{14.13}$$

The following criterion may be used to locate the pole pair:

$$1.2\omega_p \le 2.5\omega_p \quad \text{and} \quad 0.3 < \xi < 0.7 \tag{14.14}$$

where ξ is the damping ratio associated with δ_1, given by

$$\xi_\delta = \cos\left[\tan^{-1}\left(\frac{b}{a}\right)\right]. \tag{14.15}$$

It can be shown

$$\tau = \frac{2a}{b^2 + a^2 - \omega_p^2} \tag{14.16}$$

$$K_f = \frac{T_j \cos \alpha}{I(b^2 + a^2 - \omega_p^2)}. \tag{14.17}$$

3. One of the most critical design parameters of a roll/yaw controller is the minimum impulse bit ΔH; if it is too large, the sensor noise will cause severe and rapid multiple pulse firings since the vehicle is not turned away from the roll boundary sufficiently quickly. Hence, from these considerations, as a conservative estimate [4]

$$\frac{1.2 \tan \alpha}{\cos \alpha} \leq \frac{\Delta H}{\phi_D H_r} \leq \frac{2}{1 + \sin \alpha + 2 \sin(45° + \alpha/2)}. \tag{14.18}$$

One can choose the value of ΔH corresponding to the optimum design locus given in [4].

4. The effective hysteresis of the modulator corresponding to the minimum pulse width t_{min} is given by

$$h = \frac{K_f t_{min}}{\phi_D \tau}. \tag{14.19}$$

To make the modulator as noise rejectant as possible, the hysteresis should be such that $h\phi_D \geq 2.5\sigma_\psi$, where σ_ψ is the noise variance at the output of the filter.

Table 14.5 shows the values of the various roll/yaw control parameters thus chosen as initial estimates.

Roll/yaw Response

Using the design parameters the equations of motion are solved numerically under various modes of operation. Figure 14.7(a) shows the motion of the tip of the spin axis over three orbits. It can be observed that the spin axis moves in a diverging spiral to reach the dead bands ($\pm\phi_D$). In the absence of nutation, the spin axis will go into a limit cycle behavior along the dashed lines shown in Figure 14.7(a). Hence the maximum positive yaw error can only be $\pm 0.54°$. The maximum negative yaw error is found to be $-0.49°$. For axis correction, a jet torque of 0.02 N m is considered to act for 0.4 s. Figure 14.7(b)

Table 14.5 Roll/yaw controller parameters.

Reaction wheel momentum bias, H_τ	10 N ms
Roll dead band, ψ_D	± 0.4 deg
Thruster offset angle, α	3 deg
Minimum impulse bit, ΔH	0.008 N ms
Dead beat pulse	0.008 N ms
Minimum thruster on-time, t_{min}	0.4 s
Modulator feedback gain, K_t	0.0547 rad
Modulator feedback time constant, τ	12.5 s
Modulator hysteresis, h	0.25
Mutation attenuation delay time	30 s
Nutation period, T_N	50 s
Offset thruster force	0.04 N

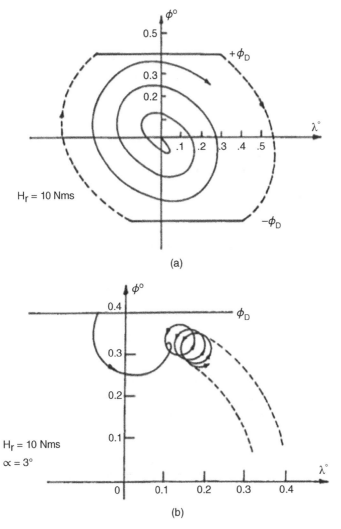

Figure 14.7 Depicted in (a) is the motion of the tip of the spin axis over three orbits, and in (b) typical corrected trace of the spin axis in the roll/yaw plane.

shows a typical trace of the spin axis in roll and yaw plane in the correction mode. Here the spin axis is assumed to reach the dead band with $\dot{\phi} = 0.0001$ rad s^{-1} and $\gamma = -0.0001$ rad s^{-1}. It can also be observed that after the firing of the nutation attenuation pulse, the nutation amplitude is reduced to 0.032°. Further, because of the orbital mode, the spin axis is carried towards the other dead band limit.

14.2.4 Summary of Results of Case Study

A design of a system for three axis attitude control of an Earth observation satellite is presented. The system consists of a momentum biased pitch reaction wheel operation on proportional plus integral scheme, a set of reaction jets for momentum dumping, a

Table 14.6 Estimation of weight requirement, based on some existing systems (Arybhata, SEO, APPLe).

Item	Approximate weight, kg
Momentum wheel	6
Compressed air	2
Gas bottles, pipelines, control valves, etc	10
Control electronics	1
Reserve	1
Total	20

pair of offset thrusters for roll/yaw control and nutation damping, horizon sensor, timer, and a speedometer. The system uses continuous power of about 1 W (peak power 6 W for the momentum wheel during momentum dumping, and some power for operating valves of the jet system two or three times per day). The total weight requirement is less than 2 kg for one year life (Table 14.6). The accuracy achieved is better than 0.5°.

14.3   Chapter Summary

In this chapter, we have covered the application of classical control theory (such as the root locus technique and a PI controller) for pitch axis control of an axi-symmetric satellite in near Earth orbit. The control system uses a pair of reaction jet thrusters for momentum dumping and another pair of offset thrusters for roll/yaw control and nutation damping.The data considered is quite realistic as the satellite parameters are those of a real world satellite launched by the Indian Space Research Organization in the early phases of Indian space program, in which this author was involved.

Bibliography

1 H. Dougherty and J. Rodden. Analysis and design of whecon – an attitude control concept. In *2nd AIAA Communications Satellite Systems Conference*, page 461, San Francisco, CA, 1968.

2 Deutsche Gesellschaft für Luft-und Raumfahrt and Deutsche Forschungs-und Versuchsanstalt für Luft. *Zeitschrift für Flugwissenschaften und Weltraumforschung: ZfW*. Springer, 1977.

3 A.L. Greensite. *Analysis and design of space vehicle flight control systems*. spartan Books, New York, 1970.

4 R.P. Iwens, A.W. Fleming, and V.A. Spector. Precision attitude control with a single body-fixed momentum wheel. In *AIAA Mechanics and Control of Flight Conference (5–9 August, Anaheim, CA)*, 1974.

5 J.K. Nor. Analysis and design of a fixed-wheel attitude stabilization system for a stationary satellite. Technical report, Institute of Aerospace Studies, University of Toronto, 1974.

6 S. Shrivastava, U.R. Prasad, V.K. Kumar, and Y. Ramakrishna. Control of an earth observation satellite using momentum wheel and offset thrusters. *Indian Journal of Technology*, 20, July 1982.

7 S.K. Shrivastava and N. Nagarajan. Three-axis control of a communication satellite. In *Fifth National Systems Conference*, Ludhiana, India, 1978.

8 J.G. Truxal, editor. *Control engineers' handbook*. McGraw-Hill, New York, 1958.

Part III

Flight Vehicle Control via Modern State Space Based Methods

Roadmap to Part III

"If you think Education is expensive, try Ignorance"
— Derek Bok, *Former President of MIT*

Part III covers Fundamentals of Flight Vehicle Control via Modern State Space Based methods and consists of Chapters 15 through 23. The basics of matrix theory and linear algebra needed for understanding this part of the subject are presented in Appendix C and form the background needed for the material in this part. Chapter 15 gives an overview of the state space representation of linear systems in continuous as well as discrete time and sampled data formulations. Examples are given both from aircraft models as well as spacecraft models, taken from the previous Part I of mathematical modeling. Chapter 16 covers the state space dynamic system response via the state transition matrix approach, including the cases of continuous time, discrete time and sampled data systems. The application of these methods for aircraft and spacecraft models is highlighted. Chapter 17 covers the first of the three structural properties of a linear state space system, namely stability. A thorough discussion of the stability of dynamic systems is presented starting with nonlinear systems, equilibrium states and the basics of the Lyapunov stability theory. Then techniques for testing the Hurwitz stability of real matrix are presented including the seldom covered Fuller's conditions along with its relationship to the Routh-Hurwitz criterion and the Lyapunov matrix equation approach. Then Chapter 18 covers the other two structural properties of a linear state space system, namely Controllability (stabilizability)/Observability (detectability). Examples are given highlighting the state space models encountered in aircraft dynamics as well as spacecraft dynamics, keeping track of the similarities as well differences between aircraft and spacecraft situations. Chapters 19 and 20 present two popular design techniques namely Pole Placement and LQR techniques respectively. Chapter 21 covers Observer based feedback control along with variable order Dynamic Compensator design and 'strong stability' concept. Finally the aspect 'spillover instability' is introduced, possibly for the first time in an Undergraduate textbook. Then Chapter 22 illustrates the application of these design methods to aircraft control problems. Finally Chapter 23 illustrates the application of these design methods to spacecrafts by using satellite formation flying problem as an example.

15

Time Domain, State Space Control Theory

15.1 Chapter Highlights

In this chapter, we briefly cover the fundamentals of state space control theory by first reviewing the techniques of representing a set of simultaneous, ordinary differential (continuous time) equations in a state space representation, along with the formal definition of state variables, control (input) variables, controlled (output) variables and measurement (sensor) variables. Then we also review the discrete time, difference equations in a state space form. We then discuss the importance of linear transformations of state space systems. Finally we present a method to obtain a linearized state space model about a nominal (or equilibrium) state from a nonlinear state space model using the Jacobian linearization process.

15.2 Introduction to State Space Control Theory

State space based control theory is a significantly different viewpoint compared to the frequency domain, Laplace transformation based (and transfer function based) control theory we covered in Part II of this book. In state space control theory, we do not rely on Laplace transformations at all to solve the differential equations of motion. Instead, we try to solve them directly in time domain, using the concept of a state transition matrix. State space based control theory is relatively new in comparison to the frequency domain, transfer function based control theory. For that reason, the literature sometimes refers to transfer function based control theory as classical control theory and the time domain, state space control theory as modern control theory. It can be said that state space based control theory revolutionized the mindset of control theorists because of its significant advantages over the classical control theory. The concept of a state variable in the dynamics of a physical system in itself is a profound concept, discovering the importance of internal dynamics. Recall that in the transfer function based control theory, the ratio of of output function over the input function (which indeed forms the transfer function) is of importance with no explicit concern for the internal dynamics, whereas in state space control theory, the state variables which embody the critical internal dynamics of the system take the center stage and the output function and input function take relatively subordinate roles. To better appreciate the importance of different roles of state variables, output variables and input variables in a state space based control theory, it is important for us to delve deep into the state space representation of

Flight Dynamics and Control of Aero and Space Vehicles, First Edition. Rama K. Yedavalli.
© 2020 John Wiley & Sons Ltd. Published 2020 by John Wiley & Sons Ltd.

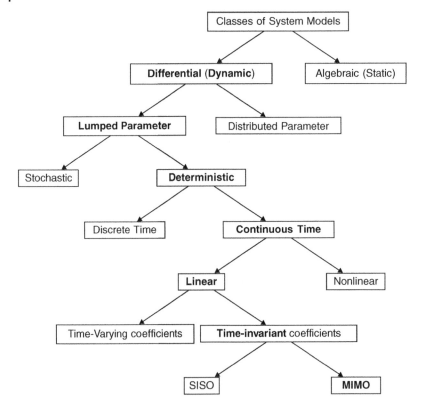

Figure 15.1 Classes of models.

dynamic systems. In this connection, it also helps to make a distinction between single input, single output (SISO) and multiple input, multiple output (MIMO) systems. Before that task, it is important to get an overview of the classes of models we focus on in obtaining the state space models. This overview is depicted pictorially in Figure 15.1. The models we focus on in this book are indicated by bold letters.

15.2.1   State Space Representation of Dynamic Systems

The state of a dynamic system is the smallest set of linearly independent variables (called state variables) such that the knowledge of these variables at $t = t_0$ together with the input at $t \geq t_0$ completely determines the behavior of the system for any time $t \geq t_0$.

When a dynamic system is modeled by ordinary differential equations, it is relatively easy to identify the set of state variables. For example, if we have a differential equation

$$\frac{d^2\theta}{dt^2} + 5\frac{d\theta}{dt} + 6\theta = e^t \tag{15.1}$$

We can re-write it as

$$\frac{d^2\theta}{dt^2} = e^t - 6\theta - 5\frac{d\theta}{dt} \tag{15.2}$$

then it is easy to observe that we need $\theta(t_0)$ and $\frac{d\theta}{dt}(t_0)$ to completely determine the behavior of $\theta(t)$ for all $t \geq t_0$. Thus $\theta(t)$ and $\frac{d\theta}{dt}(t)$ become the two state variables.

One main feature of the state space representation is that an nth order system ($n = 2$ in the above case) can be expressed in first order form with n state variables. Thus the state space representation of the above second order differential equation is obtained by first defining

$$\theta(t) = x_1(t)$$

and

$$\frac{d\theta(t)}{dt} = x_2(t)$$

and then rewriting the above equation as two first order equations in the state variables $x_1(t)$ and $x_2(t)$.

$$\dot{x}_1(t) = x_2(t)$$

and

$$\dot{x}_2(t) = \frac{d^2\theta}{dt^2}(t) = -5x_2(t) - 6x_1(t) + e^t$$

i.e.

$$\begin{bmatrix} \dot{x}_1(t) \\ \dot{x}_2(t) \end{bmatrix} = \begin{bmatrix} 0 & 1 \\ -6 & -5 \end{bmatrix} \begin{bmatrix} x_1(t) \\ x_2(t) \end{bmatrix} + \begin{bmatrix} 0 \\ 1 \end{bmatrix} e^t.$$

This is in the form

$$\dot{\vec{x}} = A\vec{x} + B\vec{u}$$

and this is the state space representation of the dynamic system represented by the equation 15.1.

In the above example, the equation considered is a linear differential equation and thus the resulting state space description is a linear state space description, but it does not have to be linear. We can have a nonlinear set of differential equations that can be put into state space form. So in general, the state space description of any dynamic system described by ordinary differential equations is given by

$$\dot{\vec{x}} = \vec{f}(\vec{x}, \vec{u}, t)$$

where $\vec{x}$ is the state vector, $\vec{u}$ is the control vector and $\vec{f}$ is a vector of nonlinear functions in x_i and u_i. Typically, we write

$$\vec{x} \in \mathbb{R}^n, \text{ i.e. } \vec{x} = \begin{bmatrix} x_1 \\ x_2 \\ \vdots \\ x_n \end{bmatrix}$$

$$\vec{u} \in \mathbb{R}^m, \text{ i.e. } \vec{u} = \begin{bmatrix} u_1 \\ u_2 \\ \vdots \\ u_m \end{bmatrix}.$$

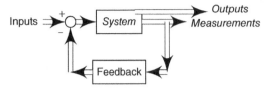

Outputs
Measurements

Figure 15.2 Typical block diagram for a dynamic system.

In such cases, the nonlinear differential equations are linearized about an equilibrium to get a linear state space representation in small motions around the equilibrium. We have already discussed the linearization of nonlinear systems about a nominal or equilibrium state in Part I of this book. Hence, going forward, in this part of the book, we focus on the resulting linear models.

15.2.2 Linear State Space Systems

Once we focus on linear state space system dynamics, we can view the development of linear state space control theory in a very systematic manner. The main idea behind this control theory is to make use of these linear state space based mathematical models to predict the behavior of physical systems, and then use these mathematical predictions to synthesize decisions (control logic) to modify the behavior of the physical system to behave in a desired way. For that we need actuators and sensors in a feedback fashion to compare the actual behavior of the output to the desired behavior continuously with the help of sensors and then apply actuation based on a control law to make the output evolve in such a way that it reduces the error between the actual behavior and the desired behavior. A typical feedback control system is pictorially shown in Figure 15.2.

The three basic steps of control theory are modeling, analysis, and design.

The variables in any generic control system can be classified into the following categories.

- Inputs, or control variables: these are variables available in the system that one can use or manipulate to effect the system's dynamic behavior. The vector of control variables is denoted by $\vec{u}$, which belongs to a real vector space of dimension $m \times 1$.
- Outputs, or controlled variables are those variables within the system's dynamics that we wish to control. Thus there is considerable difference between control variables and controlled variables. The vector of output (controlled) variables is denoted by $\vec{y}$, which also belongs to a real vector space of dimension $k \times 1$. We denote the output vector as

$$\vec{y} \in \mathbb{R}^k \text{ i.e. } \vec{y} = \begin{bmatrix} y_1 \\ y_2 \\ \vdots \\ y_k \end{bmatrix}.$$

- Measurements or sensed variables are those variables within the system's dynamics that we can sense or measure so that we can supply that information to the actuation variables. The vector of measurement variables is denoted by $\vec{z}$, which also belongs to a real vector space of dimension $l \times 1$. We denote the measurement vector as

$$\vec{z} \in \mathbb{R}^l \text{ i.e. } \vec{z} = \begin{bmatrix} z_1 \\ z_2 \\ \vdots \\ z_l \end{bmatrix}.$$

Table 15.1 Comparison of classical and modern control theory features.

Classical (transfer functions)	modern (state space)
SISO	MIMO
Linear	Possibly nonlinear
Time invariant	Possibly time-varying
Frequency domain	Time domain
Trial and error	Systematic, optimal
Initial conditions neglected	Not neglected

In some books, outputs and measurements are treated as synonymous, but in this book, we distinguish between output variables and measurement variables. Thus in our viewpoint, the number of outputs we wish to control could be quite different from the number of measurements (or sensors) we use.

15.2.3 Comparison of Features of Classical and Modern (State Space Based) Control Theory

As mentioned earlier, the very premise of state space based control theory using the concept of state variables is quite different from the input/output based classical control theory. Let us now elaborate on the features of this new state space based control theory that offers many significant advantages over the classical control theory. However, it needs to be kept in mind that the classical control theory, in spite of some glaring shortcomings, has its own advantages and thus should not be undermined or discarded. In fact, contrary to the Modern state space methods, they have stood the test of time and are so powerful that there is reluctance on the part of industries to try out these new modern state space methods despite their advantages. So a prudent control engineer needs to master both classical as well as modern control theory concepts and then use them based on the nature of the application problem in a judicious way. With that understanding, we now briefly summarize the comparison of these two viewpoints in Table 15.1. Recall that SISO stands for single input, single output while MIMO stands for multiple input, multiple output systems.

Having understood the importance of state space based control theory, in what follows, we quickly review a few basic approaches to represent the given system (linear) dynamics in a state variable representation.

15.3 State Space Representation in Companion Form: Continuous Time Systems

A class of single-input, single-output systems can be described by an nth order linear ordinary differential equation:

$$\frac{d^n y}{dt^n} + a_{n-1}\frac{d^{n-1}y}{dt^{n-1}} + \cdots + a_2\frac{d^2 y}{dt^2} + a_1\frac{dy}{dt} + a_0 y = u(t). \tag{15.3}$$

The class of systems can be reduced to the form of n first order state equations as follows. Define the state variables as

$$x_1 = y, \quad x_2 = \frac{dy}{dt}, \quad x_3 = \frac{d^2y}{dt^2}, \cdots \quad x_n = \frac{d^{n-1}y}{dt^{n-1}}. \tag{15.4}$$

These particular state variables are often called phase variables. As a direct result of this definition. $n - 1$ first order differential equations are $\dot{x}_1 = x_2, \dot{x}_2 = x_3, \cdots \dot{x}_{n-1} = x_n$. The nth equation is $\dot{x}_n = d^n y/dt^n$. Using the original differential equation and the preceding definitions gives

$$\dot{x}_n = -a_0 x_1 - a_1 x_2 - \cdots - a_{n-1} x_n + u(t) \tag{15.5}$$

so that

$$\dot{x}(t) = \begin{bmatrix} 0 & 1 & 0 & 0 & \cdots & 0 \\ 0 & 0 & 1 & 0 & \cdots & 0 \\ \vdots & & & & & \vdots \\ 0 & 0 & 0 & 0 & \cdots & 1 \\ -a_0 & -a_1 & -a_2 & -a_3 & \cdots & -a_{n-1} \end{bmatrix} x(t) + \begin{bmatrix} 0 \\ 0 \\ 0 \\ \vdots \\ 0 \\ 1 \end{bmatrix} u(t) \tag{15.6}$$

$$= Ax(t) + Bu(t).$$

The output is $y(t) = x_1(t) = [1\,0\,0 \cdots 0]\vec{x}(t) = C\vec{x}(t)$. In this case the coefficient matrix A is said to be in the phase variable canonical form.

15.4 State Space Representation of Discrete Time (Difference) Equations

We can also represent a set of difference equations, which typically describe the dynamics of a linear discrete time systems in a state variable form. In the following difference equations, we use the notation k to refer to a general discrete time $t_k \in \tau$.

$$x(k + 1) = A(k)x(k) + B(k)u(k) \tag{15.7}$$

$$y(k) = C(k)x(k) + D(k)u(k). \tag{15.8}$$

The matrices A, B, C, D have the same dimensions as in the continuous time case, but their meanings are different. The block diagram representation of equation 15.7 and equation 15.8 is show in Figure 15.3.

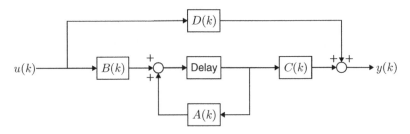

Figure 15.3 Typical block diagram for a discrete time dynamic system.

15.4.1 Companion Form for Discrete Time Systems

A comparable class of discrete time systems is described by an nth order difference equation

$$y(k + n) + a_{n-1}y(k + n - 1) + \cdots + a_2y(k + 2) + a_1y(k + 1) + a_0y(k)) = u(k). \tag{15.9}$$

Phase variable type states can be defined as

$$x_1(k) = y(k), \quad x_2(k) = y(k + 1), \quad x_3(k) = y(k + 2), \cdots \quad x_n(k) = y(k + n - 1) \tag{15.10}$$

where the discrete time points t_k are simply referred to as k. With these definitions, the first $n - 1$ state equations are of the form

$$x_i(k + 1) = x_{i+1}(k) \tag{15.11}$$

and so on. The original difference equation becomes

$$y(k + n) = x_n(k + 1) = -a_0x_1(k) - a_1x_2(k) - \cdots - a_{n-1}x_n(k) + u(k). \tag{15.12}$$

Thus we can get a companion form description for discrete time systems as well given by:

$$x(k + 1) = \begin{bmatrix} 0 & 1 & 0 & 0 & \cdots & 0 \\ 0 & 0 & 1 & 0 & \cdots & 0 \\ \vdots & & & & & \vdots \\ 0 & 0 & 0 & 0 & \cdots & 1 \\ -a_0 & -a_1 & -a_2 & -a_3 & \cdots & -a_{n-1} \end{bmatrix} x(k) + \begin{bmatrix} 0 \\ 0 \\ 0 \\ \vdots \\ 0 \\ 1 \end{bmatrix} u(k) \tag{15.13}$$

$$= Ax(k) + Bu(k).$$

A special case of discrete time system is the sampled data system, which is obtained by discretizing a continuous time system with sampling. While there are multiple variations to these sampled data systems based on the sampling rate, for simplicity, we focus on the case of a constant sampling period T. Using this sampling phenomenon, we can convert an original continuous time differential equation into an equivalent discrete time difference equation. Conceptually we are approximating the derivative of continuous function as a discrete function using either a forward difference approximation or a backward difference approximation for the derivative. The following example illustrates this procedure.

Example 15.1   A continuous time system is described by

$$\ddot{y} + 4\dot{y} + y = u(t)$$

so that $a_0 = 1$ and $a_1 = 4$. Use the forward difference approximation for derivatives

$$\dot{y}(t_k) \cong \frac{y(t_{k-1}) - y(t_k)}{T}$$

and

$$\ddot{y} \cong \frac{\dot{y}(t_{k-1}) - \dot{y}(t_k)}{T}$$

where $T = t_{k-1} - t_k$ is the constant sampling period. Find the approximate difference equation that describes the resulting sampled data system.

Solution:
It follows that

$$\ddot{y} = [y(k+2) - 2y(k+1) + y(k)]/T^2$$

so that substitution into the difference equation and regrouping terms gives

$$y(k+2) + (4T - 2)y(k+1) + (T^2 - 4T + 1)y(k) = T^2 u(k).$$

Thus the discrete coefficients are

$$a_0 = T^2 - 4T + 1$$

and

$$a_1 = 4T - 2.$$

The resulting sampled data system state space description in companion form is given by

$$x((k+1)) = \begin{bmatrix} 0 & 1 \\ -(T^2 - 4T + 1) & -(4T - 2) \end{bmatrix} x(k) + \begin{bmatrix} 0 \\ T^2 \end{bmatrix} u(k) \qquad (15.14)$$
$$= Ax(k) + Bu(k).$$

Note that in a sampled data system, the state space plant matrices A and B are functions of the constant sampling period T.

If a backward difference approximation to the derivatives is used, a very different set of coefficients will be found. Also, the time argument on the u input term will change.

15.5 State Space Representation of Simultaneous Differential Equations

The same method of defining the state variables can be applied to multiple input, multiple output systems described by several coupled differential equations if the inputs are not differentiated.

Example 15.2   A system has three inputs u_1, u_2, u_3 and three outputs y_1, y_2, y_3. The input output equations are

$$\ddot{y}_1 + a_1\ddot{y}_1 + a_2(\dot{y}_1 + \dot{y}_2) + a_3(y_1 - y_3) = u_1(t)$$
$$\ddot{y}_2 + a_4(\dot{y}_2 - \dot{y}_1 + 2\dot{y}_3) + a_5(y_2 - y_1) = u_2(t)$$
$$\dot{y}_3 + a_6(y_3 - y_1) = u_3(t).$$

Solution:
Notice that in the second equation $\dot{y}_3$ can be eliminated by using the third equation. State variables are selected as the outputs and their derivatives up to the $(n-1)$th, where n is the order of the highest derivative of a given output. Select

$$x_1 = y_1, x_2 = \dot{y}_1, x_3 = \ddot{y}_1, x_4 = y_2, x_5 = \dot{y}_2, x_6 = y_3. \text{ Then}$$

$$\dot{x}_1 = x_2, \ \dot{x}_2 = x_3, \ \dot{x}_4 = x_5$$
$$\dot{x}_1 = -a_1 x_3 - a_2(x_2 + x_5) - a_3(x_1 - x_6) + u_1$$
$$\dot{x}_5 = -a_4(x_5 - x_2 + 2x_6) - a_5(x_4 - x_1) + u_2$$
$$\dot{x}_6 = -a_6(x_6 - x_1) + u_3.$$

Eliminating $\dot{x}_6$ from the $\dot{x}$ equations leads to

$$
\begin{bmatrix} \dot{x}_1 \\ \dot{x}_2 \\ \dot{x}_3 \\ \dot{x}_4 \\ \dot{x}_5 \\ \dot{x}_6 \end{bmatrix}
=
\begin{bmatrix}
0 & 1 & 0 & 0 & 0 & 0 \\
0 & 0 & 1 & 0 & 0 & 0 \\
-a_3 & -a_2 & -a_1 & 0 & -a_2 & a_3 \\
0 & 0 & 0 & 0 & 1 & 0 \\
a_5 - 2a_4a_6 & a_4 & 0 & -a_5 & -a_4 & 2a_4a_6 \\
a_6 & 0 & 0 & 0 & 0 & -a_6
\end{bmatrix}
\begin{bmatrix} x_1 \\ x_2 \\ x_3 \\ x_4 \\ x_5 \\ x_6 \end{bmatrix}
+
\begin{bmatrix}
0 & 0 & 0 \\
0 & 0 & 0 \\
1 & 0 & 0 \\
0 & 0 & 0 \\
0 & 1 & -2a_4 \\
0 & 0 & 1
\end{bmatrix}
\begin{bmatrix} u_1 \\ u_2 \\ u_3 \end{bmatrix}.
$$

The output equation is

$$
\begin{bmatrix} y_1 \\ y_2 \\ y_3 \end{bmatrix}
=
\begin{bmatrix}
1 & 0 & 0 & 0 & 0 & 0 \\
0 & 0 & 0 & 1 & 0 & 0 \\
1 & 0 & 0 & 0 & 0 & 1
\end{bmatrix} x.
$$

When derivatives of the input appear in the system differential equation, the previous method of state variable selection must be modified in such a way that the interpretation of some of the state variables change. The new state variables are defined in such a way that they include the appropriate derivatives of the input invariables in them. A simple example illustrates this procedure. Consider

$$\ddot{y} + c_1 \dot{y} + c_2 y = u(t) + c_3 \dot{u}.$$

Rearranging these terms, we get

$$\ddot{y} - c_3 \dot{u} = -c_1 \dot{y} - c_2 y + u.$$

We then define the two state variables as $x_1 = y$, $x_2 = \dot{y} - c_3 u$, where we assume c_3 is a constant. Then we have $\dot{x}_1 = \dot{y} = x_2 + c_3 u$ and $\dot{x}_2 = -c_1[x_2 + c_3 u] - c_2 x_1 + u$, which then has the following state space representation

$$
\begin{bmatrix} \dot{x}_1 \\ \dot{x}_2 \end{bmatrix}
=
\begin{bmatrix} 0 & 1 \\ -c_2 & -c_1 \end{bmatrix}
\begin{bmatrix} x_1 \\ x_2 \end{bmatrix}
+
\begin{bmatrix} c_3 \\ (1 - c_1 c_3) \end{bmatrix}
[u].
$$

The output equation is

$$[y] = [1 \ 0].x$$

Note that in the general case there are multiple ways to arrange the terms within a given set of equations and thus the definition of state variables needs to be tailored to the given set of equations. A more systematic method involves drawing elaborate state diagrams to carefully identify the state variables, which is out of the scope of our present purpose of focusing on the specific equations related to our aerospace vehicle equations of motion. Thus we omit the discussion of state diagrams. The major conceptual point is that, in the end, with appropriate definition of state variables, most of the linear ordinary constant coefficient differential equations can be always described by a state variable representation.

15.6 State Space Equations from Transfer Functions

$$\frac{Y(s)}{U(s)} = T(s) = \frac{\beta_m s^m + \beta_{m-1} s^{m-1} + \cdots + \beta_1 s + \beta_0}{s^n + a_{n-1} s^{n-1} + \cdots + a_1 s + a_0}$$

or a Z transform transfer function

$$\frac{y(z)}{u(z)} = T(z) = \frac{\beta_m z^m + \beta_{m-1} z^{m-1} + \cdots + \beta_1 Z + \beta_0}{z^n + a_{n-1} z^{n-1} + \cdots + a_1 z + a_0}.$$

Example 15.3   Select a suitable set of state variables for system whose transfer function is

$$T(s) = \frac{s+3}{s^3 + 9s^2 + 24s + 20}$$

$$= \frac{s+3}{(s+2)^2(s+5)} = \frac{2/9}{(s+2)} + \frac{1/3}{(s+2)^2} + \frac{-2/9}{(s+5)}.$$

Solution:

There are multiple state space realizations for this transfer function. One of them is

$$\begin{bmatrix} \dot{x}_1 \\ \dot{x}_2 \\ \dot{x}_3 \end{bmatrix} = \begin{bmatrix} -9 & 1 & 0 \\ -24 & 0 & 1 \\ -20 & 0 & 0 \end{bmatrix} \begin{bmatrix} x_1 \\ x_2 \\ x_3 \end{bmatrix} + \begin{bmatrix} 0 \\ 1 \\ 3 \end{bmatrix} u$$

and

$$y = \begin{bmatrix} 1 & 0 & 0 \end{bmatrix} x.$$

Another option is

$$\begin{bmatrix} \dot{x}_1 \\ \dot{x}_2 \\ \dot{x}_3 \end{bmatrix} = \begin{bmatrix} 0 & 1 & 0 \\ 0 & 0 & 1 \\ -20 & -24 & -9 \end{bmatrix} \begin{bmatrix} x_1 \\ x_2 \\ x_3 \end{bmatrix} + \begin{bmatrix} 0 \\ 0 \\ 1 \end{bmatrix} u$$

and

$$y = \begin{bmatrix} 3 & 1 & 0 \end{bmatrix} x.$$

Using the factor form of $T(s)$ another realization is possible

$$\begin{bmatrix} \dot{x}_1 \\ \dot{x}_2 \\ \dot{x}_3 \end{bmatrix} = \begin{bmatrix} -5 & 1 & 0 \\ 0 & -2 & 1 \\ 0 & 0 & -2 \end{bmatrix} \begin{bmatrix} x_1 \\ x_2 \\ x_3 \end{bmatrix} + \begin{bmatrix} 0 \\ 1 \\ 3 \end{bmatrix} u$$

and

$$y = \begin{bmatrix} 1 & 0 & 0 \end{bmatrix} x.$$

Another realization can also be obtained from a partial fraction expansion

$$\begin{bmatrix} \dot{x}_1 \\ \dot{x}_2 \\ \dot{x}_3 \end{bmatrix} = \begin{bmatrix} -5 & 0 & 0 \\ 0 & -2 & 1 \\ 0 & 0 & -2 \end{bmatrix} \begin{bmatrix} x_1 \\ x_2 \\ x_3 \end{bmatrix} + \begin{bmatrix} 1 \\ 0 \\ 1 \end{bmatrix} u$$

and

$$y = \begin{bmatrix} \frac{-2}{9} & \frac{1}{3} & \frac{2}{9} \end{bmatrix} x.$$

15.6.1   Obtaining a Transfer Function from State and Output Equations

Given

$$\dot{\vec{x}} = A\vec{x} + Bu$$
$$y = C\vec{x} + Du$$

where y and u are of dimension 1 (single input and single output, SISO) but $\vec{x}$ is a vector of dimension n, the transfer function between output $Y(s)$ to input $U(s)$ is given by

$$G(s) = \frac{Y(s)}{U(s)}$$
$$= C(sI - A)^{-1}B + D$$
$$= \frac{N(s)}{D(s)}$$

where $N(s)$ and $D(s)$ are the numerator and denominator polynomials, respectively, of each of the transfer function $G(s)$. Note that C is a $1 \times n$ matrix and B is an $n \times 1$ matrix and D is a scalar (1×1).

Then the denominator polynomial is an nth degree polynomial. With A, B, C, and D as inputs, MATLAB has a command

```
[Num, Den]  =  ss2tf(A, B, C, D, 1)
```

to get the transfer function.

Note that we can get transfer matrices for MIMO systems, in which the elements of this matrix would be transfer functions between a given output to a given input for various combinations of those outputs and inputs. Then the last argument in the above MATLAB command is no longer 1 and needs to be changed accordingly.

Note that the transformation from state space to transfer function is unique, but the transformation from transfer function to state space is not unique. Also to get the state space from the transfer function use

```
[A, B, C, D]  =  tf2ss(Num, Den)
```

Important Observation

While we observed that there could be multiple realizations of state space matrices from a given transfer function, unfortunately MATLAB output to the above command produces only one single realization of the state space A, B, C, D matrices. It turns out that MATLAB gets this particular state space realization via the minimal realization procedure. However, this author believes that there is interesting research that can be carried out knowing all the possible multiple realizations and as such prefers not to be constrained by this single realization that is obtained from MATLAB output.

15.7   Linear Transformations of State Space Representations

Since state space representation of a linear time invariant system is not unique (in the sense that the same set of differential equations can be described by various sets of state variables), it is clear that one can transform the state space representation from one set of state variables to another set of state variables by linear matrix transformations. For example, if we have a given set of state variable vector $\vec{x}$ for the state space representation of a linear time invariant system as given by

$$\dot{\vec{x}} = A\vec{x} + B\vec{u}$$

then we can transform this system to another set of state variables $\vec{x}_n$ such that

$$\vec{x}_n = T\vec{x}$$

where T is any non-singular matrix, so that

$$\vec{x} = T^{-1}\vec{x}_n.$$

Thus, we can get a new state space representation of the same system given by

$$\dot{\vec{x}}_n = A_n\vec{x}_n + B_n\vec{u}$$

where $A_n = T^{-1}AT$ and $B_n = T^{-1}B$. While there is one-to-one correspondence between state variables of vector x and of vector x_n, the physical meaning could be lost in this new set of state variables of vector x_n. This type of linear matrix transformation with any non-singular transformation matrix T is also called a similarity transformation because the plant matrices A and A_n are similar matrices in the sense that the eigenvalues of A and A_n are the same. Put another way, the eigenvalues are invariant under a similarity transformation.

The ability to transfor one state space representation to another state space representation by a similarity transformation is a powerful concept with far reaching implications. This means that if we select the appropriate transformation matrix T, we can transform a given A matrix to various forms for the new A_n matrix. Later in the chapter, we observe that we can transform a given state space system to different forms like phase variable canonical form, controllability canonical form, observability canonical form, Jordan canonical form, and so on. Of these the Jordan canonical form (alternatively labeled the modal decomposition form) deserves special mention, which will be used extensively in later sections/chapters of this Part III. For simplicity of concept, let us assume the matrix A has distinct eigenvalues. Then the special similarity transformation matrix T, which transforms the matrix A into a pure diagonal matrix J (referred to later as the Jordan matrix) whose diagonal entries are nothing but the n eigenvalues of the matrix A, is denoted by $\mathcal{M}$ and is labeled as the modal matrix. It is important to realize that this modal matrix is nothing but the matrix formed by the eigenvectors (or normalized eigenvectors) of the matrix A. This means the modal matrix can be a complex matrix (albeit with complex conjugate entries for a real matrix A) and the new state variables in the new transformed version are called the modes of the linear system. This is an instance where even though the original state variables in vector x belonging to a real vector space are all real and have possibly some physical interpretation, the modes of the system may belong to the complex vector space! Thus modes are simply mathematical entities and that is what we mean by the new state variables losing their physical significance under a similarity transformation. In other words, similarity transformations may transform a real vector space into a complex vector space and as such the transformation matrices are not restricted to be real matrices. At this point, let us elaborate on the transformation to modal coordinates. As explained before, we observed that given a real state variable vector $\vec{x}$ for the state space representation of a linear time invariant system as given by

$$\dot{\vec{x}} = A\vec{x} + B\vec{u}$$

then we can transform this system to modal coordinates denoted by the vector another set of state variables $\vec{q}$ such that

$$\vec{q} = M\vec{x}$$

where M is the non-singular matrix, so that

$$\vec{x} = M^{-1}\vec{q}.$$

Thus, the representation in modal coordinates is given by

$$\dot{\vec{q}} = J\vec{q} + B_n\vec{u}$$

where $J = M^{-1}AM$ and $B_n = M^{-1}B$. Assuming distinct eigenvalues, we observe that the matrix J is a pure diagonal matrix with those distinct eigenvalues as the diagonal elements. However, we see that in this transformation process the Jordan matrix J and the modal matrix M could be complex matrices with the complex conjugate eigenvalues as the diagonal elements of J and the corresponding complex conjugate eigenvectors as columns of the modal matrix. However, it is still possible to keep the Jordan matrix as a real matrix and the corresponding Modal matrix also as a real matrix by splitting the real part and imaginary parts of the complex conjugate pair eigenvalues as a block diagonal real matrix as follows. Let

$$J = \begin{bmatrix} a + jb & 0 \\ 0 & a - jb \end{bmatrix}.$$

Let the corresponding complex conjugate eigenvector be

$$\vec{e} = \begin{bmatrix} v + jw & v - jw \end{bmatrix}.$$

Then the corresponding real version of the Jordan matrix is given

$$J_r = \begin{bmatrix} a & b \\ -b & a \end{bmatrix}.$$

Then we can obtain a real version of this complex eigenvector by separating out the real and imaginary parts (which by themselves are real vectors) and form the real modal matrix as follows:

$$M_r = \begin{bmatrix} v_{r11} & w_{r11} \\ v_{r21} & w_{r21} \end{bmatrix}$$

where v_{r11}, v_{r21} are components of the real vector $\vec{v}$, and w_{r11}, w_{r11} are components of the real vector $\vec{w}$.

Thus $J_r = M_r^{-1}AM_r$.

This real transformation matrix viewpoint is quite important in some applications. The following simple example illustrates this observation.

Example 15.4   Obtain the real version of the modal decomposition for the following matrix

$$A = \begin{bmatrix} 0 & 1 \\ -2 & -4 \end{bmatrix}.$$

Solution:
Obtaining the eigenvalues and eigenvectors of this matrix (by using MATLAB command [evec, eval] = eig(A)) we obtain the diagonal eigenvalues matrix J as

$$J = \begin{bmatrix} -1 + j1.7321 & 0 \\ 0 & -1 - j1.7321 \end{bmatrix}$$

and the modal matrix M as

$$M = \begin{bmatrix} 0.2236 + j0.3873 & 0.2236 - j0.3873 \\ -0.8944 & -0.8944 \end{bmatrix}.$$

Accordingly from the procedure described above, we obtain the real version of the modal decomposition as

$$J_r = \begin{bmatrix} -1 & 1.7321 \\ -1.7321 & -1 \end{bmatrix}$$

and

$$M_r = \begin{bmatrix} 0.2236 & 0.3873 \\ -0.8944 & 0 \end{bmatrix}.$$

Note that

$$M_r^{-1} = \begin{bmatrix} 0 & -1.1181 \\ 2.5820 & 0.6455 \end{bmatrix}$$

so that we get

$$M_r^{-1} A M_r = J_r.$$

We further elaborate on this similarity transformation concept at the appropriate juncture when we discuss the various other canonical forms mentioned before.

15.8 Linearization of Nonlinear State Space Systems

Until now, we have discussed the state space representation for systems that are described by linear differential equations. However, the majority of times, the original differential equations of motion for any dynamic system turn out to be nonlinear. In that case, in general, the state space description of any dynamic system described by nonlinear first order ordinary differential equations is given by

$$\dot{\vec{x}} = \vec{f}(\vec{x}, \vec{u}, t)$$

where $\vec{x}$ is the state vector, $\vec{u}$ is the control vector and $\vec{f}$ is a vector of nonlinear functions in x_i and u_i. Typically, we write

$$\vec{x} \in R^n, \text{ i.e. } \vec{x} = \begin{bmatrix} x_1 \\ x_2 \\ \vdots \\ x_n \end{bmatrix}$$

$$\vec{u} \in R^m, \text{ i.e. } \vec{u} = \begin{bmatrix} u_1 \\ u_2 \\ \vdots \\ u_m \end{bmatrix}.$$

Consider the following three classes of nonlinear systems:

1. $\dot{\vec{x}} = \vec{f}(\vec{x}, \vec{u}, t)$
2. $\dot{\vec{x}} = \vec{f}(\vec{x}, t)$
3. $\dot{\vec{x}} = \vec{f}(\vec{x})$.

Out of these, consider the second class of systems

$$\dot{\vec{x}} = \vec{f}(\vec{x}, t), \ \vec{x}(t_0) = \vec{x}_0.$$

We assume that the above equation, has a unique solution starting at the given initial condition, i.e. we have one single solution corresponding to each initial condition. Let us denote this solution as

$$\vec{x}(t; \vec{x}_0, t_0) \equiv \vec{x}(t) \text{ for simplicity}$$
$$\vec{x}(t_0; \vec{x}_0, t_0) \equiv \vec{x}_0.$$

15.8.1 Equilibrium State

In the above class of systems, a state $\vec{x}_e$ where $\vec{f}(\vec{x}_e, t) = 0$ *for all t* is called an equilibrium state of the system, i.e. the equilibrium state corresponds to the constant solution of the system. If the system is linear time invariant (i.e. $\vec{f}(\vec{x}, t) = A\vec{x}$), then there exists only one equilibrium state if A is non-singular and many equilibrium states if A is singular. For non-linear systems there may be one or more equilibrium states.

Any isolated equilibrium point can always be transferred to the origin of the coordinates

$$\text{i.e. } \vec{f}(0, t) = 0$$

by a proper coordinate transformation. So one can always take $\vec{x}_e = 0$ without any loss of generality. The origin of state space is always an equilibrium point for linear systems and for linear systems all equilibrium states behave the same way (because if $\vec{x}(t)$ is a solution $\overline{\vec{x}}(t)$ is also a solution; then $\vec{x}(t) \to \overline{\vec{x}}(t)$ is also a solution for the linear system).

In such cases, the nonlinear differential equations are linearized about an equilibrium to get a linear state space representation in small motions around the equilibrium. One such linearization process is labeled as the Jacobian method and the resulting linearized state space matrix is called the Jacobian matrix.

15.8.2 Linearizing a Nonlinear State Space Model

Consider the general nonlinear state variable model

$$\left. \begin{array}{l} \dot{\vec{x}} = \vec{f}(\vec{x}, \vec{u}, t) \\ \vec{y} = \vec{h}(\vec{x}, \vec{u}, t). \end{array} \right\} \tag{15.15}$$

The above set of nonlinear differential equations can be linearized about a constant, equilibrium solution, which can also be called the steady state solution. This is the most common type of linearization process, i.e. linearization about a given steady state condition. In a slightly different viewpoint, the nonlinear system of equations can also be linearized about a given nominal trajectory, where the nominal trajectory satisfies the original nonlinear set of differential equations. It does not have to be a constant solution. This linearization process can be referred to as the Jacobian linearization. Of course, Jacobian linearization holds good for constant, steady state equilibrium conditions. In Chapter 3 of Part I of this book, we have already covered the straightforward method of linearization about a steady state. That simple, brute force linearization process involves expanding the original nonlinear equations in terms of the steady state plus some perturbation. Then assuming those perturbations to be small, we neglect second and higher order terms, along with making a small angle approximation whenever there are trigonometric functions involved. However, in this chapter, we present the more general Jacobian linearization procedure.

In this linearization process about any given nominal trajectory, strictly speaking, the nominal trajectory needs to be the solution of the original nonlinear equation. However, the majority of the times, the given nominal trajectory is not checked to be the solution per se of the original nonlinear equation at all times, because that process itself could be very cumbersome in some nonlinear equations. Instead, the majority of the times, an engineering judgment based nominal trajectory, which most of the times happens to be the ideal or desired behavior of the system, discerned more from the physics of the problem rather than in a mathematically rigorous way.

15.8.3 Linearization About a Given Nominal Condition: Jacobian Method

A more general viewpoint of linearization is to be able linearize the given nonlinear equations about any given nominal condition, where the nominal condition need not be a steady state, constant solution of the original nonlinear equation. In other words, the nominal solution is only required to satisfy the original nonlinear differential equation with no constraint that the nominal solution be a constant solution. Thus in this viewpoint

$$\vec{\delta x}(t) = \vec{x}(t) - \vec{x}_n(t)$$
$$\vec{\delta u}(t) = \vec{u}(t) - \vec{u}_n(t)$$
$$\vec{\delta y}(t) = \vec{y}(t) - \vec{y}_n(t)$$

where $\vec{x}_n(t)$, $\vec{u}_n(t)$, and $\vec{y}_n(t)$ are such that

$$\dot{\vec{x}_n}(t) = \vec{f}(\vec{x}_n(t), \vec{u}_n(t), t).$$

Then Equation 15.15 can be written as

$$\dot{\vec{x}_n}(t) + \dot{\vec{\delta x}}(t) = \vec{f}(\vec{x}_n(t) + \vec{\delta x}(t), \vec{u}_n(t) + \vec{\delta u}(t), t)$$

$$= \vec{f}(\vec{x}_n, \vec{u}_n, t) + \left[\frac{\partial \vec{f}}{\partial \vec{x}}\right]_n \vec{\delta x} + \left[\frac{\partial \vec{f}}{\partial \vec{u}}\right]_n \vec{\delta u} + \text{higher order terms}$$

$$\vec{y}_n(t) + \vec{\delta y}(t) = \vec{h}(\vec{x}_n(t) + \vec{\delta x}(t), \vec{u}_n(t) + \vec{\delta u}(t), t)$$

$$= \vec{h}(\vec{x}_n, \vec{u}_n, t) + \left[\frac{\partial \vec{h}}{\partial \vec{x}} \right]_n \vec{\delta x} + \left[\frac{\partial \vec{h}}{\partial \vec{u}} \right]_n \vec{\delta u} + \text{higher order terms}$$

where $[\cdot]_n$ means the derivatives are evaluated at the nominal solutions. Since the nominal solutions satisfy Equation 15.15, the first terms in the preceding Taylor series expansions cancel. For sufficiently small $\vec{\delta x}$, $\vec{\delta u}$, and $\vec{\delta y}$ perturbations, the higher order terms can be neglected leaving the linear equations

$$\vec{\delta x} = \left[\frac{\partial \vec{f}}{\partial \vec{x}} \right]_n \vec{\delta x} + \left[\frac{\partial \vec{f}}{\partial \vec{u}} \right]_n \vec{\delta u}$$

$$= A\vec{\delta x} + B\vec{\delta u} \tag{15.16}$$

and

$$\vec{\delta y} = \left[\frac{\partial \vec{h}}{\partial \vec{x}} \right]_n \vec{\delta x} + \left[\frac{\partial \vec{h}}{\partial \vec{u}} \right]_n \vec{\delta u}$$

$$= C\vec{\delta x} + D\vec{\delta u}. \tag{15.17}$$

Here

$$A_{ij} = \frac{\partial f_i}{\partial x_j}\bigg|_n \quad B_{ij} = \frac{\partial f_i}{\partial u_j}\bigg|_n$$

$$\tag{15.18}$$

$$C_{ij} = \frac{\partial h_i}{\partial x_j}\bigg|_n \quad D_{ij} = \frac{\partial h_i}{\partial u_j}\bigg|_n .$$

Note that when whenever the nominal solutions $\vec{x}_n$ and $\vec{u}_n$ are time varying, the matrices A, B, C, and D could be time varying. However, if $\vec{x}_n$ and $\vec{u}_n$ are constant, then the matrices A, B, C, and D are constant. These derivative matrices are also called Jacobian matrices. When $\vec{x}_n$ and $\vec{u}_n$ are constant, these solutions are called equilibrium (or steady state) solutions. In other words, for equilibrium solutions, we have $\vec{f}(\vec{x}_n, \vec{u}_n) = \vec{f}(\vec{x}_e, \vec{u}_e) = 0$. In general there can be many equilibrium solutions. Also when the given nominal solution does not exactly satisfy the original nonlinear equation, the inaccuracies encountered in this situation are regarded as perturbations from the linearized system and its effect can be reduced by a proper control design for the linearized system.

Example 15.5 Find the equilibrium points for the system described by $\ddot{y} + (1 + y)\dot{y} - 2y + 0.5y^3 = 0$ and get the linearized state spaced model.

Solution:
Letting $x_1 = y$ and $x_2 = \dot{y}$ gives the state variable model

$$\begin{bmatrix} \dot{x}_1 \\ \dot{x}_2 \end{bmatrix} = \begin{bmatrix} x_2 \\ 2x_1 - 0.5x_1^3 - (1 + x_1)x_2 \end{bmatrix} = \vec{f}(\vec{x}) = \begin{bmatrix} f_1(\vec{x}) \\ f_2(\vec{x}) \end{bmatrix}.$$

In other words

$$f_1(\vec{x}) = x_2$$
$$f_2(\vec{x}) = 2x_1 - 0.5x_1^3 - (1 + x_1)x_2.$$

Equilibrium points are solutions of $\vec{f}(\vec{x}) = 0$, so each must have $x_2 = 0$ and $2x_1 - 0.5x_1^3 = 0$. There are three equilibrium solutions possible. They are

$$\vec{x}_{e_1} = \begin{bmatrix} 0 \\ 0 \end{bmatrix}; \vec{x}_{e_2} = \begin{bmatrix} 2 \\ 0 \end{bmatrix}; \vec{x}_{e_3} = \begin{bmatrix} -2 \\ 0 \end{bmatrix}.$$

So corresponding to each equilibruim solution, there will be a linearized state space model. The Jacobian matrix is

$$A = \begin{bmatrix} \dfrac{\partial f_1}{\partial x_1} & \dfrac{\partial f_1}{\partial x_2} \\ \dfrac{\partial f_2}{\partial x_1} & \dfrac{\partial f_2}{\partial x_2} \end{bmatrix}_{eq} = \begin{bmatrix} 0 & 1 \\ \left(2 - \dfrac{3x_1^2}{2} - x_2\right) & -(1 + x_1) \end{bmatrix}$$

a) Corresponding to $\begin{bmatrix} x_{1_e} \\ x_{2_e} \end{bmatrix} = \begin{bmatrix} 0 \\ 0 \end{bmatrix}$, the linearized state space matrix A is

$$A = \begin{bmatrix} 0 & 1 \\ 2 & -1 \end{bmatrix}$$

i.e. $\delta\dot{x} = A\delta x$.

b) Corresponding to $\begin{bmatrix} x_{1_e} \\ x_{2_e} \end{bmatrix} = \begin{bmatrix} 2 \\ 0 \end{bmatrix}$, the linearized state space matrix A is

$$A = \begin{bmatrix} 0 & 1 \\ -4 & -3 \end{bmatrix}$$

i.e. $\delta\dot{x} = A\delta x$.

c) Corresponding to $\begin{bmatrix} x_{1_e} \\ x_{2_e} \end{bmatrix} = \begin{bmatrix} 2 \\ 0 \end{bmatrix}$, the linearized state space matrix A is

$$A = \begin{bmatrix} 0 & 1 \\ -4 & 1 \end{bmatrix}$$

i.e. $\delta\dot{x} = A\delta x$.

Once the linearized state space system is obtained about the equilibrium points, our interest would then be understand the stability domain (the system's behavior as $t \to \infty$ in the neighborhood of those equilibrium points. Structural properties of linear state space systems such as stability, controllability, and observability are discussed in subsequent chapters.

15.9 Chapter Summary

This chapter is an important chapter summarizing the basics of the state space representation of a dynamic system with emphasis on linear systems. We discussed the cases of continuous time systems (governed by differential equations), as well as discrete time systems (governed by difference equations). In addition, the powerful concept of similarity transformations is explained. Finally we presented the procedures for linearizing a nonlinear set of differential equations about a nominal point to obtain linearized state space models. These fundamental concepts form the foundation for discussing various extensions in later sections/chapters of this part of the book. Fundamental concepts discussed in this chapter are also available in many excellent textbooks such as [1–7]

15.10 Exercises

Exercise 15.1. Consider the field controlled DC servomotor equations given by

$$L_f \dot{i}_f + R_f i_f = e$$
$$J\ddot{\theta} + \beta\dot{\theta} = T_c$$
$$K_T i_f = T_c$$

Here, e is the control voltage; i_f the field current; and T_c the control torque. Obtain the state space representation of the above system for the following scenarios:

(a) i_f, θ, and $\dot{\theta}$ as the state variables; the voltage e as the control variable; θ and $\dot{\theta}$ as the two output variables; and $\dot{\theta}$ as the measurement variable.

(b) T_c, θ, and $\dot{\theta}$ as the state variables; e as the control variable; θ as the output variable; and $\dot{\theta}$ and T_c as the measurement variable.

(c) θ, $\dot{\theta}$, and $\ddot{\theta}$ as the state variables; e as the control variable, $\theta + \dot{\theta}$ and $\ddot{\theta}$ as the output variables; and θ as the measurement variable.

Exercise 15.2. The equations of motion of an electromechanical system are given by

$$\ddot{x} + k_1 \dot{x} + k_2 \theta = k_3 e$$
$$\ddot{\theta} - b_1 \theta - b_2 \dot{x} = -b_3 e$$

where k_1, k_2, k_3, b_1, b_2, and b_3 are constants. Obtain the state space representation with e as the control variable; θ as the output variable; and $\dot{x}$ and $\dot{\theta}$ as the two measurement variables.

Exercise 15.3. The attitude dynamics of a rigid satellite in space is governed by the following rotational motion equations:

$$J_x \dot{\omega}_x + (J_z - J_y)\omega_y \omega_z = T_x$$
$$J_y \dot{\omega}_y + (J_x - J_z)\omega_x \omega_z = T_y$$
$$J_z \dot{\omega}_z + (J_y - J_x)\omega_x \omega_y = T_z$$

where J_x, J_y, and J_z are the principle moments of inertia about the principle axes x, y, and z respectively. Likewise, ω_x, ω_y, and ω_z are the angular velocity components. Given that the nominal (steady state) values are

$$\overline{\omega}_x = \overline{\omega}_y = 0$$
$$\overline{\omega}_z = \Omega = \text{constant}$$
$$\overline{T}_x = \overline{T}_y = \overline{T}_z = 0$$

linearize the above nonlinear equations about the given nominal values and obtain the linearized state space model.

Exercise 15.4. The orbital motion of a satellite in earth orbit is given by the equations

$$\ddot{r} - \dot{\theta}^2 r = -\frac{\mu}{r^2} + a_r$$
$$a_i = r\ddot{\theta} + 2\dot{r}\dot{\theta}$$

where a_i and a_r are the control accelerations, r and θ are the radial and transverse components of the position vector and μ is a constant given by $\mu = \omega_0^2 R^3$ for a circular orbit where ω_0 is the angular velocity and R is the radius of the circular orbit and is thus a constant. Taking the circular orbit as the nominal, obtain the linearized state space model about the nominal circular orbit with nominal accelerations being zero. Note that

$$r_{\text{nominal}} = R = \text{constant}$$

$$\dot{\theta}_{\text{nominal}} = \omega_o = \text{constant}.$$

Use the Jacobian method of linearization. After you get the state space matrices A and B analytically, determine the entries of those matrices for a specific circular orbit, namely for a geosynchronous equatorial orbit.

Exercise 15.5: Find the state space representation for the following set of equations

$$\dot{\theta}_1 + 2(\theta_1 + \theta_2) = u_1$$

$$\ddot{\theta}_2 + 3\dot{\theta}_2 + 4\theta_2 = u_2.$$

Exercise 15.6: Find the state space representation for the following set of equations

$$\ddot{\theta}_1 + 3\dot{\theta}_1 + 2(\theta_1 - \theta_2) = u_1 + \dot{u}_2$$

$$\dot{\theta}_2 + 3(\theta_2 - \theta_1) = u_2 + 2\dot{u}_1.$$

Bibliography

1 T.E Fortmann and K.L Hitz. *An Introduction to Linear Control Systems.* Mercel-Dekker, 1977.

2 G.F Franklin, J.D Powell, and A Emami-Naeni. *Feedback Control of Dynamic Systems.* Pearson-Prentice-Hall, 2006.

3 B. Friedland. *Control system design: An introdcution to state-space methods.* McGraw-Hill, 1986.

4 H. Kwakernaak and R. Sivan. *Linear Optimal Control Systems.* Wiley Interscience, 1972.

5 R.E. Skelton. *Dynamic Systems Control: Linear Systems Analysis and Synthesis.* Wiley, New York, 1 edition, 1988.

6 J.G. Truxal, editor. *Control engineers' handbook.* McGraw-Hill, New York, 1958.

7 W.L. Brogan. *Modern Control Theory.* Prentice Hall, 1974.

16

Dynamic Response of Linear State Space Systems (Including Discrete Time Systems and Sampled Data Systems)

16.1 Chapter Highlights

In this chapter, we present the solution to the system of differential equations represented by a state space description. That is, we present expressions for the state variable trajectories as a function of time for both unforced and forced cases, using the concept of the state transition matrix (STM). We cover the cases of both continuous time STM as well as the discrete time STM and then introduce the state space representation of a sampled data system, which is a special case of a general discrete time system. Methods to determine the STMs using the Caley–Hamilton theorem are presented and the important properties of these STMs are highlighted.

16.2 Introduction to Dynamic Response: Continuous Time Systems

We have seen that in general, the state space representation of a dynamic system is given by

$$\dot{\vec{x}} = \vec{f}(\vec{x}, \vec{u}) \tag{16.1}$$

with a given initial condition $\vec{x}(0) = \vec{x}_0$. Given this, our objective now is to solve this set of differential equations, by which we mean that we need to get the state trajectories (i.e. all state variables as an explicit function of time). When the control vector is zero, we label that response as homogeneous response or uncontrolled response or unforced response, all of these phrases being synonymous with each other. Similarly when the control vector is non-zero (i.e. the forcing function is present) but given or known, then we label the corresponding system response as a non-homogeneous response or controlled response or forced response, all of these phrases again being synonymous with each other corresponding to that particular input (control) function vector $\vec{u}(t)$. For a nonlinear system, this dynamic response is highly dependent on various quantities such as the initial time t_0, initial state $\vec{x}(t_0)$ and, of course, in the case of forced response, even on the control vector $\vec{u}(t)$. Because of the difficulty in obtaining analytical solutions for a nonlinear system, typically, we resort to numerical techniques to obtain the dynamic response for a nonlinear system. However, it turns out that even for a linear system, when

Flight Dynamics and Control of Aero and Space Vehicles, First Edition. Rama K. Yedavalli.
© 2020 John Wiley & Sons Ltd. Published 2020 by John Wiley & Sons Ltd.

the plant matrices are time varying, as represented by the state space representation of a linear time varying system, of the following form

$$\dot{x} = A(t)x + B(t)u \tag{16.2}$$

it is still not a trivial task to obtain the dynamic response in an analytical form. It turns out that we can get a nice, general analytical form of solution only for a linear time invariant system, i.e for a state space system in which the plant matrices are constant matrices. However, before we specialize the dynamic response situation for a linear time invariant (LTI) system, we can still consider the general case of linear time varying system and discuss the concept of the state transition matrix (STM) and the role it plays in getting the state trajectories.

16.2.1 The State Transition Matrix and its Properties

Whether it is a time varying system or a time invariant system, once it is a linear system, the evolution of the state trajectories from a given time instant τ to any other time instant t, is captured by the special matrix labeled the STM. For the general linear time varying system, we can write (temporarily ignoring the vector notation),

$$x(t) = \vec{\Phi}(t, \tau)x(\tau). \tag{16.3}$$

The STM for a linear system satisfies the following properties.

- $\vec{\Phi}(t,t) =$ identity matrix I, for any t
- $x(t_3) = \vec{\Phi}(t_3, t_1)x(t_1)$ for any t_3 and t_1
- $x(t_3) = \vec{\Phi}(t_3, t_2)x(t_2)$ for any t_3 and t_2
- $x(t_2) = \vec{\Phi}(t_2, t_1)x(t_1)$ for any t_2 and t_1
- $\vec{\Phi}(t_3, t_1) = \vec{\Phi}(t_3, t_2)\vec{\Phi}(t_2, t_1)$
- $\vec{\Phi}(t, \tau) = [\vec{\Phi}(\tau, t)]^{-1}$ for any τ and t.

The above properties are quite simple to derive and play a very important role in the understanding of the behavior of a linear system's state trajectories. Note that for the special case of a linear time invariant system, i.e. if the matrix $\vec{A}$ is constant, then we observe that

$$\vec{\Phi}(t, \tau) = e^{(t-\tau)\vec{A}}. \tag{16.4}$$

Keep in mind that both sides of the above expression are matrices, not just scalars, for a multi-variable state space system. We are highlighting this fact by denoting the matrix A as $\vec{A}$.

There is no such simple expression for the STM of a linear time varying system. Its determination very much depends on the specificity of the time varying matrix $\vec{A}(t)$. For some special, simple time varying matrices, it is sometimes possible to determine the STM in a relatively easier fashion but in general it is not a simple task. It is also important to realize that for a linear time varying system, the STM depends both on the initial given time τ as well as the present time t, whereas for a linear time invariant system, it is only the difference $t - \tau$, that matters. Thus in a linear LTI system, there is no loss of generality in taking the initial time τ or t_0 to be zero. Hence, going forward, we present the dynamic response expressions for LTI systems.

16.3   Solutions of Linear Constant Coefficient Differential Equations in State Space Form

Consider the linear state space system

$$\dot{\vec{x}} = A\vec{x} + B\vec{u} \tag{16.5}$$

$$\vec{x}(t_0) = \vec{x}_0.$$

16.3.1   Solution to the Homogeneous Case

The simplest form of the general constant coefficient differential equation is the homogenous, i.e. uncontrolled function

$$\dot{\vec{x}} = A\vec{x} \tag{16.6}$$

where A is a constant $n \times n$ matrix. Fortunately, the solution to this homogenous set of equations mimics the scalar version and is thus given by

$$\vec{x(t)} = e^{At}\vec{c} \tag{16.7}$$

where $\vec{c}$ is the constant vector of integration that depends on the initial condition. For simplicity, omitting the vector notation, we write

$$\vec{x(t)} = e^{A(t-t_0)}\vec{x(t_0)}. \tag{16.8}$$

Note that $e^{A(t-t_0)}$ is an $n \times n$ matrix as well and is labeled the STM, since it helps the transition of the state trajectories from $t(0)$ to any other arbitrary time instant t. It is important to keep in mind that this STM plays an extremely important role in understanding many structural properties of a linear state space system such as stability, controllability, observability and others. In a later section, we discuss various methods of determining this STM matrix in an analytical form so that considerable insight can be gained on the evolution of the state trajectories as a function of time.

16.3.2   Solution to the Non-homogeneous (Forced) Case

Now adding the control variable (going forward, for simplicity and reducing clutter, we omit the vector notation for $x(t)$ and $u(t)$ hoping that its context is fully understood and not cause any confusion by now), we can get the forced response by the expression

$$x(t) = e^{A(t-t_0)}x(t_0) + \int_{t_0}^{t} e^{A(t-\lambda)}Bu(\lambda)\,\mathrm{d}\lambda. \tag{16.9}$$

Note that in the above expression, we assume that the control function $u(t)$ is known or given for all time t, starting from the initial time $t(0)$. Also, now that we are considering multi-variable systems and keeping in mind that, in matrix theory, the product AB is not same as the product BA, it is important to pay attention to the order of multiplication in all the terms inside the integral in the above expression. The dimensional compatibility needs to be strictly adhered to. No mixing up of those terms is allowed. Keep in mind that the STM is an $n \times n$ matrix and the B matrix is an $n \times m$ matrix and the control vector $u(t)$ is an $m \times 1$ vector.

Thus it is reiterated that only in the case of a linear time invariant system can we get a closed form analytical expression for the dynamic response of the state space system.

In order to gain better insight into the state trajectory evolution as a function of time, it helps to determine the STM by analytical means. Since the exponential function is an analytic function of the matrix A, we can get an analytical expression for the STM using the powerful Caley–Hamilton theorem, which is elaborated next.

16.4 Determination of State Transition Matrices Using the Cayley–Hamilton Theorem

Let $f(x)$ be a function that is analytic in the complex plane and let A be an $n \times n$ matrix whose eigenvalues λ_i belong to the complex plane. Then $f(x)$ has a power series representation

$$f(x) = \sum_{k=0}^{\infty} \alpha_k x^k. \tag{16.10}$$

It is possible to regroup the infinite series for $f(x)$ so that

$$f(x) = \Delta(x) \sum_{k=0}^{\infty} \beta_k x^k + R(x). \tag{16.11}$$

Here R is the remainder, $\Delta(x)$ is an nth degree polynomial in x, and $\Delta(x) = 0$ at $x = \lambda_i$, $i = 1, 2, \ldots, n$. The remainder $R(x)$ will have degree $\leq n - 1$. The analytic function of a square matrix A is defined by the same series as its scalar counterpart but with A replacing x. Thus, when $f(A)$ is any analytic function of A,

$$f(A) = \Delta(A) \sum (\beta_k) A^k + R(A). \tag{16.12}$$

but using the Cayley–Hamilton theorem $\Delta(A) = 0$. Therefore

$$f(A) = R(A) \tag{16.13}$$

where $R(A)$ is a polynomial in A of degree $n - 1$. In general, $R(A)$ can be written as

$$R(A) = \alpha_0 I + \alpha_1 A + \alpha_2 A^2 + \cdots + \alpha_{n-1} A^{n-1}. \tag{16.14}$$

Note that in the scalar case when $x = \lambda_i$, we get

$$R(\lambda_i) = \alpha_0 + \alpha_1 \lambda_i + \alpha_2 \lambda_i^2 + \cdots + \alpha_{n-1} \lambda_i^{n-1}. \tag{16.15}$$

Thus

$$f(\lambda_i) = R(\lambda_i) \quad \text{where} \quad i = 1, 2, \ldots, n \tag{16.16}$$

and

$$f(A) = R(A). \tag{16.17}$$

Equations 16.16 and 16.17 can be used to find any analytic function of A. For example for our purposes $f(A) = e^{At}$.

Procedure

1. Find eigenvalues of A (assume they are distinct).
2. Form

$$e^{At} = \alpha_0 I + \alpha_1 A + \alpha_2 A^2 + \cdots + \alpha_{n-1} A^{n-1}. \tag{16.18}$$

3. Form the n equations

$$e^{\lambda_i t} = \alpha_0 + \alpha_1 \lambda_i + \alpha_2 \lambda_i^2 + \ldots + \alpha_{n-1} \lambda_i^{n-1} \quad \text{where} \quad i = 1, 2, \ldots, n. \tag{16.19}$$

From these n equations, solve for $\alpha_0, \alpha_1, \ldots, \alpha_{n-1}$.
4. Substitute these expressions for α_i in Equation 16.18.

Example 16.1 Find e^{At} when

$$A = \begin{bmatrix} -1 & \frac{1}{2} \\ 0 & 1 \end{bmatrix}. \tag{16.20}$$

Solution

Since the matrix A is triangular, we can immediately see its eigenvalues are

$$\lambda_1 = -1$$
$$\lambda_2 = 1$$

and are obviously distinct.

$$e^{At} = \alpha_0 I + \alpha_1 A (\text{note } n = 2)$$

$$= \alpha_0 \begin{bmatrix} 1 & 0 \\ 0 & 1 \end{bmatrix} + \alpha_1 \begin{bmatrix} -1 & \frac{1}{2} \\ 0 & 1 \end{bmatrix}$$

$$= \begin{bmatrix} \alpha_0 & 0 \\ 0 & \alpha_0 \end{bmatrix} + \begin{bmatrix} -\alpha_1 & \frac{1}{2}\alpha_1 \\ 0 & \alpha_1 \end{bmatrix}$$

$$= \begin{bmatrix} \alpha_0 - \alpha_1 & \frac{1}{2}\alpha_1 \\ 0 & \alpha_0 + \alpha_1 \end{bmatrix}.$$

The first equation from the above system (i.e. the second row of the matrix equation) can be written algebraically as

$$e^{\lambda_1 t} = \alpha_0 + \alpha_1 \lambda_1$$

which, after substituting λ_1, becomes

$$e^{-t} = \alpha_0 + \alpha_1(-1)$$
$$= \alpha_0 - \alpha_1.$$

The second equation from above is written:

$$e^{\lambda_2 t} = \alpha_0 + \alpha_1 \lambda_2$$

which, after substituting λ_2 becomes

$$e^t = \alpha_0 + \alpha_1(1)$$
$$= \alpha_0 + \alpha_1.$$

Then the two equations, with λ_1 and λ_2 substituted in, are

$$\alpha_0 - \alpha_1 = e^{-t}$$
$$\alpha_0 + \alpha_1 = e^t$$

which can be solved to obtain

$$\alpha_0 = \frac{1}{2}(e^t + e^{-t})$$

$$\alpha_1 = \frac{1}{2}(e^t - e^{-t}).$$

Substituting these values α_i in the e^{At} expression, we obtain

$$e^{At} = \begin{bmatrix} e^{-t} & \frac{1}{4}(e^t - e^{-t}) \\ 0 & e^t \end{bmatrix}.$$

16.4.1 For Repeated Roots

If λ_i is repeated m_i times, then you can get additional independent equations as follows:

$$f(\lambda_i) = R(\lambda_i)$$

$$\left.\frac{df(\lambda)}{d\lambda}\right|_{\lambda_i} = \left.\frac{dR(\lambda)}{d\lambda}\right|_{\lambda_i}$$

$$\left.\frac{d^2f(\lambda)}{d\lambda^2}\right|_{\lambda_i} = \left.\frac{d^2R(\lambda)}{d\lambda^2}\right|_{\lambda_i}$$

$$\vdots$$

$$\left.\frac{d^{m_i-1}f(\lambda)}{d\lambda^{m_i-1}}\right|_{\lambda_i} = \left.\frac{d^{m_i-1}R(\lambda)}{d\lambda^{m_i-1}}\right|_{\lambda_i}.$$

Example 16.2 Find e^{At} when $A = \begin{bmatrix} -3 & 2 \\ 0 & -3 \end{bmatrix}$.

Solution
The eigenvalues are

$$\lambda_1 = \lambda_2 = -3 \text{ (repeated twice, i.e. } m = 2).$$

Note that $n = 2$.

$$e^{At} = \alpha_0 I + \alpha_1 A$$

$$= \begin{bmatrix} \alpha_0 - 3\alpha_1 & 2\alpha_1 \\ 0 & \alpha_0 - 3\alpha_1 \end{bmatrix}.$$

Now

$$e^{\lambda_1 t} = \alpha_0 + \alpha_1 \lambda_1$$

which, upon substitution of $\lambda_1 = -3$, can be written

$$e^{-3t} = \alpha_0 - 3\alpha_1.$$

The second equation is obtained by

$$\frac{d}{d\lambda}(e^{\lambda t})\Big|_{\lambda=\lambda_1=-3} = \left[\frac{d}{d\lambda}(\alpha_0 + \alpha_1\lambda)\right]_{\lambda=\lambda_1=-3}$$

$$te^{\lambda_1 t} = te^{-3t} = \alpha_1$$

$$\text{i.e. } \alpha_1 = te^{-3t}.$$

∴ The two equations to be solved are

$$\alpha_0 - 3\alpha_1 = e^{-3t}$$

$$\alpha_1 = te^{-3t}.$$

Therefore

$$e^{At} = \begin{bmatrix} e^{-3t} & 2te^{-3t} \\ 0 & e^{-3t} \end{bmatrix}.$$

Notice that when you have a repeated eigenvalue a t appears in the STM, which has ramifications in the stability of that system.

It is left to the reader to make sure that all the properties of the the general STM discussed before are valid for the special case of the matrix $e^{\vec{A}t}$. However, care needs to be exercised in noticing that in general

$$e^{\vec{A}t}e^{\vec{B}t} \neq e^{(\vec{A}+\vec{B})t} \tag{16.21}$$

unless matrices $\vec{A}$ and $\vec{B}$ commute, i.e. $\vec{A}\vec{B} = \vec{B}\vec{A}$, which seldom occurs.

Summarizing, the following methods are applicable for finding $\vec{\Phi}$ for linear LTI systems.

1. Using the Laplace transform

$$\vec{\Phi}(t, \tau) = \mathcal{L}[\vec{I}s - \vec{A}^{-1}]$$

   $\vec{\Phi}(t, \tau)$ is then found by replacing t by $t - \tau$, since

$$\vec{\Phi}(t, \tau) = \vec{\Phi}(t - \tau, 0)$$

   when A is constant.
2. Using the equation

$$\vec{\Phi}(t, \tau) = \alpha_0\vec{I} + \alpha_1\vec{A} + \cdots + \alpha_{n-1}\vec{A}^{n-1}$$

   where

$$e^{\lambda_i(t-\tau)} = \alpha_0 + \alpha_1\lambda_i + \cdots + \alpha_{n-1}\lambda_i^{n-1}$$

   and, if some eigenvalues are repeated, derivatives of the above expression with respect to λ must be used.
3. Making use of the Jordan form,

$$\vec{\Phi}(t, \tau) = \vec{M}e^{\vec{J}(t-\tau)}\vec{M}^{-1}$$

   where $\vec{J}$ is the Jordan form (or the diagonal matrix $\vec{\Lambda}$), and $\vec{M}$ is the modal matrix.

4. Using an infinite series

$$\vec{\Phi}(t,\tau) \cong \vec{I} + \vec{A}(t-\tau) + \frac{1}{2}\vec{A}^2(t-\tau)^2 + \frac{1}{3!}\vec{A}^3(t-\tau)^3 + \ldots .$$

This infinite series can be truncated after a finite number of terms to obtain an approximation for the transition matrix.

16.5 Response of a Constant Coefficient (Time Invariant) Discrete Time State Space System

Consider the homogeneous case

$$x(k+1) = Ax(k) \quad \text{with} \quad x(0) = x_o \tag{16.22}$$

where the coefficient matrix A is a constant and initial conditions x_o are known, i.e.

$$x(1) = Ax(0) \tag{16.23}$$

$$x(1) = Ax(0) \tag{16.24}$$

$$x(2) = Ax(1) \tag{16.25}$$

$$= A^2x(0) \tag{16.26}$$

$$x(3) = Ax(2) \tag{16.27}$$

$$= A^3x(0). \tag{16.28}$$

At an arbitrary time t_k

$$x(k) = A^k x(0) \tag{16.29}$$

which can also be written as

$$x(k) = A^k x(0). \tag{16.30}$$

When the matrix A is constant the discrete transition matrix is

$$\phi(k,j) = A^{k-j}. \tag{16.31}$$

Hence the state transition matrix STM for the discrete time state space system is given by

$$\text{STM} = A^k. \tag{16.32}$$

Next, consider the non-homogeneous case

$$x(k+1) = Ax(k) + B(k)u(k). \tag{16.33}$$

In this case we are given the series of input vectors $u(k)$ as well as the initial conditions x_o. Then

$$x(1) = Ax(0) + B(0)u(0) \tag{16.34}$$

$$x(2) = Ax(1) + B(1)u(1) \tag{16.35}$$

$$= A^2x(0) + AB(0)u(0) + B(1)u(1) \tag{16.36}$$

$$x(3) = Ax(2) + B(2)u(2) \tag{16.37}$$

$$= A^3 x(0) + A^2 B(0)u(0) + AB(1)u(1) + B(2)u(2). \tag{16.38}$$

At an arbitrary time t_k

$$x(k) = A^k x(0) + \sum_{j=0}^{k-1} A^{k-1-j} B(j)u(j) \tag{16.39}$$

which can also be written as

$$x(k) = A^k x(0) + \sum_{j=1}^{k} A^{k-j} B(j-1)u(j-1). \tag{16.40}$$

Notice that, in above analysis, we assumed A to be a constant matrix (independent of k) but we allowed B to be time varying (i.e. B is a function of k). Also, in the discrete time system, the convolution integration present in continuous systems is replaced by a discrete summation.

Example 16.3   Given a matrix

$$A = \begin{bmatrix} 0.368 & 0 \\ 0.632 & 1 \end{bmatrix}$$

find A^k.

Solution
We know

$$\lambda_1 = 1$$
$$\lambda_2 = 0.368$$

and

$$A^k = \alpha_0 I + \alpha_1 A$$
$$= \begin{bmatrix} \alpha_0 + 0.368\alpha_1 & 0 \\ 0.632\alpha_1 & \alpha_0 + \alpha_1 \end{bmatrix}.$$

The first equation from this system of equations is

$$(\lambda_1)^k = \alpha_0 + \alpha_1 \lambda_1$$

which, after substituting in $\lambda_1 = 1$, can be rearranged into an equation in two unknowns α_0 and α_1

$$1 = \alpha_0 + \alpha_1.$$

The second equation from the system of equations is

$$(\lambda_2)^k = \alpha_0 + \alpha_1 \lambda_2$$

which, after substituting in $\lambda_2 = 0.368$, can also be rearranged into an equation in two unknowns α_0 and α_1 as such

$$0.368^k = \alpha_0 + 0.368\alpha_1.$$

Now armed with two equations, we can solve for the two unknowns α_0 and α_1 to obtain

$$A^k = \begin{bmatrix} (0.368)^k & 0 \\ [1 - (0.368)^k] & 1 \end{bmatrix}.$$

Example 16.4   Find $A^k = \begin{bmatrix} 1 & -1 & 1 \\ 0 & 1 & 1 \\ 0 & 0 & 1 \end{bmatrix}^k$.

Solution

It is easy to see that this is a repeated eigenvalue case, with

$$\lambda_1, \lambda_2, \lambda_3 = 1$$

repeated three times. Then

$$A^k = \alpha_0 I + \alpha_1 A + \alpha_2 A^2$$
$$= \begin{bmatrix} \alpha_0 + \alpha_1 + \alpha_2 & -\alpha_1 - 2\alpha_2 & \alpha_1 + \alpha_2 \\ 0 & \alpha_0 + \alpha_1 + \alpha_2 & \alpha_1 + 2\alpha_2 \\ 0 & 0 & \alpha_0 + \alpha_1 + \alpha_2 \end{bmatrix}.$$

For the first eigenvalue λ_1,

$$(\lambda_1)^k = \alpha_0 + \alpha_1 \lambda_1 + \alpha_2 \lambda_1$$
$$= \alpha_0 + \alpha_1(1) + \alpha_2(1)^2$$
$$= \alpha_0 + \alpha_1 + \alpha_2.$$

However, because this is a repeated eigenvalue case,

$$\left. \frac{d(\lambda^k)}{d\lambda} \right|_{\lambda=\lambda_1} = k\lambda^{k-1}|_{\lambda=\lambda_1=1} = k = \alpha_1 + 2\alpha_2 \tag{16.41}$$

$$\left. \frac{d^2(\lambda^k)}{d\lambda^2} \right|_{\lambda=\lambda_1} = k(k-1)\lambda^{k-2}|_{\lambda=\lambda_1=1} = k(k-1) = 2\alpha_2. \tag{16.42}$$

Therefore,

$$A^k = \begin{bmatrix} 1 & -k & \dfrac{k(3-k)}{2} \\ 0 & 1 & k \\ 0 & 0 & 1 \end{bmatrix}$$

and so for $k = 1000$

$$A^{1000} = \begin{bmatrix} 1 & -1000 & \dfrac{1000(3-1000)}{2} \\ 0 & 1 & 1000 \\ 0 & 0 & 1 \end{bmatrix}.$$

The above example illustrates the power of the Cayley–Hamilton theorem. Observe that the above expression for A^{1000} is obtained as a simple analytical expression rather than getting the matrix A multiplied a thousand times.

16.6   Discretizing a Continuous Time System: Sampled Data Systems

Consider the continuous time system

$$\dot{x} = Ax + Bu \tag{16.43}$$

which we wish to solve using a computer. Say our computer samples the continuous inputs u at time instants

$$t = kT \quad \text{where} \quad k = 0, 1, 2 \dots. \tag{16.44}$$

Then the discrete time representation of Equation 16.43 is

$$x((k+1)T) = G(T)x(kT) + H(T)u(kT). \tag{16.45}$$

Here, the matrices G and H depend upon the sampling period T; obviously they are constant matrices if the sampling period is fixed. To determine these matrices, first recall the solution to Equation 16.43 is

$$x(t) = e^{At}x(0) + e^{At} \int_0^{\tau} e^{-At}Bu(\tau)d\tau \tag{16.46}$$

assuming the components of $u(t)$ are constant over the interval between any two consecutive sampling instants, i.e. $u(t) = u(kT)$ for the kth sampling period. Since

$$x((k+1)T) = e^{A(k+1)T}x(0) + e^{A(k+1)T} \int_0^{(k+1)T} e^{-At}Bu(\tau)d\tau \tag{16.47}$$

and

$$\begin{aligned} x(kT) &= e^{AkT}x(0) + e^{AkT} \int_{kT}^{(k+1)\tau} e^{-A\tau}Bu(\tau)d\tau \\ &= e^{At}x(kT) + e^{AT} \int_0^T e^{-At}Bu(kT)dt \\ &= e^{AT}x(kT) + \int_0^T e^{A\lambda}Bu(kT)d\lambda \end{aligned} \tag{16.48}$$

where

$$\lambda \equiv T - t. \tag{16.49}$$

So then

$$G(T) = e^{AT} \tag{16.50}$$

$$H(T) = \left(\int_0^T e^{At}dt \right) B. \tag{16.51}$$

Note that we used the special letters G and H to denote the plant matrices for the sampled data system, so that we can highlight the fact that these G and H matrices are functions of the continuous time plant matrices A and B and are also functions of the constant sampling period T. Thus, finally, a sampled data system is always denoted by the state space representation

$$x((k+1)T) = G(T)x(kT) + H(T)u(kT) \tag{16.52}$$

Example 16.5 Obtain a discrete time state space representation of the following continuous time system:

$$\begin{bmatrix} \dot{x}_1 \\ \dot{x}_2 \end{bmatrix} = \begin{bmatrix} 0 & 1 \\ 0 & -2 \end{bmatrix} \begin{bmatrix} x_1 \\ x_2 \end{bmatrix} + \begin{bmatrix} 0 \\ 1 \end{bmatrix} [u]. \tag{16.53}$$

Solution

The discrete time system representation will be in the form

$$x((k+1)T) = G(T)x(kT) + H(T)u(kT) \tag{16.54}$$

where

$$\begin{aligned} G(T) &= e^{AT} \\ &= \begin{bmatrix} 1 & \frac{1}{2}(1 - e^{-2T}) \\ 0 & e^{-2T} \end{bmatrix} \end{aligned} \tag{16.55}$$

and

$$\begin{aligned} H(T) &= \left(\int_0^T e^{At} dt \right) B \\ &= \left\{ \int_0^T \begin{bmatrix} 1 & \frac{1}{2}(1 - e^{-2T}) \\ 0 & e^{-2T} \end{bmatrix} \right\} \begin{bmatrix} 0 \\ 1 \end{bmatrix} \\ &= \begin{bmatrix} \frac{1}{2} \left(T + \frac{e^{-2T} - 1}{2} \right) \\ \frac{1}{2}(1 - e^{-2T}) \end{bmatrix}. \end{aligned} \tag{16.56}$$

Therefore

$$\begin{aligned} \begin{bmatrix} x_1((k+1)T) \\ x_2((k+1)T) \end{bmatrix} &= \begin{bmatrix} 1 & \frac{1}{2}(1 - e^{-2T}) \\ 0 & e^{-2T} \end{bmatrix} \begin{bmatrix} x_1(kT) \\ x_2(kT) \end{bmatrix} \\ &\quad + \begin{bmatrix} \frac{1}{2} \left(T + \frac{e^{-2T} - 1}{2} \right) \\ \frac{1}{2}(1 - e^{-2T}) \end{bmatrix} u(kT). \end{aligned} \tag{16.57}$$

Then a given sampling period T can be substituted into the general Equation 16.58 to obtain an equation describing the specific system.

For example, with $T = 1$ s, the sampled data state space system is given by

$$\begin{bmatrix} x_1((k+1)) \\ x_2((k+1)) \end{bmatrix} = \begin{bmatrix} 1 & 0.432 \\ 0 & 0.135 \end{bmatrix} \begin{bmatrix} x_1(k) \\ x_2(k) \end{bmatrix} + \begin{bmatrix} 0.284 \\ 0.432 \end{bmatrix} u(k). \tag{16.58}$$

Example 16.6 Discretize the following system, assuming a sampling period $T = 0.1$ s:

$$\dot{x} = \begin{bmatrix} -3 & 1 \\ 0 & -2 \end{bmatrix} x + \begin{bmatrix} 1 \\ 1 \end{bmatrix} u.$$

Solution

To find the sampled data system plant matrix G, which is equal to e^{AT}, we first need to find the matrix e^{At} for the given A matrix of the continuous time system. Following the procedure given in the discussion of e^{At} matrix determination using the Cayley–Hamilton theorem we obtain

$$e^{At} = \begin{bmatrix} e^{-3t} & -e^{-3t} + e^{-2t} \\ 0 & e^{-2t} \end{bmatrix}.$$

Then $G(T) = e^{AT}$ for $T = 0.1$ s is given by

$$G(T) = e^{AT}$$
$$= \begin{bmatrix} e^{-3T} & -e^{-3T} + e^{-2T} \\ 0 & e^{-2T} \end{bmatrix}$$
$$= \begin{bmatrix} 0.741 & 0.0779 \\ 0 & 0.819 \end{bmatrix}.$$

Then to calculate the matrix H

$$H(T) = \left(\int_0^T e^{At} dt \right) B$$
$$= \left(\int_0^T \begin{bmatrix} e^{-3t} & -e^{-3t} + e^{-2t} \\ 0 & e^{-2t} \end{bmatrix} dt \right) \begin{bmatrix} 1 \\ 1 \end{bmatrix}$$
$$= \frac{1}{2} \begin{bmatrix} 1 - e^{-2T} \\ 1 - e^{-2T} \end{bmatrix}$$
$$= \begin{bmatrix} 0.0906 \\ 0.0906 \end{bmatrix}.$$

And so the discretized approximation of the given continuous system is

$$x(k+1) = \begin{bmatrix} 0.741 & 0.07779 \\ 0 & 0.819 \end{bmatrix} x(k) + \begin{bmatrix} 0.0906 \\ 0.0906 \end{bmatrix} u(k).$$

16.7 Chapter Summary

This chapter discussed the important concept of getting the dynamic response (state trajectories) of linear state space systems using the concept of the STM. It is shown that for the special case of linear continuous time LTI systems, this STM takes on a simple form as the e^{At} matrix, which can be determined analytically using the Caley–Hamilton theorem. We then discussed the STM for linear discrete time LTI systems, which turns out to be the A_d^k matrix where A_d is the plant matrix of the discrete time system. Finally we covered the case sampled data systems, which are discrete time systems obtained by sampling a continuous time system with a constant sampling period T. For this case, it is shown that the plant matrix for the sampled data system is $A_d = e^{\tilde{A}T}$ where $\tilde{A}$ is the plant matrix for the continuous time system. The important properties of STMs are discussed. Fundamental concepts such as those discussed in this chapter are also available in a more consolidated way in many excellent textbooks dedicated to state space control systems such as [1–8]

16.8 Exercises

Exercise 16.1. Consider the continuous time system

$$\begin{bmatrix} \dot{x}_1 \\ \dot{x}_2 \end{bmatrix} = \begin{bmatrix} -1 & 0.5 \\ 0 & 1 \end{bmatrix} \begin{bmatrix} x_1 \\ x_2 \end{bmatrix} + \begin{bmatrix} 5 \\ 1 \end{bmatrix} \tag{16.59}$$

and the initial condition

$$x_0 = \begin{bmatrix} 1 \\ 2 \end{bmatrix}. \tag{16.60}$$

Given that

$$e^{At} = \begin{bmatrix} e^{-t} & \frac{1}{4}(e^t - e^{-t}) \\ 0 & e^t \end{bmatrix}. \tag{16.61}$$

Obtain the analytical expressions and state trajectories for $x_1(t)$ and $x_2(t)$ when
(a) $u \equiv 0$
(b) $u \equiv e^{-t}$.

Exercise 16.2. The following matrices are the state space matrices for an aircraft longitudinal dynamics at a given flight condition.

$$\begin{bmatrix} \dot{u} \\ \dot{\alpha} \\ \dot{\theta} \\ \dot{q} \\ \dot{h} \end{bmatrix} = \begin{bmatrix} -0.0451 & 6.348 & -32.2 & 0 & 0 \\ -0.0021 & -2.0244 & 0 & 1 & 0 \\ 0 & 0 & 0 & 1 & 0 \\ 0.0021 & -6.958 & 0 & -3.0757 & 0 \\ 0 & -176 & 176 & 0 & 0 \end{bmatrix} \begin{bmatrix} u \\ \alpha \\ \theta \\ q \\ h \end{bmatrix} + \begin{bmatrix} 0 \\ -0.160 \\ 0 \\ -11.029 \\ 0 \end{bmatrix} \delta_e \tag{16.62}$$

where the control input δ_e is the perturbation elevator deflection (in radians). Use MATLAB to plot the open loop system ($u \equiv 0$) trajectories for the outputs

$$y = x = \begin{bmatrix} u \\ \alpha \\ \theta \\ q \\ h \end{bmatrix}. \tag{16.63}$$

Exercise 16.3. Given the continuous time system state space model

$$\dot{x} = \begin{bmatrix} 0 & 1 \\ 0 & -1 \end{bmatrix} x + \begin{bmatrix} 0 \\ 1 \end{bmatrix} u \tag{16.64}$$

obtain the corresponding discretized (sampled data) system state space matrices G and H assuming a sampling interval $T = 1$ s.

Hint:

1. $e^{At} = \begin{bmatrix} 1 & 1 - e^{-t} \\ 0 & e^{-t} \end{bmatrix}.$

Exercise 16.4. Obtain the STM for the discrete time system state space model given by

$$x(k+1) = \begin{bmatrix} 0 & 1 \\ -3 & -4 \end{bmatrix} x(k). \tag{16.65}$$

Exercise 16.5. Obtain the STM for the continuous time system state space model given by

$$\dot{x} = \begin{bmatrix} 0 & 1 \\ -3 & -4 \end{bmatrix} x. \tag{16.66}$$

Bibliography

1 T.E Fortmann and K.L Hitz. *An Introduction to Linear Control Systems.* Mercel-Dekker, 1977.

2 G.F Franklin, J.D Powell, and A Emami-Naeni. *Feedback Control of Dynamic Systems.* Pearson-Prentice-Hall, 2006.

3 B. Friedland. *Control system design: An introdcution to state-space methods.* McGraw-Hill, 1986.

4 H. Kwakernaak and R. Sivan. *Linear Optimal Control Systems.* Wiley Interscience, 1972.

5 K. Ogata and Y. Yang. *Modern control engineering.* Prentice-Hall, Englewood Cliffs, NJ, 1970.

6 R.E. Skelton. *Dynamic Systems Control: Linear Systems Analysis and Synthesis.* Wiley, New York, 1 edition, 1988.

7 J.G. Truxal, editor. *Control engineers' handbook.* McGraw-Hill, New York, 1958.

8 W.L. Brogan. *Modern Control Theory.* Prentice Hall, 1974.

17

Stability of Dynamic Systems with State Space Representation with Emphasis on Linear Systems

17.1 Chapter Highlights

In this chapter, we thoroughly discuss the various concepts related to the stability of dynamic systems represented by state space models. We first consider the nonlinear state space models and briefly discuss their stability using Lyapunov stability concepts. Then we specialize these notions to linear state space models and in particular to linear time invariant (LTI) state space systems. Within this linear systems stability discussion, we first consider the continuous time system case, then the discrete time system case, treating sampled data systems as part of discrete time systems. The distinction between continuous time systems and discrete time systems is necessary and important because the stability regions are different. For continuous time systems, the open left half of the complex plane forms the stability region (i.e negative real part eigenvalue criterion) whereas for discrete time systems, the unit circle within the complex plane with origin as the center of that circle forms the stability region (i.e. magnitudes of the eigenvalues being ≤ 1 as the criterion). Going forward, we refer to the continuous time system stability as the Hurwitz stability and the discrete time system stability as the Schur stability.

17.2 Stability of Dynamic Systems via Lyapunov Stability Concepts

We have noticed that, in general, the state space description of any dynamic system described by nonlinear first order ordinary differential equations is given by

$$\dot{\vec{x}} = \vec{f}(\vec{x}, \vec{u}, t)$$

where $\vec{x}$ is the state vector, $\vec{u}$ is the control vector and $\vec{f}$ is a vector of nonlinear functions in x_i and u_i. Typically, we write

$$\vec{x} \in R^n, \text{ i.e. } \vec{x} = \begin{bmatrix} x_1 \\ x_2 \\ \vdots \\ x_n \end{bmatrix}$$

$$\vec{u} \in R^m, \text{ i.e. } \vec{u} = \begin{bmatrix} u_1 \\ u_2 \\ \vdots \\ u_m \end{bmatrix}.$$

Flight Dynamics and Control of Aero and Space Vehicles, First Edition. Rama K. Yedavalli.
© 2020 John Wiley & Sons Ltd. Published 2020 by John Wiley & Sons Ltd.

Consider the following three classes of nonlinear systems:

1. $\dot{\vec{x}} = \vec{f}(\vec{x}, \vec{u}, t)$
2. $\dot{\vec{x}} = \vec{f}(\vec{x}, t)$
3. $\dot{\vec{x}} = \vec{f}(\vec{x})$.

Out of these, consider the second class of systems.

$$\dot{\vec{x}} = \vec{f}(\vec{x}, t), \; \vec{x}(t_0) = \vec{x}_0.$$

We assume that the above equation, has a unique solution starting at the given initial condition, i.e. we have one single solution corresponding to each initial condition. Let us denote this solution as

$$\vec{x}(t; \vec{x}_0, t_0) \equiv \vec{x}(t) \text{ for simplicity}$$
$$\vec{x}(t_0; \vec{x}_0, t_0) \equiv \vec{x}_0.$$

17.2.1 Equilibrium State

In the above class of systems, a state $\vec{x}_e$ where $\vec{f}(\vec{x}_e, t) = 0$ for all t is called an equilibrium state of the system, i.e. the equilibrium state corresponds to the constant solution of the system. If the system is linear time invariant (i.e. $\vec{f}(\vec{x}, t) = A\vec{x}$), then there exists only one equilibrium state if A is non-singular and many equilibrium states if A is singular. For nonlinear systems there may be one or more equilibrium states.

Any isolated equilibrium point can always be transferred to the origin of the coordinates

$$\text{i.e. } \vec{f}(0, t) = 0$$

by a proper coordinate transformation. So one can always take $\vec{x}_e = 0$ without any loss of generality. The origin of state space is always an equilibrium point for linear systems and for linear systems all equilibrium states behave the same way (because if $\vec{x}(t)$ is a solution $\overline{\vec{x}}(t)$ is also a solution; then $\vec{x}(t) \to \overline{\vec{x}}(t)$ is also a solution for the linear system).

For nonlinear systems, the stability of the system is defined in terms of the stability of its equilibrium points via the Lyapunov stability notions. Let us elaborate on this important concept (see Figure 17.1).

For a continuous time system with an equilibrium state $\vec{x}_e$, the following definitions of stability are widely used.

- **Definition 1**
  The equilibrium state $\vec{x}_e$ is said to be stable in the sense of Lyapunov if for any t_0 and for any $\epsilon > 0$ there exists a $\delta(\epsilon, t_0) > 0$ such that if $||\vec{x}(t_0) - \vec{x}_e|| \leq \delta$, then $||\vec{x}(t_0) - \vec{x}_e|| \leq \epsilon$ for all $t \geq t_0$. Note that it is necessarily true that $\delta \leq \epsilon$.
- **Definition 2**
  The equilibrium state $\vec{x}_e$ is said to be asymptotically stable if
  (a) it is stable in the sense of Lyapunov
  (b) for all t_0, there exists a $\rho(t_0) > 0$ such that $||\vec{x}(t_0) - \vec{x}_e|| < \rho$ implies $||\vec{x}(t) - \vec{x}_e|| \to 0$ as $t \to \infty$.

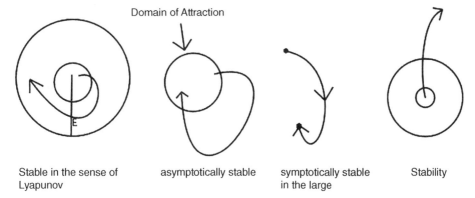

Figure 17.1 Pictorial representation of stability of equilibrium states.

Asymptotic stability ensures that all solutions converge to the equilibrium state as $t \to \infty$ provided the initial deviation is less than ρ.

- **Definition 3**
  The equilibrium state $\vec{x}_e$ is said to be globally asymptotically stable or asymptotically stable in the large if
  1. it is stable in the sense of Lyapunov
  2. for any $\vec{x}(t_0)$ and any t_0 $||\vec{x}(t) - \vec{x}_e|| \to 0$ as $t \to \infty$.

If, in the above definitions of stability, δ and P are independent of t_0 then we say uniformly stable and uniformly asymptotically stable respectively.

Remark 1.   Thus, these definitions of stability are basically concerned with whether neighboring solutions remain neighboring or not.

Remark 2.   Also, the concept of stability involves the behavior as $t \to \infty$ and also on the behavior of the trajectories (motions) perturbed from a fixed motion (equilibrium state).

In Figure 17.2 points A, C, E, F, and G are equilibrium points (solutions), of which A, E, F, and G are isolated equilibrium points. A and F are unstable equilbrium points. E and G are stable equilibrium points. C is neutrally stable.

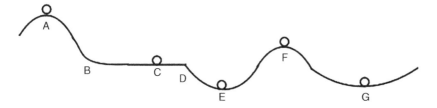

Figure 17.2 Stability analogy.

The amount of initial deviation of the ball is analogous to δ and ρ in Definitions 1 and 2. If the initial deviation is small, then the ball at G would return to G. If it is too big it may not return. Thus the stability definition involves the amount of initial perturbation.

The point C, loosely speaking, corresponds to stability in the sense of Lyapunov.

17.2.2 Lyapunov Method to Determine Stability

The second method of Lyapunov is based on the fact that if the system has an asymptotically stable equilibrium state, then the stored energy of the system displaced within the domain of attraction decays with increasing time until it finally assumes its minimum value at the equilibrium state.

A Lyapunov function is constructed in terms of the state variables x_i of the system under consideration. It is a scalar which is positive (mathematical equivalent of energy).

Theorem 17.1   Let us consider the autonomous (free and stationary; that is, not explicitly dependent on time) system

$$\vec{x} = \vec{f}(\vec{x})$$

where the origin is an equilibrium point, i.e. $\vec{f}(0) = 0$ for all t. Then the origin is stable in the sense of Lyapunov if there exists

1. a scalar Lyapunov function $V(\vec{x}) > 0$ having continuous first partial derivatives
2. $\dot{V}(\vec{x})$ is negative semi-definite (≤ 0).

Theorem 17.2   The origin is asymptotically stable if

1. $V(\vec{x}) > 0$
2. $\dot{V}(\vec{x}) < 0$.

Theorem 17.3   The origin is globally asymptotically stable (i.e. asymptotically stable in the large) if

1. $V(0) = 0$
2. $V(\vec{x}) > 0$ for all $\vec{x} \neq 0$
3. $\dot{V}(\vec{x}) < 0$ for all $\vec{x} \neq 0$, and
4. $V(\vec{x}) \to \infty$ as $||\vec{x}|| \to \infty$.

Remark 3.   The Lyapunov function is not unique.

Remark 4.   The Lyapunov stability theorems are sufficient conditions for stability.

Remark 5.   The inability to find a satisfactory Lyapunov function does not mean the system is unstable.

17.2.3 Lyapunov Stability Analysis for Linear Time Invariant Systems

Consider

$$\dot{\vec{x}} = A\vec{x}(t)$$
$$\vec{x}(0) = \vec{x}_0$$
$$\vec{x} \in R^n$$

and let A be non-singular so that there is only one equilibrium state, which is the origin ($\vec{x} \equiv 0$).

Let us choose a Lyapunov function $V(\vec{x})$ as

$$V(\vec{x}) = \vec{x}^T P \vec{x} > 0 \tag{17.1}$$

where P is a symmetric positive definite matrix. Then

$$\dot{V}(\vec{x}) = \dot{\vec{x}}^T P \vec{x} + \vec{x}^T P \dot{\vec{x}}$$
$$= (A\vec{x})^T P \vec{x} + \vec{x}^T P (A\vec{x})$$
$$= \vec{x}^T (A^T P + PA)\vec{x}.$$

Since $V(\vec{x})$ is chosen positive, we require for asymptotic stability that $\dot{V}(\vec{x})$ be negative definite, i.e.

$$\dot{V} = -\vec{x}^T Q \vec{x}$$

where

$$A^T P + PA = -Q \tag{17.2}$$

with Q being any positive definite matrix. Thus if we start with a P as in Equation 17.1, we require Q given by Equation 17.2 to be positive definite for asymptotic stability. Instead of starting with P, we can specify a positive definite Q and solve Equation 17.2 for P and check whether it is positive definite or not.

The equation can also be written as

$$\boxed{A^T P + PA + Q = 0} \tag{17.3}$$

and is called the matrix Lyapunov equation. Note that it is the linear matrix equation that can be solved for P given A and Q.

Theorem 17.4 For a given positive definite Q and an A such that $\lambda_j + \lambda_k \neq 0$ for all $j, k = 1, 2, \ldots n$ the solution to Equation 17.2 is unique and is a symmetric matrix.

Theorem 17.5 The real parts of the eigenvalues of a constant matrix A are negative (i.e. the matrix A is an asymptotically stable matrix) if and only if for any given symmetric positive definite matrix Q there exists a symmetric positive definite matrix P that is the unique solution of the Lyapunov equation (Equation 17.3).

This important theorem tells us that for a LTI system, we have two ways of checking the stability, either by solving the matrix Lyapunov equation and checking the positive definiteness of that solution matrix or by simply computing the eigenvalues of the state space matrix A of the linear system. Thus both of these methods are equivalent to each other and both serve as a necessary and sufficient condition for the stability of a linear time invariant system.

Example 17.1   Determine the stability of the following system by the Lyapunov method.

$$\begin{bmatrix} \dot{x}_1 \\ \dot{x}_2 \end{bmatrix} = \begin{bmatrix} 0 & 1 \\ -1 & -1 \end{bmatrix} \begin{bmatrix} x_1 \\ x_2 \end{bmatrix}.$$

Solution
Since A is non-singular, solve the Lyapunov equation with say $Q = I_2$, i.e.

$$\begin{bmatrix} 0 & -1 \\ 1 & -1 \end{bmatrix} \begin{bmatrix} P_{11} & P_{12} \\ P_{12} & P_{22} \end{bmatrix} + \begin{bmatrix} P_{11} & P_{12} \\ P_{12} & P_{22} \end{bmatrix} \begin{bmatrix} 0 & 1 \\ -1 & -1 \end{bmatrix} = \begin{bmatrix} -1 & 0 \\ 0 & -1 \end{bmatrix}$$

which is simply the set of three equations in three unknowns

$$-2P_{12} = -1 \tag{17.4}$$

$$P_{11} - P_{12} - P_{22} = 0 \tag{17.5}$$

$$P_{12} - 2P_{22} = -1. \tag{17.6}$$

Solving, we find the matrix P as

$$P = \begin{bmatrix} \frac{3}{2} & \frac{1}{2} \\ \frac{1}{2} & 1 \end{bmatrix}.$$

Then check the definiteness of this matrix. Using the principal minor (Sylvester's criterion) test

$$\Delta_1 = \frac{3}{2} > 0$$

$$\Delta_2 = \begin{vmatrix} \frac{3}{2} & \frac{1}{2} \\ \frac{1}{2} & 1 \end{vmatrix} > 0.$$

Therefore P is positive definite and hence A is asymptotically stable.

17.3   Stability Conditions for Linear Time Invariant Systems with State Space Representation

Going forward, we now focus our attention on stability conditions for general LTI systems with state space representation. We have noted that within this LTI framework, the discussion of stability revolves around the stability of the plant matrix A of that LTI system. Within these LTI systems, we have continuous time systems, discrete time systems, and sampled data systems that are considered as a special case of discrete time systems. Accordingly, we address the stability analysis of continuous time LTI systems separately from the discrete time LTI systems. Recall that the stability region of continuous time LTI systems is the left half of the complex plane (labeled as the Hurwitz stability) whereas that of the discrete time system is the unit circle in the complex plane with origin as the center of that unit circle (labeled as the Schur stability).

17.3.1 Continuous Time Systems: Methods for Checking the Hurwitz Stability of a Real matrix

We have learned from Appendix C, as well as the previous chapter on the dynamic response of linear time invariant state space systems, that the state trajectories asymptotically converge to zero when the state space plant matrix A has negative real part eigenvalues. Thus the necessary and sufficient condition for the (Hurwitz) stability of linear state space matrix A is that it has negative real part eigenvalues. However, computing the eigenvalues of a real matrix, especially for large order matrices, is a computationally (or numerically) intensive task. Hence, even several decades back, there were efforts to determine the stability of a matrix without actually computing the eigenvalues of the matrix. One such criterion is the popular Routh–Hurwitz criterion. Recall that we alluded to the Routh–Hurwitz criterion for stability in Part II of this book when dealing with transfer function techniques in which we needed to check whether a given nth degree polynomial with real coefficients has all its roots with negative real parts or not. As proposed originally, the Routh–Hurwitz criterion is thus a polynomial stability test.

17.3.1.1 Method 1. Checking Stability via the Routh–Hurwitz Criterion

It is clear that we can easily apply the original Routh–Hurwitz polynomial stability test to determine the stability of a real matrix because it turns out that roots of the nth degree polynomial obtained by the operation $Det(\lambda I - A) = 0$ of an $n \times n$ real matrix A are nothing but the eigenvalues of that matrix A. This special polynomial obtained via this operation is called the characteristic polynomial of the A matrix. Thus by applying the Routh–Hurwitz criterion to the characteristic polynomial of a real matrix A, we can determine the stability of that matrix.

Recall that the Routh–Hurwitz criterion states that a necessary condition for the roots of the characteristic polynomial

$$a_0 s^n + a_1 s^{n-1} + \ldots a_{n-1}s + a_n = 0 \tag{17.7}$$

to have negative real parts is that all the coefficients must be non-zero and be of the same sign. Since we covered the Routh–Hurwitz criterion in Part II of this book, we do not intend to repeat it here. However, it is worth mentioning here that an alternative way of stating the Routh- Hurwitz criterion is that the following Hurwitz determinants all be positive. Note that the square matrices whose determinants we are calculating are formed in a specific pattern with the coefficients in the characteristic polynomial as follows. Thus the equivalent necessary and sufficient condition for the stability of a real A matrix whose characteristic polynomial is given by

$$a_0 s^n + a_1 s^{n-1} + \ldots a_{n-1}s + a_n = 0 \tag{17.8}$$

is that all the Hurwitz Determinants Δ_i be positive, where Δ_i are given by

$$\Delta_1 = a_1 > 0 \tag{17.9}$$

$$\Delta_2 = \begin{vmatrix} a_1 & a_0 \\ a_3 & a_2 \end{vmatrix} > 0 \qquad \Delta_3 = \begin{vmatrix} a_1 & a_0 & 0 \\ a_3 & a_2 & a_1 \\ a_5 & a_4 & a_3 \end{vmatrix} > 0 \tag{17.10}$$

$$\Delta_4 = \begin{vmatrix} a_1 & a_o & 0 & 0 \\ a_3 & a_2 & a_1 & a_o \\ a_5 & a_4 & a_3 & a_2 \\ a_7 & a_6 & a_5 & a_4 \end{vmatrix} > 0 \tag{17.11}$$

$$\Delta_5 = \begin{vmatrix} a_1 & a_o & 0 & 0 & 0 \\ a_3 & a_2 & a_1 & a_o & 0 \\ a_5 & a_4 & a_3 & a_2 & a_1 \\ a_7 & a_6 & a_5 & a_4 & a_3 \\ a_9 & a_8 & a_7 & a_6 & a_5 \end{vmatrix} \tag{17.12}$$

$$\Delta_i = \begin{vmatrix} a_1 & a_o & 0 & 0 & \cdots \\ a_3 & a_2 & a_1 & a_o & \cdots \\ a_5 & a_4 & a_3 & a_2 & \cdots \\ \vdots & \vdots & \vdots & \vdots & \cdots \\ a_{si-1} & a_{2i-2} & \cdots & \cdots & \cdots & a_i \end{vmatrix}. \tag{17.13}$$

17.3.1.2 Method 2. Via the Positive Definiteness of the Lyapunov Equation Solution Matrix

We covered this test in the previous section as part of the general Lyapunov stability analysis for nonlinear systems. However, now, we restate the Lyapunov stability condition for completeness sake, treating it as another available method for testing the Hurwitz stability of a real matrix A.

Theorem 17.6 The given real matrix A is (Hurwitz) stable if and only if, for a given symmetric, positive definite matrix Q, there exists a symmetric, positive definite matrix P as solution to the Lyapunov Matrix Equation $PA + A^T P + Q = 0$.

Important Observation. Note that MATLAB has a Lyapunov matrix equation solver routine called Lyap with a command $X = Lyap(A, C)$, which solves a general Lyapunov matrix equation (with possibly a complex A matrix) of the form $A^* X + XA^T = -C$. This is slightly different from the above Lyapunov matrix equation we discussed with a real A matrix. So care needs to be exercised in using this MATLAB routine for the above case. We need to replace the A matrix in the MATLAB command with the A^T matrix of our real matrix A so that the MATLAB routine actually solves the above intended Lyapunov matrix equation with a real A matrix.

Inspired by the Routh–Hurwitz criterion as well as the Lyapunov equation criterion, Fuller [7] gave another interesting set of conditions for the stability of a real matrix A. The concept behind Fuller's conditions is to realize that every real (constant) matrix has eigenvalues, and, if complex, happen to occur in complex conjugate pairs. Hence, Fuller converted the Hurwitz stability testing problem into a non-singularity testing problem for a new, higher dimensional matrix, derived from the original A matrix (whose stability we are after) that has the property that its eigenvalues are the pairwise summations of the eigenvalues of the A matrix. These new, higher dimensional matrices are formed by the so-called Kronecker operations on the original matrix A. This is done using three classes of Kronecker based matrices of various dimensions, namely (i) the Kronecker sum matrix $D = K[A]$ of dimension n^2 where n is the dimension of the original state space matrix A, (ii) the Lyapunov matrix, $\mathcal{L} = L[A]$m which is of dimension $n(n+1)/2$ and finally (iii) the bi-alternate sum matrix $\mathcal{G} = G[A]$, which is of dimension $n(n-1)/2$. We now present the results of Fuller [7].

17.3.1.3 Methods 3 to 5. Via Fuller's Conditions of Non-singularity

First we present Fuller's method of building higher order matrices based on Kronecker operations.

Fuller's Kronecker Operation based matrices, We now briefly review the method of building few higher order matrices via Kronecker based operations, which have some special properties in terms their eigenvalues. Most of the following material is adopted from Fuller[7].

Definition 17.1 Let A be an n dimensional matrix $[a_{ij}]$ and B an m dimensional matrix $[b_{ij}]$. The mn dimensional matrix C defined by

$$\begin{bmatrix} a_{11}B & \cdots & a_{1n}B \\ a_{21}B & \cdots & a_{2n}B \\ \vdots & \vdots & \vdots \\ a_{m1}B & \cdots & a_{mn}B \end{bmatrix} \tag{17.14}$$

is called the Kronecker product of A and B and is written

$$A \times B = C. \tag{17.15}$$

Corollary 17.1 Let the characteristic roots of matrices A and B be $\lambda_1, \lambda_2, \ldots \lambda_n$, and $\mu_1, \mu_2, \ldots \mu_m$, respectively. Then the characteristic roots of the matrix

$$\sum_{p,q} h_{pq} A^p \times B^q \tag{17.16}$$

are the mn values $\sum_{p,q} h_{pq} \lambda_i^p \times \mu_j^q$, $i = 1, 2, \ldots, n$ and $j = 1, 2, \ldots, m$.

Corollary 17.2 The characteristic roots of the matrix $A \oplus B$ where

$$A \oplus B = A \times I_m + I_n \times B \tag{17.17}$$

are the mn values $\lambda_i + \mu_j$, $i = 1, 2, \ldots, n$ and $j = 1, 2, \ldots, m$.

The matrix $A \oplus B$ is called the Kronecker sum of A and B.

Now we specialize the above operation to build another special matrix labeled as the Kronecker sum matrix $\mathcal{D} = K[A]$, which is nothing but the Kronecker sum matrix A with itself.

Case I: Kronecker sum matrix $\mathcal{D} = K[A]$: Kronecker Sum of A with itself: let $\mathcal{D}$ be the matrix of dimension $k = n^2$, defined by

$$\mathcal{D} = A \times I_n + I_n \times A. \tag{17.18}$$

Corollary 17.3 The characteristic roots of $\mathcal{D}$ are $\lambda_i + \lambda_j$, $i = 1, 2, \ldots, n$ and $j = 1, 2, \ldots, n$. Henceforth, we use an operator notation to denote $\mathcal{D}$. We write $\mathcal{D} = K[A]$.

Example 17.2 For

$$A = \begin{bmatrix} a_{11} & a_{12} \\ a_{21} & a_{22} \end{bmatrix}$$

with λ_1, and λ_2 as eigenvalues, the previous D matrix is given by

$$D = \begin{bmatrix} 2a_{11} & a_{12} & a_{12} & 0 \\ a_{21} & a_{11} + a_{22} & 0 & a_{12} \\ a_{21} & 0 & a_{22} + a_{11} & a_{12} \\ 0 & a_{21} & a_{21} & 2a_{22} \end{bmatrix}$$

with eigenvalues $2\lambda_1$, $\lambda_1 + \lambda_2$, $\lambda_2 + \lambda_1$, and $2\lambda_2$.

MATLAB has a computer routine to build the matrix D from the given matrix A.

17.3.1.4 Method 3. Stability Condition I (for the A Matrix to be Hurwitz Stable) in Terms of the Kronecker Sum Matrix $D = K[A]$)

Theorem 17.7 For the characteristic roots of A to have all of their real parts negative (i.e., for A to be asymptotically stable), it is necessary and sufficient that in the characteristic polynomial

$$(-1)^k |K[A] - \lambda I_k| \tag{17.19}$$

the coefficients of λ, $i = 0, 1, 2, \ldots, k - 1$ should all be positive.

Fuller's stability condition II for the Hurwitz stability of a real matrix A via Lyapunov matrix $\mathcal{L} = L[A]$. We now define another Kronecker related matrix $\mathcal{L}$ called the Lyapunov matrix and state a stability theorem in terms of this matrix.

Definition 17.2 Lyapunov matrix $\mathcal{L}$: the elements of the Lyapunov matrix $\mathcal{L}$ of dimension $l = \frac{1}{2}[n(n+1)]$ in terms of the elements of the matrix A are given as follows. For $p > q$:

$$\mathcal{L}_{pq,rs} = \begin{vmatrix} a_{ps} & \text{if} & r - q \text{ and } s < q \\ a_{pr} & \text{if} & r \geq q, r \neq p, s = q \\ a_{pp} + a_{qq} & \text{if} & r = p \text{ and } s = q \\ a_{qs} & \text{if} & r = p \text{ and } s \leq p, s \neq q \\ a_{qr} & \text{if} & r > p \text{ and } s = p \\ 0 & & \text{otherwise} \end{vmatrix} \tag{17.20}$$

and for $p = q$:

$$\mathcal{L}_{pq,rs} = \begin{vmatrix} 2a_{ps} & \text{if} & r = p \text{ and } s < p \\ 2a_{pp} & \text{if} & r = p \geq q, \text{ and } s = p \\ 2a_{pr} & \text{if} & r = p \text{ and } s = q \\ 0 & & \text{otherwise} \end{vmatrix}. \tag{17.21}$$

Corollary 17.4 The characteristic roots of $\mathcal{L}$ are $\lambda_1 + \lambda_2$, $i = 1, 2, \ldots, n$ and $j = 1, 2, \ldots, i$.

Example 17.3 If

$$A = \begin{bmatrix} a_{11} & a_{12} \\ a_{21} & a_{22} \end{bmatrix} \tag{17.22}$$

with eigenvalues λ_1 and λ_2, then the Lyapunov matrix is given by

$$\mathcal{L} = \begin{bmatrix} 2a_{11} & 2a_{12} & 0 \\ a_{21} & a_{11} + a_{22} & a_{12} \\ 0 & 2a_{21} & 2a_{22} \end{bmatrix}$$

with eigenvalues $2\lambda_1, \lambda_1 + \lambda_2$, and $2\lambda_2$. We observe that, when compared with the eigenvalues of the Kronecker sum matrix $\mathcal{D}$, the eigenvalues of $\mathcal{L}$ omit the repetition of eigenvalues $\lambda_1 + \lambda_2$. Again, for simplicity, we use operator notation to denote $\mathcal{L}$. We write $\mathcal{L} = L[A]$. A method to form the $\mathcal{L}$ matrix from the matrix $\mathcal{D}$ is given by Jury in [3]. We include a MATLAB code for building this matrix in Appendix C.

17.3.1.5 Method 4. Stability Condition II for A in Terms of the Lyapunov Matrix $\mathcal{L} = L[A]$

Theorem 17.8 For the characteristic roots of A to have all of their real parts negative (i.e. for A to be an asymptotically stable matrix), it is necessary and sufficient that in the characteristic polynomial

$$(-1)^l |L[A] - \lambda I_l| \tag{17.23}$$

the coefficients of λ_i $i = 1, 2, \ldots, l-1$ should all be positive.

Clearly, Theorem 17.8 is an improvement over Theorem 17.7, since the dimensions of $\mathcal{L}$ are less than that of $\mathcal{D}$.

Fuller's Stability Condition III for a Real Matrix A via Bialternate Sum matrix $\mathcal{G} = G[A]$
Finally, there is another matrix, called the bi-alternate sum matrix, of reduced dimension $m = \frac{1}{2}[n(n-1)]$ in terms of which a stability theorem like that given earlier can be stated.

Definition 17.3 Bi-alternate sum matrix $\mathcal{G}$: the elements of the bialternate sum matrix $\mathcal{G}$ of dimension $m = \frac{1}{2}[n(n-1)]$ in terms of the elements of the matrix A are given as follows:

$$\mathcal{G} = \begin{vmatrix} -a_{ps} & \text{if} & r = q \text{ and } s < q \\ a_{pr} & \text{if} & r \neq p, s = q \\ a_{pp} + a_{qq} & \text{if} & s = p \text{ and } s = q \\ a_{qs} & \text{if} & r = p \text{ and } s \neq q \\ -a_{qr} & \text{if} & s = p \\ 0 & & \text{otherwise} \end{vmatrix} . \tag{17.24}$$

Note that $\mathcal{G}$ can be written as $\mathcal{G} = A \cdot I_n + I_n \cdot A$ where $\cdot$ denotes the bi-alternate product (see [3] for details on the bi-alternate product). Again, we use operator notation to denote $\mathcal{G}$. We write $\mathcal{G} = G[A]$.

Corollary 17.5 The characteristic roots of $\mathcal{G}$ are $\lambda_i + \lambda_j$, for $i = 2, 3, \ldots, n$ and $j = 1, 2, \ldots, i-1$.

In [3] a simple computer amenable methodology is given to form the G matrix from the given matrix A.

Example 17.4 For

$$A = \begin{bmatrix} a_{11} & a_{12} \\ a_{21} & a_{22} \end{bmatrix}$$

with λ_1 and λ_2 as eigenvalues, the bialternate sum matrix G is given by the scalar

$$G = [a_{22} + a_{11}]$$

where the characteristic root of G is $\lambda_1 + \lambda_2 = a_{11} + a_{22}$.

Example 17.5 When $n = 3$, for the matrix

$$A = \begin{bmatrix} a_{11} & a_{12} & a_{13} \\ a_{21} & a_{22} & a_{23} \\ a_{31} & a_{32} & a_{33} \end{bmatrix}$$

with λ_1, λ_2, and λ_3 as eigenvalues, the bialternate sum matrix G is given by

$$G = \begin{bmatrix} a_{22} + a_{11} & a_{23} & -a_{13} \\ a_{12} & a_{33} + a_{11} & a_{12} \\ -a_{31} & a_{21} & a_{33} + a_{22} \end{bmatrix}$$

with eigenvalues $\lambda_1 + \lambda_2, \lambda_2 + \lambda_3 \, \lambda_3 + \lambda_1$.

Note that, when compared with the eigenvalues of D and L, the eigenvalues of G omit the eigenvalues of the type $2\lambda_i$.

17.3.1.6 Method 5. Stability Condition III for a real Matrix A in Terms of the Bialternate Sum Matrix $G[A]$)

Theorem 17.9 For the characteristic roots of A to have all of their real parts negative, it is necessary and sufficient that in $(-1)^n$ times the characteristic polynomial of A, namely,

$$(-1)^n |[A] - \lambda I_n| \tag{17.25}$$

and in $(-1)^m$ times the characteristic polynomial of G namely,

$$(-1)^m |G[A] - \mu I_m| \tag{17.26}$$

the coefficients of λ ($i = 0, \ldots, n - 1$) and $\mu (j = 1, 2, \ldots, m - 1)$ should all be positive.

This theorem improves somewhat on Theorems 17.2 and 17.3, since the dimensions of G are less than the dimensions of D and L, respectively.

One important consequence of the fact that the eigenvalues of D, L, and G include the sum of the eigenvalues of A is the following fact, which is stated as a lemma to emphasize its importance.

Lemma 17.1

$$\det K[A] = 0$$
$$\det L[A] = 0$$
$$\det G[A] = 0$$

if at least one complex pair of the eigenvalues of A is on the imaginary axis and

$$\det A = 0$$

if and only if at least one of the eigenvalues of A is at the origin of the complex plane. It is important to note that $\det K[A]$, $\det L[A]$, and $\det G[A]$ represent the constant coefficients in the corresponding characteristic polynomials mentioned earlier. It may also be noted that the previous lemma explicitly takes into account the fact that the matrix A is a real matrix and hence has eigenvalues in complex conjugate pairs.

17.3.2 Connection between the Lyapunov Matrix Equation Condition and Fuller's Condition II

It is interesting and somewhat not surprising that the Lyapunov matrix equation $PA + A^T P + Q = 0$ and the Kronecker operation based Lyapunov matrix $L[A]$ are related. Let $\vec{p}$ and $\vec{q}$ be the vectors formed from the (symmetric) matrices P and Q of the Lyapunov matrix equation as follows: they are the elements on and below the leading diagonals of the symmetric matrices P and Q respectively. Thus $\vec{p}$ and $\vec{q}$ are given by

$$\vec{p} = \begin{bmatrix} p_{11} \\ p_{12} \\ p_{22} \\ p_{13} \ \vdots \\ p_{nn} \end{bmatrix}$$

$$\vec{q} = \begin{bmatrix} q_{11} \\ q_{12} \\ q_{22} \\ q_{13} \ \vdots \\ q_{nn} \end{bmatrix}.$$

Then Fuller's Lyapunov matrix $L[A]$ is nothing but the coefficient matrix in the algebraic set of simultaneous equation

$$Hp = q \tag{17.27}$$

where $H = L[A]$, i.e. $L[A]p = q$. The most interesting thing is the revelation that the matrix $L[A]$ has eigenvalues that are pairwise summations of the eigenvalues of A, avoiding specific redundancy (in indices $i \neq j$).

Important Remark. It is the belief of this author that Fuller's conditions have not received the attention and admiration they deserve. It is the author's opinion that this is the first time in a textbook that all these various conditions for stability testing of a real matrix are put together at one place and presented in a unifying framework showing the connection between these seemingly different (yet equivalent) conditions. It is not surprising that they are all connected since, in the end, all of them serve as necessary and sufficient conditions for the (Hurwitz) stability of a real matrix.

17.3.3 Alternate Stability Conditions for Second Order (Possibly Nonlinear) Systems [1]

This discussion is borrowed from [1].

An alternative criterion for the equilibrium of second order differential equations of the form

$$\ddot{x} = f(x) \tag{17.28}$$

results when $f(x_o) = 0$ where x_o is the equilibrium condition. A small variation δ from equilibrium

$$x = x_o + \delta \tag{17.29}$$

results in

$$\begin{aligned}
\ddot{x} &= \ddot{\delta} \\
&= f(x_o + \delta) \\
&= f(x_o) + \delta f'(x_o) + \tfrac{1}{2}\delta f''(x_o) + \ldots
\end{aligned} \tag{17.30}$$

by Taylor expansion about the equilibrium where dots denote differentiation with respect to time t and primes with respect to space x. For example,

$$f'(x_o) = \left.\frac{df(x)}{dx}\right|_{x=x_o}. \tag{17.31}$$

For small perturbations δ the perturbation equation becomes

$$\ddot{\delta} - f'(x_o)\delta = 0 \tag{17.32}$$

where the stability of δ (and therefore equilibrium) is assured when $f'(x_o) < 0$ and instability when $f'(x_o) > 0$. For example, consider a pendulum equation of the form

$$\ddot{\theta} + \frac{g}{l}\sin\theta = 0 \tag{17.33}$$

where θ is the deviation from the bottom rest, l is the pendulum length, and g is the gravitational acceleration. Therefore,

$$\begin{aligned}
\ddot{\theta} &= \frac{g}{l}\sin\theta \\
&= f(\theta)
\end{aligned} \tag{17.34}$$

and

$$f'(\theta) = -\frac{g}{l}\cos\theta. \tag{17.35}$$

From the position of the bottom rest:

$$\theta(0) = 0 \quad f'(0) = -\frac{g}{l} < 0. \tag{17.36}$$

This corresponds to a small displacement or velocity perturbation with a small oscillation about bottom rest, which will continue indefinitely in the absence of damping. The oscillation can be made as small as desired by keeping the perturbation sufficiently small. The position of bottom rest (equilibrium) is therefore Lyapunov and also Lagrange stable since it is bounded. The addition of damping results in asymptotic stability of the solution. For the position of the top rest:

$$\theta(\pi) = 0 \tag{17.37}$$

$$f'(\pi) = \frac{g}{l} > 0. \tag{17.38}$$

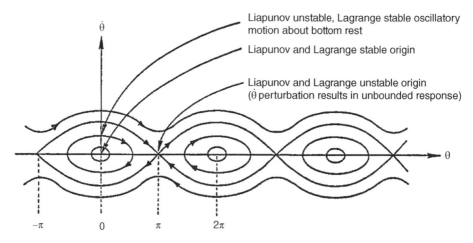

Liapunov unstable, Lagrange stable oscillatory motion about bottom rest

Liapunov and Lagrange stable origin

Liapunov and Lagrange unstable origin ($\dot\theta$ perturbation results in unbounded response)

Figure 17.3 Pendulum phase plane diagram.

The position of the top rest is Lyapunov unstable because a non-zero deviation in attitude (θ) or attitude rate ($\dot\theta$) produces a dramatic change in θ no matter how small the initial perturbation. Since θ is unbounded for any initial value of $\dot\theta$ (from the position of the top rest), the top rest position is also Lagrange unstable. A phase diagram of the pendulum is available in Figure 17.3.

17.4 Stability Conditions for Quasi-linear (Periodic) Systems

Systems described by equations with periodic coefficients (quasi-linear systems) may be described by a matrix equation of the type

$$\frac{dx}{dt}(\tau) = A(\tau)x(\tau) \tag{17.39}$$

where $x(\tau)$ is a $n \times 1$ column vector and $A(\tau)$ is an $n \times n$ matrix of known periodic coefficients with period T. Here τ is a time dependent variable (e.g. $\tau = \omega t$). A numerical procedure (Floquet theory) can be used to determine the stability of the zero (trivial) solution of Equation 17.39 for the special case $A(\tau) = A(\tau + T)$ when each element of $A(\tau)$ is either periodic (with period T) or constant. Second order differential equations of the type

$$\frac{d^2x}{d\tau^2} + q(\tau)\frac{dx}{d\tau} + r(\tau)x = 0 \tag{17.40}$$

where $q(\tau + T) = q(\tau)$ and $r(\tau + T) = r(t)$ can be transformed using

$$x(\tau) = y(\tau)e^{-\frac{1}{2}\int_0^\tau q(\alpha)d\alpha} \tag{17.41}$$

to yield Hill's equation

$$\frac{d^2y}{d\tau^2} + p(\tau)y = 0 \tag{17.42}$$

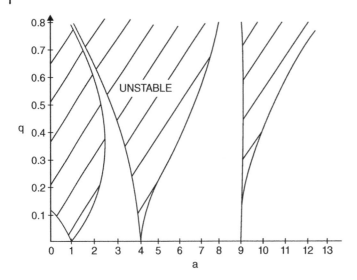

Figure 17.4 Solution stability for Mathieu's equation.

where

$$p(\tau) = [r(\tau) - \frac{1}{2}\frac{dq(\tau)}{d\tau} - \frac{1}{4}q^2(\tau)]$$

$$= p(\tau + T).$$

(17.43)

A special case of Hill's equation is the Mathieu equation

$$\frac{d^2y}{d\tau^2} + (a + 16q\cos 2\tau)y = 0$$

(17.44)

where a and q are real. Solutions of the Mathieu equation are known as Mathieu functions. Consider the solution to Equation 17.44

$$y = e^{\pi\tau}\phi(\tau)$$

(17.45)

where $\phi(\tau)$ is periodic in τ with a period π or 2π. Since this solution is unchanged if $-\tau$ is written for τ,

$$y = e^{-\mu\tau}\phi(-\tau)$$

(17.46)

is another independent solution. Therefore, the general solution is

$$y = c_1 e^{\mu t}\phi(\tau) + c_2 e^{-\mu t}\phi(\tau)$$

(17.47)

where c_1 and c_2 are arbitrary constants. The solution is stable if μ is imaginary and unstable if μ is real. The stability of solutions is shown in Figure 17.4

17.5 Stability of Linear, Possibly Time Varying, Systems

Given a generic linear system with time varying coefficients and the initial conditions

$$\vec{x}(t) = A(t)\vec{x}(t) + B(t)\vec{u}(t)$$

(17.48)

$$\vec{x}(t_0) = \vec{x}_0.$$

Let us consider the uncontrolled system (i.e. $\vec{u} \equiv 0$). Then

$$\dot{\vec{x}}(t) = A(t)\vec{x}(t) \tag{17.49}$$
$$\vec{x}(t_0) = \vec{x}_0.$$

Corresponding to each initial condition $\vec{x}_0$, there is a unique solution $\vec{x}(t, \vec{x}_0, t_0)$.

17.5.1 Equilibrium State or Point

In Equation 17.49 a state $\vec{x}_e$ where

$$A(t)\vec{x}_e = 0 \text{ for all } t \tag{17.50}$$

is called an equilibrium state of the system, i.e. the equilibrium state corresponds to the constant solution of the system in Equation 17.49.

There may be more than one equilibrium state for linear systems. For example, if $A(t)$ is constant and non-singular there is only one equilibrium state whereas if A is singular there are more than one equilibrium states. Note that $\vec{x}_e \equiv 0$, i.e. the origin of the state space is always an equilibrium state for a linear system. The concept of stability involves the behavior of the state $\vec{x}(t)$ as $t \to \infty$ when perturbed from the equilibrium state at $t = t_0$, i.e.

(a) Stability is a concept involving asymptotic behavior ($t \to \infty$).
(b) It involves the concept of neighborhood, i.e. when perturbed by some amount at $t = t_0$ from the equilibrium state, whether the state $\vec{x}(t)$ remains in the neighborhood of the equilibrium state $\vec{x}_e$ (Figure 17.5).

The origin of the state space is always an equilibrium point for linear systems. We can talk about the stability of the origin without any loss of generality. For linear systems all equilibrium states (other than zero) also behave in the same way as the zero solution because if $\vec{x}_{e_1} \neq 0$ is an equilibrium state and $\vec{x}_{e_2}$ is also an equilibrium state, then $\vec{x}_{e_1} - \vec{x}_{e_2}$ is also an equilibrium state because

$$A(t)\vec{x}_{e_1} = 0$$
$$A(t)\vec{x}_{e_2} = 0.$$

Figure 17.5 Lyapunov neighborhood.

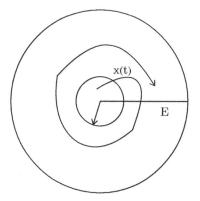

Therefore

$$A(t)(\vec{x}_{e_1} - \vec{x}_{e_2}) = 0$$

so $\vec{x}_{e_1}$ is also an equilibrium state. Since all equilibrium states behave the same way (this holds only for linear systems) we can therefore study the behavior of the null solution (or origin).

Definition 17.4   The origin of the state space is stable in the sense of Lyapunov if for any t_0 and any $\epsilon > 0$, there exists a $\delta(t, t_0) > 0$ such that

$$||\vec{x}(t_0)|| \le \delta \rightarrow ||\vec{x}(t)|| \le \epsilon \text{ for all } t \ge t_0.$$

Definition 17.5   The origin of the state space is said to be asymptotically stable if

(a) it is stable in the sense of Lyapunov, and
(b) for all t_0, there exists a $\delta(t_0) > 0$ such that $||\vec{x}(t_0)|| < \delta$ implies that $||\vec{x}(t)|| \rightarrow 0$ as $t \rightarrow \infty$.

Definition 17.6   The origin of the state space is said to be globally asymptotically stable or asymptotically stable in the large if

(a) it is stable in the sense of Lyapunov or
(b) for any $\vec{x}(t_0)$ and any t_0, $||\vec{x}(t)|| \rightarrow 0$ as $t \rightarrow \infty$.

Consider

$$\dot{\vec{x}}(t) = A(t)\vec{x}(t)$$
$$\vec{x}(t_0) = \vec{x}_0 \vec{x}(t) = \Phi(t, t_0)x(t_0)$$
$$||\vec{x}(t)|| = ||\Phi(t, t_0)x(t_0)|| \le ||\Phi(t, t_0)|| \; ||x(t_0)||.$$

The system is stable in the sense of Lyapunov iff

$$||\Phi(t, t_0)|| < N(t_0).$$

The system is asymptotically stable iff

$$||\Phi(t, t_0)|| < N(t_0)$$
$$\text{and } ||\Phi(t, t_0)|| \rightarrow 0 \text{ as } t \rightarrow \infty.$$

Thus, for linear systems, asymptotic stability does not depend on $\vec{x}(t_0)$.

Note that for linear systems, the origin is asymptotically stable if it is asymptotically stable in the large, i.e. the two previous definitions collapse into one and thus can be combined into one concept. Hence, for linear systems, we simply talk about either

(a) stability in the sense of Lyapunov or
(b) asymptotic stability.

Remark 6. For linear systems (because the origin is always an equilibrium state and all equilibrium states behave the same way, and the behavior of the origin of the state space implies the behavior of $||\vec{x}(t)||$), we can simply talk about stability of the system, rather than the stability of the origin or equilibrium state.

Important remark about the stability of linear time varying systems

Note that determination of stability, even in the case of linear systems, can be difficult if it happens to be a time varying system, i.e. when the state space matrix A has time varying elements, i.e. when A is $A(t)$.

It is important to keep in mind that the stability of linear time varying systems cannot be inferred from the ad hoc logic that it can be approximated as a linear time invariant system at each time instant and thus that the given linear time varying system is stable when it is known to be stable at each time instant. This logic is erroneous and is a common misconception as can be seen by the example below.

The matrix

$$A(t) = \begin{bmatrix} -1 + 1.5 \cos^2 t & 1 - 1.5 \cos t \sin t \\ -1 - 1.5 \cos t \sin t & -1 + \sin^2 t \end{bmatrix}$$

with initial conditions of $x_1(t_0) = 1$ and $x_2(t_0) = 0$ is seen to be stable when the above matrix is frozen for each time instant (and thereby treating it as a time invariant matrix at that instant) but it can be easily seen that the actual time trajectory of the state variable x_1 has on unbounded oscillatory (unstable) response. It can be shown that the STM for this system is given by [4]

$$\Phi(t, 0) = \begin{bmatrix} e^{t/2} \cos t & e^{-t} \sin t \\ -e^{t/2} \sin t & e^{-t} \cos t. \end{bmatrix}$$

Thus the actual time varying system is unstable.

Another useful result [2, 10] with regard to stability of linear time varying systems is that if the eigenvalues satisfy

$$\text{Re} \lambda_i(t) \leq -k < 0 \tag{17.51}$$

for all t (i.e eigenvalues of the system frozen at each time instant have a relative stability degree of k) *and* if the time variation of $A(t)$ is sufficiently slow, then the system is stable.

17.5.2 Review of the Stability of Linear Time Invariant Systems in Terms of Eigenvalues

17.5.2.1 Continuous Time Systems: Hurwitz Stability

In this case

$$\vec{x}(t) = e^{A(t-t_0)}\vec{x}(t_0).$$

If A has an eigenvalue λ_i, then by the Frobenius theorem $e^{A(t-t_0)}$ has eigenvalues $e^{\lambda_i(t-t_0)}$.

Thus $||\vec{x}(t)||$ behavior depends on the matrix $e^{\lambda_i(t-t_0)}$, which in turn depends on λ_i. For continuous time systems, the stability of the system is thus given by the following criterion: let $\lambda_i = \beta_i + j\omega_i$.

(a) The system is unstable if β_i (real part of λ_i) > 0 for any distinct root or $\beta_i \geq 0$ for any repeated root (Figure 17.6).

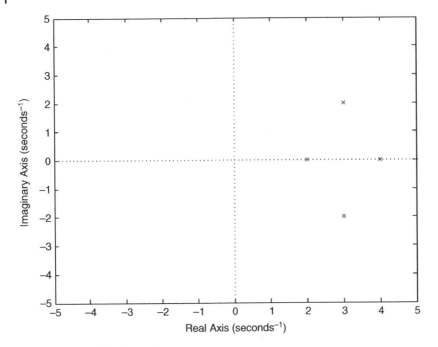

Figure 17.6 Unstable eigenvalues.

(b) The system is stable in the sense of Lyapunov if $\beta_i \leq 0$ for all distinct roots and $\beta_i < 0$ for all repeated roots, i.e. there are no multiple poles on the imaginary axis and all distinct poles are in the left half of the complex plane (Figure 17.7).

(c) The system is asypmtotically stable if $\beta_i < 0$ for all roots (Figure 17.8).

Continuous Time Systems

$$\vec{x}(t) = A\vec{x}(t)$$
$$\vec{x}(t) = e^{A(t-t_0)}\vec{x}(t_0)$$
$$\text{Eigenvalues of } A : \lambda_i = \beta_i \pm j\omega_i$$

Table 17.1 summarizes the Hurwitz stability conditions.

Table 17.1 Hurwitz stability criteria for continuous time systems.

Unstable	If $\beta_i > 0$ for any single (or distinct) root or $\beta_i \geq 0$ for any repeated root
Stable in the sense of Lyapunov or neutrally stable	If $\beta_i \leq 0$ for all distinct roots and $\beta_i < 0$ for all repeated roots
Asymptotically stable	If $\beta_i < 0$ for all roots

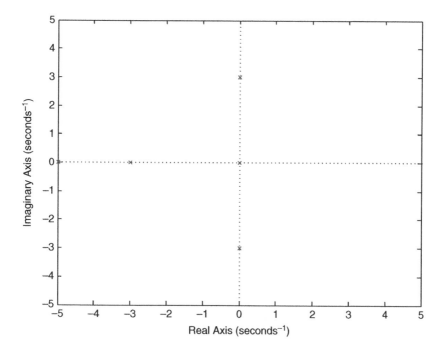

Figure 17.7 Eigenvalues stable in the sense of Lyapunov.

Figure 17.8 Asymptotically stable.

Domain of Attraction

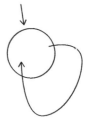

17.5.2.2   Discrete Time Systems (Schur Stability)

$$\vec{x}(k+1) = A\vec{x}(k)$$
$$\vec{x}(k) = A^k\vec{x}(0)$$

Eigenvalues of A : λ_i

For discrete time systems, it is the unit circle in the complex plane that determines stability (Schur Stability). If all the eigenvalue *magnitudes* lie inside the unit circle then the discrete time system is stable (Figure 17.9).

Table 17.2 summarizes the Schur stability conditions.

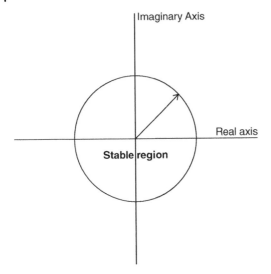

Imaginary Axis

Figure 17.9 Eigenvalues of a discrete time system.

Real axis

Stable region

Table 17.2 Schur stability criteria for discrete time systems.

Unstable	If $|\lambda_i| > 1$ for any distinct root or $|\lambda_i| \geq 1$ for any repeated root
Stable in the sense of Lyapunov or neutrally stable	If $|\lambda_i| \leq 1$ for all distinct roots and $|\lambda_i| < 1$ for all repeated roots
Asymptotically stable	If $|\lambda|_i < 1$ for all roots

17.6 Bounded Input–Bounded State Stability (BIBS) and Bounded Input–Bounded Output Stability (BIBO)

Once we have an input forcing function, the stability discussion involves the bounded-ness issue of the input function, leading to the concept of the bounded input–bounded state (BIBS) stability, or bounded input–bounded output (BIBO) stability. Consider

$$\dot{\vec{x}} = A\vec{x} + B\vec{u}$$
$$\vec{y} = C\vec{x}$$
$$\vec{x}(t_0) = \vec{x}_0.$$

Definition 17.7 If there is a fixed, finite constant K such that $||\vec{u}|| \leq K$ for every t, then the input is said to be bounded. If for every bounded input and for arbitrary initial conditions $\vec{x}(t_0)$, there exists a scalar $\delta > 0$ such that $||\vec{x}|| \leq \delta$, then the system is BIBS stable.

Definition 17.8 If $||\vec{u}|| \leq K$, then if $||\vec{y}|| \leq \alpha K_m \bar{\delta}$, then the system is BIBO stable.

Note that BIBS stability requires the open loop system matrix A to be asymptotically stable. However, it is possible to achieve BIBO stability even if the open loop system matrix A is not asymptotically stable, because the outputs may not be linked to the unstable part of the states. This brings in the issues of controllability and observability of state space systems, which is the topic of the next chapter.

17.6.1 Lagrange Stability

Lagrange stability is a boundedness concept applicable to linear systems. If a small deviation from some equilibrium point remains bounded, then the motion is said to be Lagrange (or infinitesimal) stable. There is a subtle difference between Lagrange stability and stability in the sense of Lyapunov. Note that stability in the sense of Lyapunov requires the solution of the differential equation, beginning sufficiently close to the equilibrium (origin), and that the resulting trajectory must remain arbitrarily close to itself after perturbation. This concept is applicable to the solutions of both linear and nonlinear differential equations.

However, Lagrange stability requires only the solution (trajectory) to remain within a finite distance from equilibrium while Lyapunov stability requires arbitrarily small deviations (perturbation) from equilibrium. Therefore, Lyapunov stability implies Lagrange stability, but the opposite is not true. A linear time invariant system is said to be stable if and only if the roots of its characteristic equation have negative real parts. If any root has a real part that is positive, the system is unstable. For an asymptotically stable system, the state variables are not only expected to remain in the neighborhood of equilibrium but eventually even return to the equilibrium state when the system is given an initial small perturbation. If any root has a real part that is zero, the system is unstable.

17.7 Chapter Summary

This long chapter presents a thorough treatment on the most important structural property of a state space system, namely stability. In particular, we thoroughly covered the various methods to assess the Hurwitz stability of a real matrix, which is treated as the plant matrix of linear time invariant state space system. Stability, controllability and observability are designated as the three important structural properties one should be thoroughly familiar with to master the subject of linear state space systems. We covered the stability conditions for both continuous time systems (Hurwitz stability) as well as discrete time systems (Schur stability). In the next chapter we cover the other two structural properties, namely controllability and observability, and related issues. Fundamental concepts such as those discussed in this chapter are also available in a more consolidated way in many excellent textbooks dedicated to state space control systems such as [4–6, 8, 9, 11–13]

17.8 Exercises

Exercise 17.1. Find the Hurwitz stability nature of the following matrix using the Lyapunov matrix equation method.

$$A = \begin{bmatrix} -1 & -1.5 \\ 0.8 & -0.9 \end{bmatrix}$$

For the same matrix above, determine its Schur stability nature.

Exercise 17.2. Find the Hurwitz stability nature of the following matrix

$$A = \begin{bmatrix} 0 & 0 & -1 & 2 \\ -16 & 0 & 1 & 0 \\ 4 & -1 & -6 & 0 \\ 0 & 1 & 0 & -1 \end{bmatrix}$$

using the Routh–Hurwitz criterion, Fuller's method, and the Lyapunov matrix equation method.

Exercise 17.3. Find the Hurwitz stability nature of the following matrix

$$A = \begin{bmatrix} -3 & 1 & 2 & 4 \\ -1 & -3 & -1 & -2 \\ -2 & 1 & -3 & -4 \\ -4 & 2 & 4 & -3 \end{bmatrix}$$

using the Routh–Hurwitz criterion, Fuller's method, and the Lyapunov matrix equation method.

Exercise 17.4. Find the Hurwitz stability nature of the following matrix

$$A = \begin{bmatrix} -1 & 0 & -1 & 2 \\ -16 & -6 & 1 & 0 \\ 4 & -1 & -1 & 0 \\ 0 & 1 & 0 & 0 \end{bmatrix}$$

using the Routh–Hurwitz criterion, Fuller's method, and the Lyapunov matrix equation method.

Exercise 17.5. Find the Hurwitz stability nature of the following matrix

$$A = \begin{bmatrix} 4.3881 & -2015.1 & -62.986 \\ 0.0049482 & -2.2724 & 0.92897 \\ 2.4217 & -425.91 & -34.761 \end{bmatrix}$$

using the Routh–Hurwitz criterion, Fuller's method, and the Lyapunov matrix equation method.

Bibliography

1 Vladimir A. Chobotov. *Spacecraft Attitude Dynamics and Control.* Krieger Publishing Company, 1991.

2 C. A. Desoer. Slowly varying system x=a(t)x. 1969.

3 E.I. Jury. Inners and Stability of Dynamic Systems, volume, New York. 1974.

4 T.E Fortmann and K.L Hitz. *An Introduction to Linear Control Systems.* Mercel-Dekker, 1977.

5 G.F Franklin, J.D Powell, and A Emami-Naeni. *Feedback Control of Dynamic Systems.* Pearson-Prentice-Hall, 2006.

6 B. Friedland. *Control system design: An introdcution to state-space methods.* McGraw-Hill, 1986.

7 A.T. Fuller. Conditions for a matrix to have only characteristic roots with negative real parts. *Journal of Mathematical Analysis and Applications*, 23:71–98, 1968.

8 H. Kwakernaak and R. Sivan. *Linear Optimal Control Systems*. Wiley Interscience, 1972.

9 K. Ogata and Y. Yang. *Modern control engineering*. Prentice-Hall, Englewood Cliffs, NJ, 1970.

10 H. H. Rosenbrock. The stability of linear time-dependent control systems. 1963.

11 R.E. Skelton. *Dynamic Systems Control: Linear Systems Analysis and Synthesis*. Wiley, New York, 1 edition, 1988.

12 J.G. Truxal, editor. *Control engineers' handbook*. McGraw-Hill, New York, 1958.

13 W.L. Brogan. *Modern Control Theory*. Prentice Hall, 1974.

18

Controllability, Stabilizability, Observability, and Detectability

18.1 Chapter Highlights

In this chapter, the two other structural properties of linear state space systems, namely controllability and observability concepts are discussed. As the name implies, the ability to control the dynamic system is termed controllability. Thus it answers the question as to whether there exists a control vector $u(t)$, consisting of m control variables, that can take the state vector, consisting of n variables (at a given initial time) to any other arbitrary value of the state vector in a finite amount of time. If such a control (forcing) function exists, then we say the system is (state) controllable. Obviously, before we try to design a control system, it is important to make sure that the system is indeed controllable. In this chapter, we develop conditions of controllability for a linear state space system. Not surprisingly, controllability conditions involve the nature of matrix pair (A, B), where A is the $n \times n$ plant matrix and B is the $n \times m$ control distribution matrix. In other words, the controllability question explores the relationship between the control (or actuation) variables and the state variables. After getting conditions of controllability, we also discuss the related issue of stabilizability for uncontrollable systems. The observability concept is very similar and is thus dual to the concept of controllability. In this, we ask the question of whether the entire state vector, consisting of n variables (at a given time), can be observed or not from only a set of l output (or measurement or sensor) variables. If it does, then we say the system is completely observable. Accordingly, the observability concept explores the relationship between sensor (measurement) variables and state variables and thus involves the matrix pair (A, M), where A is the $n \times n$ plant matrix and M is the $l \times n$ measurement matrix. After getting conditions of observability, we also discuss the related issue of detectability, for unobservable systems. Note that this observability concept can be used in the context of a set of k output variables and their relationship to the state variables, in which case, we use the matrix pair (A, C) in the conditions, where C is the $k \times n$ output matrix.

18.2 Controllability of Linear State Space Systems

A linear, possibly time varying, system

$$\dot{x} = A(t)x + B(t)u \tag{18.1}$$

$$y = C(t)x \tag{18.2}$$

is said to be output controllable over the interval $t\epsilon[t_0, t_1]$ if the output $y(t)$ can be taken to an arbitrary value $y(t_1)$ from an arbitrary initial condition $y(t_0)$. Mathematically speaking, a linear system is output controllable over the interval $t\epsilon[t_0, t_1]$ if there exists $u(t)$ in the same interval such that

$$y(t_1) = C(t_1) \int_{t_0}^{t_1} \phi(t_1, \sigma)B(\sigma)u(\sigma)d\sigma \qquad \text{where} \qquad y\epsilon R^k. \tag{18.3}$$

Accordingly, a possibly time varying, linear system

$$\dot{x} = A(t)x + B(t)u \tag{18.4}$$

$$\tag{18.5}$$

is said to be state controllable over the interval $t\epsilon[t_0, t_1]$ if the state vector $x(t)$ can be taken to an arbitrary value $x(t_1)$ from an arbitrary initial condition $x(t_0)$. Mathematically speaking, a linear system is state controllable over the interval $t\epsilon[t_0, t_1]$ if there exists $u(t)$ in the same interval such that

$$x(t_1) - x(t_0) = \int_{t_0}^{t_1} \phi(t_1, \sigma)B(\sigma)u(\sigma)d\sigma \qquad \text{where} \qquad x\epsilon R^n. \tag{18.6}$$

For general, linear time varying systems, there is an additional notion of reachability wherein there is a difference between going from the origin to some given state $x(t_1)$ in some finite time interval, which could be different from going from a given state $x(t_1)$ to the origin. In other words, controllable subspace could be different from reachable subspace. However, for continuous time, linear time invariant systems (LTI systems), these two notions coincide, and thus we can simply discuss only the controllability notion, not worrying about the reachability notion. Going forward, we focus on the conditions for controllability for continuous time LTI systems.

For time invariant systems (i.e. systems with constant coefficient matrices)

$$\phi(t, \sigma) = e^{A(t-\sigma)}$$
$$= \alpha_0(t - \sigma)I + \alpha_1(t - \sigma)A + \ldots \alpha_{n-1}A^{n-1} \;. \tag{18.7}$$

Substituting Equation 18.7 into the general case Equation 18.6 and integrating, (we are omitting vector notation, but realize that x and u are vectors of dimension n and m respectively) we have

$$\tilde{x} = \int_{t_0}^{t_1} [\alpha_0 I + \alpha_1 A + \ldots \alpha_{n-1}A^{n-1}]Bu(\sigma)d\sigma$$

$$= [B, AB, A^2B, \ldots A^{n-1}B] \left\{ \begin{array}{c} \int_{t_0}^{t_1} \alpha_0(t_1 - \sigma)u(\sigma)d\sigma \\ \vdots \\ \int_{t_0}^{t_1} \alpha_{n-1}(t_1 - \sigma)u(\sigma)d\sigma \end{array} \right\}. \tag{18.8}$$

Then

$$\tilde{x} = Bu_1 + ABu_2 + A^2Bu_3, \ldots A^{n-1}Bu_m \tag{18.9}$$

and so an arbitrary $\tilde{x} \epsilon R^n$ can be written as a linear combination of h^i only if the set of h^i spans R^n. Thus, it can be proved that a time invariant system is state controllable if and only if the matrix

$$[B \quad AB \quad \cdots \quad A^{n-1}B] \equiv H_c \tag{18.10}$$

has rank n. The columns of matrix 18.10 are those subspaces along which the state x may be moved by our control input u.

Hence, in summary, the condition for (state) controllability for linear time invariant state space systems, is that the controllability matrix denoted by H_c,

$$H_c = \begin{bmatrix} B & AB & \cdots & A^{n-1}B \end{bmatrix}$$

has rank n.

Note that the controllability matrix H_c is a rectangular matrix of dimensions $n \times mn$. Thus controllability requires that out of the mn columns there should exist at least n columns that are linearly independent. If this condition is satisfied, then we say the system is completely (state) controllable.

Conceptually speaking, this controllability condition is exploring the way the control variables are linked to the state variables and thus the connection between them. Since controllability is also a structural property of a linear state space system, just like the stability property, it is also invariant under similarity transformation. Hence, the majority of times, it is much easier to investigate the controllability condition in modal coordinates, rather than in the original state variables, because (barring special cases) modal coordinates are completely decoupled from each other (like in the distinct eigenvalue case), and thus controllability investigation simply involves whether each mode is directly connected to the control variables or not. Hence in the next section, we provide conditions of controllability via modal decomposition.

18.3 State Controllability Test via Modal Decomposition

Consider $\dot{\vec{x}} = A\vec{x} + B\vec{u}$. Transform this system to modal coordinates $\vec{q}$, i.e. $\dot{\vec{q}} = Jq + B_n\vec{u}$. Note that when there are repeated eigenvalues J may not be completely diagonal. So let us consider two cases.

18.3.1 Distinct Eigenvalues Case

J is completely diagonal. The condition for controllability then is that the elements of each row of B_n (that correspond to each distinct eigenvalue) are not all zero.

Example 18.1 Assess the controllability of the system

$$\begin{bmatrix} \dot{q}_1 \\ \dot{q}_2 \end{bmatrix} = \begin{bmatrix} -1 & 0 \\ 0 & -2 \end{bmatrix} \begin{bmatrix} q_1 \\ q_2 \end{bmatrix} + \begin{bmatrix} 2 \\ 0 \end{bmatrix} \vec{u}.$$

Solution
The above system is not completely controllable, because every element in the last row of B_n is zero. If we change the B_n matrix to, say,

$$B_n = \begin{bmatrix} 2 \\ 1 \end{bmatrix}$$

then it is completely controllable.

18.3.2 Repeated Eigenvalue Case

Then J has Jordan blocks J_i, i.e. $J = \begin{bmatrix} J_1 & & & 0 \\ & J_2 & & \\ & & \ddots & \\ 0 & & & J_p \end{bmatrix}$. The corresponding B_n matrix can be

written as

$$B_n = \begin{bmatrix} \leftarrow & B^T_{n_1} & \rightarrow \\ \leftarrow & B^T_{n_2} & \rightarrow \\ & \vdots & \\ \leftarrow & B^T_{n_p} & \rightarrow \end{bmatrix}.$$

Let $B^T_{n_i,l}$ denote the last row of $B^T_{n_i}$. Then the system is completely controllable if:

(a) The elements of $B^T_{n_i,l}$ (corresponding to the Jordan block J_i) are not all zero.
(b) The last rows $B^T_{n_i,l}$ of the r Jordan blocks associated with the same eigenvalue form a linearly independent set.

Example 18.2   Assess the controllability of the following system

$$\begin{bmatrix} \dot{q}_1 \\ \dot{q}_2 \\ \dot{q}_3 \end{bmatrix} = \begin{bmatrix} -1 & 1 & 0 \\ 0 & -1 & 0 \\ \hline 0 & 0 & -2 \end{bmatrix} \begin{bmatrix} q_1 \\ q_2 \\ q_3 \end{bmatrix} + \begin{bmatrix} 4 & 2 \\ 0 & 0 \\ 3 & 0 \end{bmatrix} \begin{bmatrix} u_1 \\ u_2 \end{bmatrix}.$$

Solution
The system is not completely controllable because of zero entries in the last row of the B_n matrix corresponding to the J_1 matrix.

$$\begin{bmatrix} \dot{q}_1 \\ \dot{q}_2 \\ \dot{q}_3 \end{bmatrix} = \begin{bmatrix} -1 & 1 & 0 \\ 0 & -1 & 0 \\ \hline 0 & 0 & -1 \end{bmatrix} \begin{bmatrix} q_1 \\ q_2 \\ q_3 \end{bmatrix} + \begin{bmatrix} 4 & 2 \\ 2 & 3 \\ 4 & 6 \end{bmatrix} \begin{bmatrix} u_1 \\ u_2 \end{bmatrix}$$

is not completely controllable because the eigenvalue -1 has two Jordan blocks and the last rows of B_n corresponding to these Jordan blocks associated with the same eigenvalue [2 3] and [4 6] form a linearly dependent set.

18.4   Normality or Normal Linear Systems

Consider $\dot{\vec{x}} = A\vec{x} + B\vec{u}$, given $\vec{u} \in \mathbb{R}^m, \vec{x} \in \mathbb{R}^n$, where $B = [b_1 b_2 b_3 \cdots b_m]$ and $b_i \in \mathbb{R}^{n \times 1}$ is a column. We say the system is normal if each of the systems

$$\vec{x}(t) = A\vec{x}(t) + b_1 u_1(t)$$
$$\vec{x}(t) = A\vec{x}(t) + b_2 u_2(t)$$
$$\vdots$$
$$\vec{x}(t) = A\vec{x}(t) + b_m u_m(t)$$

is completely controllable, i.e. normality involves controllability from each component of $\vec{u}$. Clearly a normal system is always completely controllable.

Caution: A completely controllable system is not necessarily normal.

Now, we switch our attention to systems that are not completely controllable. Once the system is known to be uncontrollable, the question then is to worry about what the nature of the uncontrollable states is. So the moment the system is known to be uncontrollable, the first task is to separate out the controllable subspace from the uncontrollable subspace. This is done by employing an appropriate similarity transformation matrix. These details are discussed next.

18.5 Stabilizability of Uncontrollable Linear State Space Systems

The controllable subspace of the linear time invariant system

$$\vec{x}(t) = A\vec{x}(t) + B\vec{u}(t)$$

is the linear subspace consisting of the states that can be reached from the zero state within a finite time.

Theorem 18.1 The controllable subspace of the n dimensional linear time invariant system

$$\vec{x}(t) = A\vec{x}(t) + B\vec{u}(t) \tag{18.11}$$

is the linear subspace spanned by the columns of the controllability matrix

$$H_c = \begin{bmatrix} B & AB & \cdots & A^{n-1}B \end{bmatrix}.$$

Let the dimension of the controllable subspace be $n_c < n$. (Note that if the system is completely controllable, i.e. when $n_c = n$, there is no need to go any further.)

Then with a similarity transformation, one can transform Equation 18.11 into controllability canonical form given by

$$\vec{x}_c(t) = \begin{bmatrix} A_{11c} & A_{12c} \\ 0 & A_{22c} \end{bmatrix} \vec{x}_c(t) + \begin{bmatrix} B_{11c} \\ 0 \end{bmatrix} \vec{u}(t) \tag{18.12}$$

where A_{11c} is an $n_c \times n_c$ matrix and the pair $\{A_{11c}, B_{11c}\}$ is completely controllable.

Note: $\vec{x}_c = T^{-1}\vec{x}(t)$ or $\vec{x}_c = T\vec{x}(t)$.

Definition 18.1 The system of Equations 18.11 or 18.12 is stabilizable if and only if the uncontrollable subspace is stable (or the unstable subspace is controllable), i.e. if and only if A_{22c} is an asymptotically stable matrix.

Remarks:

1) Note that stability and stabilizability are two entirely different concepts. Of course stabilizability involves the concept of stability (and uncontrollability).
2) If a system is unstabilizable, then obviously the engineer has to go back to the drawing board (i.e. the open loop plant) and reorient the actuators and plant in such a way that either the system is controllable or at least stabilizable.

18.5.1 Determining the Transformation Matrix T for Controllability Canonical Form

Consider $H_c = \begin{bmatrix} B & AB & \cdots & A^{n-1}B \end{bmatrix}$. Determine the rank of H_c; let it be n_c. Obviously $n_c < n$. (As mentioned earlier if $n_c = n$, there is no interest in going to controllability canonical form anyway.) So consider $n_c < n$. Let the n_c columns of H_c that are linearly independent be also the columns of T. Then make up the rest of the $n - n_c$ columns of T such that T is non-singular, i.e.

$$T = \begin{bmatrix} \uparrow & \uparrow & & \uparrow & \uparrow & & \uparrow \\ h_{c1} & h_{c2} & \cdots & h_{cn_c} & d_1 & \cdots & d_{n-n_c} \\ \downarrow & \downarrow & & \downarrow & \downarrow & & \downarrow \end{bmatrix}$$

where the arbitrary columns d_i are such that T is non-singular.

Example 18.3 Determine the controllability and stabilizability of the system

$$\{A, B\} = \left\{ \begin{bmatrix} -3 & 1 \\ -2 & \frac{3}{2} \end{bmatrix}, \begin{bmatrix} 1 \\ 4 \end{bmatrix} \right\}.$$

Note that the eigenvalues of A are 1 and -2.5.

Solution

$$H_c = \begin{bmatrix} 1 & 1 \\ 4 & 4 \end{bmatrix}$$

$n_c = 1$.

This system is not completely controllable, so we need to go to the controllability canonical form. The transformation matrix

$$T = \begin{bmatrix} 1 & \otimes_1 \\ 4 & \otimes_2 \end{bmatrix}$$

where $\otimes_1$ and $\otimes_2$ are such that T is non-singular. So let

$$T = \begin{bmatrix} 1 & 1 \\ 4 & 2 \end{bmatrix}$$

so that

$$T^{-1} = \begin{bmatrix} -1 & \frac{1}{2} \\ 2 & -\frac{1}{2} \end{bmatrix}.$$

Then

$$B_c = T^{-1}B$$

$$A_c = T^{-1}AT$$

and so

$$\{A_c, B_c\} = \left\{ \begin{bmatrix} 1 & \frac{1}{2} \\ 0 & \frac{-5}{2} \end{bmatrix}, \begin{bmatrix} 1 \\ 0 \end{bmatrix} \right\}.$$

Since the uncontrollable subspace (i.e. the A_{22c} matrix, which in this example is simply the scalar -2.5), is stable, the above system is stabilizable.

Remark 2. Notice that without going to the controllability canonical form we could not have ascertained which of those eigenvalues corresponds to the controllable subspace and which to the uncontrollable subspace. Thus the beautiful property of the controllability canonical form is that it clearly separates out all the controllable subspace states into one set and all the uncontrollable subspace states into a distinctly separate set.

Remark 3. It is also important to realize that when a system is uncontrollable, it implies that a linear combination of the original (physically meaningful) state variables is uncontrollable, not necessarily each individual state variable.

Remark 4. Notice that, for linear time invariant systems, since time does not enter into the conditions of controllability/stabilizability, these conditions and concepts are valid for both continuous time as well as discrete time linear state space systems.

Next we switch our attention to another structural property of a linear state space system, namely a property dual to the property of controllability, labeled observability. While the controllability property focuses on the connection between the state variables and control variables (i.e the property of the pair (A, B)), the observability property concerns the connection between the state variables and either the output variables (involving the pair (A, C) matrices) or the measurement variables (involving the pair (A, M) matrices). Let us elaborate on this next.

18.6 Observability of Linear State Space Systems

Again, take a generic linear state space system

$$\begin{aligned} \dot{x} &= A(t)x + B(t)u \\ y &= C(t)x + D(t)u. \end{aligned} \tag{18.13}$$

We ask now: is every motion of the state visible in the output? If not, there may exist some internal instabilities in the system not visible in the output. An observability problem relates to the problem of determining the initial value of the state vector knowing only its output y over some interval of time. The output of the system in Equation 18.13 can be expressed as

$$y(t) = C(t)\phi(t, t_o)x(t_o) + \int_{t_o}^{t_1} [\phi(t, \sigma)B(\sigma) + D\delta(t - \sigma)]u(\sigma)d\sigma. \tag{18.14}$$

Keep in mind the second term on the right hand side of Equation 18.14 is a known quantity.

We may study the homogeneous system

$$\begin{aligned} \dot{x} &= A(t)x \\ y &= C(t)x. \end{aligned} \tag{18.15}$$

Now

$$\begin{aligned} y(t) &= G(t, t_o)x_o \\ &= C(t)\phi(t_1 t_o)x_o \\ &= G(t, t_o)x_o. \end{aligned} \tag{18.16}$$

Observability requires the right hand side of Eq. 18.16

$$G(t, t_1)x_o \equiv 0 \tag{18.17}$$

over the interval $t\epsilon[t_o, t_1]$ only if

$$x_o = 0. \tag{18.18}$$

Now premultiply Equation 18.16 by $[C(t)\phi(t, t_o)]^T$ to get

$$\phi^T(t, t_o)C^T(t)y(t) = \phi^T(t, t_o)C^T(t)C(t)\phi(t, t_o)x_o. \tag{18.19}$$

Integrating,

$$\int_{t_o}^{t_1} \phi^T(t, t_o)C^T(t)y(t)dt = W(t_1, t_o)x_o \tag{18.20}$$

and therefore

$$x_o = W(t_1, t_o)^{-1} \int_{t_o}^{t_1} \phi^T(t, t_o)C^T(t)y(t)dt. \tag{18.21}$$

Therefore the initial state vector can be obtained uniquely if

$$W(t_1, t_o) \equiv \int_{t_o}^{t_1} G^T(t, t_o)G(t, t_o)dt \tag{18.22}$$

is a non-singular matrix.

For general, linear time varying systems, there is an additional notion of constructibility (also referred to as reconstructibility in some books) wherein there is a difference between constructing the state from some given future outputs $y(t_1)$ in some finite time interval, which could be different from constructing the state from some given given past outputs $y(t_1)$. In other words, observable subspace could be different from constructible (or reconstuctible) subspace. However, for continuous time, linear time invariant systems (LTI systems), these two notions coincide, and thus we can simply discuss only the observability notion, not worrying about the constructibility (or reconstructibility) notion. Going forward, we focus on the conditions for observability for continuous time LTI systems. Following the same logic employed for the controllability case, it can be proven that a time invariant linear system is observable if and only if the matrix with the rows

$$\begin{bmatrix} C \\ CA \\ CA^2 \\ \vdots \\ CA^{n-1} \end{bmatrix}$$ (18.23)

has rank n.

18.7 State Observability Test via Modal Decomposition

Consider

$$\dot{\vec{x}} = A\vec{x}$$ (18.24)
$$\vec{y} = C\vec{x}.$$ (18.25)

Note Equation 18.25 could just as well have been replaced by an equation relating measurement variables

$$\vec{z} = M\vec{x}$$ (18.26)

but we will proceed here with the output equation as written in Equation 18.14. Transform this to modal coordinates $\vec{q}$, $\vec{x} = M\vec{q}$, i.e.

$$\dot{\vec{q}} = J\vec{q}$$
$$\vec{y} = C_n\vec{q}.$$

Note that when these are repeated eigenvalues J may not be completely diagonal. So let us consider two cases.

18.7.1 The Distinct Eigenvalue Case

J is completely diagonal. The condition for observability is that the elements of each column of each column of C_n (that corrsponds to each distinct eigenvalue) are not all zero.

Example 18.4   Assess the observability of the system

$$\begin{bmatrix} \dot{q}_1 \\ \dot{q}_2 \end{bmatrix} = \begin{bmatrix} -1 & 0 \\ 0 & -2 \end{bmatrix} \begin{bmatrix} q_1 \\ q_2 \end{bmatrix}$$

$$y = \begin{bmatrix} 2 & 0 \end{bmatrix} \begin{bmatrix} q_1 \\ q_2 \end{bmatrix}.$$

Solution
The system is not completely observable because of the presence of 0 (i.e. the C_{12} element). If we change the C_n matrix to say [2 1], then it is completely observable.

18.7.2 Repeated Eigenvalue Case

Then J has Jordan blocks J_i, i.e. $J = \begin{bmatrix} J_1 & & & 0 \\ & J_2 & & \\ & & \ddots & \\ 0 & & & J_p \end{bmatrix}$. The corresponding C_n matrix can

be written as

$$C_n = \begin{bmatrix} \uparrow & \uparrow & & \uparrow \\ C_{n1} & C_{n2} & \cdots & C_{np} \\ \downarrow & \downarrow & & \downarrow \end{bmatrix}.$$

Let B_{ni1} denote the first column of B_{ni}. Then the system is completely observable if:

(a) The elements of C_{ni1} (corresponding to the Jordan block J_i) are not all zero.
(b) The first columns C_{ni1} of the r Jordan blocks associated with the same eigenvalue form a linearly independent set.

Example 18.5   Consider the system

$$\begin{bmatrix} \dot{q}_1 \\ \dot{q}_2 \\ \dot{q}_3 \end{bmatrix} = \begin{bmatrix} -1 & 1 & 0 \\ 0 & -1 & 0 \\ 0 & 0 & -2 \end{bmatrix} \begin{bmatrix} q_1 \\ q_2 \\ q_3 \end{bmatrix}$$

$$\begin{bmatrix} y_1 \\ y_2 \end{bmatrix} = \begin{bmatrix} 4 & 0 & 3 \\ 2 & 0 & 0 \end{bmatrix} \begin{bmatrix} q_1 \\ q_2 \\ q_3 \end{bmatrix}.$$

Solution

The system is completely observable. However, if $C_n = \begin{bmatrix} 0 & 4 & 3 \\ 0 & 2 & 0 \end{bmatrix}$ then the system is not completely observable because the first column is all zeroes. Similarly the system

$$\begin{bmatrix} \dot{q}_1 \\ \dot{q}_2 \\ \dot{q}_3 \end{bmatrix} = \begin{bmatrix} -1 & 1 & 0 \\ 0 & -1 & 0 \\ 0 & 0 & -1 \end{bmatrix} \begin{bmatrix} q_1 \\ q_2 \\ q_3 \end{bmatrix}$$

$$\begin{bmatrix} y_1 \\ y_2 \end{bmatrix} = \begin{bmatrix} 2 & 4 & 4 \\ 3 & 2 & 6 \end{bmatrix} \begin{bmatrix} q_1 \\ q_2 \\ q_3 \end{bmatrix}$$

is not completely observable because the eigenvalue -1 has two Jordan blocks and the first columns of C_{ni} corresponding to these two Jordan blocks $\begin{bmatrix} 2 \\ 3 \end{bmatrix}$ and $\begin{bmatrix} 4 \\ 6 \end{bmatrix}$ form a linearly dependent set.

18.8 Detectability of Unobservable Linear State Space Systems

Definition 18.2   The unobservable (or unreconstructible) subspace of the linear time invariant system

$$\vec{x}(t) = A\vec{x} + B\vec{u}$$
$$\vec{y}(t) = C\vec{x}$$

(18.27)

is the linear subspace consisting of the state $\vec{x}(t_0) = \vec{x}_0$ for which

$$\vec{y}(t; \vec{x}_0, t_0, 0) = 0$$
$$t \geqslant t_0.$$

Theorem 18.2 The observable subspace of the n dimensional linear time invariant system

$$\vec{x}(t) = A\vec{x}(t) + B\vec{u}(t)$$

has the dimensions equal to the rank of the observability matrix

$$H_\phi = \begin{bmatrix} C^T & A^T C^T & A^{2^T} C^T \cdots & A^{n-1^T} C^T \end{bmatrix}.$$

Let the dimensions of the controllable subspace be $n_\phi < n$. Then with a similarity transformation, one can transform Equation 18.27 into observability canonical form

$$\vec{x}_\phi(t) = \begin{bmatrix} A_{11\phi} & 0 \\ A_{21\phi} & A_{22\phi} \end{bmatrix} \vec{x}_\phi(t) + \begin{bmatrix} B_{1\phi} \\ B_{2\phi} \end{bmatrix} \vec{u}(t)$$
$$\vec{y}(t) = \begin{bmatrix} C_{11\phi} & 0 \end{bmatrix}$$
$$\vec{x}_\phi(t_0) = \begin{bmatrix} 0 \\ x_{110} \end{bmatrix}$$

(18.28)

where $A_{11\phi}$ is an $n_\phi \times n_\phi$ matrix and the pair $\{A_{11\phi}, C_{11\phi}\}$ is completely observable.

Definition 18.3 The system of Equations 18.27 or 18.28 is detectable if and only if the unobservable subspace is stable (or the unstable subspace is observable), i.e. if and only if $A_{22\phi}$ is an asymptotically stable matrix.

Remarks:

1) Note that detectability involves the concept of stability. When the measurement matrix M is involved, this concept is referred to as reconstructibility.
2) If a system is undetectable (or unreconstructible), then obviously the engineer has to go back to the open loop system and rearrange or reorient the sensors (and the definitions of outputs) in such a way that either the system is completely observable or at least detectable.

18.8.1 Determining the Transformation Matrix T for Observability Canonical Form

Let

$$\vec{x} = A\vec{x}$$
$$\vec{y} = C\vec{x}.$$

Let us transform this to observability canonical form by the transformation matrix T given by $\vec{x}_\phi = T\vec{x}$ so that $\vec{x} = T^{-1}\vec{x}_\phi$. Then

$$\dot{\vec{x}}_\phi = TAT^{-1}\vec{x}_\phi \qquad\qquad = A_\phi \vec{x}_\phi$$
$$\text{and } \vec{y} = CT^{-1}\vec{x}_\phi \qquad\qquad = C_\phi \vec{x}_\phi.$$

When the system is not completely observable $n_\phi < n$. Then the transformation matrix T is built as follows. Consider

$$H_\phi^T = \begin{bmatrix} C \\ CA \\ \vdots \\ CA^{n-1} \end{bmatrix}. \qquad\qquad (18.29)$$

Let the n_ϕ rows of H_c, which are linearly independent, be the n_ϕ columns of T. Then make up the rest of $n - n_\phi$ rows of T such that T is non-singular, i.e.

$$T = \begin{bmatrix} \leftarrow & h_{\phi_1}^T & \rightarrow \\ \leftarrow & h_{\phi_2}^T & \rightarrow \\ & \vdots & \\ \leftarrow & h_{\phi n_\phi}^T & \rightarrow \\ \leftarrow & d_1^T & \rightarrow \\ \leftarrow & d_2^T & \rightarrow \\ & \vdots & \\ \leftarrow & d_{n-n_\phi}^T & \rightarrow \end{bmatrix}$$

where d_i^T are the rows of T such that T is non-singular.

Example 18.6 Express the system

$$A = \begin{bmatrix} -3 & -2 \\ 1 & \frac{3}{2} \end{bmatrix}$$
$$C = \begin{bmatrix} 1 & 4 \end{bmatrix}$$

in observability canonical form.

Solution

$$H_\phi = [C^T A^T C^T]$$
$$= \begin{bmatrix} 1 & 1 \\ 4 & 4 \end{bmatrix}$$

so, since the the rank of H_ϕ is less than 2, the system is not observable. The rank $n_\phi = 1$. So let $T = \begin{bmatrix} 1 & 4 \\ \otimes_1 & \otimes_2 \end{bmatrix}$ where $\otimes_1$ and $\otimes_2$ are any two elements which make T non-singular. So let the second row of T be say [12] so that

$$T = \begin{bmatrix} 1 & 4 \\ 1 & 2 \end{bmatrix} \qquad\qquad (18.30)$$

and

$$T^{-1} = \begin{bmatrix} 1 & 4 \\ \frac{1}{2} & -\frac{1}{2} \end{bmatrix}.$$ (18.31)

Then it can be seen that

$$
\begin{aligned}
A_\phi &= TAT^{-1} \\
&= \begin{bmatrix} 1 & 0 \\ -\frac{1}{2} & -\frac{5}{2} \end{bmatrix} \\
C_\phi &= CT^{-1} \\
&= \begin{bmatrix} 1 & 0 \end{bmatrix}.
\end{aligned}
$$

Since the unobservable subspace is stable, this system is detectable.

Remark 6. Notice that without going to the observability canonical form we could not have ascertained as to which of those eigenvalues corresponds to the observable subspace and which to the unobservable subspace. Thus the beautiful property of the observability canonical form is that it clearly separates out all the observable subspace states into one set and all the unobservable subspace states into a distinctly separate set.

Remark 7. It is also important to realize that when a system is unobservable, it implies that a linear combination of the original (physically meaningful) state variables is unobservable, not necessarily each individual state variable.

Remark 8. Notice that, for linear time invariant systems, since time does not enter into the conditions of observability/detectability, these conditions and concepts are valid for both continuous time as well as discrete time linear state space systems.

Now that we understand the important concepts of both controllability (stabilizability) as well as of observability (detectability), it is clear that both of these concepts together play an extremely important role in the feedback control systems design for linear state space systems. It is interesting and important to keep in mind that both of these concepts are dual to each other. Hence in what follows, we examine the implications and importance of these two concepts together.

18.9 Implications and Importance of Controllability and Observability

(1) They serve as sufficient conditions for designing controllers for regulating the outputs.
(2) Quite often, whenever a system is not completely controllable and observable, it means that the state space model has more state variables than necessary. This means there is an avenue available for reducing the number of state variables in the model, which is a model reduction problem!

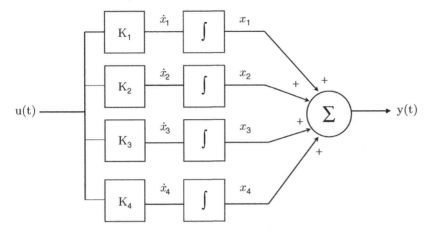

Figure 18.1 Model reduction.

(3) Also, whenever a system is uncontrollable and unobservable, it may mean that the actuators (B matrix elements) and sensors (M or C matrix elements) are not placed at the appropriate locations. In other words, these concepts together prompt the engineer to look into the aspect of optimization of actuator and sensor locations.

(4) Notice that the norm of the controllability matrix, H_c, relates to the degree of controllability and the norm of the observability matrix, H_ϕ, relates to the degree of observability. Hence, these measures can be used to estimate the amount of control effort needed to accomplish a control task and similarly to estimate the amount of richness of the measurement data to assess the quality of measurements.

For example, consider the following situation: the system in Figure 18.1 has the state variable description

$$\dot{\vec{x}} = \vec{0}\vec{x}(t) + \begin{bmatrix} k_1 \\ k_2 \\ k_3 \\ k_4 \end{bmatrix} u(t)$$

where

$$\vec{0} = \begin{bmatrix} 0 & & & 0 \\ & 0 & & \\ & & 0 & \\ 0 & & & 0 \end{bmatrix}$$

$$\vec{x} = \begin{bmatrix} x_1 \\ x_2 \\ x_3 \\ x_4 \end{bmatrix}$$

$$y = \begin{bmatrix} 1 & 1 & 1 & 1 \end{bmatrix} \begin{bmatrix} x_1 \\ x_2 \\ x_3 \\ x_4 \end{bmatrix}.$$

Obviously this system is neither controllable nor observable. However if we define $\dot{\xi} = k\vec{u}(t)$ when $= k = \sum k_i$, where ξ is a scalar and k is assumed to be non-zero, then the scalar system

$$\begin{matrix} \dot{\xi} = ku(t) \\ y(t) = \xi(t) \end{matrix} \quad \text{i.e.} \quad \begin{matrix} \dot{\xi} = 0\xi + ku(t) \\ y(t) = \xi(t) \end{matrix}$$

is both controllable and observable (even though the A matrix, which in this case is a scalar, is zero!). This means that we do not need four state variables x_1, x_2, x_3, x_4 but instead need only one state variable ξ to describe the system dynamics!

(5) Also, finally, whenever a system is uncontrollable (or unobservable), it means that there is some pole-zero cancellation in the corresponding transfer function.

Example 18.7 The lateral/directional dynamics of an aircraft at a particular flight condition (Figure 18.2) is described by the approximate linear model

$$\begin{bmatrix} \dot{p} \\ \dot{r} \\ \dot{\beta} \\ \dot{\phi} \end{bmatrix} = \begin{bmatrix} -10 & 0 & -10 & 0 \\ 0 & -0.7 & 9 & 0 \\ 0 & -1 & -0.7 & 0 \\ 1 & 0 & 0 & 0 \end{bmatrix} \begin{bmatrix} p \\ r \\ \beta \\ \phi \end{bmatrix} + \begin{bmatrix} 20 & 2.8 \\ 0 & -3.13 \\ 0 & 0 \\ 0 & 0 \end{bmatrix} \begin{bmatrix} \delta_A \\ \delta_R \end{bmatrix}. \tag{18.32}$$

(a) Verify if the system is controllable with both inputs operable.

(b) Now suppose a malfunction prevents manipulation of the rudder, i.e. input δ_R. Is the system controllable using only ailerons (i.e. δ_A alone)?

(c) If we outfit the aircraft with only a rate gyro, which senses the roll rate p, are the dynamics of the system observable?

(d) If we instead install a bank angle indicator that measures ϕ, is the system now observable?

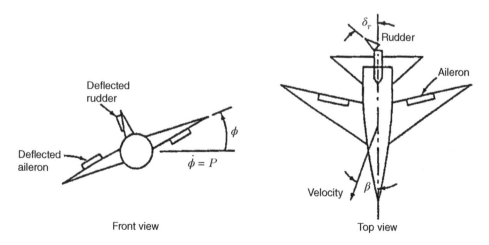

Figure 18.2 Aircraft lateral dynamics.

Solution

(a) To verify the system is indeed controllable with both controllers working, we form the concatenated matrix with the columns:

$$[B \ AB \ A^2B \ A^3B] =$$

$$\begin{bmatrix} 20 & 2.8 & -200 & -28 & 2000 & 248.7 & -20000 & 2443.2 \\ 0 & -3.13 & 0 & 2.191 & 0 & 26.63 & 0 & -58.08 \\ 0 & 0 & 0 & 3.13 & 0 & -4.382 & 0 & -23.569 \\ 0 & 0 & 20 & 2.8 & -200 & -28 & 2000 & 248.7 \end{bmatrix}.$$

Note here the B matrix is the full 4×2 matrix defined in the problem statement because both the aileron and rudder are working. The rank of this matrix is 4 and so the airplane is indeed controllable when both rudder and ailerons are working.

(b) Now with the rudder malfunctioning, we again form the concatenated matrix with the columns

$$[B \ AB \ A^2B \ A^3B] = \begin{bmatrix} 20 & -200 & 2000 & 20000 \\ 0 & 0 & 0 & 0 \\ 0 & 0 & 0 & 0 \\ 0 & 20 & -200 & 2000 \end{bmatrix}$$

noting that this time the B matrix is only the first column of the 4×2 matrix specified in the lateral/direction state space equation. This matrix has rank 2, which is less than $n = 4$ and so we conclude that the system is not completely controllable using only ailerons.

(c) If we are measuring only first state variable p, with a roll rate gyro, the corresponding M matrix as such

$$M = [1 \ 0 \ 0 \ 0] \qquad (18.33)$$

and then form the concatenated matrix with the columns

$$[M^T \ A^T M^T \ (A^2)^T M^T \ (A^3)^T M^T] = \begin{bmatrix} 1 & -10 & 100 & -1000 \\ 0 & 0 & 10 & -114 \\ 0 & -10 & 107 & -984.9 \\ 0 & 0 & 0 & 0 \end{bmatrix} \qquad (18.34)$$

the rank of which is 3, less than $n = 4$, and so we conclude that the system is not completely observable using just a rate gyro.

(d) If the fourth state variable ϕ is the only measurement, we build M as follows

$$M = [0 \ 0 \ 0 \ 1] \qquad (18.35)$$

which leads to the matrix with the following concatenated columns:

$$[M^T \ A^T M^T \ (A^2)^T M^T \ (A^3)^T M^T] = \begin{bmatrix} 0 & 1 & -10 & 100 \\ 0 & 0 & 0 & 10 \\ 0 & 0 & -10 & 107 \\ 1 & 0 & 0 & 0 \end{bmatrix} \qquad (18.36)$$

the rank of which is $4 = n$, and so we can conclude that the entire system is observable with a bank angle indicator as the only sensor used.

18.10 A Display of all Three Structural Properties via Modal Decomposition

For conceptual ease, we consider only the distinct eigenvalue case.

It is clear that every linear state space system

$$\dot{\vec{x}} = A\vec{x} + B\vec{u}$$
$$\vec{y} = C\vec{x}$$

can be visualized as consisting of some controllable as well as observable set of state variables, some controllable but unobservable set of state variables, some observable but uncontrollable set of state variables, and finally neither controllable nor observable set of state variables. This division or distribution of state variables is more visible or transparent in the modal coordinates, especially in the distinct eigenvalue case. This is illustrated below. The Jordan canonical form is given by

$$\dot{\vec{q}} = J\vec{q} + \mathcal{M}^{-1}B\vec{u}$$
$$= J\vec{q} + B_n\vec{u}$$
$$\vec{y} = C\mathcal{M}\vec{q}$$
$$= C_n\vec{q}.$$

We can rearrange the modal vector $\vec{q}(t)$ such that

$$\vec{q}(t) = \begin{bmatrix} q_{co}(t) \\ q_c(t) \\ q_o(t) \\ \tilde{q}(t) \end{bmatrix}$$

where

$$\begin{bmatrix} \dot{q}_{co} \\ \dot{q}_c \\ \dot{q}_o \\ \dot{\tilde{q}} \end{bmatrix} = \begin{bmatrix} J_{co} & & & \\ \hline & J_c & & \\ \hline & & J_o & \\ \hline & & & J \sim \end{bmatrix} \begin{bmatrix} q_{co} \\ q_c \\ q_o \\ \tilde{q} \end{bmatrix} + \begin{bmatrix} B_{co} \\ B_c \\ 0 \\ 0 \end{bmatrix} \vec{u}$$

$$\vec{y} = \begin{bmatrix} C_{co} & | & 0 & | & C_o & | & 0 \end{bmatrix} \begin{bmatrix} q_{co} \\ q_c \\ q_o \\ \tilde{q} \end{bmatrix}$$

where the modes in each of the four categories discussed before are easily understood and visible, by carefully looking at the Jordan matrix, the control distribution matrix and the corresponding output matrix.

18.11 Chapter Summary

In this chapter, we considered the other two structural properties of a linear state space system, namely controllability and observability. We have learned that for time varying systems, there are additional concepts such as reachability and constructibility. However, we focused on continuous time LTI systems for which it suffices to discuss only controllability and observability, which turn out to be dual to each other. Then within systems which are uncontrollable, we have learnt how to decompose that uncontrollable system into a controllable subspace and an uncontrollable subspace. If the uncontrollable subspace is stable, we called that system a stabilizable system. Similarly we have seen that an unobservable system can be decomposed into an observable subspace and an unobservable subspace. If the unobservable subspace is stable, we called that system a detectable system. These notions of controllability (stabilizabilty) and observability (detectability) are extremely important to check before we embark on designing controllers for linear state space systems. Fundamental concepts such as those discussed in this chapter are also available in a more consolidated way in many excellent textbooks dedicated to state space control systems such as [1–9]

18.12 Exercises

Exercise 18.1. Given the state space system

$$
\begin{bmatrix} \dot{x}_1 \\ \dot{x}_2 \\ \dot{x}_3 \\ \dot{x}_4 \end{bmatrix} = \begin{bmatrix} -2 & 1 & 0 & 0 \\ 0 & -2 & 0 & 0 \\ 0 & 0 & 3 & 0 \\ 0 & 0 & 0 & -4 \end{bmatrix} \begin{bmatrix} x_1 \\ x_2 \\ x_3 \\ x_4 \end{bmatrix} + \begin{bmatrix} 2 & 6 \\ 0 & 6 \\ 1 & 1 \\ 5 & 3 \end{bmatrix} \begin{bmatrix} u_1 \\ u_2 \end{bmatrix} \tag{18.37}
$$

and that only u_1 or u_2 is available for implementation, which control would you pick and why?

Exercise 18.2. Test the controllability of the following systems. For the systems which are uncontrollable test the stabilizability.

(a) $A = \begin{bmatrix} -1 & 0 \\ 0 & -2 \end{bmatrix} ; B = \begin{bmatrix} 2 \\ 0 \end{bmatrix}$

(b) $A = \begin{bmatrix} -1 & 0 \\ 0 & 4 \end{bmatrix} ; B = \begin{bmatrix} 5 \\ 0 \end{bmatrix}$

(c) $A = \begin{bmatrix} -1 & 0 \\ 0 & -6 \end{bmatrix} ; B = \begin{bmatrix} 0 \\ 5 \end{bmatrix}$

(d) $A = \begin{bmatrix} -5 & 0 \\ 0 & -7 \end{bmatrix} ; B = \begin{bmatrix} 4 \\ 6 \end{bmatrix}$

(e) $A = \begin{bmatrix} -1 & 1 \\ 0 & -2 \end{bmatrix} ; B = \begin{bmatrix} 2 \\ 3 \end{bmatrix}$

(f) $A = \begin{bmatrix} -5 & 0 \\ 0 & -5 \end{bmatrix} ; B = \begin{bmatrix} 2 \\ 3 \end{bmatrix}.$

Exercise 18.3. Ascertain if the following systems are already in the controllability canonical form and if they are, investigate the stabilizability.

(a) $A = \begin{bmatrix} -5 & 1 & 1 \\ 0 & 4 & 2 \\ 0 & 0 & 5 \end{bmatrix} ; B = \begin{bmatrix} 1 \\ 1 \\ 0 \end{bmatrix}$

(b) $A = \begin{bmatrix} -5 & 1 & 0 \\ 0 & 4 & 0 \\ 0 & 0 & -3 \end{bmatrix} ; B = \begin{bmatrix} 1 \\ 1 \\ 0 \end{bmatrix}$

(c) $A = \begin{bmatrix} -2 & 1 & 5 \\ 0 & -2 & 4 \\ 0 & 0 & -6 \end{bmatrix} ; B = \begin{bmatrix} 3 \\ 0 \\ 0 \end{bmatrix}$

(d) $A = \begin{bmatrix} -5 & 1 & 0 & 1 \\ 0 & 4 & 0 & 2 \\ 0 & 0 & -2 & 0 \\ 0 & 0 & 0 & -3 \end{bmatrix} ; B = \begin{bmatrix} 1 \\ 1 \\ 0 \\ 0 \end{bmatrix}.$

Exercise 18.4. Test the observability of the following systems. For the systems which are unobservable test the detectability.

(a) $A = \begin{bmatrix} -1 & 0 \\ 0 & -2 \end{bmatrix} ; C^T = \begin{bmatrix} 2 \\ 0 \end{bmatrix}$

(b) $A = \begin{bmatrix} -1 & 0 \\ 0 & 4 \end{bmatrix} ; C^T = \begin{bmatrix} 5 \\ 0 \end{bmatrix}$

(c) $A = \begin{bmatrix} -1 & 0 \\ 0 & -6 \end{bmatrix} ; C^T = \begin{bmatrix} 0 \\ 5 \end{bmatrix}$

(d) $A = \begin{bmatrix} -5 & 0 \\ 0 & -7 \end{bmatrix} ; C^T = \begin{bmatrix} 4 \\ 6 \end{bmatrix}$

(e) $A = \begin{bmatrix} -1 & 0 \\ 1 & -2 \end{bmatrix} ; C^t = \begin{bmatrix} 2 \\ 3 \end{bmatrix}$

(f) $A = \begin{bmatrix} -5 & 0 \\ 0 & -5 \end{bmatrix} ; C^T = \begin{bmatrix} 2 \\ 3 \end{bmatrix}.$

Exercise 18.5. Ascertain if the following systems are already in the observability canonical form and if they are, investigate the detectability.

(a) $A = \begin{bmatrix} -5 & 0 & 0 \\ 1 & 4 & 0 \\ 1 & 2 & 5 \end{bmatrix} ; C^T = \begin{bmatrix} 1 \\ 1 \\ 0 \end{bmatrix}$

(b) $A = \begin{bmatrix} -5 & 0 & 0 \\ 1 & 4 & 0 \\ 0 & 0 & -3 \end{bmatrix} ; C^T = \begin{bmatrix} 1 \\ 1 \\ 0 \end{bmatrix}$

(c) $A = \begin{bmatrix} -2 & 0 & 0 \\ 1 & -2 & 0 \\ 5 & 4 & -6 \end{bmatrix} ; C^T = \begin{bmatrix} 3 \\ 0 \\ 0 \end{bmatrix}$

(d) $A = \begin{bmatrix} -5 & 0 & 0 & 0 \\ 1 & 4 & 0 & 0 \\ 0 & 0 & -2 & 0 \\ 1 & 2 & 0 & -3 \end{bmatrix} ; C^T = \begin{bmatrix} 1 \\ 1 \\ 0 \\ 0 \end{bmatrix}.$

Bibliography

1 T.E Fortmann and K.L Hitz. *An Introduction to Linear Control Systems*. Mercel-Dekker, 1977.

2 G.F Franklin, J.D Powell, and A Emami-Naeni. *Feedback Control of Dynamic Systems*. Pearson-Prentice-Hall, 2006.

3 B. Friedland. *Control system design: An introdcution to state-space methods*. *McGraw-Hill*, 1986.

4 T. Kailath. *Linear Systems*. Prentice Hall, 1980.

5 H. Kwakernaak and R. Sivan. *Linear Optimal Control Systems*. Wiley Interscience, 1972.

6 K. Ogata and Y. Yang. *Modern control engineering*. Prentice-Hall, Englewood Cliffs, NJ, 1970.

7 R.E. Skelton. *Dynamic Systems Control: Linear Systems Analysis and Synthesis*. Wiley, New York, 1 edition, 1988.

8 J.G. Truxal, editor. *Control engineers' handbook*. McGraw-Hill, New York, 1958.

9 W.L. Brogan. *Modern Control Theory*. Prentice Hall, 1974.

19

Shaping of Dynamic Response by Control Design: Pole (Eigenvalue) Placement Technique

19.1 Chapter Highlights

In this first chapter on control design for linear state space systems, we present a very important and popular control design method, widely known as the pole placement method. Notice the rather casual name pole placement inspired by a transfer function based viewpoint. Strictly speaking in our discussion on linear state space systems, we need to label it as the eigenvalue placement method. However, because of our familiarity with frequency domain, transfer function based methods as well the need for us to be conversant with both classical control theory as well as modern control theory, we do not mind calling this method the pole placement method. However, when we combine the objective of placing the eigenvalues at some desired locations in the complex plane as well as placing a subset of eigenvectors also with some desired directions, then we label that design method as the eigenstructure assignment method. Note that eigenvectors in time domain based methods are equivalent to zeroes of the transfer function in the classical control theory. In this chapter, we also note that various trade-offs between different controller structures as full state feedback, measurement (or output) feedback and finally the observer (or estimator) based feedback.

19.2 Shaping of Dynamic Response of State Space Systems using Control Design

Finally, we are at a stage where we are ready to design a controller $\vec{u}(t)$ such that the output $\vec{y}(t)$ behaves the way we want. This is the control design phase. Recall that when the state space system

$$\dot{\vec{x}} = A\vec{x} + B\vec{u} \qquad \vec{x}(0) = \vec{x}_0$$
$$\vec{y} = C\vec{x} + D\vec{u} \tag{19.1}$$
$$\vec{z} = M\vec{x} + N^0\vec{u}$$

is completely controllable and completely observable (both from outputs as well as measurements), we are in a position to design a controller $\vec{u}(t)$ as a function of the measurements $\vec{z}(t)$ to make the output $\vec{y}(t)$ behave the way we want.

Flight Dynamics and Control of Aero and Space Vehicles, First Edition. Rama K. Yedavalli.
© 2020 John Wiley & Sons Ltd. Published 2020 by John Wiley & Sons Ltd.

Now let us write $\vec{u}(t) = \vec{f}(\vec{z}(t))$ in general, i.e. we can build a general nonlinear controller if we have to. However, for simplicity, we first attempt to design linear controllers, i.e.

$$\begin{array}{ccc} \vec{u}(t) & = & K(t) \quad \vec{z}(t) \\ m \times 1 & & m \times l \quad l \times 1. \end{array} \tag{19.2}$$

The control law in Equation 19.2 is called a measurement feedback control law. Here the control gain K is simply a matrix of dimension $m \times l$ whose entries are yet to bet determined. These gains can be time varying. However, again for simplicity, we attempt to design a linear constant control gain K, i.e.

$$\vec{u}(t) = K\vec{z}(t). \tag{19.3}$$

Now, the next question we ask is: how do we want our $\vec{y}(t)$ to behave? That is, what is the expected desired behavior of the output? Typically the outputs $\vec{y}(t)$ (and $\vec{x}(t)$) are always modeled as some errors from steady state and therefore very frequently we want $\vec{y}(t)$ as well as $\vec{x}(t)$ to go to zero so that the system when perturbed from the steady state is required to go back to the steady state and we use feedback control to achieve this objective. Desiring $\vec{y}(t)$ and $\vec{x}(t)$ to go to zero is called the regulation problem. If instead $\vec{y}(t)$ and $\vec{x}(t)$ are required to reach some desired output $\vec{y}_d(t)$ and $\vec{x}_d(t)$ then we call it a tracking problem. So in a way the regulation problem is a special case of the tracking problem because in that case $\vec{y}_d(t)$ and $\vec{x}_d(t)$ are simply zero. Let us assume we are designing a controller for regulation.

Now when the controller $\vec{u}(t) = K\vec{z}(t)$ is controlling the system, the closed loop system is given by

$$\begin{aligned} \dot{\vec{x}} &= (A + BKM) \quad \vec{x}(0) = \vec{x}_0 \\ \vec{y} &= (C + DKM)\vec{x}. \end{aligned} \tag{19.4}$$

Obviously, if we want $\vec{x}(t)$ and then eventually $\vec{y}(t)$ to go to zero, we require the closed loop system matrix

$$A_c = (A + BKM)$$

to be an asymptotically stable matrix, i.e. it should have eigenvalues with negative real parts. Unfortunately designing a measurement feedback control gain K such that the closed loop system matrix A_c is asymptotically stable is quite difficult and there may not even exist a gain K that can do this. Thus while measurement feedback is the most practical and simple situation, it is also most inadequate, i.e.

$$\vec{u}(t) = K\vec{z}(t) \begin{cases} \text{simple and practical} \\ \text{but} \\ \text{inadequate or unacceptable performance} \end{cases}.$$

So we look for the other extreme. We assume that *all* state variables are available for measurement! That is, we assume $\vec{z}(t) = \vec{x}(t) \Rightarrow M = I_n$. In this case we have

$$\vec{u}(t) = K\vec{x}(t) \tag{19.5}$$

and here K is an $m \times n$ control gain matrix and this is called a full state feedback controller. Then the closed loop system is given by

$$\dot{\vec{x}} = (A + BK) \quad \vec{x}(0) = \vec{x}_0$$
$$\vec{y} = (C + DK)\vec{x}.$$

Fortunately for this full state feedback case, assuming complete controllability and observability, it is always possible to design a control gain K such that $A_c = (A + BK)$ is an asymptotically stable matrix.

Thus the full state feedback control

$$\vec{u}(t) = K\vec{x}(t) \begin{cases} \text{idealistic and may be impractical} \\ \text{but} \\ \text{is the best we can have from a performance point of view} \end{cases}.$$

Of course, there is a compromise, a via media way of designing a controller that is practical and yet yields a reasonably good performance, and that is given by an observer based controller, i.e. in this case, we build an estimator or observer of the entire state $\vec{x}$, called $\hat{\vec{x}}$ still using only the available measurements $\vec{z}(t)$ and then use these estimated states $\hat{\vec{x}}$ in the controller, i.e.

$$\vec{u} = \vec{f}(\hat{\vec{x}})$$

where

$$\hat{\vec{x}} = \vec{g}(\vec{z}(t)).$$

In what follows, we attempt to examine all these cases one by one.

In that connection, we first present a control design algorithm for full state feedback controller $\vec{u} = K\vec{x}$, because it is known to yield the best performance. Even though it could be impractical, knowing what this controller can achieve in terms of performance would allow us to know the reference or benchmark the performance and then use this knowledge to assess how close any other controller we design is to this best performance we aspire.

Typically, the performance of the closed loop system can be measured from two perspectives, namely from desired time response/relative stability point of view or from regulation/tracking measure point of view. Out of which, the desired time response point of view can be accomplished by our ability to place the eigenvalues in some desired locations in the complex plane as well as our ability to place the eigenvectors in some desired directions as well. This is because, as we have seen from the previous discussions, eigenvalues dictate the nature of the time response and eigenvectors dictate the amplitude of the time response. Hence to get some desired time response in the closed loop system, we need to specify the desired eigenvalues as well as the desired eigenvectors together. A control design method that accomplishes this is labeled as the eigenstructure assignment method. However, complete independent assignment of eigenvalues as well as eigenvectors is not always possible because of the inherent constrained relationship between the eigenvalues and the corresponding eigenvectors. So practical eigenstructure assignment is necessarily constrained and cannot be arbitrary. Since an eigenstructure assignment (even the constrained one) control design is an advanced subject requiring detail beyond the scope of an undergraduate course, what we attempt in this book is a simplified version of this ambitious task, and that is simple, arbitrary desired eigenvalue assignment problem. When we specify the stable eigenvalues of the closed loop system matrix $A_c = A + BK$ as desired closed loop eigenvalues that are supposedly known or given to us, then it is easy to determine the gain K such that $(A + BK)$ has those desired eigenvalues. This process of designing gain K such that $(A + BK)$ has the known desired eigenvalues λ_{d_i} is called control design by pole placement or control design by eigenvalue placement or control design by pole (or eigenvalue) assignment. In what follows, we present control design methods for arbitrary eigenvalue placement. The algorithms for this objective differ based on the number of control inputs, m. When $m = 1$ (i.e. a single input case), the resulting control gain is unique, whereas when we have multiple inputs ($m \geqslant 2$) we could have multiple (non-unique) gains to achieve that objective. So we consider both of these

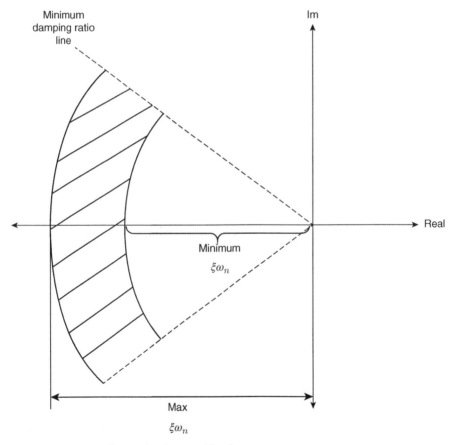

Figure 19.1 Pole placement region in the complex plane.

cases. First, we present the control design method for the single input case. The typical pole placement region in the complex plane is shown in Figure 19.1.

19.3 Single Input Full State Feedback Case: Ackermann's Formula for Gain

Given

$$\begin{rcases} \dot{\vec{x}} = A\vec{x} + B\vec{u} & A \text{ is } n \times n \\ & B \text{ is } n \times 1 \\ \vec{u} = K\vec{x} & K \text{ is } 1 \times n \end{rcases}. \tag{19.6}$$

Our aim is to design the full state feedback control gain K such that the closed loop system matrix $A + BK$ has eigenvalues places at the given desired locations. Let the desired eigenvalues $\lambda_{d1}, \lambda_{d2}, \cdots, \lambda_{dn}$ (with the assumption that complex ones are specified in complex conjugate pairs).

Steps for Full State Feedback Control Gain Determination:

Step 1. From these desired eigenvalues, first compute the characteristic polynomial of the desired closed loop system and let this polynomial be

$$\alpha_d(s) = s^n + \alpha_1 s^{n-1} + \alpha_2 s^{n-2} + \cdots + \alpha_{n-1} s + \alpha_n. \tag{19.7}$$

Step 2. Form the matrix

$$\alpha_d(A) = A^n + \alpha_1 A^{n-1} + \alpha_2 A^{n-2} + \cdots + \alpha_{n-1} A + \alpha_n I. \tag{19.8}$$

Step 3. Form the matrix

$$C_x(A) = \begin{bmatrix} B & AB & A^2 B & \cdots & A^{n-1} B \end{bmatrix}. \tag{19.9}$$

Note that C_x will be an $n \times n$ matrix (so is $\alpha_d(A)$).

Step 4. Finally

$$K = -e_n^T C_x^{-1} \alpha_d(A) \tag{19.10}$$

where

$$e_n^T = \begin{bmatrix} 0 & 0 & \cdots & 0 & 1 \end{bmatrix}. \tag{19.11}$$

Step 5. Compute the eigenvalues of $A_c = A + BK$ and make sure the eigevvalues of A_c are indeed $\lambda_{d1}, \lambda_{d2}, \cdots, \lambda_{dn}$.

Example 19.1 Given the state space system

$$\dot{x} = \begin{bmatrix} -4 & 0 \\ 2 & -2 \end{bmatrix} + \begin{bmatrix} 4 \\ 0 \end{bmatrix} u \tag{19.12}$$

find a gain matrix K such that the closed loop eigenvalues are

$$\lambda_{d1} = -5 \tag{19.13}$$

$$\lambda_{d2} = -12 \tag{19.14}$$

$$\lambda_{d1} = -5; \lambda_{d2} = -12.$$

Solution:

Step 1. So $\alpha_d(s) = (s+5)(s+12) = s^2 + 17s + 60$.

Step 2. Now

$$\alpha_d(A) = A^2 + \alpha_1 A + \alpha_2 I_2$$

$$= A^2 + 17A + 60I_2$$

$$= \begin{bmatrix} 16 & 0 \\ -12 & 4 \end{bmatrix} + \begin{bmatrix} -68 & 0 \\ 34 & -34 \end{bmatrix} + \begin{bmatrix} 60 & 0 \\ 0 & 60 \end{bmatrix}$$

$$= \begin{bmatrix} 8 & 0 \\ 22 & 30 \end{bmatrix}.$$

Step 3.

$$C_x(A) = \begin{bmatrix} B & AB \end{bmatrix} = \begin{bmatrix} 4 & -16 \\ 0 & 8 \end{bmatrix}.$$

Step 4.

$$K = -\begin{bmatrix} 0 & 1 \end{bmatrix} \begin{bmatrix} 4 & -16 \\ 0 & 8 \end{bmatrix}^{-1} \begin{bmatrix} 8 & 0 \\ 22 & 30 \end{bmatrix}$$

$$= -\begin{bmatrix} 0 & 1 \end{bmatrix} \begin{bmatrix} \frac{1}{4} & \frac{1}{2} \\ 0 & \frac{1}{8} \end{bmatrix} \begin{bmatrix} 8 & 0 \\ 22 & 30 \end{bmatrix} = -\begin{bmatrix} \frac{22}{8} & \frac{30}{8} \end{bmatrix}$$

$$= \begin{bmatrix} -\frac{11}{4} & -\frac{15}{4} \end{bmatrix}.$$

Step 5.

$$A + BK = -\begin{bmatrix} -4 & 0 \\ 2 & -2 \end{bmatrix} + \begin{bmatrix} 4 \\ 0 \end{bmatrix} \begin{bmatrix} -\frac{11}{4} & -\frac{15}{4} \end{bmatrix}$$

$$= \begin{bmatrix} -15 & -15 \\ 2 & -2 \end{bmatrix}$$

which has $\lambda_1 = -5$ and $\lambda_2 = -12$.

Recall that in a single input case, the control gain is unique.

Typically, the locations of all the n desired eigenvalues within the complex plane, needed for the control gain determination, are determined based on the speed of response specifications such as the desired damping ratios and natural frequencies of all the complex conjugate pairs based on the desired rise time, settling time, and time constant characteristics in comparison to the open loop situation. Thus the n desired eigenvalues are selected to reside in a bounded region in the left half of the complex plane, where the dominant eigenvalue pair has a real part with sufficient stability degree. The upper bound on the real part of the eigenvalues farthest from the imaginary axis are selected to avoid saturation of the actuators. Note that the further the real parts from imaginary axis, the larger the control gains. Thus the desired pole locations need to be selected judiciously considering the trade-off between acceptable speed of response specifications as well as the limits on the control effort.

19.4 Pole (Eigenvalue) Assignment using Full State Feedback: MIMO Case

In this section, we consider the MIMO case. Accordingly the full state feedback control gain K is an $m \times n$ matrix where this time $m > 1$. We present a technique with which we can construct a matrix K that places the eigenvalues of the closed loop matrix

$$(A + BK) \text{ (where } \vec{u} = K\vec{x})$$

in any desired set of locations specified a priori

$$\lambda_i \in \Omega, \ \Omega = \{\lambda_{d1}, \lambda_{d2}, \cdots, \lambda_{dn}\}$$

(where λ_{di} denotes the desired closed loop eigenvalues with the complex eigenvalues being specified in pairs). The eigenvalues of $A + BK$ are the roots of

$$\Delta'(\lambda) = |\lambda I_n - (A + BK)| = |\lambda I_n - A - BK| \qquad = 0$$

$$= |[\lambda I_n - A][I_n - (\lambda I_n - A)^{-1}BK]| \qquad = 0$$

$$= |\lambda I_n - A||I_n - (\lambda I_n - A)^{-1}BK| \qquad = 0$$

where the characteristic polynomial for the open loop is

$$\Delta(\lambda) = |\lambda I_n - A|.$$

Define

$$\Phi(\lambda) \equiv (\lambda I_n - A)^{-1}$$

$$\Delta'(\lambda) = \Delta(\lambda)|I_n - \Phi(\lambda)BK|.$$

However,

$$|I_n - \Phi(\lambda)BK| = |I_m - K\Phi(\lambda)B|$$

from the identity $|I_n + MN| = |I_m + NM|$ where M is $n \times m$ and N is $m \times n$.

The matrix K must be selected so that $\Delta'(\lambda_{di}) = 0$ for each specified λ_{di}. This will be accomplished by forcing

$$|I_m - K\Phi(\lambda)B| = 0 \text{ for all } \lambda_{di} \in \Omega$$

where

$$\Delta'(\lambda) = \Delta(\lambda)|I_m - K\Phi(\lambda)B|.$$

The determinant will be zero if a column of the above matrix is zero. Thus we will choose K to force a column of $I_m - K\Phi(\lambda)B$ to zero. Let the ith column of I_m be $e^i \in \mathbb{R}^m$ and let the ith column of $I_m - K\Phi(\lambda)B \equiv \Psi(\lambda)$ be Ψ^i. Thus $\lambda_\alpha \in \Omega$ is a root of $\Delta'(\lambda)$ if

$$e^i - K\Psi^i(\lambda_\alpha) = 0$$

$$\text{or } K\Psi^i(\lambda_\alpha) = e^i \qquad\qquad (19.15)$$

for $i = 1, 2, \cdots n$ and $\alpha = 1, 2, \cdots n$.

If an independent equation like Equation (19.15) can be found for every $\lambda_\alpha \in \Omega$, then G can be obtained by combining the independent equations as follows.

$$\begin{bmatrix} K\Psi^i(\lambda_1) & & & = e^i \\ & K\Psi^j(\lambda_2) & & = e^j \\ & & \ddots & \\ & & K\Psi^r(\lambda_n) & = \cdots \end{bmatrix}$$

into one, i.e.

$$\begin{bmatrix} K \\ \downarrow \\ m \times n \end{bmatrix} \underbrace{\begin{bmatrix} \Psi^i(\lambda_1) & \Psi^j(\lambda_2) & \cdots & \Psi^r(\lambda_n) \end{bmatrix}}_{\overline{\Psi} \atop n \times n} = \underbrace{\begin{bmatrix} e^i & e^j & \cdots & e^r \end{bmatrix}}_{E \atop m \times n}$$

yielding

$$\boxed{K = E\overline{\Psi}^{-1}}.$$

If all the desired λ_α are distinct, it will always be possible to find n linearly independent columns of the $n \times mn$ matrix $\begin{bmatrix} \Psi(\lambda_1) & \Psi(\lambda_2) & \cdots & \Psi(\lambda_n) \end{bmatrix}$ if the rank of each $\Psi(\lambda_i)$ is m.

Note: Controllability of the pair (A, B) is a necessary and sufficient condition for arbitrary eigenvalue placement. Recall that in the MIMO case, the control gain K is not unique for a given set of desired eigenvalues. That non-uniqueness comes from the choice we have in selecting the linearly independent columns from the columns of Ψ matrix in the procedure described earlier. It may be noted that some choices may yield complex gains (especially when the desired eigenvalues include complex conjugate pairs) and we discard those complex gains.

Let us illustrate the concept of the existence of multiple gains in the multiple input case that can all place the closed loop eigenvalues at the given desired locations by a simple example below.

Example 19.2   Let $A = \begin{bmatrix} 0 & 2 \\ 0 & 3 \end{bmatrix}$, $B = \begin{bmatrix} 0 \\ 1 \end{bmatrix}$. The open loop system is unstable. Controllability is easily verified. Then

$$\Phi = \frac{\begin{bmatrix} \lambda - 3 & 2 \\ 0 & \lambda \end{bmatrix}}{\lambda(\lambda - 3)}, \Psi(\lambda) = \Psi_1(\lambda) = \frac{\begin{bmatrix} 2 \\ \lambda \end{bmatrix}}{\lambda(\lambda - 3)}.$$

If the desired poles are $\lambda_{d1} = -3$, $\lambda_{d2} = -4$, then $\Psi_1(\lambda_1) = \begin{bmatrix} -1/9 & -1/6 \end{bmatrix}^T$ and $\Psi_1(\lambda_2) = \begin{bmatrix} 1/14 & -1/7 \end{bmatrix}^T$ are linearly independent. We get $K = -\begin{bmatrix} 1 & 1 \end{bmatrix} \begin{bmatrix} 36 & 1 \\ -42 & -28 \end{bmatrix} = \begin{bmatrix} 6 & 10 \end{bmatrix}$. This feedback gain matrix gives closed-loop eigenvalues at $\lambda = -3$ and $\lambda = -4$

Example 19.3   A system is described as $A = \begin{bmatrix} -2 & 1 & 0 \\ 0 & -2 & 0 \\ 0 & 0 & 4 \end{bmatrix}$, $b = \begin{bmatrix} 0 & 0 \\ 0 & 1 \\ 1 & 0 \end{bmatrix}$. Find a constant state feedback matrix K that yields closed-loop pole $\Gamma = -2, -3, -4$. Inverting $\lambda I - A$

leads to

$$\overline{\Phi} = \begin{bmatrix} 0 & 1/(\lambda + 2)^2 \\ 0 & 1/(\lambda + 2) \\ 1/(\gamma - 4) & 0 \end{bmatrix}.$$ (19.16)

A non-singular G matrix can be obtained as

$$\overline{\Psi} = [\Psi_1(-2)\Psi_1(-3)\Psi_1(-4)] = \begin{bmatrix} 0 & 1 & 1/4 \\ 0 & -1 & -1/2 \\ -1/6 & 0 & 0 \end{bmatrix}.$$ (19.17)

Corresponding to this choice, $x = [e_1 e_2 e_2] = \begin{bmatrix} 1 & 0 & 0 \\ 0 & 1 & 1 \end{bmatrix}$.

Example 19.4 A system is described as $A = \begin{bmatrix} -2 & 1 & 0 \\ 0 & -2 & 0 \\ 0 & 0 & 4 \end{bmatrix}$, $b = \begin{bmatrix} 0 & 0 \\ 0 & 1 \\ 1 & 0 \end{bmatrix}$. Find a constant state feedback matrix K that yields closed-loop pole $\Gamma = -2, -3, -4$. Inverting $\lambda I - A$ leads to

$$\overline{\Phi} = \begin{bmatrix} 0 & 1/(\lambda + 2)^2 \\ 0 & 1/(\lambda + 2) \\ 1/(\gamma - 4) & 0 \end{bmatrix}.$$ (19.18)

A non-singular G matrix can be obtained as

$$\overline{\Psi} = [\Psi_1(-2)\Psi_1(-3)\Psi_1(-4)] = \begin{bmatrix} 0 & 1 & 1/4 \\ 0 & -1 & -1/2 \\ -1/6 & 0 & 0 \end{bmatrix}.$$ (19.19)

Corresponding to this choice, $x = [e_1 e_2 e_2] = \begin{bmatrix} 1 & 0 & 0 \\ 0 & 1 & 1 \end{bmatrix}$.

Example 19.5

$$\dot{x} = \begin{bmatrix} -0.01 & 0 \\ 0 & -0.02 \end{bmatrix} x(t) + \begin{bmatrix} 1 & 1 \\ -0.25 & 0.75 \end{bmatrix} u(t).$$

Here we have two inputs. Let the controller be given by the full state feedback controller of the type,

$$u(t) = \begin{bmatrix} K_{11} & K_{12} \\ K_{21} & K_{22} \end{bmatrix} x(t).$$

Let us now build this gain K such that the desired closed loop characteristic polynomial given by $s^2 + 0.2050s + 0.01295$ with the desired eigenvalues being $-0.1025 \pm j0.04944$. In this simple case, we can analytically form the closed-loop characteristic equation as a function of the control gains and, by equating that with the desired characteristic equation given above, we can determine all the possible gains that place the closed-loop

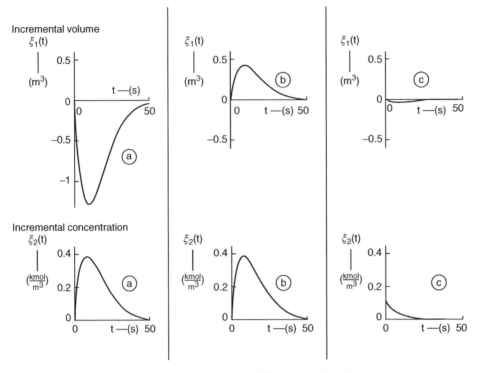

Figure 19.2 Closed-loop responses corresponding to different initial conditions and gain matrices.

eigenvalues at the given locations. It turns out three gains exist that can do that job. They are given by

$$
K_a = \begin{bmatrix} 1.1 & 3.7 \\ 0 & 0 \end{bmatrix}, K_b = \begin{bmatrix} 0 & 0 \\ 1.1 & -1.2333 \end{bmatrix}, K_c = \begin{bmatrix} 0.1 & 0 \\ 0 & 0.1 \end{bmatrix}.
$$

The state trajectories for each of closed loop systems formed by these gains are shown in the Figure 19.2. It can be seen that those state trajectories look markedly different from each other even though all closed loop system matrices have the same set of closed loop eigenvalues. Could you explain why? (See Exercise 19.3).

Important observation: While we noted that there could be multiple gains existing in the MIMO pole placement design that all place the closed loop eigenvalues at the specified desired locations, unfortunately MATLAB's in-built place command for determining the gain for the pole placement problem in the MIMO case still produces only one single gain and gives it as the output. It turns out that MATLAB gets this particular single gain by a different method than the one we discussed in this chapter. However, this author believes that there is interesting research that can be carried out knowing all the possible multiple gains and as such prefers not to be constrained by this single realization that is obtained from MATLAB output. Steps need to be taken to produce all the possible choices for this MIMO pole placement problem.

This non-uniqueness of the control gains for the MIMO case allows us to look for alternative performance measures that demand the control gain that achieves the performance objective in the best possible way, i.e in an optimal way where a performance index defines what that optimal way is. This leads to the subject of the optimal quadratic regulation problem which forms the topic of the next chapter.

19.5 Chapter Summary

In this chapter, we addressed the issue of designing controllers that can make the output behave the way we want. We examined the different controller structures, such as measurement feedback, full state feedback, and observer based feedback, along with the trade-offs between them. Then we defined what we mean by the desired behavior for an output in terms of desired time response or desired relative stability or desired regulation/tracking points of view. Finally, we examined the control design aspect for achieving the desired relative stability requirements via arbitrary eigenvalue placement capability and discussed control design methods for the same, one for the single input case and the other for multiple input case. In the next chapter, we attempt to do the same for regulation/tracking as the desired behavior. Fundamental concepts such as those discussed in this chapter are also available in a more consolidated way in many excellent textbooks dedicated to state space control systems such as [1–8]

19.6 Exercises

Exercise 19.1. An airplane is found to have poor short-period flying qualities in a particular flight regime. To improve the flying qualities, a stability augmentation system using state feedback is to be employed. Determine the feedback gains so that the airplane's short-period characteristics are

$$\lambda_{sp} = -2.1 \pm 2.14i. \tag{19.20}$$

Assume the original short-period dynamics are given by

$$\begin{bmatrix} \Delta\dot{\alpha} \\ \Delta\dot{q} \end{bmatrix} = \begin{bmatrix} -0.334 & 1.0 \\ -2.52 & -0.387 \end{bmatrix} \begin{bmatrix} \Delta\alpha \\ \Delta q \end{bmatrix} + \begin{bmatrix} -0.027 \\ -2.6 \end{bmatrix} \begin{bmatrix} \Delta\delta_e \end{bmatrix}. \tag{19.21}$$

For this problem, use both Ackermann's formula as well as the MIMO gain information and check if they give the same answer or not.

Exercise 19.2. In the satellite attitude control problem, where the dynamics are given by

$$\begin{bmatrix} \dot{\omega}_1 \\ \dot{\omega}_2 \end{bmatrix} = \begin{bmatrix} 0 & -10 \\ 10 & 0 \end{bmatrix} \begin{bmatrix} \omega_1 \\ \omega_2 \end{bmatrix} + \begin{bmatrix} 1 & 0 \\ 0 & 1 \end{bmatrix} \begin{bmatrix} T_1 \\ T_2 \end{bmatrix} \tag{19.22}$$

design a full state feedback control gain such that the closed loop system has a natural frequency $\omega_n = 10$ rad s^{-1} and a damping ratio of $\xi = 0.7$.

Exercise 19.3. Consider a linear multi-variable system with multiple inputs, i.e.

$$\dot{x} = Ax + Bu \qquad u \in \mathbb{R}^m, \quad m > 1$$
$$x \in \mathbb{R}^4 \tag{19.23}$$

It is desired to place the eigenvalues of the closed loop system with a full state feedback gain K at

$$\lambda_{d1,d2} = -0.3 \pm 1.2i$$
$$\lambda_{d3,d4} = -1.2 + 2.5i, \tag{19.24}$$

Three engineers come up with three different gains K_a, K_b, and K_c such that all three achieve closed loop systems with the desired eigenvalues.

(a) Is it possible for the engineers to come up with different gains such that all achieve desired closed loop eigenvalues?

(b) Then let us take each of these gains and plot the time responses of $x_i(t)$ for some given initial conditions. It turns out that the time histories look completely different for each of the three gains even though all the time histories asymptotically go to zero. Can you explain why the time histories were different even though they supposedly have the same closed loop eigenvalues?

Exercise 19.4. The linearized longitudinal motion of a helicopter near hover (see Figure 19.3) can be modeled by the normalized third order system

$$\begin{bmatrix} \dot{q} \\ \dot{\theta} \\ \dot{u} \end{bmatrix} = \begin{bmatrix} -0.4 & 0 & -0.01 \\ 1 & 0 & 0 \\ 1.4 & 9.8 & -0.02 \end{bmatrix} \begin{bmatrix} q \\ \theta \\ u \end{bmatrix} + \begin{bmatrix} 6.3 \\ 0 \\ 9.8 \end{bmatrix} \delta$$

where
q = pitch rate
θ = pitch angle of fuselage
u = horizontal velocity
δ = rotor tilt angle (control variable).
Suppose our sensor measures the horizontal velocity u as the output; that is, $y = u$.

(a) Find the open-loop pole locations.

(b) Is the system controllable?

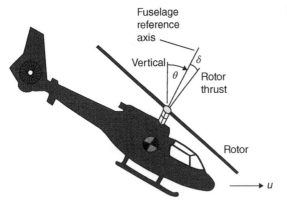

Fuselage reference axis

Vertical

δ

θ

Rotor thrust

Rotor

u

Figure 19.3 Helicopter problem.

(c) Find the feedback gain that places the poles of the system at $s = -1 \pm 1j$ and $s = -2$. Note that this is a measurement feedback.

(d) Now assume full state feedback and find all the possible control gains that take place the clopsed loop eigenvalues at the above mentioned locations.

Notes: In all these problems, plot all the state trajectories and output trajectories, assuming a non-zero initial condition in the state. Be sensible in selecting the initial conditions $\underline{x}(0)$. Plot the open loop trajectories first and then plot the closed loop trajectories larger and defend the final control design you suggest to the customer.

Bibliography

1 T.E Fortmann and K.L Hitz. *An Introduction to Linear Control Systems.* Mercel-Dekker, 1977.

2 G.F Franklin, J.D Powell, and A Emami-Naeni. *Feedback Control of Dynamic Systems.* Pearson-Prentice-Hall, 2006.

3 B. Friedland. *Control system design: An introduction to state-space methods.* McGraw-Hill, 1986.

4 H. Kwakernaak and R. Sivan. *Linear Optimal Control Systems.* Wiley Interscience, 1972.

5 K. Ogata and Y. Yang. *Modern control engineering.* Prentice-Hall, Englewood Cliffs, NJ, 1970.

6 R.E. Skelton. *Dynamic Systems Control: Linear Systems Analysis and Synthesis.* Wiley, New York, 1 edition, 1988.

7 J.G. Truxal, editor. *Control engineers' handbook.* McGraw-Hill, New York, 1958.

8 W.L. Brogan. *Modern Control Theory.* Prentice Hall, 1974.

20

Linear Quadratic Regulator (LQR) Optimal Control

20.1 Chapter Highlights

We have seen in the previous chapter that for a standard multiple input/multiple-output system, the pole placement control design technique described produces multiple gains, which all achieve the desired closed loop locations but with possibly different closed loop state and output time response trajectories. This is because in that technique, we focused only on eigenvalue placement with no attention paid to the resulting eigenvectors, which are different for different gains, thereby yielding various time response trajectories. This indicates that there is an avenue available to demand more than just desired eigenvalue placement, which is a measure of relative stability, with emphasis on achieving a desired damping ratio and natural frequencies in the closed loop trajectories. Instead of relative stability, the other measure of performance, as mentioned earlier, is the regulation/tracking behavior. Just as full eigenstructure assignment control design requires knowledge of an advanced nature, so does tracking control design. However, regulation is a special case of the tracking problem and hence, commensurate with the scope of this book aimed at an undergraduate student body, in this chapter, we present a highly useful and well received control design method for regulation as the control objective. This design method is popularly known as the LQR (linear quadratic regulator) technique. This technique is labeled as an optimal control method because the problem formulation involves minimizing a quadratic performance index that reflects our desire to take the output as closely to zero (regulate) as possible with minimal control effort. Hence the label optimal control.

20.2 Formulation of the Optimum Control Problem

The dynamic process considered here, as elsewhere in this text, is, as usual, characterized by the vector matrix differential equation

$$\dot{x} = Ax + Bu \qquad y = Cx \qquad (20.1)$$

where x is the process state, u is the control input, y is the output and matrices A, B and C are known (given) matrices. Again, as before, we seek a linear control law

$$u(t) = -Kx(t) \qquad (20.2)$$

Flight Dynamics and Control of Aero and Space Vehicles, First Edition. Rama K. Yedavalli.
© 2020 John Wiley & Sons Ltd. Published 2020 by John Wiley & Sons Ltd.

where K is a suitable gain matrix. Here, however, instead of seeking a gain matrix to achieve specified closed-loop pole locations, we now seek a gain to minimize a specified performance criterion V (or cost function) expressed as the integral of a quadratic form in the state x plus a second quadratic form in in the control u

$$V = \int_t^T [y'(\tau)\overline{Q(\tau)}y(\tau) + u'(\tau)Ru(\tau)]d\tau \tag{20.3}$$

where $\overline{Q}$ and R are symmetric positive definite matrices. They represent the weightings we impose on the various output and control variables in the regulation problem. The above performance index can then be rewritten as

$$V = \int_t^T [x'(\tau)Q(\tau)x(\tau) + u'(\tau)Ru(\tau)]d\tau \tag{20.4}$$

where Q is a symmetric, possibly positive semi-definite matrix and R is always a symmetric, positive definite matrix.

Some explanatory remarks about this performance criterion are in order before we attempt to find the optimum gain matrix K.

First, we note that minimization of V also minimizes ρV where ρ is any positive constant.

Second, regarding the limits on the integral, the lower limit t is identified as the present time, and the upper limit T is the terminal time, or final time. The time difference $T - t$ is the control interval, or time-to-go. If the terminal time T is finite and fixed, the time-to-go keeps decreasing to zero, at which time the control process ends. This situation is characteristic of missile guidance problems, as will be discussed in an example below. The more customary case, however, is that in which the terminal time is infinite. In this case we are interested in the behavior of the process "from now on", including the steady state. This is precisely the case addressed by pole placement, and is the case that will receive the major portion of our attention subsequently.

Finally, consider the weighting matrices Q and R. These are often called the state weighting matrix and control weighting matrix, respectively. We are about to derive a recipe for finding the control gain matrix G in terms of these weighting matrices. In other words, we can plug the matrices Q and R, along with the matrices A and B that define the dynamic process, into a computer program and direct it to find K.

The weighting matrix Q specifies the importance of the various components of the state vector relative to each other. For example, suppose that x_1 represents the system error, and that $x_2, \ldots, x_k$ represent successive derivatives, i.e.

$$\dot{x}_2 = \dot{x}$$
$$\dot{x}_3 = \ddot{x}$$
$$\ldots$$
$$\dot{x}_k = x^{(k-1)}.$$

If only the error and none of its derivatives are of concern, then we might select a state weighting matrix

$$Q = \begin{bmatrix} 1 & 0 & \ldots & 0 \\ 0 & 0 & \ldots & 0 \\ \ldots & \ldots & \ldots & \ldots \\ 0 & 0 & \ldots & 0 \end{bmatrix} \tag{20.5}$$

which will yield the quadratic form

$$x'Qx = x_1^2.$$ (20.6)

However, the choice of (20.6) as a state weighting matrix may lead to a control system in which the velocity $x_2 = i$ is larger than desired. To limit the velocity, the performance integral might include a velocity penalty, i.e.

$$x'Qx = x_1^2 + c^2 x_2^2$$ (20.7)

which would result from a state weighting matrix

$$Q = \begin{bmatrix} 1 & 0 & \cdots & 0 \\ 0 & c^2 & \cdots & 0 \\ \cdots & \cdots & \cdots & \cdots \\ 0 & 0 & \cdots & 0 \end{bmatrix}.$$ (20.8)

Another possible situation is one in which we are interested in the state, only through its influence on the system output

$$y = Cx.$$ (20.9)

In other words, we are interested in regulating the output $y(t)$, not just the state $x(t)$. For example, for a system with a single output

$$y = c'x$$ (20.10)

a suitable performance criterion might be

$$y^2 = x'cc'x.$$ (20.11)

So in this case

$$Q = cc'a$$ (20.12)

where a is any positive scalar. It should, by now, be obvious that the choice of the state weighting matrix Q depends on what the system designer is trying to achieve.

20.3 Quadratic Integrals and Matrix Differential Equations

When the control law Equation 20.14 is used to control the dynamic process in Equataion 20.13, the closed-loop dynamic behavior is given by

$$\dot{x} = Ax - BKx = A_c x$$ (20.13)

where

$$A_c = A - BK$$ (20.14)

is the closed-loop dynamics matrix. In most cases considered in this text, we are interested in the case in which A, B, and K are constant matrices, but there is really no need to restrict them to be constant; in fact, the theoretical development is much easier if we do not assume that they are constant. Thus, we permit the closed-loop matrix A, to vary with time. Since A, may be time varying we cannot write the solution to Equation 20.13 as a matrix exponential. However, the solution to Equation 20.13 can be written in terms of the general state transition matrix

$$x(\tau) = \Phi_c(\tau, t)x(t)$$ (20.15)

where Φ_c is the state transition matrix corresponding to A_c. Using Equaton (20.15), the performance index can be expressed as a quadratic form in the initial state $x(t)$. In particular

$$V = \int_t^T [x'(\tau)Qx(\tau) + x'(\tau)K'RKx(\tau)]d\tau$$

$$= \int_t^T x'(t)\Phi_c(\tau, t)Q + K'RK\Phi(\tau, t)x(t)d\tau.$$

The initial state $x(t)$ can be moved outside the integral to yield

$$V = x'(t)P(t, T)x(t) \tag{20.16}$$

where

$$P(t, T) = \int_t^T \Phi_c'(\tau, t)\{Q + K'RK\}\Phi_c(\tau, t)d\tau. \tag{20.17}$$

For purposes of determining the optimum gain, i.e. the matrix K that results in the closed-loop dynamics matrix $A_c = A - BK$, which minimizes the resulting integral, it is convenient to find a differential equation satisfied. For this purpose, we note that V is a function of the initial time t. Thus we can write (20.16) as

$$V(t) = \int_t^T x'(\tau)L(\tau)x(\tau)d\tau \tag{20.18}$$

where

$$L = Q + K'RK. \tag{20.19}$$

(Note that L is not restricted to be constant.) Thus, by the definition of an integral

$$\frac{dV}{dt} = -x'(\tau)Lx(\tau)|_{\tau=1} = -x'(t)Lx(t) \tag{20.20}$$

However,

$$\frac{dV}{dt} = \dot{x}'(t)P(t, T)x(t) + x'(t)\dot{P}(t, T)x(t) + x'(t)P(t, T)\dot{x}(t). \tag{20.21}$$

(The dot over P denotes differentiation with respect to t, that is,

$$\dot{P}(t, T) = \partial P(t, T)/\partial t).$$

We obtain

$$\frac{dV}{dt} = x'(t)[A_c'(t)P(t, T) + \dot{P}(t, T) + P(t, T)A_c(t)]x(t). \tag{20.22}$$

We thus have two expressions for $\frac{dV}{dt}$. Both are quadratic forms in the initial state $x(t)$, which is arbitrary. The only way two quadratic forms in x can be equal for any (arbitrary) x is if the matrices underlying the forms are equal. Thus we have found that the matrix P satisfies the differential equation

$$-L = A_c'P + \dot{P} + PA_c. \tag{20.23}$$

One should not forget that

$$P = P(t, T) \quad A_c = A_c(t) \quad L = L(t). \tag{20.24}$$

We have already determined

$$P(t, T) = \int_t^T \Phi_c'(\tau, t)L(t)\Phi_c(\tau, t)d\tau \tag{20.25}$$

where

$$P(t, T) = 0 \tag{20.26}$$

is the required condition.

20.4 The Optimum Gain Matrix

When any gain matrix K is chosen to close the loop, the corresponding closed-loop performance has been shown to be given by

$$V(t) = x'(t)P(t, T)x(t). \tag{20.27}$$

In terms of the matrices A, B, and K, Q, and R become

$$-\dot{P} = P(A - BK) + (A' - K'B')P + Q + K'RK. \tag{20.28}$$

Our task now is to find the matrix K that makes the solution to Equation (20.28) as small as possible. What does it mean for one matrix to be smaller than another? We are really interested in the quadratic forms resulting from these matrices, and thus we are seeking the matrix P for which the quadratic form

$$\hat{V} = x'\hat{P}x < x'Px \tag{20.29}$$

for any arbitrary initial state $x(t)$ and any matrix $P \neq \hat{P}$. The problem of finding an optimum gain matrix can be approached by a number of avenues.

Now the minimizing matrix P that results from the minimizing gain K,

$$-\dot{\hat{P}} = \hat{P}(A - B\hat{K}) + (A' - \hat{K}'B')\hat{P} + Q + \hat{K}'\hat{K}. \tag{20.30}$$

Any non-optimum gain matrix K and the corresponding matrix P can be expressed in terms of these matrices:

$$P = \hat{K} + N$$
$$K = \hat{K} + Z$$

$$-(\dot{\hat{P}} + \dot{N}) = (\hat{P} + N)[A - B(\hat{K} + Z)] + (A' - (\hat{K}' + Z')B')(\hat{P} + N)$$
$$+ Q + (\hat{K}' + Z')R(\hat{K} + Z). \tag{20.31}$$

On subtracting (20.30) from (20.31) we obtain the following differential equation for N

$$-\dot{N} = NAc + Ac'N + (\hat{K}'R - \hat{P}B)Z + Z'(R\hat{K} - B'\hat{P}) + Z'RZ \tag{20.32}$$

where $A_c = A - BK = A - B(\hat{K} + Z)$. The differential equation (20.32) is exactly in the form of (9.16) with L in the latter being given by

$$L = (\hat{K}'R - \hat{P}B)Z + Z'(\hat{K}'R - \hat{P}B)' + Z'RZ. \tag{20.33}$$

Using (20.25) we see that the solution to (20.32) is of the form

$$N(t, T) = \int_t^T \Phi_c'(T, t)L\Phi_c(T, t)\mathrm{d}\tau. \tag{20.34}$$

Now if $\hat{V}$ is minimum, then we must have

$$x'\hat{P}x \le x'(\hat{P} + N)x = x'\hat{P}x + x'Nx \tag{20.35}$$

which implies that the quadratic form $x'Nx$ must be positive-definite, or at least positive semi-definite. Now look at L as given by (20.33). If Z is sufficiently small the linear terms dominate the quadratic term $Z'RZ$.

We conclude that for the control law $\hat{K}$ to be optimum, we must have

$$R\hat{K} - B'P = 0 \tag{20.36}$$

or, since the control weighting matrix R is always non-singular,

$$\hat{K} = R^{-1}B'\hat{P}. \tag{20.37}$$

This gives the optimum gain matrix in terms of the solution to the differential equation (20.30) that determines $\hat{P}$. When (20.37) is substituted into Equation (20.30) the following differential equation results for $\hat{P}$:

$$-\dot{\hat{P}} = \hat{P}A + A'\hat{P} - \hat{P}BR^{-1}B'\hat{P} + Q. \tag{20.38}$$

One obvious method of solving is the numerical integration of (20.38). Since P is symmetric, there are $k(k+1)/2$ coupled, scalar equations to be integrated. It should be noted that these equations are integrated backward in time, because the condition that must be satisfied is

$$\hat{P}(T, T) = 0 \tag{20.39}$$

and we are interested in $\hat{P}(T, T)$ for $t < T$.

20.5 The Steady State Solution

In an application in which the control interval is finite, the gain matrix K will generally be time varying even when the matrices A, B, Q, and R are all constant, because the solution matrix $P(t, T)$ of the matrix Riccati equation will not be constant. However, suppose the control interval is infinite. We want a control gain K that minimizes the performance integral

$$V_\infty = \int_t^\infty (x'Qx + u'Ru)\mathrm{d}\tau. \tag{20.40}$$

In this case the terminal time T is infinite, so the integration (backward in time) will converge to a constant matrix $\bar{P}$.

For an infinite terminal time

$$V_\infty = x'Px \tag{20.41}$$

here $\bar{P}$ satisfies the algebraic Riccati equation (ARE)

$$0 = \bar{P}A + A'\bar{P} - \bar{P}BR^{-1}B'\bar{P} + Q \tag{20.42}$$

and the optimum gain in the steady state is given by

$$\overline{P} = R^{-1}B'\overline{P}. \tag{20.43}$$

For most design applications the following facts about the solution of (20.42) will suffice:

1. if the system is asymptotically stable, or
2. if the system defined by the matrices (A, B) is controllable, and the system defined by (A, C) where $C'C = Q$, is observable,

then the ARE has a unique, positive definite solution $\overline{P}$.

20.6 Disturbances and Reference Inputs

When there are exogenous input functions (such as disturbances or even some reference inputs) in the system, we need to consider a more general model for the state space system, as given by

$$\dot{x} = Ax + Bu + Ew \tag{20.44}$$

where $w(t)$ is the exogenous (disturbance or reference variable) vector. Assume that $w(t)$ satisfies a differential equation

$$\dot{w}(t) = A_0 w(t). \tag{20.45}$$

Hence the entire (meta)state satisfies the differential equation

$$\dot{\xi} = A_{meta}\xi + B_{meta}u. \tag{20.46}$$

Here

$$\xi = \begin{bmatrix} x \\ w \end{bmatrix} \tag{20.47}$$

$$A_{meta} = \left[\begin{array}{c|c} A & E \\ \hline 0 & A_0 \end{array} \right] \tag{20.48}$$

and

$$B_{meta} = \begin{bmatrix} B \\ 0 \end{bmatrix}, \tag{20.49}$$

The exogenous state $w(t)$ is not controllable. Hence we need to assume that the meta state is stabilizable. Then an appropriate performance integral would be

$$V = \int_t^T (\xi' Q_{meta}\xi + u'Ru)d\tau. \tag{20.50}$$

The weighting matrix for the metastate is of the form

$$Q_{meta} = \left[\begin{array}{c|c} Q & 0 \\ \hline 0 & 0 \end{array} \right]. \tag{20.51}$$

For $x = \dot{x} = 0$, the required steady state control $\overline{u}$ and the steady state exogenous input $\overline{w}$ must satisfy

$$B\overline{u} + E\overline{w} = 0. \tag{20.52}$$

Express the total control u as the sum of the steady state control and a corrective control v:

$$u(t) = \bar{u} + v(t).$$ (20.53)

Then

$$\dot{\xi} = A_{meta}\xi + B_{meta}v.$$ (20.54)

The corrective control v does tend to zero and minimizes to

$$\overline{V} = \int_t^\infty (\xi' Q_{meta}\xi + v'Rv)d\tau$$ (20.55)

thereby, finally, minimizing the quadratic form

$$\overline{V} = \int_t^\infty (x'Qx + v'Rv)d\tau.$$ (20.56)

Thus eventually the optimal controller is $v(t)$ where

$$v(t) = u(t) - \bar{u}$$ (20.57)

where $u(t)$ is the LQR optimal control for the system without the exogenous input. However, there are some issues with this approach and the assumptions we made. Even if the meta system is stabilizable, the steady state control $\bar{u}$ may not exist. Even if it exists, it may not be unique. Also minimizing the index V with control v in it is not same as minimizing $\overline{V}$ with control u in it because of the presence of cross terms appearing in the index $\overline{V}$ with u in it (because $u = \bar{u} + v$). For issues related to these, as well as for unstabilizable meta systems, the reader is encouraged to consult books such as [3].

Example 20.1 Let us illustrate the LQR design method by the following simple example:

$$\begin{bmatrix} \dot{x}_1 \\ \dot{x}_2 \end{bmatrix} = \begin{bmatrix} -1.1969 & 2.0 \\ 2.4763 & -0.8581 \end{bmatrix} \begin{bmatrix} x_1 \\ x_2 \end{bmatrix} + \begin{bmatrix} -0.1149 \\ -14.1249 \end{bmatrix} u.$$

The weighting matrices are chosen as follows:

$$Q = \begin{bmatrix} 1 & 0 \\ 0 & 10 \end{bmatrix}$$

$$R = 5.$$

(a) Find the optimal gain K that minimizes the quadratic cost function.
(b) Plot the state trajectory of all two state variables to the following initial conditions:

$$x_o = \begin{bmatrix} 1 & 1 \end{bmatrix}^T.$$ (20.58)

This problem can be solved using MATLAB.

(a) After defining the matrices A, B, Q, and R we simply call upon the in-built command

```
K=lqr (A,B,Q,R)
```

which yields the optimal gain

$$K = \begin{bmatrix} -0.2779 & -1.3814 \end{bmatrix}.$$ (20.59)

(b) Then, we define state space system and follow its trajectory for the first 5 s after perturbation

```
sys=ss(A-B*K,[],eye(2),[])
initial(sys,x0,5)
```

Note here C is simply a 2×2 identity matrix. The output of this command (the state trajectories) is pictured in Figure 20.1

In the above example, we specified a given set of weighting matrices Q and R and obtained an optimal LQR gain corresponding to those weighting matrices and analyzed the closed loop system state trajectories. However, if the closed loop system state trajectories are not satisfactory, we need to change those weighting matrices and redo the optimal control problem for the new set of weighting matrices. Instead of going through this process in an iterative way with multiple runs, it is more prudent to select the weightings also in an optimal way in the sense that the closed loop system delivers acceptable performance with these optimal weightings. Conceptually, we know that in this LQR problem, a smaller state regulation cost is achieved at the expense of larger control regulation cost because we are minimizing the sum of these two costs. So if we can generate a trade-off curve between the state regulation cost and the control effort (reflected via the control gain elements) as a function of a scalar weighting parameter, then we can

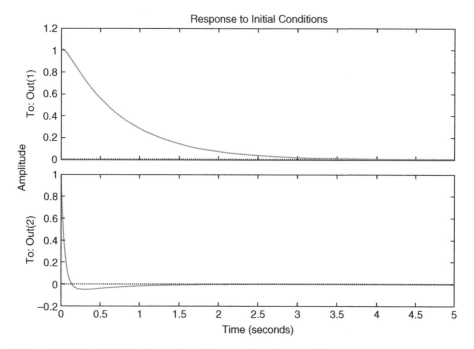

Figure 20.1 The solution to the system of two equations in two unknowns.

select that optimal weighting parameter based on the trade-off between the two costs. Since the weighting matrices Q and R appear in the Riccati equation as a ratio, we can fix one of the weighting matrices, either Q or R and vary the other. Since as a matrix, each of these matrices have too many design variables, we propose to fix the state weighting matrix Q and select a scalar design parameter ρ within the weighting matrix R. Thus we take the overall control weighting matrix R as $R = \rho R_o$, where we now fix the weighting matrix R_o, so that ρ then becomes our single design variable, in the sense that for each ρ we get an optimal LQR gain. Thus the objective of this optimal LQR problem is to select the best value of ρ, which in turn allows us to select the best LQR gain based on the trade-off information between the state regulation cost, and the control effort. The steps involved in getting this trade-off curve are explained next.

20.7   Trade-Off Curve Between State Regulation Cost and Control Effort

We observed that in the LQR problem formulation, we are minimizing the total regulation cost comprising the sum of state regulation cost and the control regulation cost as given by

$$V = \int_t^\infty (x'Qx + u'Ru)d\tau. \tag{20.60}$$

Intuitively, it is clear that the more control effort we expend, the less the state regulation cost will be. In other words, we do not obtain any insight into the trade-off between the state regulation cost and the control effort just by looking at the total minimized cost. So, we need to quantify control effort, which is not exactly same as the total control regulation cost. For this, let us take out one of the design variables out of the total regulation cost. Let the control weighting matrix R be written as $R = \rho R_o$ where R_o is, as usual, a symmetric positive definite matrix and the scalar design variable ρ is positive which we vary during the LQR optimization procedure. Thus we write the total performance cost index as

$$V = \int_t^\infty (x'Qx + \rho u'R_o u)d\tau. \tag{20.61}$$

We then write

$$V_x = \int_t^\infty (x'Qx)d\tau \tag{20.62}$$

and

$$V_u = \int_t^\infty (u'R_o u)d\tau. \tag{20.63}$$

Thus we have,

$$V = V_x + \rho V_u. \tag{20.64}$$

We now label the cost V_x as the state regulation cost and the cost V_u as the control effort. Once we compute these two costs separately for a given value of the design variable ρ, we can note the state regulation cost for a given control effort. Then we could draw the trade-off curve between the state regulation cost and the corresponding control effort as a function of the design variable ρ. This requires us to know a technique that allows us to compute these individual regulation costs separately. In what follows, we present a generic method to compute the quadratic cost of any linear asymptotically stable linear state space system.

20.7.1 Method to Evaluate a Quadratic Cost Subject to a Linear (Stable) State Space System

Consider $\dot{x} = Ax$, $x(0) = x_0$, $A \in R^{n \times n}$, where A is assumed to be an $n \times n$ asymptotically stable matrix. Also consider the cost function,

$$J = \int_0^\infty x^T Q x dt \quad Q \text{ is symmetric, } PD > 0. \tag{20.65}$$

Then the value of the cost function J can be easily calculated as follows:

$$J = x_0^T P x_0 \tag{20.66}$$

where P is the solution of the Lyapunov equation,

$$PA + A^T P + Q = 0. \tag{20.67}$$

Note that P is a symmetric, positive definite matrix. Note that a Lyapunov equation solution exists only when A is an asymptotically stable matrix.

Now we can apply this technique in our LQR problem to evaluate the state reputation cost and control effort separately as follows:

Consider the LQR problem with the index,

$$J = \int_0^\infty (y^T \overline{Q} y + u^T R u) dt = \int_0^\infty (x^T Q x + u^T R u) dt. \tag{20.68}$$

Let $Q \ (= C^T \overline{Q} C)$ be fixed and $R = \rho I_m$ so that

$$J = \int_0^\infty (x^T Q x + \rho u^T u) dt$$

$$= \int_0^\infty (x^T Q x dt) + \rho \int_0^\infty (u^T u) dt \tag{20.69}$$

$$= J_x + \rho J_u.$$

We call J_x the state reputation cost, J_u the control effort cost (or simply control effort).

Remember that we have,

$$u = Kx \quad K = -R^{-1} B^T \overline{P} \tag{20.70}$$

where $\overline{P}$ is the symmetric positive definite solution of the algebraic Riccati equation,

$$\overline{P} A + A^T \overline{P} - \overline{P} B R^{-1} B^T \overline{P} + Q = 0. \tag{20.71}$$

The closed loop system is,

$$\dot{x} = (A - BR^{-1} B^T \overline{P}) \quad x = A_{Cl} x, \qquad x(0) = x_0 \tag{20.72}$$

where A_{Cl} is asymptotically stable. Obviously we can easily find,

$$J_x = \int_0^\infty (x^T Q x dt) = x_0^T P_x x_0 \tag{20.73}$$

where P_x comes from the solution to the Lyapunov equation,

$$P_x A_{Cl} + A_{Cl}^T P_x + Q = 0. \tag{20.74}$$

Similarly we can find,

$$\begin{aligned} J_u &= \int_0^\infty (u^T u) dt \\ &= \int_0^\infty x^T K^T K x\, dt \\ &= x_0^T P_u x_0 \end{aligned} \tag{20.75}$$

where P_u comes from the solution of the Lyapunov equation,

$$P_u A_{Cl} + A_{Cl}^T P_u + K^T K = 0. \tag{20.76}$$

Note that the optimal cost, J, is given by,

$$\begin{aligned} J &= x_0^T \overline{P} x_0 \\ &= J_x + \rho J_u. \end{aligned} \tag{20.77}$$

Obviously J_u can also be found out by knowing J, J_x, and ρ, because for a given ρ, we see that,

$$J_u = \frac{(J - J_x)}{\rho}. \tag{20.78}$$

So the procedure to find the most practical, optimal control gain, K^*, is to vary ρ in increments and for each ρ solve the LQR problem and find the costs J_x and J_u and plot them, which will look as in Figure 20.2.

Then knowing the limits of J_x and J_u, we can zero in the most desirable ρ^* and in turn the optimal control gain K^* (corresponding to that ρ^*). In other words, there is a mini-optimization within the overall LQR optimization problem. Thus, finally, the utility of the LQR optimal control problem is completely realized only after we obtain this trade-off curve. Without this trade-off curve, there is not much we accomplish by simply solving the original LQR control design problem. In other words, obtaining this trade-off curve gives closure to the LQR optimal control design procedure.

Figure 20.2 Desirable region for ρ, and thus for K^*.

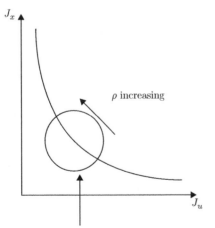

J_x

ρ increasing

J_u

Note that the eigenvalues of the closed loop system matrix A_{Cl} are determined by the optimal control gain, K, and we can get some control over them by treating the entire matrix elements of Q and R as design variables, which could be unwieldy in a practical situation, but the point is to know that we do have many design variables at our disposal to come up with a meaningful and useful control gain that meets the closed loop system specifications at the end of the design exercise.

20.8 Chapter Summary

In this chapter, we have covered a very important and popular control design method called the LQR (linear quadratic regulator) technique. In this method, the control objective is formulated in an optimal control framework in which a quadratic performance index is minimized with respect to the control function subject to a linear dynamic system description serving as an equality constraint. This optimization method yields the optimal control function as a full state feedback controller. Hence this method belongs to the class of functional optimization methods rather than the parameter optimization methods in which the controller structure is pre-supposed and the optimization is done on the pre-supposed controller gains as parameters. Of course, parameter optimization methods allow flexibility in specifying the controller structure but it is difficult to obtain a solution to the optimization problem, whereas functional optimization methods (to which LQR formulation belongs) offer no control over the structure of the optimal controller but it is easier to find the solution to the posed optimization problem. It is also important to realize that the optimality requirement begets stability of the closed loop system as a nice by-product. In other words, the LQR method, which produces the optimal control gain via a Riccati equation solution, automatically guarantees the stability of the closed loop system. Of course, the difference from pole placement design is that in the pole placement design case, we have control over the locations of the closed loop eigenvalues, whereas in an LQR method, while guaranteeing a stable closed loop system, does not give us any a priori knowledge about the locations of those closed loop eigenvalues. However, it is known and proved that the closed loop eigenvalues of an LQR designed closed loop system follow a pattern called the Butterworth pattern.

Fundamental concepts such as those discussed in this chapter are also available in a more consolidated way in many excellent textbooks dedicated to state space control systems such as [1–8]

20.9 Exercises

Exercise 20.1. An airplane is found to have poor short-period flying qualities in a particular flight regime. To improve the flying qualities, a stability augmentation system using state feedback is to be employed. Assume the original short-period dynamics are given by

$$\begin{bmatrix} \Delta\dot{\alpha} \\ \Delta\dot{q} \end{bmatrix} = \begin{bmatrix} -0.334 & 1.0 \\ -2.52 & -0.387 \end{bmatrix} \begin{bmatrix} \Delta\alpha \\ \Delta q \end{bmatrix} + \begin{bmatrix} -0.027 \\ -2.6 \end{bmatrix} \begin{bmatrix} \Delta\delta_e \end{bmatrix}. \tag{20.79}$$

For this problem, use the LQR method to design a full state feedback control. Select the state weighting matrix Q to be an identity matrix and the control weighting to be a positive scalar ρ. Plot the trade-off curve, and finally select an appropriate scalar ρ and the corresponding optimal control gain. Analyze the resulting closed loop eigenvalue scenario and compare it with the pole placement design that was done for this problem in the previous chapter.

Exercise 20.2. In the satellite attitude control problem, the dynamics are given by

$$\begin{bmatrix} \dot{\omega}_1 \\ \dot{\omega}_2 \end{bmatrix} = \begin{bmatrix} 0 & -10 \\ 10 & 0 \end{bmatrix} \begin{bmatrix} \omega_1 \\ \omega_2 \end{bmatrix} + \begin{bmatrix} 1 & 0 \\ 0 & 1 \end{bmatrix} \begin{bmatrix} T_1 \\ T_2 \end{bmatrix}. \tag{20.80}$$

For this problem, use the LQR method to design a full state feedback control. Select the state weighting matrix Q to be an Identity matrix and the control weighting matrix to be a positive scalar ρ times the identity matrix. Plot the trade-off curve, and finally select an appropriate scalar ρ and the corresponding optimal control gain. Analyze the resulting closed loop eigenvalue scenario and compare it with the pole placement design that was done for this problem in the previous chapter.

Bibliography

1 T.E Fortmann and K.L Hitz. *An Introduction to Linear Control Systems.* Mercel-Dekker, 1977.

2 G.F Franklin, J.D Powell, and A Emami-Naeni. *Feedback Control of Dynamic Systems.* Pearson-Prentice-Hall, 2006.

3 B. Friedland. *Control system design: An introdcution to state-space methods.* McGraw-Hill, 1986.

4 H. Kwakernaak and R. Sivan. *Linear Optimal Control Systems.* Wiley Interscience, 1972.

5 K. Ogata and Y. Yang. *Modern control engineering.* Prentice-Hall, Englewood Cliffs, NJ, 1970.

6 R.E. Skelton. *Dynamic Systems Control: Linear Systems Analysis and Synthesis.* Wiley, New York, 1 edition, 1988.

7 J.G. Truxal, editor. *Control engineers' handbook.* McGraw-Hill, New York, 1958.

8 W.L.Brogan. *Modern Control Theory.* Prentice Hall, 1974.

21

Control Design Using Observers

21.1 Chapter Highlights

In this chapter, we present a design method, which, in some sense, combines the best features of the two control design methodologies, namely it achieves reasonably good performance but is more practical in control law implementation in that the control law still uses only the available measurements without requiring all the state variables be measured for feedback purposes. In this method, we use the measurements information to construct an estimate of the entire state, and feed those estimated state variables to the actuator. Hence the name observer based feedback. Finally we also look at few other possible controller structures labeled dynamic compensators of reduced order. Finally, the possibility of spillover instability that may occur when we drive higher order systems with lower order controllers is presented.

21.2 Observers or Estimators and Their Use in Feedback Control Systems

We have learnt that pure measurement feedback, while it is simple and practical, cannot guarantee the stability of the closed loop system. In the other extreme, full state feedback can guarantee arbitrary pole (i.e. eigenvalue) placement (under the assumption of complete controllability) but it is impractical because it is expensive to measure all the state variables. So we need a compromise. That compromise is offered by observer or estimator based controllers. In these, we take the measurements and build an estimator or observer for the entire state as a function of these measurements and then we use these estimated states in the full state feedback control scheme. Consider

$$\dot{\vec{x}} = A\vec{x} + B\vec{u} \quad \longrightarrow \text{state } \vec{x} \in^n \tag{21.1}$$

$$\vec{z} = M\vec{x} \quad \longrightarrow \text{measurements } \vec{z} \in^l. \tag{21.2}$$

Let $\hat{\vec{x}}$ be the state estimate. Obviously we want $\hat{\vec{x}}$ to be as close to the actual state $\vec{x}$ as possible. So let us build a model to generate $\hat{\vec{x}}$. So let us write

$$\dot{\hat{\vec{x}}} = A\hat{\vec{x}} + B\vec{u} + F\vec{z} \tag{21.3}$$

Flight Dynamics and Control of Aero and Space Vehicles, First Edition. Rama K. Yedavalli.
© 2020 John Wiley & Sons Ltd. Published 2020 by John Wiley & Sons Ltd.

where $F\vec{\tilde{z}}$ is the forcing function term for the estimator. Here $\vec{\tilde{z}}$ is called the measurement residual because we want to drive the estimator with known measurements $\vec{z}$ and we have yet to determine what $\vec{\tilde{z}}$ is and what F is.

F is the filter or estimator or observer gain matrix and it is of dimensions $n \times l$, because $\vec{\tilde{z}}$ is of the same dimensions as $\vec{z}$. Now, to determine F and $\vec{\tilde{z}}$, let us calculate them from Equation 21.1.

$$\dot{\vec{x}} - \dot{\vec{\hat{x}}} = A(\vec{x} - \vec{\hat{x}}) - F\vec{\tilde{z}}.$$

To make the right-hand side contain $(\vec{x} - \vec{\hat{x}})$, let us write

$$\vec{\tilde{z}} = \vec{z} - \vec{\hat{z}} = M\vec{x} - M\vec{\hat{x}} = M(\vec{x} - \vec{\hat{x}}). \tag{21.4}$$

Then

$$(\dot{\vec{x}} - \dot{\vec{\hat{x}}}) = A(\vec{x} - \vec{\hat{x}}) - FM(\vec{x} - \vec{\hat{x}})$$
$$= (A - FM)(\vec{x} - \vec{\hat{x}}).$$

If we define $(\vec{x} - \vec{\hat{x}})$ to be the estimation error $\vec{e}$, then you see that

$$\dot{\vec{e}} = (A - FM)\vec{e}$$
$$\vec{e}(0) = \vec{x}(0) - \vec{\hat{x}}(0)$$
$$= \vec{x}(0) - 0$$

where we deliberately take $\vec{\hat{x}}(0) = 0$. Obviously the estimation error $\vec{e}$ can be made asymptotically zero if we choose the F matrix such that $(A - FM)$ is an asymptotically stable matrix. Then with $\vec{\tilde{z}}$ given by Equation 21.4, the final form of the estimator dynamics is given by

$$\dot{\vec{\hat{x}}} = A\vec{\hat{x}} + B\vec{u} + FM(\vec{x} - \vec{\hat{x}})$$

$$\boxed{\dot{\vec{\hat{x}}} = (A - FM)\vec{\hat{x}} + B\vec{u} + F\vec{z}.} \tag{21.5}$$

In other words, the estimator or observer is given by Equation 21.5, and as promised, the estimator is driven by the measurements $\vec{z}$ and has a plant matrix $(A - FM)$. So as long as we make sure that the estimator plant matrix $(A - FM)$ is asymptotically stable, then the estimated states $\vec{\hat{x}}$ generated by Equation 21.5 are such that the estimation error $\vec{e} \to 0$ asymptotically.

Thus the main task now is to build the estimator gain matrix F. For the present assume that we can build an estimator or observer gain matrix F such that $(A - FM)$ is an asymptotically stable matrix.

Then the controller, i.e. the observer-based controller, can be given by

$$\vec{u} = K\vec{\hat{x}} \tag{21.6}$$

where K is the $m \times n$ controller gain matrix. If we now use this controller in the system, then the closed loop system is given by

$$\dot{\vec{x}} = A\vec{x} + BK\vec{\hat{x}}$$

$$\dot{\vec{\hat{x}}} = FM\vec{x} + (A - BK - FM)\vec{\hat{x}}$$

i.e. $\begin{bmatrix} \dot{\vec{x}} \\ \dot{\vec{\hat{x}}} \end{bmatrix} = \begin{bmatrix} A & BK \\ FM & A + BK - FM \end{bmatrix} \begin{bmatrix} \vec{x} \\ \vec{\hat{x}} \end{bmatrix}$

$\begin{bmatrix} \vec{x}(0) \\ \vec{\hat{x}}(0) \end{bmatrix} = \begin{bmatrix} \vec{x}_0 \\ 0 \end{bmatrix}.$

Thus the closed loop system with an observer based controller is a $2n$th order system with the closed loop system matrix of order $2n \times 2n$.

Quite interestingly, it can be shown that the eigenvalues of the overall closed loop system Equation 21.6 are the union of the eigenvalues of $(A + BK)$ and $(A - FM)$, i.e. the eigenvalues of the closed-loop system as though the whole state is measured and the eigenvalues of the observer!

Then if we can build a controller gain matrix K such that $(A + BK)$ is asymptotically stable and a filter gain matrix $(A - FM)$ is asymptotically stable, then we can see that the overall estimation based controller closed loop system can be made asymptotically stable.

Caution: Even though the matrices $(A + BK)$ and $(A - FM)$ can be individually asymptotically stable, the matrix $(A + BK - FM)$ need not be stable! The matrix $(A + BK - FM)$ is called the controller closed loop system matrix and if K and F are such that $(A + BK - FM)$ is also asymptotically stable, then we say the closed loop system is strongly stable and this concept is called strong stability. Figure 21.1 shows the typical regions of placement of eigenvalues for the controller and estimator.

Example 21.1 A broom balancer is a control system that balances a broom on an arbitrary object, such as a moving cart. Such a system is pictured in Figure 21.2. This problem was considered in [2]. Note the uncontrolled system here is fundamentally unstable, as the broom would fall down if left uncontrolled. The governing equation describing this system is:

$$(M + m)\ddot{z} + ml\ddot{\theta} \cos \theta - ml\dot{\theta}^2 \sin \theta = u \tag{21.7}$$

where

- $M = 8$kg is the mass of the cart
- $m = 2$kg is the mass of the broom
- $l = 2$m is the length of the broomstick
- $g \approx 10$m/s^2 for convenience.

(a) Linearize the above system with nominal values of

$$z_n(t) = 0$$
$$\dot{z}_n(t) = 0$$
$$\theta_n(t) = 0$$
$$\dot{\theta}_n(t) = 0$$
$$u_n(t) = 0$$

and obtain a linear state space model with horizontal distance $z(t)$ as the output(controlled) variable as well as the measurement variable.

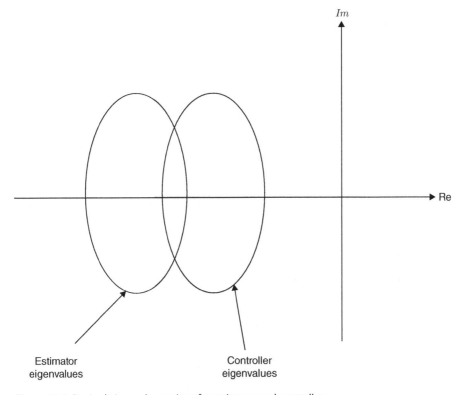

Figure 21.1 Desired eigenvalue regions for estimator and controller.

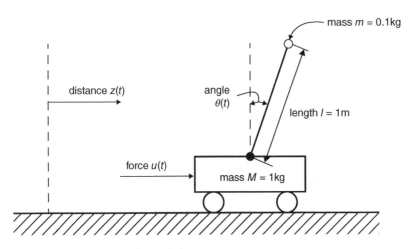

Figure 21.2 An image of a broom balancer.

(b) Using the A, B, C, and M matrices found from part (a), investigate the following
  (i) open loop stability
  (ii) controllability
  (iii) observability.
(c) Design a full state feedback controller such that the controlled variable $z(t)$ is regulated reasonably well. Use the LQR method and select the output and control weightings such that the dominant eigenvalue (that eigenvalue closest to the imaginary axis) has a real part of at least -1.
(d) Now design an observer by determining the observer gain matrix F such that the observer closed loop matrix $A - FM$ has closed loop eigenvalues at

$$\lambda_{e_1} = -2$$
$$\lambda_{e_2} = -3$$
$$\lambda_{e_{3,4}} = -2 \pm i.$$

Hint: The eigenvalues of a generic matrix Z are the same as those of its transpose Z^T.
(e) After obtaining the controller gain matrix K and the filter gain matrix F, form the overall observer based controller driven closed loop system and investigate if the system is strongly stable.
(f) Plot the output variable trajectory and the control trajectory for an initial condition of $\theta(0) = 5°$ and all others zero. Keep in mind the control is a feedback function of the state estimate $\hat{x}$.

Solution
(a) The linearized state space model is as follows:

$$\begin{bmatrix} \dot{z} \\ \ddot{z} \\ \dot{\theta} \\ \ddot{\theta} \end{bmatrix} = \begin{bmatrix} 0 & 1 & 0 & 0 \\ 0 & 0 & -1 & 0 \\ 0 & 0 & 0 & 1 \\ 0 & 0 & 11 & 0 \end{bmatrix} \begin{bmatrix} z \\ \dot{z} \\ \theta \\ \dot{\theta} \end{bmatrix} + \begin{bmatrix} 0 \\ 1 \\ 0 \\ -1 \end{bmatrix} u$$

$$y = z = \begin{bmatrix} 1 & 0 & 0 & 0 \end{bmatrix} \begin{bmatrix} z \\ \dot{z} \\ \theta \\ \dot{\theta} \end{bmatrix}.$$

Since we are measuring the controlled variable, we may say $y = z$.
(b) (i) Three of the four eigenvalues of the open loop system matrix A have non-negative real parts and therefore the open-loop system is unstable. The physical interpretation of this is that we do not expect a broom to balance itself without any forcing function supplied.
  (ii) Using the given A and B matrices we form the matrix with the columns

$$\begin{bmatrix} B & AB & A^2B & A^3B \end{bmatrix} = \begin{bmatrix} 0 & 1 & 0 & 1 \\ 1 & 0 & 1 & 0 \\ 0 & -1 & 0 & -11 \\ -1 & 0 & -11 & 0 \end{bmatrix}$$

the rank of which is 4, equal to the order of the A matrix. We therefore conclude the system is controllable.

(iii) To comment on the observability of the system we form the concatenated matrix with the rows

$$\begin{bmatrix} C \\ CA \\ CA^2 \\ CA^3 \end{bmatrix} = \begin{bmatrix} 1 & 0 & 0 & 0 \\ 0 & 1 & 0 & 0 \\ 0 & 0 & -1 & 0 \\ 0 & 0 & 0 & -1 \end{bmatrix}$$

the rank of which is 4, equal to the order of the A matrix. We therefore conclude the system is observable.

(c) We will form a gain K using the `lqr` command in MATLAB. To do this, first we must choose the weighting matrix Q and control effort R. Since z is the only state variable of interest to us, we will weight the matrix Q as follows:

$$Q = \begin{bmatrix} 25 & 0 & 0 & 0 \\ 0 & 0 & 0 & 0 \\ 0 & 0 & 0 & 0 \\ 0 & 0 & 0 & 0 \end{bmatrix}.$$

We choose the control effort

$$R = 5$$

which achieves the requirement on the dominant eigenvalue. which outputs the gain

$$K = \begin{bmatrix} -2.236 & -3.584 & -39.734 & -12.245 \end{bmatrix}.$$

The closed loop matrix $A_{cl} = A - BK$ formed using the gain K found above has the eigenvalues

$$\lambda_{1,2} = -3.3112 \pm 0.0298i$$
$$\lambda_{3,4} = -1.0188 \pm 1.0006i$$

which meet the requirement that the dominant eigenvalue has a real part greater in magnitude than unity.

(d) Recognize the matrix property given in the hint:

$$(A - FM)^T = A^T - M^T F^T \tag{21.8}$$

where

$$M = \begin{bmatrix} 1 & 0 & 0 & 0 \end{bmatrix} \tag{21.9}$$

because the horizontal distance z is the only state variable we wish to observe. Then, the right-hand side of Equation 21.8 is in the form $A - BK$ and we can therefore trick the `place` command into finding a filter gain with the desired poles p as follows:

```
transp_F = place(A',M',p)
```

Realize the output of this command is in fact the transpose of the filter matrix. So we can transpose the output to obtain the filter gain:

$$F = \begin{bmatrix} 9 \\ 42 \\ -148 \\ -492 \end{bmatrix}. \tag{21.10}$$

You may confirm the eigenvalues of $A - FM$ are indeed the desired eigenvalues

$$\lambda_{1,2} = -2 \pm i$$
$$\lambda_3 = -2$$
$$\lambda_4 = -3.$$

(e) Keep in mind the state variables in the observer based system are $\begin{bmatrix} x & \hat{x} \end{bmatrix}^T$. Then the overall closed-loop observer based controller is given by

$$u = K\hat{x}$$
$$= \begin{bmatrix} 0 & 0 & 0 & 0 & -2.236 & -3.585 & -39.734 & -12.245 \end{bmatrix} \tag{21.11}$$

where the first four 0 entries correspond to the x part of the state variables and the last four (non-zero) entries correspond to the observer based controller $\hat{x}$ as defined in Equation 21.11, and the closed loop 4×8 concatenated matrix matrix A_{obv} as follows

$$A_{obv} = \begin{bmatrix} A & -BK \\ FM & A - BK - FM \end{bmatrix}$$

$$= \begin{bmatrix} 0 & 1 & 0 & 0 & 0 & 0 & 0 & 0 \\ 0 & 0 & -1 & 0 & 2.236 & 3.585 & 39.734 & 12.245 \\ 0 & 0 & 0 & 1 & 0 & 0 & 0 & 0 \\ 0 & 0 & 11 & 0 & -2.236 & -3.585 & -39.734 & -12.245 \\ 9 & 0 & 0 & 0 & -9 & 1 & 0 & 0 \\ 42 & 0 & 0 & 0 & -39.764 & 3.585 & 38.734 & 12.245 \\ -148 & 0 & 0 & 0 & 148 & 0 & 0 & 1 \\ -492 & 0 & 0 & 0 & 489.764 & -3.585 & -28.734 & -12.245 \end{bmatrix}$$

the eigenvalues of which are

$$\lambda_{1,2} = -2 \pm i$$
$$\lambda_3 = -2$$
$$\lambda_4 = -3$$
$$\lambda_{5,6} = -3.3112 \pm 0.0298i$$
$$\lambda_{7,8} = -1.0188 \pm 1.0006i.$$

Notice these are the eigenvalues of $A - BK$ and $A - FM$. Though these eigenvalues are stable, those of the matrix $A - BK - FM$ are not, and we must therefore conclude this is not a strongly stable system.

(f) The output variable trajectory is found using the 4×8 state coefficient matrix A_{obv} and we build the C matrix as follows

$$C_{obv} = \begin{bmatrix} 1 & 0 & 0 & 0 & 0 & 0 & 0 & 0 \end{bmatrix} \tag{21.12}$$

because the output variable z is the first state variable as defined in the state space system given. Keep in mind here that the first four entries of the C_{obv} matrix relate to the four state variables, and the last four entries to the *estimates* of those four state variables. Then

```
out = ss(A_obv, [], C_obv, [])
      initial(out, x0)
```

where x0 is the initial conditions as described in the problem:

$$x_0 = \begin{bmatrix} 0 & 0 & 5 & 0 & 0 & 0 & 0 & 0 \end{bmatrix}^T \tag{21.13}$$

which outputs the output trajectory shown in Figure 21.3

We are also interested in observing the time response of the angle θ to confirm our designed controller actually balances the broomstick (Figure 21.4).

To output this trajectory we choose the C_θ matrix as follows

$$C_\theta = \begin{bmatrix} 0 & 0 & 1 & 0 & 0 & 0 & 0 & 0 \end{bmatrix} \tag{21.14}$$

because θ is the third state variable.

To obtain the control trajectory, we use the

$$C = u = K\hat{x} \tag{21.15}$$

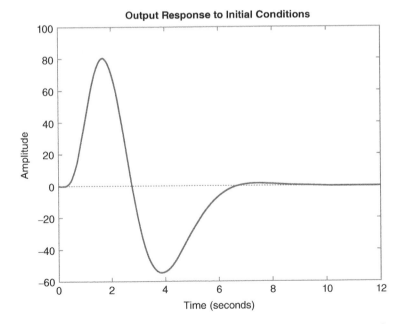

Figure 21.3 Output variable trajectory for the observer based controller example.

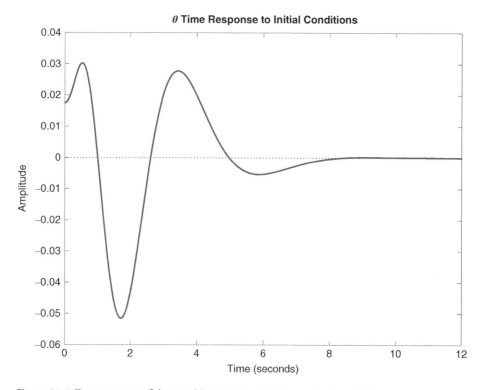

Figure 21.4 Time response of the variable θ given an initial perturbation of 1°.

matrix as specified in Equation 21.11. Then we use the following lines of code

```
cntrl = ss(A_obv, [], u, [])
        initial(cntrl, x0)
```

to output the control trajectory, displayed in Figure 21.5.

While the above mentioned observer based feedback control is a good way to build a practical controller, it is clear that the feedback control law and the resulting closed loop system order becomes large when the number of state variables (and thus the number of state estimate variables needed in the control law) is large. Hence attempts were made to examine if we can build reduced order controllers to still achieve reasonably satisfactory stability and performance characteristics. This leads us to the concept of dynamic compensators of given order as alternative controller structures. We expand on that notion in the next section.

21.3 Other Controller Structures: Dynamic Compensators of Varying Dimensions

In the previous section we presented a method for designing an estimator based feedback controller. In that method, the estimator is a full state estimator with the same dimensions as the original state it is trying to estimate. The closed loop system was seen to be of order $2n$, which could be deemed high in some application problems. So there

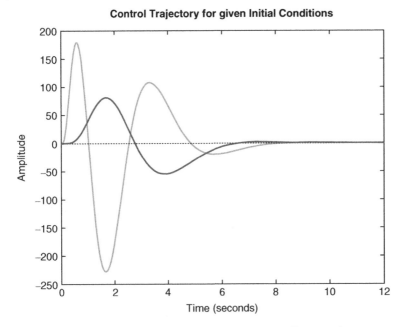

Figure 21.5 Control trajectory for the observer based controller example.

is some incentive to look for reduced order controllers. This leads us to the concept of dynamic compensators of varying dimensions. Let us elaborate on this.

As before let us consider the standard linear state space system given by

$$\dot{x} = Ax + Bu \tag{21.16}$$

$$y = Cx \rightarrow \quad \text{measurements or outputs.} \tag{21.17}$$

Let us revisit the various types of feedback controllers we have attempted to design until now.

1. Full state feedback:

$$u = Kx \rightarrow \text{with closed loop system matrix given by} \tag{21.18}$$

$$A_c = (A + BK). \tag{21.19}$$

2. Output feedback:

$$u = Ky \rightarrow \text{with closed loop system matrix given by} \tag{21.20}$$

$$A_c = (A + BKC). \tag{21.21}$$

3. Observer based feedback:

$$u = K\hat{x} \tag{21.22}$$

$$\dot{\hat{x}} = (A - FC)\hat{x} + Fy + Bu = (A - FC)\hat{x} + FCx + BK\hat{x}$$

$$= (A - BK - FC)\hat{x} + FCx.$$

Closed loop system:

$$\begin{bmatrix} A & BK \\ FC & A + BK - FC \end{bmatrix}\cdot \qquad (21.23)$$

Now, we introduce the dynamic compensator of varying dimensions as follows. Let s be the dimension of the dynamic compensator state ρ. We want the control u to be a function of the measurements y and the dynamic compensator state ρ. Accordingly we write dynamic compensator of order, s as:

$$u = Ky + H\rho \qquad (21.24)$$

$$\dot{\rho} = J\rho + Fy = Fy + J\rho \qquad (21.25)$$

$$\rho \in R^s \quad s \to \text{dynamic compensator state dimension.} \qquad (21.26)$$

We label the vector ρ as the dynamic compensator state because it is generated by another state space model with the measurement vector y as its input. Then we augment the original state with dynamic compensator state to get the following closed loop system,

$$\dot{x} = Ax + BKy + H\rho = (A + BKC)x + H\rho \qquad (21.27)$$

$$\dot{\rho} = J\rho + Fy = FCx + J\rho. \qquad (21.28)$$

Thus we have,

$$\begin{bmatrix} \dot{x} \\ \dot{\rho} \end{bmatrix} = \begin{bmatrix} A + BKC & H \\ FC & J \end{bmatrix}\begin{bmatrix} x \\ \rho \end{bmatrix}$$

$$= \left\{ \begin{bmatrix} A & 0 \\ 0 & 0 \end{bmatrix} + \begin{bmatrix} B & 0 \\ 0 & I \end{bmatrix}\begin{bmatrix} K & H \\ F & J \end{bmatrix}\begin{bmatrix} C & 0 \\ 0 & I \end{bmatrix} \right\}\begin{bmatrix} x \\ \rho \end{bmatrix}$$

$$\dot{\zeta} = (\hat{A} + \hat{B}\hat{K}\hat{C})\zeta. \qquad (21.29)$$

Thus the new gain matrix is,

$$\hat{K} = \begin{bmatrix} K & H \\ F & J \end{bmatrix}. \qquad (21.30)$$

We can then pose an LQR problem with output feedback formulation where the new augmented state is ζ, and the closed loop system matrix is,

$$A_{cl} = (\hat{A} + \hat{B}\hat{K}\hat{C}). \qquad (21.31)$$

This closed loop system matrix is similar in structure to the standard output feedback closed loop control system matrix. Thus we can follow the LQR output feedback problem formulation to design the gain matrix $\hat{K}$ via parameter optimization methods. Note that $\hat{A}, \hat{B}$, and $\hat{C}$ are known and $\hat{K}$ is the only unknown matrix. Also notice that we now have control over the dimensions of the controller. Thus in practice, depending on the physics of the application, we could try out dynamic compensators of various dimensions starting from $s = 1$ and keep increasing them until we arrive at a dynamic compensator gain matrix that achieves satisfactory stability and performance characteristics in the overall closed loop system matrix. The detailed control design of dynamic compensators is clearly out of the scope of an undergraduate controls course but is discussed here to convey the basic concept for completeness purposes.

Finally, we conclude this chapter with yet another interesting conceptual topic in control design, namely spillover instability in linear state space dynamic systems.

21.4 Spillover Instabilities in Linear State Space Dynamic Systems

In many applications, especially in the state space modeling of the dynamics of large flexible space structures [1,3–9], we end up with a very high dimensional linear state space system in the open loop, i.e. the state variable vector $\vec{x}$ is of very high dimensions (possibly running into tens or even hundreds). Since on-board computers (on a large flexible structure space vehicle) have constraints on their memory requirements, it is impractical to design a controller of such high dimensions. Hence, for practicality, the control designer, is forced to consider a much lower order model for control design purposes. Thus we have a two model scenario, a very large order open loop model, which we label as the evaluation model (because that model is supposed to represent the real model of the flexible space structure), and a low order control design model. Let us denote the following as the evaluation model for simulation purposes.

$$\dot{\vec{x}} = A\vec{x} + B\vec{u} \quad A \in R^{N \times N} \quad B \in R^{N \times m} \tag{21.32}$$

where N is very large and m $<<<$ N.

For practical purposes we have to reduce this model for control design purposes. Let us write,

$$\dot{\vec{x}} = \begin{bmatrix} \dot{\vec{x}}_R \\ \dot{\vec{x}}_T \end{bmatrix} = \begin{bmatrix} A_R & A_{RT} \\ A_{TR} & A_T \end{bmatrix} \begin{bmatrix} \vec{x}_R \\ \vec{x}_T \end{bmatrix} + \begin{bmatrix} B_R \\ B_T \end{bmatrix} \vec{u} \tag{21.33}$$

where we simply take the reduced model (which is the top half of the above model) for control design purposes, i.e.,

$$\dot{\vec{x}}_R = A_R \vec{x}_R + B_R \vec{u} \rightarrow \text{control design model.} \tag{21.34}$$

Here,

$$\vec{x}_R \in R^{n_R} \quad n_R \ll N. \tag{21.35}$$

Let us now design a full state feedback controller for this reduced order model. Then $\vec{u} = G_R \vec{x}_R$ and the closed loop system for the reduced order control design model is,

$$\dot{\vec{x}}_R = (A_R + B_R G_R)\vec{x}_R. \tag{21.36}$$

From a previous chapter discussion, we know that under the assumption of complete controllability, we can always design a full state feedback gain G_R such that $(A_R + B_R G_R)$ is a stable matrix. However, when we now drive the real evaluation model with a controller designed for a reduced order model, we get the closed loop system as

$$\begin{bmatrix} \dot{\vec{x}}_R \\ \dot{\vec{x}}_T \end{bmatrix} = \begin{bmatrix} A_R + B_R G_R & A_{RT} \\ A_{TR} + B_R G_R & A_T \end{bmatrix} \begin{bmatrix} \vec{x}_R \\ \vec{x}_T \end{bmatrix}. \tag{21.37}$$

Even if $(A_R + B_R G_R)$ is stable (by design) and A_T is stable, it turns out that the overall closed loop system matrix of the above system could be unstable because of the interaction of the unmodeled dynamics with the controller, reflected by the submatrix, $A_{TR} + B_R G_R$. This phenomenon, i.e. instability due to the unmodeled dynamics interacting with the controller, is called spillover instability.

This is especially critical in low damped flexible space structure control design.

This simply means that the controller is exciting the higher order modes (flexible) and pumping energies into the higher order modes which are neglected in the control design. For this reason selecting the right type of reduced order control design model is very critical (to avoid spillover instability, see Figure 21.6). Suppose now that the reduced order controller is an observer based controller (we omit vector notation for now, in this discussion), i.e.

$$u = G_R \hat{x}_R \tag{21.38}$$

where

$$\dot{\hat{x}}_R = A_R \hat{x}_R + B_R u + F_R(z - M_R \hat{x}_R) \tag{21.39}$$

and

$$z = Mx = \begin{bmatrix} M_R & M_T \end{bmatrix} \begin{bmatrix} x_R \\ x_T \end{bmatrix} \tag{21.40}$$

is the actual measurement but the estimator, out of necessity, feeds only the measurement residual $(z - M_R \hat{x}_R)$. Then the resulting closed loop system is,

$$\begin{bmatrix} \dot{x}_R \\ \dot{\hat{x}}_R \\ \dot{x}_T \end{bmatrix} = \begin{bmatrix} A_R & B_R G_R & A_{RT} \\ F_R M_R & A_R + B_R G_R - F_R M_R & F_R M_T \\ A_{TR} & B_T G_R & A_T \end{bmatrix} \begin{bmatrix} x_R \\ \hat{x}_R \\ x_T \end{bmatrix}. \tag{21.41}$$

Figure 21.6 Spillover instability.

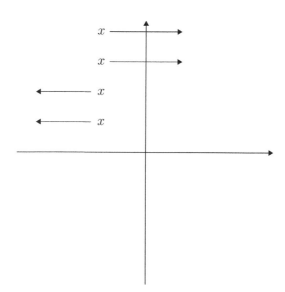

So again the above closed loop system can become unstable even if A_{RT} and A_{TR} are zero and A_T is stable because of the terms $B_T G_R$ and $F_R M_T$ in the off-diagonal entries. This possible instability is caused by the terms $B_T G_R$ and/or $F_R M_T$. We designate:

the effect due to $B_T G_R \rightarrow$ control spillover
the effect due to $F_R M_T \rightarrow$ observation spillover.

So in a realistic control design, there is a chance of control and observation spillover instabilities. For more on this exciting topic, literature in the area of dynamic and control of flexible space structures, such as [3, 4] is recommended as further reading. Clearly, the details of this research area are beyond the scope of an undergraduate book like this, but the concept of spillover instabilities is clearly very easy to understand, important, and interesting.

21.5 Chapter Summary

This chapter has covered more important and useful material on a practical method of designing controllers for linear state space dynamic systems, via the concept of a observer based feedback controller that combines the features of measurement feedback as well as the full state feedback in an interesting way, making use of the concept of a state estimator (or state observer). In addition, we introduced an additional controller structure, namely dynamic compensators of varying dimensions. Finally, the interesting and important concept of spillover instabilities in a two model framework is explained.

21.6 Exercises

Exercise 21.1. Consider again the spacecraft attitude control problem where the model is given by

$$\begin{bmatrix} \dot{\omega}_x \\ \dot{\omega}_y \end{bmatrix} = \begin{bmatrix} 0 & -1 \\ 1 & 0 \end{bmatrix} \begin{bmatrix} \omega_x \\ \omega_y \end{bmatrix} + \begin{bmatrix} 1 \\ 0 \end{bmatrix} T_x \tag{21.42}$$

and suppose only ω_x is measured. Then the measurement state space equation $z(t)$ is

$$z = \begin{bmatrix} 1 & 0 \end{bmatrix} \begin{bmatrix} \omega_x \\ \omega_y \end{bmatrix}. \tag{21.43}$$

(a) Determine the observer gain matrix such that the eigenvalues of $(A - FM)$ are located a little left of the open loop system.
(b) Then form an optimal closed loop system matrix and investigate the stability of the compensator matrix. Is the system strongly stable?

Bibliography

1 M. Balas. Feedback control of flexible systems. *IEEE Transactions on Automatic Control*, 23(4).

2 T.E Fortmann and K.L Hitz. *An Introduction to Linear Control Systems.* Mercel-Dekker, 1977.

3 Juang J. N, K. B. Lim, and J. L. Junkins. Robust eigenstructure assignment for flexible structure. *J. Guid Cont. Dyn.*, 12:381–387, 1989.

4 L. Meirovitch and I. Tuzcu. Unified theory for the dynamics and control of maneuvering flexible aircraft. *AIAA Journal*, 42(4):714–727, 2004.

5 R.E. Skelton. *Dynamic Systems Control: Linear Systems Analysis and Synthesis.* Wiley, New York, 1 edition, 1988.

6 M.R. Waszak and D.K. Schmidt. Flight dynamics of aeroelastic vehicles. *Journal of Aircraft*, 25(6):563–571, 1988.

7 B. Wie, J.A. Lehner, and C.T. Plescia. Roll/yaw control of a flexible spacecraft using skewed bias momentum wheels. *Journal of Guidance, Control, and Dynamics*, 8(4):447–451, 1985.

8 B. Wie and C.T. Plescia. Attitude stabilization of flexible spacecraft during station-keeping maneuvers. *Journal of Guidance, Control, and Dynamics*, 7(4):430–436, 1984.

9 R. K. Yedavalli and R. E. Skelton. Determination of critical parameters in large flexible space structures with uncertain model data. *ASME Journal of Dynamic Systems, Measurement and Control*, pages 238–244, December 1983.

22

State Space Control Design: Applications to Aircraft Control

22.1 Chapter Highlights

In this chapter, we apply the state space control design methods we learnt in the previous chapters to problems in the control of atmospheric (aero) vehicles.

22.2 LQR Controller Design for Aircraft Control Application

Example 22.1 The aircraft BRAVO at a particular flight condition has the following equations of longitudinal motion:

$$\begin{bmatrix} \dot{u} \\ \dot{\alpha} \\ \dot{q} \\ \dot{\theta} \end{bmatrix} = \begin{bmatrix} -0.007 & 0.012 & 0 & -9.81 \\ -0.128 & -0.54 & 1 & 0 \\ 0.064 & 0.96 & -0.99 & 0 \\ 0 & 0 & 1 & 0 \end{bmatrix} \begin{bmatrix} u \\ \alpha \\ q \\ \theta \end{bmatrix} + \begin{bmatrix} 0 \\ -0.036 \\ -12.61 \\ 0 \end{bmatrix} u.$$

The weighting matrices are chosen as follows:

$$Q = \begin{bmatrix} 1 & 0 & 0 & 0 \\ 0 & 10 & 0 & 0 \\ 0 & 0 & 50 & 0 \\ 0 & 0 & 0 & 1 \end{bmatrix}$$

$R = 5$.

This choice means q_{22} and q_{23} penalize any persistent transient motion motion of the angle of attack and pitch rate, and the weighting factor of 5 on the control ensures only moderate deflection of the elevator results.

(a) Find the optimal gain K that minimizes the quadratic cost function.
(b) Plot the state trajectory of all four state variables to the following initial conditions:

$$x_o = \begin{bmatrix} 0 & 1 & 0 & 0 \end{bmatrix}^T. \tag{22.1}$$

Flight Dynamics and Control of Aero and Space Vehicles, First Edition. Rama K. Yedavalli.
© 2020 John Wiley & Sons Ltd. Published 2020 by John Wiley & Sons Ltd.

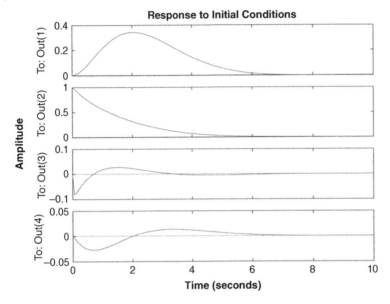

Figure 22.1 Longitudinal state variable trajectories (i.e. the solution to the system of four equations in four unknowns).

Solution
This problem can be solved using MATLAB.

(a) After defining the matrices A, B, Q, and R we simply call upon the in-built command

```
K=lqr(A,B,Q,R)
```

which yields the optimal gain

$$K = \begin{bmatrix} 0.475 & -0.3869 & -3.2253 & -5.35 \end{bmatrix}. \tag{22.2}$$

(b) Then, we define state space system and follow its trajectory for the first 25 s after perturbation

```
sys=ss(A−B*K,[],eye(4),[])
        initial(sys,x0,25)
```

Note here C is simply a 4×4 identity matrix. The output of this command (the state trajectories) is pictured in Figure 22.1.

22.3 Pole Placement Design for Aircraft Control Application

The state space description of the lateral/directional mode of an aircraft is

$$\dot{x} = \begin{bmatrix} -0.2543 & 0.1830 & 0 & -1 \\ 0 & 0 & 1 & 0 \\ -15.982 & 0 & -8.402 & 2.193 \\ 4.495 & 0 & -0.3498 & -0.7605 \end{bmatrix} x + \begin{bmatrix} 0 & 0.0708 \\ 0 & 0 \\ 28.984 & 2.548 \\ -0.2218 & -4.597 \end{bmatrix} u \tag{22.3}$$

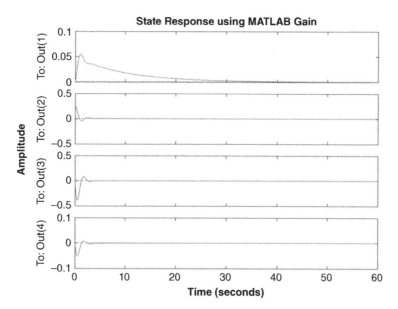

Figure 22.2 State variable response to MATLAB gain.

and we would like the closed loop system to have the following eigenvalues:

$$\begin{bmatrix} \lambda_{d1} & \cdots & \lambda_{d4} \end{bmatrix} = \begin{bmatrix} -8.5 & -0.1 & -1.25 + 2.5i & -1.25 - 2.5i \end{bmatrix}. \tag{22.4}$$

Given the state space description and the desired eigenvalues, we can use MATLAB's in-built `place` command, which outputs the gain:

$$K_{\text{MATLAB}} = \begin{bmatrix} -0.5033 & 0.2791 & -0.2254 & 0.2017 \\ -1.0233 & -0.1950 & 0.2667 & -1.6649 \end{bmatrix}. \tag{22.5}$$

The state response to the MATLAB gain is depicted in Figure 22.2. Alternatively, we can use MATLAB's symbolic math toolbox to implement Brogan's algorithm and find the other possible gains K that allow us to achieve the desired closed-loop poles. Recall Brogan's algorithm for a general $n \times n$ A matrix and $m \times n$ B matrix,

$$\phi = (\lambda \mathbf{I}_{n \times n} - \mathbf{A})^{-1} \tag{22.6}$$

$$\psi = \phi \mathbf{B} \tag{22.7}$$

$$\overline{\psi} = [\psi_1(\lambda_1) \quad \cdots \quad \psi_m(\lambda_n)] \tag{22.8}$$

$$\mathbf{K} = -\mathbf{E}\,\overline{\psi}^{-1} \tag{22.9}$$

$$\mathbf{A}_{\text{cl}} = \mathbf{A} - \mathbf{B}\mathbf{K}. \tag{22.10}$$

In MATLAB, we start by defining the matrices A and B and the symbol x, which we will use for ease of coding in place of λ in the above equations, as follows:

```
A = [-0.2543 0.183 0 -1;0 0 1 0;-15.982 0 -8.402 2.193;
            ... 4.495 0 -0.3498 -0.7605];
B = [0 0.0708;0 0;28.984 2.548;-0.2218 -4.597];
            syms x
```

Then, from Equation (22.6),

```
phi = (x*eye(length(A))-A)  ^-1;
```

and from Equation (22.7),

```
psi = phi*B;
```

The above part of the code will remain the same for all gains K that we find. It is the different possibilities for $\bar{\psi}$ and therefore $\mathbf{E}$ that result in multiple gains K that still achieve the desired closed loop output. For instance, if we build $\bar{\psi}$ using the columns $[\psi_1(\lambda_1) \quad \psi_1(\lambda_2) \quad \psi_2(\lambda_3) \quad \psi_2(\lambda_4)]$, i.e.

```
psibar = horzcat(subs(psi(:,1),x,despol(1))
subs(psi(:,1),x,... despol(2)),subs(psi(:,2),
x,despol(3)),subs(psi(:,2),x,... despol(4)));
```

then we have to define $\mathbf{E}$ as the 2×4 matrix

$$E = \begin{bmatrix} 1 & 1 & 0 & 0 \\ 0 & 0 & 1 & 1 \end{bmatrix} \tag{22.11}$$

and from Equation (22.9)

```
K = -E*psibar^-1;
```

which turns out to be

$$K = \begin{bmatrix} -0.0039 & 0.0301 & 0.0055 & 0.0081 \\ 0.4221 & 0.0493 & 0.0154 & -0.3138 \end{bmatrix}. \tag{22.12}$$

Using the gain K calculated above and substituting into Equation (22.10),

```
Acl = A-B*K;
```

which is

$$A_{cl} = \begin{bmatrix} -0.2842 & 0.1795 & -0.0011 & -0.9778 \\ 0 & 0 & 1 & 0 \\ -16.9455 & -0.9985 & -8.6020 & 2.7589 \\ 6.4361 & 0.22 & -0.28 & -2.2138 \end{bmatrix}. \tag{22.13}$$

The state response using this gain is given in Figure 22.3. This is not the only possible gain K we can use. If instead we build $\bar{\psi}$ using the columns $[\psi_1(\lambda_1) \quad \psi_2(\lambda_2) \quad \psi_2(\lambda_3) \quad \psi_2(\lambda_4)]$, i.e.

```
psibar =horzcat(subs(psi(:,1),x,despol(1)),
subs(psi(:,2),x,... despol(2)),subs(psi(:,2),x,
despol(3)),subs(psi(:,2),x,... despol(4)));
```

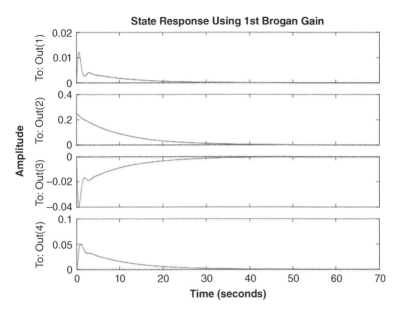

Figure 22.3 State response to first possible gain, as given by Brogan's algorithm.

then we define **E** as

$$E = \begin{bmatrix} 1 & 0 & 0 & 0 \\ 0 & 1 & 1 & 1 \end{bmatrix} \tag{22.14}$$

and this time, the gain is found to be

$$K = \begin{bmatrix} 0.005 & -0.0002 & 0.0023 & 0.0003 \\ 0.45 & -0.0459 & 0.0051 & -0.3402 \end{bmatrix} \tag{22.15}$$

with its corresponding closed loop matrix

$$A_{cl} = \begin{bmatrix} -0.2862 & 0.1863 & -.0004 & -0.9759 \\ 0 & 0 & 1 & 0 \\ -17.274 & 0.1214 & -8.4805 & 3.0694 \\ 6.5626 & -0.2111 & -0.3268 & -2.3334 \end{bmatrix}. \tag{22.16}$$

The state response for this gain is given in Figure 22.4. These are still not the only possible gains K we can use. If instead we build $\overline{\psi}$ using the columns $[\psi_2(\lambda_1) \quad \psi_1(\lambda_2) \quad \psi_2(\lambda_3) \quad \psi_2(\lambda_4)]$, i.e.

```
psibar = horzcat(subs(psi(:,2),x,despol(1)),
subs(psi(:,1),x,... despol(2)),subs(psi(:,2),x,
despol(3)),subs(psi(:,2),x,... despol(4)));
```

then **E** is defined in MATLAB as

$$E = \begin{bmatrix} 0 & 1 & 0 & 0 \\ 1 & 0 & 1 & 1 \end{bmatrix} \tag{22.17}$$

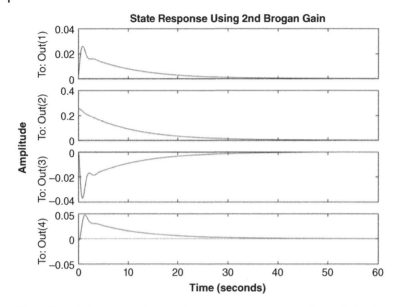

Figure 22.4 State response to second possible gain, as given by Brogan's algorithm.

and the gain in this case is found to be

$$K = \begin{bmatrix} -0.0086 & 0.0299 & 0.0033 & 0.0083 \\ 0.4649 & 0.0508 & 0.0354 & -0.3159 \end{bmatrix} \tag{22.18}$$

with its corresponding closed loop system

$$A_{cl} = \begin{bmatrix} -0.2872 & 0.1794 & -0.0025 & -0.9776 \\ 0 & 0 & 1 & 0 \\ -16.9181 & -0.9975 & -8.5893 & 2.7575 \\ 6.6342 & 0.2269 & -0.1877 & -2.2235 \end{bmatrix}. \tag{22.19}$$

The state response to this gain is given in Figure 22.5. Still other possible gains K such that the desired closed loop poles are achievable. If we build $\overline{\psi}$ using the columns $[\psi_2(\lambda_1) \quad \psi_2(\lambda_2) \quad \psi_1(\lambda_3) \quad \psi_1(\lambda_4)]$, i.e.

```
psibar = horzcat(subs(psi(:,2),x,despol(1)),
subs(psi(:,2),x,... despol(2)),subs(psi(:,1),x,
despol(3)),subs(psi(:,1),x,... despol(4)));
```

then we define **E** as

$$E = \begin{bmatrix} 0 & 0 & 1 & 1 \\ 1 & 1 & 0 & 0 \end{bmatrix}. \tag{22.20}$$

This yields the gain

$$K = \begin{bmatrix} 0.1027 & 0.1610 & 0.0445 & -0.9487 \\ -0.009 & -0.0434 & 0.0218 & -0.1173 \end{bmatrix} \tag{22.21}$$

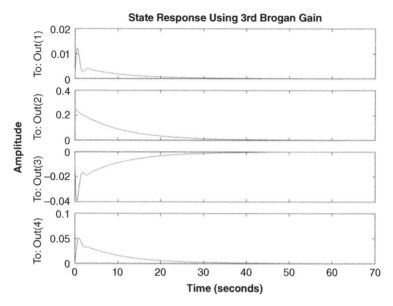

Figure 22.5 State response to third possible gain, as given by Brogan's algorithm.

and its corresponding closed loop system

$$
A_{cl} = \begin{bmatrix} -0.2537 & 0.1861 & -0.0015 & -0.9917 \\ 0 & 0 & 1 & 0 \\ -18.9366 & -4.5556 & -9.7481 & 29.9881 \\ 4.431 & -0.2353 & -0.2596 & -1.0982 \end{bmatrix}.
$$ (22.22)

The state response to this gain is given in Figure 22.6. Yet other possible gains K are available. For example, we can build $\overline{\psi}$ using the columns $[\psi_2(\lambda_1) \quad \psi_1(\lambda_2) \quad \psi_1(\lambda_3) \quad \psi_1(\lambda_4)]$, i.e.

```
psibar = horzcat(subs(psi(:,2),x,despol(1)),
    subs(psi(:,1),x,... despol(2)),subs(psi(:,1),x,
    despol(3)),subs(psi(:,1),x,... despol(4)));
```

for which **E** is defined

$$
E = \begin{bmatrix} 0 & 1 & 1 & 1 \\ 1 & 0 & 0 & 0 \end{bmatrix}
$$ (22.23)

and the gain K is

$$
K = \begin{bmatrix} 0.1897 & 0.1992 & 0.048 & -0.9308 \\ 0.1266 & 0.0162 & 0.0272 & -0.0895 \end{bmatrix}
$$ (22.24)

with its corresponding closed loop system

$$
A_{cl} = \begin{bmatrix} -0.2633 & 0.1819 & -0.0019 & -0.9937 \\ 0 & 0 & 1 & 0 \\ -21.8021 & -5.8158 & -9.8625 & 29.4001 \\ 5.035 & 0.0303 & -0.2355 & -0.9743 \end{bmatrix}
$$ (22.25)

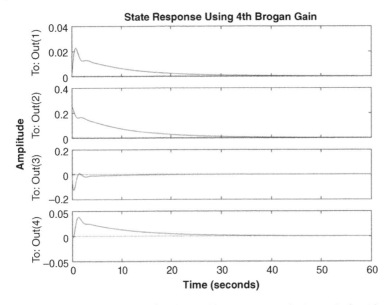

Figure 22.6 State response to fourth possible gain, as given by Brogan's algorithm.

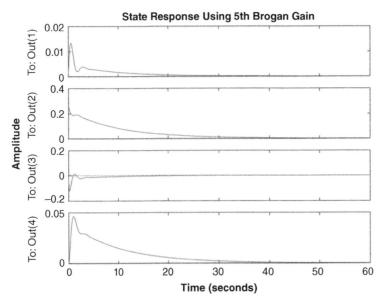

Figure 22.7 State response to fifth possible gain, as given by Brogan's algorithm.

and the state response to this gain is given in Figure 22.7. You can easily verify the eigenvalues of all of the above closed loop matrices are indeed the desired eigenvalues.

22.4 Chapter Summary

In this chapter, we learnt how to apply the LQR and pole placement control design methods to aircraft control problems.

22.5 Exercises

Exercise 22.1. In the example problem of using the LQR method for designing the optimal control gain K, we used a fixed state weighting matrix Q and a fixed control weighting matrix (which in this single input case was a scalar, which was taken as 5). In this exercise, treat the control weighting scalar as a design variable ρ (which is then a positive scalar) keeping the state weighting matrix fixed as it was in the worked out example. With ρ as a design variable, now obtain the trade-off curve discussed in this chapter for this problem and select the best ρ that achieves the most reasonable trade-off (which will be the ρ on the curved part of the trade-off curve) and obtain the optimal control gain K for that finalized optimal ρ [1]. Then plot the state trajectories and the control trajectory as well. Analyze the resulting closed loop system in terms of its closed loop eigenvalue locations.

Exercise 22.2. In the example problem of using the pole placement method for designing the control gain K, for desired pole locations, we had six different closed loop system matrices, all of them having the same desired closed loop locations. These six different gains include MATALB gain and five other gains that we labeled as Brogan gains 1, 2, 3, 4, and 5 [2]. Analyze the closed loop state trajectories very thoroughly for each of these gains and explain why these state trajectories look markedly different from each other. In this exercise, also plot the control variable trajectories and explain the behavior of these control trajectories. After observing the control trajectories, which are nothing but the control surface deflections of the aileron angle and the rudder angle, do you recommend a particular control gain among these six gains as more desirable gain than any other with the logic explained for your selection? Once you make your recommendation, analyze the eigenvector situation for that particular closed loop system with the gain you selected.

Bibliography

1 H. Kwakernaak and R. Sivan. *Linear Optimal Control Systems.* Wiley Interscience, 1972.

2 W.L. Brogan. *Modern Control Theory.* Prentice Hall, 1974.

23

State Space Control Design: Applications to Spacecraft Control

23.1 Chapter Highlights

In this chapter, we apply LQR control design method to the satellite formation control problem, which highlights the importance of all the concepts we have learned in the theory. This application problem was the result of a sponsored project in satellite formation control, with interesting conclusions of practical interest.

23.2 Control Design for Multiple Satellite Formation Flying

Here we consider the problem of several satellites residing in a given orbit to maintain their formation pattern relative to each other (see Figure 23.1). Thus the dynamics of relative motion in orbital mechanics are considered as a starting point. Naturally, because of various external perturbation torques experienced by the member satellites, their relative positions within the formation get disturbed. Thus the control objective is to control each satellite's position within the formation by using control torques so that the desired formation pattern is maintained. To keep the problem formulation simple and easily understandable, we simply assume one leader satellite and one follower satellite, and consider the relative position equations corresponding to the two satellites. These relative motion equations in orbital mechanics area are known as Clohessy and Wiltshire relative motion equations.

Definition of the Relative Coordinate Frame

The motion of a single member satellite relative to the leader (or formation origin) is described by a nonlinear system of differential equations in x, y, z, r, and ω, where x, y and z are the relative frame coordinates, r is the time-dependent orbital radius, and ω is the time dependent angular rate of the leader. If the inertial orbit of the leader is circular, then the orbital radius and angular rate become time invariant. Clohessy and Wiltshire noted that for this special case, the dynamical equations for the follower satellite have only very weak nonlinear couplings so long as the distance between the follower and the leader is much less than the orbital radius [4]. Therefore the nonlinear dynamics of a formation, which is by definition limited in spatial extent, is often well represented by

Flight Dynamics and Control of Aero and Space Vehicles, First Edition. Rama K. Yedavalli.
© 2020 John Wiley & Sons Ltd. Published 2020 by John Wiley & Sons Ltd.

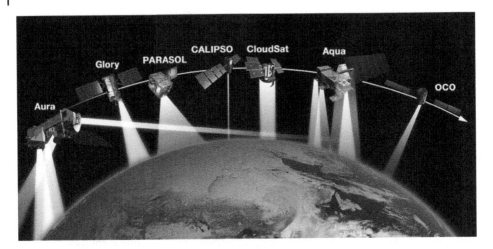

Figure 23.1 An example of satellite formation. Courtesy of NASA.

a linearization about the formation origin. The linearized system of equations is

$$\ddot{x} = 2\omega\dot{y} + F_x \tag{23.1}$$

$$\ddot{y} = -2\omega\dot{x} + 3\omega^2 x + F_y \tag{23.2}$$

$$\ddot{z} = -\omega^2 z + F_z \tag{23.3}$$

where $+x$ is anti-parallel to the leader's velocity, $+y$ is parallel to the outward directed orbital radius vector, $+z$ is parallel to the angular momentum direction, and F_x, F_y, and F_z are, respectively, the x, y, and z accelerations due to external forces. It has been noted that the motion parallel to the orbital plane, $x(t)$ and $y(t)$, is decoupled from the out-of-plane motion $z(t)$ and that the out-of-plane motion is a bounded input, bounded output (BIBO) stable, simple harmonic oscillator. Based on these observations, we will treat only the in-plane dynamics, i.e. Equations 23.1 and 23.2 in this section.

State Space Representation

The state space representation of the linear range dynamics of the relative motion of two satellites in a circular orbit formation flying with a constant angular velocity ω is given by the state space equation

$$\begin{bmatrix} \dot{x} \\ \ddot{x} \\ \dot{y} \\ \ddot{y} \end{bmatrix} = \begin{bmatrix} 0 & 0 & 1 & 0 \\ 0 & 0 & 0 & 1 \\ 0 & 0 & 0 & 2\omega \\ 0 & 3\omega^2 & -2\omega & 0 \end{bmatrix} \begin{bmatrix} x \\ \dot{x} \\ y \\ \dot{y} \end{bmatrix} + \begin{bmatrix} 0 & 0 \\ 0 & 0 \\ 1 & 0 \\ 0 & 1 \end{bmatrix} \begin{bmatrix} T_x \\ T_y \end{bmatrix}. \tag{23.4}$$

Notice we have control authority in the x and y directions. In other words, we can physically apply control torques on this system in the x and y directions. To separate these control directions, it is useful to define the columns of the input coefficient matrix above as

$$B_x = \begin{bmatrix} 0 & 0 & 1 & 0 \end{bmatrix}^T \tag{23.5}$$

$$B_y = \begin{bmatrix} 0 & 0 & 0 & 1 \end{bmatrix}^T. \tag{23.6}$$

The pair (A, B) is completely controllable for all possible values of ω and some authors have used it to design LQR controllers for this problem [2]. The pair (A, B_x) is also completely controllable but the pair (A, B_y) is uncontrollable and unstabilizable.

Design Concept: Elimination of the Radial Control Axis

The controllability characteristics for the linearized satellite formation suggest that the radial y axis of control may be eliminated altogether. One control design has used a bang–bang algorithm to maintain the formation using differential drag as its only means of actuation; this design necessarily precluded the possibility of using control forces parallel to the radial axis.

There are several benefits to be obtained from a control design that employs thrust in only one of the two orthogonal directions in the orbital plane. Such an algorithm could be used to simplify the propulsive apparatus or to reduce the need for redundancy by acting as a failure mode. Also, any target orbit may be achieved from a coplanar starting orbit by applying thrusts only in the orbit tangent, and sometimes this is the most fuel efficient method of changing orbit, as in Hohmann transfers [1]. So by limiting control thrust inputs to the x direction the controller may better use orbital dynamics to advantage.

In designing a controller that excludes radial thrust, we non-dimensionalize the state and time variables to reduce errors in calculation and to make the result more broadly applicable. Substituting typical values of $\omega < 10^{-3}$ into the above state matrix yields a state matrix too ill-conditioned to use in readily available linear algebra tools. So to improve the accuracy of calculations, the angular rate was normalized to unity [2]. Non-dimensionalizing the orbital radius makes these results applicable to all circular, Keplerian orbits. Note that after these conversions are performed, the orbital velocity $v = \omega r$ is also unity. Thus all lengths and times throughout the simulations were non-dimensionalized in the same manner.

Then, with $Q = I_{4\times4}$ and $R = 100$ we obtain a reduced input LQR gain of

$$K_{red} = \begin{bmatrix} -0.100 & -2.597 & 1.347 & -0.501 \end{bmatrix} \tag{23.7}$$

and the eigenvalues of the closed loop system

$$A_{cl} = A - B_x K_{lqr} \tag{23.8}$$

are

$$\lambda_{1,2} = -0.26 \pm 1.08i \tag{23.9}$$

$$\lambda_{3,4} = -0.42 + 0.26i. \tag{23.10}$$

Simulations

The simulation of single, large scale maneuvers is a suitable first step in testing the efficacy of this design, since such maneuvers might be used in establishing and modifying formations. Most of the results presented here correspond to a maneuver in which the follower satellite was initialized at a point on the x axis and then regulated to the origin. The non-dimensionalized initial state of the satellite is

$$\bar{x}_o = \begin{bmatrix} -1.5 \times 10^{-5} & 0 & 0 & 0 \end{bmatrix}^T. \tag{23.11}$$

Then the LQR derived gain matrix K_{red} was employed in a feedback loop to calculate a thrust trajectory that would move the follower satellite to the origin. This maneuver was simulated on several non-dimensionalized initial positions on the x axis. As would be expected from a linear system, the state and control trajectories resulting from these different initial separations were similar in the geometric sense. This observations allows the results to be condensed for all the initial positions simulated:

$$5 \times 10^{-6} \leq \frac{x_o}{r} \leq 5 \times 10^{-4}. \tag{23.12}$$

The simulations were performed entirely within the relative coordinate frame using a proprietary fourth/fifth order Runge–Kutta ODE solver to propagate both the full nonlinear equations and the linearized equations modeling the continuous time relative dynamics of the lead–follow pair. When compared, the results using the nonlinear pair were virtually indistinguishable from those using the linearized version for the formation sizes tested. Therefore the results are only applicable to formations in the linear regime defined by

$$x, y, z \ll r. \tag{23.13}$$

Results

We will now consider some measures not directly addressed by the cost function J. Compare the proposed reduced input controller K_{red} to a controller with control authority along both the x and y axes. For this full input system, the weighting matrix is

$$R = \rho \begin{bmatrix} 1 & 0 \\ 0 & 1 \end{bmatrix} \tag{23.14}$$

where ρ is a scalar and a measure of control effort. It is a good exercise for the reader to acquire the corresponding gain and closed-loop system. Here, we skip these steps and instead present the closed-loop trajectories for both the reduced and full input systems in Figure 23.2. Clearly, the reduced input trajectory is very similar to the full input trajectory. Overall, the performance of the two is very similar. In essence, the reduced input controller uses 20% less fuel, though it requires 10% more time to maneuver the same distance [3].

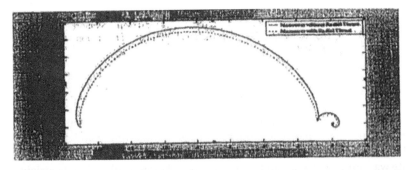

Figure 23.2 Trajectories of two follower spacecraft, one excluding radial thrust (solid), and the other including radial thrust (dashed). The spacecraft were simulated maneuvering from a point on the relative frame x axis to the origin.

Interpretation of Results

So which controller should we use? The answer depends on the mission at hand. In maintaining and changing orbits, fuel efficiency is often best obtained by limiting thrust to either parallel to the orbital velocity (x axis) or perpendicular to the orbital plane (z axis). However, there certainly are cases where fuel efficiency is not the highest priority, as in rendezvous encounters. Therefore the seeming inefficiency of thrust in the radial direction as compared to the along orbit direction should not be construed as an overwhelming reason to eliminate the radial axis of control. It does imply that, in order to get the best performance from the thrust applied in the radial direction, radial thrust should be applied judiciously.

Mathematically, we may achieve a nice balance between the two controllers discussed above by using a full input B matrix, but weighting the R matrix more heavily towards T_y, as such

$$R = \rho \begin{bmatrix} 1 & 0 \\ 0 & 4 \end{bmatrix}. \tag{23.15}$$

Here we chose $R_{22} = 4R_{11}$ after some experimentation because it seems to give the tightest trajectories and best fuel efficiency. While assessing fuel efficiency is beyond the scope of this book, it is a good exercise for the reader to experiment with the weighting of the R matrix and its effect on the closed-loop trajectory in MATLAB.

23.2.1 Pole Placement Design for the above problem

We could also design a full state feedback controller of the form

$$u(t) = K_p x(t) \tag{23.16}$$

where the gain matrix K_p can be obtained using the pole placement algorithm. Using MATLAB's in-built command `place` with the following closed-loop eigenvalues:

$$\lambda_{1,2} = -1.2265 \pm 2.9053i \tag{23.17}$$
$$\lambda_{3,4} = -0.02736 \pm 0.04638i \tag{23.18}$$

we obtain the corresponding gain matrix K_p

$$K_p = \begin{bmatrix} -0.29 & 0 & -0.5473 & -2 \\ 0 & -12.9445 & 2 & -2.4527 \end{bmatrix}. \tag{23.19}$$

In the exercise given below, the reader is asked to compare the pole placement controller behavior with the LQR controller where the LQR design weighting matrices are adjusted such that the pole placement design control effort (J_u) is as close to the LQR controller so that the state regulation cost (J_x) can then be compared for both of these designs.

23.3 Chapter Summary

In this chapter, we presented the application of LQR and pole placement control design methods to spacecraft control problems. We illustrated the use of the LQR method to the innovative application problem of satellite formation control, which uses the relative motion equations from orbital mechanics field known as Clohessy and Wiltshire equations. The conclusions drawn for this application are quite useful and insightful. Then the pole placement technique is also illustrated for the same problem.

23.4 Exercises

Exercise 23.1. Consider the pole placement design controller discussed above. Now compute the individual performance costs J_x and J_u for this pole placement controller. Then consider the LQR design method for the same system. Vary the weightings (possibly the control weighting, to keep the iterations as simple as possible) and design an LQR controller such that its control effort is as close to the control effort value you got for the pole placement controller. That is $J_{u\text{pole}}$ is almost equal to $J_{u\text{lqr}}$. Then compare the corresponding state regulation cost J_x for both of these designs.

Bibliography

1 R.R. Bate, D.D. Mueller, and J.E. White. *Fundamentals of astrodynamics*. Dover, 1971.
2 R.H. Vassar and R.B. Sherwood. Formationkeeping for a pair of satellites in a circular orbit. *AIAA Journal of Guidance,Control and Dynamics*, 8:235, 1985.
3 S. Starin, R.K. Yedavalli, and A. Sparks. Design of a lqr controller of reduced inputs for multiple spacecraft formation flying. *Proceedings of the American Control Conference*, page 1327, June 2001.
4 W.H. Clohessy and R.S. Wiltshire. Terminal guidance system for satellite rendezvous. *Journal of Aerospace Sciences*, 27:653, 1960.

Part IV

Other Related Flight Vehicles

Roadmap to Part IV

"If you want the Rainbow, you have to put up with the Rain"

– Anonymous

In this part IV of the book, our objective is to present some fundamental material on flight vehicles other than the standard fixed wing aircraft and the satellites as representing an aero and a space flight vehicle respectively. It is important for the reader to realize that there are few related flight vehicles whose dynamics and control problems are possibly quite different from the two traditional flight vehicles we have discussed so far in the previous parts of the book. Few of those non-traditional flight vehicles are (i) rotor-craft vehicles such as helicopters and quad-copters (ii) missiles, and (iii)

hypersonic vehicles, among many other such vehicles. To keep the book content concise, we focus our attention on only the vehicles mentioned above. We treat this subject matter again from a conceptual viewpoint, emphasizing the possible similarities and differences between dynamics and control specific features of these vehicles and the traditional vehicles (fixed wing aircraft and satellites).

24

Tutorial on Aircraft Flight Control by Boeing

24.1 Tutorial Highlights

This tutorial on aircraft flight control systems is based on Boeing's Model 777 flight control system [1]. The Model 777 is Boeing's first commercial airplane, and employs complete electronic control of the primary flight control system. However, the system design is such that the airplane looks and feels similar to other Boeing jet transports, and at the same time, employs the latest technology to ease the pilot's workload and the long term maintenance of the system. Before the first flight of the Model 777 on 12 June 1994, the primary flight control system had undergone many hours of testing over an eight month period. Individual components were tested, and continue to be tested, in stand-alone laboratories. The system as a whole was tested in the flight control test rig (FCTR), where all electrical and mechanical components were installed in a simulated airplane environment and operated in concert. The primary flight control system was also tested in conjunction with other airplane systems in the new systems integration laboratory (SIL). As a result of these tests, there was a high level of confidence before the system was installed in the first airplane. The initial flight testing of the airplane has shown few unexpected characteristics.

24.2 System Overview

Conventional primary flight controls systems employ hydraulic actuators and control valves controlled by cables that are driven by the pilot controls. These cables run the length of the airframe from the cockpit area to the surfaces to be controlled. This type of system, while providing desirable airplane handling characteristics, does have some very distinct drawbacks. The cable controlled system comes with a weight penalty due to the long cable runs, pulleys, brackets, and supports needed. The system requires periodic maintenance, such as lubrication and adjustments due to cable stretch. In addition, systems such as the yaw damper, which provide enhanced control of surfaces, require dedicated actuators, wiring, and controllers. This adds to the overall system weight and increases the number of components in the system.

In the Model 777, the cable control of the primary flight control surfaces has been removed. Rather, the actuators are controlled electronically. At the heart of the Model 777 primary flight control system are electronic computers. These computers convert

Flight Dynamics and Control of Aero and Space Vehicles, First Edition. Rama K. Yedavalli.
© 2020 John Wiley & Sons Ltd. Published 2020 by John Wiley & Sons Ltd.

electrical signal sent from transducers attached to the pilot controls into commands that are sent to the actuators. Because of these changes to the system, the following design features have been made possible:

1. Full time surface control employs advanced control laws. The aerodynamic surfaces are sized to afford the required airplane response during critical flight conditions. The reaction time of the control laws is much faster than that of an alert pilot. Therefore the size of the flight control surfaces could be made smaller than those needed for a conventionally controlled airplane. This results in a reduction in the overall weight of the system.
2. Retention of the desirable flight characteristics of a conventional system and the removal of the undesirable ones. This is discussed later in Section 24.4.
3. Integration of functions such as the yaw damper into the basic surface control. This allows that separate components normally used for these functions to be removed.
4. Improved system reliability and maintainability.

24.2.1 Design Philosophy

The design philosophy of the Model 777 primary flight control system maintains system operation that is consistent with a pilot's past training and experience. What is meant by this is that however different the actual system architecture is from previous Boeing airplanes, the presentation to the pilot is that of a conventional mechanical system. The Model 777 retains the conventional control column, wheel, and rudder pedals, whose operation is identical to the controls employed on other Boeing transport airplanes. The flight deck controls of the Model 777 are very similar to those of the Model 747-400.

Because the system is controlled electronically, there is an opportunity to include system control augmentation and envelope protection features that would have been difficult to provide in a conventional mechanical system. The Model 777 primary flight control system has made full use of the capabilities of this architecture by including such features as:

1. bank angle protection
2. turn compensation
3. stall and overspeed protection
4. pitch control and stability augmentation
5. thrust asymmetry compensation.

More will be said of these specific features later. What should be noted, however, is that none of these features limit the actions of the pilot. The Model 777 design utilizes envelope protection rather than envelope limiting as a deterrent. Protection deters pilot inputs. For example, the bank angle protection feature will significantly increase the wheel force a pilot encounters when attempting to roll the airplane past a predefined bank angle. However, if necessary, the pilot may override this protection by exerting a greater force on the wheel than is being exerted by the backdrive actuator. The intent is to inform the pilot that the command being given would put the airplane outside of its normal operating envelope, but the ability to do so is not precluded. This concept is central to the design of the Model 777 primary flight control system.

24.2.2 System Architecture and Redundancy

The Model 777 primary flight control system incorporates several layers of redundancy. To fully understand the amount of redundancy in the system and how it is managed, it is necessary to explore the types of components in the system and how they are used.

24.2.3 Flight Deck Controls

The 777 is equipped with standard flight deck controls. Instead of the control columns, wheel, and pedals driving quadrants and cables, they are attached to electrical transducers that convert mechanical displacement into electrical signals. Multiple transducers installed on each pilot controller ensure that the functionality of that control remains intact in the event of a single transducer failure.

A gradient control actuator is attached to the two control column feel units. These units prove the tactile feel of the control column by proportionally increasing the amount of column force a pilot experiences during a maneuver with an increase in airspeed. In addition, pilot controller backdrive actuators move the control column, wheel, and pedal in response to autopilot commands when the autopilot is engaged. This provides visual feedback to the pilot during autopilot operations.

24.2.4 System Electronics

There are two types of electronic computers used in the Model 777 primary flight control system; the actuator control electronics (ACE), which is primarily an analog device, and the primary flight computer (PFC), which utilizes digital technology. There are four ACEs and three PFCs employed in the system. The function of the ACE is to interface with the pilot control transducers and to control the actuators with analog servo loops. The role of the PFC is the calculation of control laws that convert the pilot controller position into surface commands, which are then transmitted back to the ACEs. The PFC also contains ancillary functions, such as system monitoring, crew announcements, and all the system on-board maintenance capabilities.

Four identical ACEs are used in the system, referred to as L1, L2, C, and R. These designations correspond roughly to the left, center, and right hydraulic systems on the airplane. The flight control functions are distributed among the four ACEs, such that a total failure of a single ACE will leave the major functionality of the flight control system intact. An ACE failure of this nature will have much of the same impact to the primary flight control system as that of a hydraulic system failure.

The ACEs decode the signals received from the transducers used in the pilot controls and the primary actuation. The ACEs convert the transducer position into a digital value and then transmit that value over the ARINC 629 busses for use by the PFCs. The PFCs use these pilot control positions and surfaces positions to calculate the surface commands. The PFCs then transmit the surface commands over the same ARINC 629 busses back to the ACEs, which converts them into analog commands for each actuator. There are three PFCs in the system, called L, C, and R. Where the redundancy of the ACEs lies in functional distribution, the redundancy of the PFCs is in the number of calculating elements. Each of the three PFCs, referred to as channels, are identical in design and perform identical calculations.

Internal to each PFC are three independent sets of microprocessors, ARINC 629 interfaces, and power supplies, which are referred to as lanes. All lanes perform identical calculations. Failure of a single lane internal to a PFC will cause only that lane to be shut down. That channel will continue to operate normally on two lanes with no loss of system functionality. Any subsequent failure of a channel that is already operating on two lanes, however, will cause that channel to be shut down, as a channel is not allowed to operate on a single lane. The airplane is designed to be operated indefinitely with one lane of nine failed. The proposed master minimum equipment list (MMEL) allows the airplane to be dispatched with two lanes failed out of the nine (as long as they are not within the same channel) for ten days and for a single day with one PFC channel inoperative.

24.2.5   ARINC 629 Data Bus

The ACEs and PFCs communicate with each other, as well as with other systems on the airplane, via triplex, bi-directional ARINC 629 flight control data busses, referred to as L, C, and R. The connection from the electronics unit to the data bus is an ARINC 629 coupler. Each coupler can be removed from that data bus and replaced individually without disturbing the integrity of the data bus itself.

24.2.6   Interfaces to Other Airplane Systems

The primary flight control system receives data from other airplane systems by two different methods. The air data inertial reference unit (ADIRU), standby attitude and air data reference unit (SAARU), and the autopilot flight director computers (AFDC) transmit on the ARINC 629 flight control data busses where the PFCs are able to read their data directly. Other systems, such as the flap slat electronics units (FSEU), proximity switch electronics unit (PSEU), and engine data interface unit (EDIU), among others, transmit their data on ARINC 629 systems data busses. The PFCs receive data from these systems through the airplane information management system (AIMS) data conversion gateway (DCG) function. The DCG supplies data from the systems data busses onto the flight controls data busses. This gateway between the two main set of ARINC 629 busses maintains separation between the critical flight control bus and the essential systems bus but still allows data to be passed back and forth.

24.3   System Electrical Power

There are three power sources dedicated to the primary flight control system, which are referred to as the flight controls direct current (FCDC) power system. Each of the three power systems is driven by an FCDC power supply assembly (PSA). The FCDC system is supplied by two dedicated permanent magnet generators (PMG) on each engine. Each PSA converts the PMG alternating current into 28 V DC for use by the electronics modules in the primary flight control system. Alternative power sources for the PSAs include the airplane ram air turbine (RAT), the 28 V DC airplane busses, the airplane hot battery bus, and dedicated five ampere-hour FCDC batteries. During flight, the PSAs draw power from the PMGs. For on-ground engines off operations or for in-flight failures of the PMGs, the PSAs draw from any available source.

24.3.1 Control Surface Actuation

The control surfaces of the system are controlled by hydraulically powered actuators. The elevators, ailerons, and flaperons are controlled by two actuators on each surface, the rudder is controlled by three. Each spoiler panel is powered by a single actuator. The horizontal stabilizer is positioned by two hydraulic motors driving the stabilizer jackscrew. On the Model 777, the primary flight control surfaces are actuated through the ACE sources which in turn command the hydraulic actuators.

The actuators on the elevators, ailerons, flaperons, and rudder have several operational modes. These modes, and the surfaces that each are applicable to, are defined below:

1. Active: normally all actuators on the elevators, ailerons, flaperons, and rudder receive commands from their respective ACEs and position the surfaces accordingly. The actuators will remain in the active mode until commanded into another mode by the ACEs.
2. Bypassed: in this mode, the actuator does not respond to commands from the ACE. The actuator is allowed to move freely, so that the redundant actuators on a given surface may drive the surface without any loss of authority. This mode is present on aileron, flaperons, and rudder actuators.
3. Damped: in this mode, the actuator does not respond to commands from the ACE. The actuator is allowed to move, but at a restricted rate which provides flutter damping. This mode allows the other actuator or actuators on the surface to continue to operate the surface at a rate sufficient for airplane control. This mode is present on elevator and rudder actuators.
4. Blocked: in this mode, the actuator does not respond to commands from the ACE, and is not allowed to move. When both actuators on a surface controlled by two actuators have failed, they both enter the blocked mode to provide a hydraulic lock on the surface. This mode is present on the elevator and aileron actuators.

An example using the elevator surface illustrates how these modes are used. If the inboard actuator on an elevator fails, the ACE controlling that actuator will place the actuator in the damped mode. This allows the surface to move at a limited rate under the control of the outboard actuator. Concurrent with this action, the ACE also arms the blocking mode on the outboard actuator on the same surface. If a subsequent failure occurs, which will cause the outboard actuator to be placed in the damped mode by its ACE, both actuators will then be in the damped mode and have their blocking modes armed. An elevator actuator in this configuration enters the blocking mode, which hydraulically locks the surface in place for flutter protection.

The Model 777 primary flight control test rig facility allowed complete integrated testing of all the components of the flight control system prior to being installed in the actual airplane.

24.3.2 Mechanical Control

Spoiler panels and the alternate stabilizer pitch trim system are controlled mechanically, rather than electronically. Spoilers are driven directly from control wheel deflections via a control cable. The alternate horizontal stabilizer control is accomplished by using the pitch trim levers on the aisle stand. Electrical switches actuated by the trim levers allow the PFCs to determine when alternate trim is being commanded so that appropriate commands can be given to the pitch control laws.

Spoiler panels are also used as speed brakes. The speed brake function for this spoiler pair has two positions: retracted and fully extended. The speed brake commands for spoilers are electrical in nature, with an ACE giving an extend or retract command via a solenoid operated valve in each of the actuators. Once that spoiler pair has been deployed by a speed brake command, there is no control wheel to speed brake command mixing.

24.3.3 System Operating Modes

The primary flight control system has three operating modes: normal, secondary, and direct. These modes are defined below:

1. Normal: in normal mode, the PFCs supply all commands to the ACEs. Full functionality is provided including all enhanced performance, envelope protection, and ride quality features.
2. Secondary: in this mode the PFCs supply all commands to the ACEs, just as in the normal mode. However, functionality of the system is reduced. For example, the envelope protection functions may not be active in the secondary mode. The PFCs enter this mode automatically from the normal mode when there are sufficient failures in the system or interfacing systems such that normal mode is not supported. An example of a set of failures that will automatically drop the system into secondary is total loss of airplane air data from the ADIRU and SAARU. The airplane is quite capable of being flown for a long period in secondary mode if required.
3. Direct: in the Direct mode, the ACEs do not process commands from the PFCs. Instead, each ACE decodes pilot commands directly from the pilot controller transducers and uses them for the closed loop control of the actuators. This mode will be entered automatically due to total failures of the PFCs, failures internal to the ACEs, loss of the flight control ARINC 629 data busses, or some combination of these. It may also be selected manually via the PFC disconnect Switch on the overhead panel in the flight deck. The airplane handling characteristics in the direct mode closely match those of the secondary mode.

24.4 Control Laws and System Functionality

The design philosophy employed on the Model 777 primary flight control system control laws stresses aircraft operations consistent with a pilot's past training and experience. The combination of electronic control of the system and this philosophy provides the feel of a conventional airplane, but with improved handling characteristics.

24.4.1 Pitch Control

Pitch control is accomplished through a maneuver demand control law. It is referred to as a C*U control law. C* (pronounced C-star) is a term that is used to describe the blending of the airplane pitch rate and load factor (the amount of acceleration felt by an occupant of the airplane during a maneuver). At low air speeds, the pitch rate of the airplane is the controlling factor. At high air speeds, the load factor predominates. The U term refers to the change in the airspeed away from a referenced trim speed. This introduces an element of speed stability into the airplane pitch control. The result is that the airplane is trimmed to a particular airspeed, and any deviation from that airspeed will cause a pitch change in order to return to that referenced airspeed. However, airplane configuration changes, such as a change in trailing edge flap settings, will not cause airplane trim change. Thus, the major advantage of using a maneuver demand control law is that nuisance handling characteristics found in a conventional system that increase pilot workload are minimized, while the desirable characteristics are maintained.

While in flight, the pitch trim switches on the pilot's and first officer's control wheels do not directly control the horizontal stabilizer. When the trim switches are used in flight, the pilot is actually requesting a new referenced trim speed. The airplane will pitch nose-up or nose-down in response to that change in order to achieve that new airspeed. The stabilizer will automatically trim to offload the elevator surface when necessary. When the airplane is on the ground, the pitch trim switches are used to trim the horizontal stabilizer directly. While the alternate trim levers move the stabilizer directly, even in flight, the act of doing so will also change the C*U referenced trim speed such that the net effect is the same as would have been achieved if the pitch trim switches had been used. As on a conventional airplane, trimming is required to reduce any column forces that are being held by the pilot.

The pitch control law incorporates several additional features. One is called landing flare compensation. This function provides handling characteristics during flare and landing consistent with that of a conventional airplane, which would have otherwise been altered significantly by the C*U control law. The pitch control law also incorporates stall and overspeed protection. These functions will not allow the referenced trim speed to be set below a predefined minimum value or above the maximum operating speed. They also significantly increase the column force that the pilot must hold in order to fly above or below these speeds. An additional feature incorporated into the pitch control law is turn compensation, which enables the pilot to maintain a constant altitude with minimal column input during a banked turn.

The unique Model 777 implementation of maneuver demand and speed stability in the pitch control law means that:

1. An established flight path remains unchanged unless the pilot changes it through a column input or the airspeed changes and the speed stability function takes effect.
2. Trimming is required only for airspeed changes.

24.4.2 Yaw Control

The yaw control law contains the standard functions used on other Boeing jet liners, such as the yaw damper and the rudder ratio changer functions. However, there are no separate actuators and linkages in the Model 777 for these functions as were used in previous Boeing airplanes. Rather, the commands for these functions are calculated in the PFCs and included as part of the normal rudder command to the main rudder actuators.

This reduces weight, complexity, maintenance, and spares required to be stocked. The yaw control law also incorporates several additional features. The gust suppression system reduces airplane tail wag by sensing wind gusts on the vertical fin and applying a rudder command to oppose the movement that would have been generated by the gust. Another feature is the wheel-rudder crosstie function, which reduces sideslip by using small amounts of rudder during banked turns.

One important feature in the yaw control is thrust asymmetry compensation, or TAC. This function automatically applies a rudder input for any thrust asymmetry between the two engines exceeding approximately 10% of the rate thrust. This is intended to cancel the yawing moment associated with an engine failure. TAC operates at all airspeeds above 80 knots; even on the ground during the take-off phase. It will not operate when thrust reversers are deployed.

24.4.3 Roll Control

The roll control law in the Model 777 is fairly conventional. The outboard ailerons and spoiler panels 5 and 10 are locked out in the faired position when the airspeed exceeds a value that is dependent upon speed and altitude. It roughly corresponds to flaps up. As with the yaw damper function, this function does not have a separate actuator, but is part of the normal aileron commands. The bank angle protection feature in the roll control law has been discussed previously.

24.4.4 757 Test Bed

The control laws and features discussed here were incorporated into a modified Model 757 and flown in the summer of 1992. The captain's controls remained connected to the normal Model 757 flight controls system. The Model 777 control laws were flown through the First Officer's controls. After the initial checkout and validation phase, pilots from several airlines and regulatory agencies were invited to fly the modified Model 757. The feedback from the pilots was very positive and enthusiastic. The initial flights of the Model 777 indicate that the flight characteristics of the Model 757 demonstrator were very close to those of the Model 777.

24.4.5 Primary Flight Control System Displays and Announcements

The primary displays for the primary flight control system are the engine indication and crew altering system (EICAS) display and the multi-function display (MFD). The EICAS display is very similar to that used in the Model 747-400. It displays the engine parameters, as well as the warning, caution, and advisory messages used by the flight crew. The MFD displays the status level message, which is used to determine the health of the various systems, and whether the airplane is able to be dispatched. The MFD also can display, when requested, the flight control synoptic page that shows the position of all the flight control surfaces.

This tutorial on aircraft flight control systems is based on a publicly available internal report by Boeing, which is gratefully acknowledged and referenced.

24.4.6   Glossary

ACE   - actuator control electronics
ADIRU   - air data inertial reference unit
ADM   - air data module (static and total pressure)
AFDC   - autopilot flight director computer
AIMS   - airplane information management system
ARINC   - Aeronautical Radio Inc. (industry standard)
C   - center
C*U   - pitch control law utilized in the primary flight computer
CMC   - central maintenance computer function of AIMS
DCGF   - data conversion gateway function of AIMS
EDIU   - engine data interface unit
EICAS   - engine indication and crew alerting system
ELMS   - electrical load management system
FCDC   - flight control direct current (power system)
FCTR   - flight control test rig
FSEU   - flap slat electronics unit
L   - left
L1   - left 1
L2   - left 2
LRRA   - low range radio altimeter
LRU   - line replaceable unit
MAT   - maintenance access terminal
MEL   - minimum equipment list
MFD   - multi-function display
MOV   - motor operated valve
PCU   - power control unites, actuators
PFC   - primary flight computer
PMG   - permanent magnet generator
PSA   - power supply assembly
PSEU   - proximity switch electronics unit
R   - right
RAT   - ram air turbine
SAARU   - standby attitude and air data unit
SIL   - systems integration laboratory
TAC   - thrust asymmetry compensation
WEU   - warning electronics unit

24.5   Tutorial Summary

The Model 777 primary flight control system utilizes new technology to provide significant benefits over a conventional primary flight control system. These benefits include a reduction in overall weight of the airplane, superior handling characteristics, and improved maintainability of the system. At the same time, the control of the airplane is accomplished using traditional cockpit controls thereby allowing the pilot to fly the airplane without any specialized training. The technology utilized by the Model 777 primary flight control system has earned its way onto the airplane, and is not just technology for technology's sake.

Bibliography

1 G. Bartley. Model 777 primary flight control system. *Boeing Airliner*, Oct-Dec 1994.

25

Tutorial on Satellite Control Systems

25.1 Tutorial Highlights

In this chapter, we summarize the essential features of satellite attitude control systems in a tutorial fashion. Most of the material is necessarily borrowed from the existing literature [1, 2] for the benefit of the reader so that this basic fundamental material on satellite control systems can be juxtaposed with the aircraft control systems tutorial presented in the previous chapter. This means the reader can easily compare and contrast the similarities and differences in the various features between aircraft and spacecraft control systems analysis and design.

25.2 Spacecraft/Satellite Building Blocks

A spacecraft typically consists of a payload and a number of subsystems. Partitioning the entire spacecraft in terms of these subsystems (building blocks) is helpful from an overview point of view. These building blocks could be generically classified as (i) structural subsystem, (ii) power subsystem, (iii) attitude and orbit control subsystem, (iv) telemetry, tracking, command, and communications subsystem, (v) thermal subsystem, (vi) propulsion/reaction jet control subsystem, (vii) payload, (viii) sensors and actuators, and such. In this tutorial our emphasis is on attitude and orbit control subsystems along with the propulsion/reaction jet control subsystem and the needed sensors and actuators for designing a feedback control system.

25.2.1 Attitude and Orbit Control

If a satellite orbit is a low Earth orbit (LEO) and if it is launched directly into this orbit, then the satellite will possess a low spin rate, which is imparted by the launch vehicle. So for this case, the overall objective of our control system is for the satellite to achieve its final desired attitude. Thus it is an attitude control system. However, if the satellite orbit is a geosynchronous Earth orbit (GEO) then we need few orbit control tasks to be performed. In this case, the initial low Earth elliptical orbit needs to be circularized using propulsion/reaction jet thrusters, then the satellite is put into a highly elliptical orbit with its apogee altitude equal to its final circular orbit altitude. Then again by firing the apogee kick motor at apogee, the orbit is circularized. Then once the satellite is in the desired orbit, the remaining attitude control actions need to be taken to achieve the

Flight Dynamics and Control of Aero and Space Vehicles, First Edition. Rama K. Yedavalli.
© 2020 John Wiley & Sons Ltd. Published 2020 by John Wiley & Sons Ltd.

desired attitude in its final orbit. Thus for GEO satellites we need both the attitude and orbit control actions done together.

With this backdrop, let us briefly review the sensors and actuators needed to accomplish all these tasks.

25.2.2 Attitude Control Sensors

Naturally, the attitude sensors needed for attitude control depend on the overall mission of the satellite and the accuracy requirements of the satellite mission. Earth/horizon sensors, sun sensors, star sensors, magnetometers, and gyroscopes are some of the typical sensors used for these purposes.

25.2.2.1 Earth/Horizon Senors

These are used to scan across the Earth, measuring rotation angles to define the spacecraft's attitude relative to the Earth from satellite's altitude. A few of the sensors are labeled (i) horizon crossing indicator (HCI), (ii) steerable horizon crossing indicator (SHCI), (iii) conical Earth sensor, (iv) boresight limb sensor, and such. For a more detailed account of the operation of these sensors, the reader is encouraged to consult the references given at the end of this tutorial.

25.2.2.2 Sun Sensors and Star Sensors

For most applications, the Sun can be treated as a point source. A simple Sun sensor can be used to detect a Sun reference as the Sun is relatively very bright. The basic sensing element in these sensors is the silicon solar cell. Some of these Sun sensors are labeled as Sun presence sensors, analog Sun sensors and digital Sun sensors. Sun sensors have been developed with fields of view ranging from several square arc minutes to $128 \times 128°$ and resolutions of less than an arc-second to several degrees.

In the same vein, various star sensors are labeled as (i) star scanner, (ii) gimbaled star tracker, (iii) fixed head star tracker, and such. A star tracker tracks stars within a designed field of view (FOV) and over a visual spectral magnitude range.

25.2.2.3 Magnetometers

These measure the magnetic field to milligauss accuracy. They can be used to obtain both the magnitude and direction of the magnetic field. Since Earth's magnetic field estimation (prediction) is subject to considerable uncertainty, in general, magnetometer measurements are not that accurate and their use is typically restricted to spacecraft below $1000\,\text{km s}^{-1}$. Earth's magnetic field strength in LEO is about 0.5 G.

25.2.2.4 Gyroscopes

A gyroscope is an instrument that uses a rapidly spinning mass to sense the inertial orientation of its spin axis. Rate gyros and rate integrating gyros are attitude sensors used to measure changes in the spacecraft orientation. As the names imply, rate gyros measure satellite angular rates whereas rate integrating gyros measure angular displacements directly.

For a more detailed account of the operation of these sensors, the reader is encouraged to consult the references given at the end of this tutorial.

Next we switch our attention to attitude actuators.

25.3   Attitude Actuators

Attitude actuators are employed to control the attitude of the spacecraft. Several attitude actuators are available that can be regarded as passive or active based on their functions. Momentum exchange devices such as momentum wheels [labeled as control moment gyros (CMGs) in some literature], and reaction wheels, reaction control jets (thrusters) and nutation dampers (passive) are some of the standard actuators.

25.3.1   Momentum Wheels (CMGs) and Reaction Wheels

25.3.1.1   Momentum Wheel
This is a flywheel designed to operate at a biased non-zero momentum. It provides a variable momentum storage capability about its rotation axis and is usually fixed in the spacecraft. It is especially common for dual-spin stabilized spacecraft.

25.3.1.2   Reaction Wheel
This is a flywheel designed to operate at zero bias. It consists of a two phase alternating current (AC) servomotor. It exhibits a relatively constant torque versus speed curve. It also contains a tachometer to measure its speed. momentum. It provides a variable momentum storage capability about its rotation axis and is usually fixed in the spacecraft. It is especially common for dual-spin stabilized spacecraft.

However, there does not seem to be a strict, rigorous adherence to the above definitions in the existing literature where in some accounts of some books asserting reaction wheels also have some nominal, non-zero biased angular momentum.

Finally, magnetic torques and nutation Dampers are also used as actuators for passive or active control of the satellite.

25.4   Considerations in Using Momentum Exchange Devices and Reaction Jet Thrusters for Active Control

The above mentioned sensors and actuators are used for active and passive control of spacecraft. As discussed in Part I of this book, some passive control means are the spin stabilization, dual spin stabilization, gravity gradient stabilization, Magnetic torque, etc. The standard active control means is the three axis control using reaction jet thrusters and momentum exchange devices.

A pictorial representation of a three axis stabilized satellite is shown Figure 25.1.

The basic advantages of a three axis active control system are: fast response, good pointing accuracy, possibility of using Sun oriented arrays (solar cells), and non-inertial pointing capability. The disadvantages are: limited life due to fuel expenditure, scanning sensors, and no graceful performance degradation in the event of gas jet failure.

The addition of momentum exchange devices prolongs the life of the system at the expense of increased complexity.

For economy of operation, the jet thrust should be as small as possible. This is not always possible, however, especially if large initial angular velocities are imparted to the satellite. The larger the inertia of the satellite, the larger the required jet size as can be seen by equating initial angular momentum to impulse available.

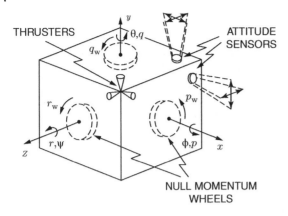

Figure 25.1 Three-axis satellite control.

Similarly, the limit cycle gas consumption is directly proportional to the vehicle inertia for the case of the zero or very small external disturbance. In this case the time average propellant consumption rate is obtained by finding the ratio of propellant expended during one limit cycle to the limit cycle period.

25.4.1 On-Orbit Operation via Pure Jet Control Systems

Control torques in active attitude control systems are generally obtained from cold or hot gas and/or electric propulsion. The simplest cold gas systems use an inert gas stored in a high-pressure vessel with initial pressures up to 400 atmospheres (Figure 25.2). Normally the gas is passed through one or more pressure regulators so

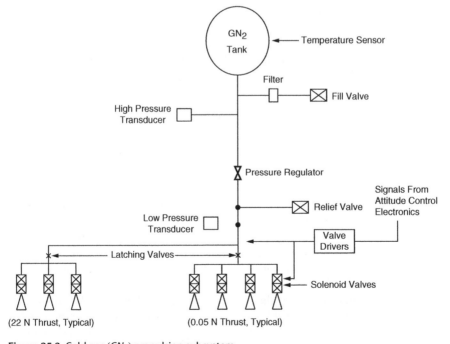

Figure 25.2 Cold gas (GN_2) propulsion subsystem.

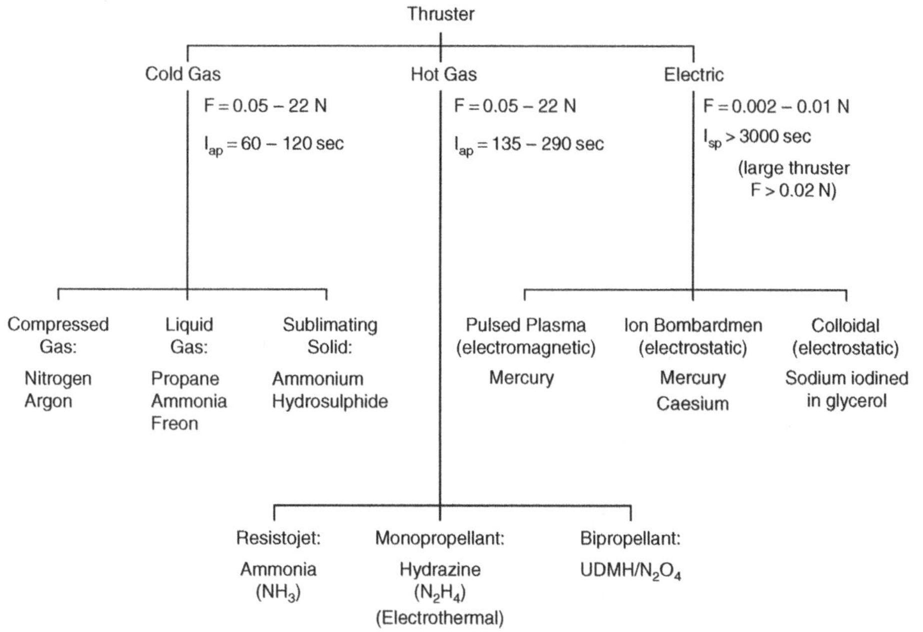

Figure 25.3 Mass expulsion thrusters.

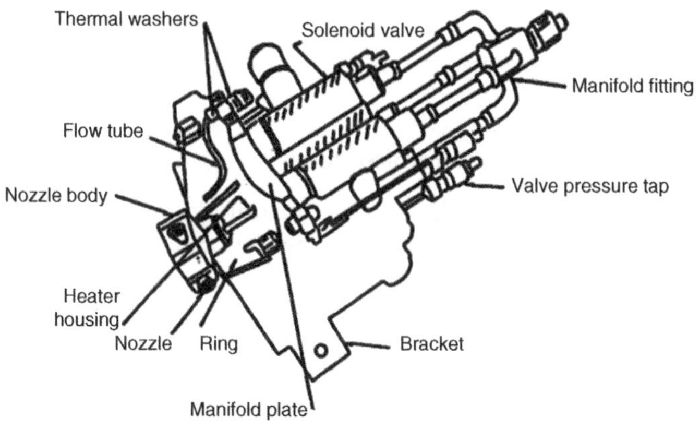

Figure 25.4 An example of a low-level thruster assembly.

that the thrusters operate at nearly constant pressure. Thrust range is typically between 0.05 to 22 N (see Figure 25.3). Efficient operation can be achieved with pulse durations of less than 10 ms to several seconds. The specific impulse I_{sp} can vary from 60 to 290 s or more, depending on the type of gas used. An example of a low-level thruster assembly is shown in Figure 25.4. An example three axis (or body) stabilized spacecraft is shown in Figure 25.5.

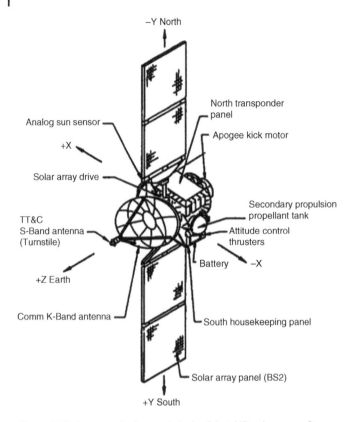

Figure 25.5 An example three axis (or body) stabilized spacecraft.

25.4.2 Recommended Practice for Active Control Systems

The following recommendations apply to the design of active attitude control systems. Specific recommendations with regard to the thrusting maneuvers and spacecraft with structural flexibility are included.

25.4.2.1 General Considerations

1. Ensure that all closed loop control systems exhibit acceptable transient response.
2. Control system torque capability must be sufficiently large to correct initial conditions errors and maintain.
3. The control logic must be consistent with the minimum impulse size and lifetime specification of the thrusters.
4. Evaluate system performance incorporating as many hardware elements in a simulation as possible.
5. Combine the normal tolerances statistically with the beginning and end-of-life center of mass location and moment of inertia characteristics.

25.4.2.2   Thrusting Maneuvers

1. Maximize the distance between the center of mass and the point of application for the control force.
2. Determine that propellant expenditure is consistent with system constraints and performance. Determine performance in the event of thruster failure.
3. Design limit cycling characteristics to avoid excessive error, excitation of the spacecraft flexible modes, excessive propellant expenditure, and excessive thruster wear.
4. Consider the effect of thrust impingement on thermal design and thrust degradation.
5. Keep the residual conditions at end of maneuver within the control capability of the control system to prevent excessive expenditure of propellants.
6. Provide capability to verify control system operation before thrust initiation.
7. Maintain current and accurate mass properties of the spacecraft and alternate configurations.
8. Consider all sources of thrust misalignment, including:
    (a) Thrust vector to thruster misalignment.
    (b) Thruster mechanical misalignment.
    (c) Thruster support structure compliance.

25.4.2.3   Structural Flexibility

1. Provide adequate separation of the rigid body and flexible mode frequencies.
2. Choose the control band pass so that it excludes the structural resonant frequencies by stiffening of the structure. Consider the use of special inner control loops, or the use of a notch filter.
3. Make the damping ratio of each flexible appendage as large as possible. Artificial stiffening or damping by means of separate control loops should be considered.
4. Under saturation, all loops should degrade gracefully.

25.5   Tutorial Summary

In this chapter, we provided a brief overview of satellite attitude control systems in a tutorial fashion to complement the aircraft control systems tutorial in the previous chapter. Most of the material in this tutorial is taken from the existing literature, which are gratefully acknowledged and referenced.

Bibliography

1 Vladimir A. Chobotov. *Spacecraft Attitude Dynamics and Control*. Krieger Publishing Company, 1991.
2 Anton H. de Ruiter, Christopher Damaren, and James R. Forbes. *Spacecraft Attitude Dynamics and Control: An Introduction*. Wiley, 2013.

26

Tutorial on Other Flight Vehicles

26.1 Tutorial on Helicopter (Rotorcraft) Flight Control Systems

26.1.1 Highlights of the Tutorial on Helicopter Flight Vehicles

In this chapter, we briefly introduce and analyze another popular atmospheric flight vehicle, namely the helicopter. Interestingly, even at the detailed mathematical modeling level, the methodology that was followed for aircraft dynamics and control holds good for these vehicles also. However, at the specificity of the vehicle, conceptually there are significant differences from the aircraft principles warranting a separate discussion of this important type of atmospheric flight vehicle. The fundamental difference is that these are rotorcraft vehicles as opposed to the fixed wing aircraft we considered before. Thus in this chapter we give a tutorial tour focused on these vehicles, highlighting both the similarities and differences with the fixed wing aircraft type along the way. A detailed account of helicopter dynamics and control is available in excellent textbooks dedicated to this subject such as [5, 6].

26.1.2 Introduction

Helicopters are a type of aircraft known as rotorcraft, for they produce the lift needed to sustain flight by means of a rotating wing, the rotor. Because rotors are powered directly, helicopters can fly at zero forward speed: they hover. They can also fly backwards, of course. Since they do not need long runways to get the needed lift, they are essentially vertical take off and landing (VTOL) vehicles. At present, there are two main kinds of helicopter: those which use a single main rotor and a small tail rotor, and those which have two main rotors in tandem.

In the single main rotor type, the rotor produces vertical thrust. By inclining this lift vector, a helicopter can be accelerated in both the fore and aft, and the lateral directions. This main rotor is usually shaft-driven and, as a result, its torque has to be countered, usually by a small tail rotor mounted at the end of the tail boom. Yaw control is achieved by varying the thrust developed by this tail rotor.

In the USA and UK the main rotor rotates counterclockwise (viewed from above); in France, they use clockwise rotation. This has some significance in relation to the use of the tail rotor. To approach some point at which to hover, the pilot of a helicopter

Flight Dynamics and Control of Aero and Space Vehicles, First Edition. Rama K. Yedavalli.
© 2020 John Wiley & Sons Ltd. Published 2020 by John Wiley & Sons Ltd.

must make his aircraft flare stop. Since it is customary for helicopter pilots to sit in the right-hand seat in the cockpit, the external view can be restricted in this flare maneuver, and, thus, often a sidewards flare is executed, which requires the pilot to apply more pressure to the left pedal in order to sideslip to the right, but this increased left pedal deflection demands a greater trimming moment from the tail rotor, which has to be achieved by an increase in the thrust of that rotor. Pilots flying French helicopters do not have so great a problem in carrying out this maneuver.

The two rotors of the tandem helicopters are normally arranged to be at the top and the front and rear of the fuselage. These rotors rotate in opposite directions, thereby ensuring that the torque is self-balancing. There is normally a significant overlap between the rotor the rotor discs, however, the hub of the rear rotor being raised above the hub of the rotor at the front. The resulting aerodynamic interference causes a loss of power, but the amount lost, being about 8–10 %, is almost the same as that lost in driving a single tail rotor.

Every rotor has blades of high aspect ratio that are very flexible. These rotors are either articulated, in which case they use hinges at the root of the blades to allow free motion of the blades in directions normal to, and in the plane of, the rotor disc. At the blade hinge, the bending moment is zero; no moment is transmitted, therefore, through the root of the blade to the fuselage of the helicopter. Recent designs have eliminated hinges; these are referred to as hingeless, or rigid, rotors.

The out-of-plane motion of the blade, perpendicular to its radial direction, is referred to as its flapping motion. Motion about the vertical hinge causes the blade to deflect in the plane of the disc and such motion is referred to as lagging motion. In hingeless rotors, flapping and lagging motion are defined as the out-of-phase and the in-phase bending, respectively.

To control a rotor means that the pitch angles of its blades can be altered to cause a change in the blade's angle of attack, thereby controlling the corresponding aerodynamic forces. On a hinged blade, the pitch bearing is usually outboard of both the flapping and lagging hinges, but on a hingeless rotor the bearing may be found either in or outboard of the major bending moment at the blade root.

While the state variables in helicopter flight are still similar to the aircraft state variables, the major difference in helicopter control is with the definition of control variables available within the helicopter dynamics. With any type of rotor, there will be an azimuthal variation of lift as the rotor rotates. Such variation affects the degree of flapping motion and, consequently, the direction of the average thrust vector of the rotor. A cyclic variation of lift can be effected, therefore, by changing a rotor blade's pitch as the blade is being rotated. This altering of blade pitch is termed the cyclic pitch control; when it causes a pitching moment to be applied to the helicopter it is called longitudinal cyclic, usually denoted by δ_B. If the applied moment is about the roll axis, the control is called the lateral cyclic, denoted by δ_A. Yaw is controlled by changing, by the same amount, the pitch angle of all the blades of the tail rotor; such a collective deflection of the blades of the tail rotor is denoted by δ_T. When the pitch angles of all the blades of the main rotor are changed by an identical amount at every point in azimuth, a change is caused in the total lift being provided by the rotor. This type of control is called collective pitch control, denoted by $\delta_{\theta 0}$. Direct control of translational motion is done by means of the collective control, since it is the means by which the direction of the thrust vector can be controlled.

Thus as can be seen from this discussion, the control variables in helicopter dynamics are significantly different from the aircraft control variables.

The importance of the collective to helicopter flight cannot be overemphasized; it is the direct lift control that allows the helicopter's vertical motion to be controlled quickly and precisely. Since there is considerable energy stored when the rotor rotates (as a result of its angular momentum) only small changes in the collective setting are needed to change vertical motion without any accompanying exchange of height for airspeed. Moreover, for small collective inputs, the ability of the helicopter's engine (or engines) to change speed is not of great concern. However, this simple means of controlling height makes it a little more difficult to control a helicopter's horizontal speed; to slow down, it is necessary to pitch a helicopter nose-up. Thus, a pilot achieves deceleration by means of pitch attitude, while maintaining his helicopter's height with the collective, which requires the pilot to demonstrate greater control coordination. It is characteristic of helicopters during the approach to hover, and at hover, that any changes in the vehicle's speed require some adjustment of the collective, which, in turn, causes a change in the helicopter's yawing motion, thereby resulting in the development of significant sideslip. These coupled motions subsequently result (in the absence of immediate and effective pilot action) in the helicopter rolling and pitching. This complex dynamic response is of particular concern when considering a helicopter's approach on the glide slope, for it can lead to deviation from the desired flight path.

Hence another major difference from fixed wing aircraft is the ever-present multitude of cross couplings between all degrees of freedom along the longitudinal, lateral, and directional axes.

With tandem rotors, matters are different. If both rotors are tilted, a change is caused in both the forward force and the pitching moment. If differential collective pitch between the rotors is used, it is possible only to produce pitching motion; yaw control is provided by tilting the rotors in opposite directions. If the cg of the helicopter is not located exactly midway between the rotors, then use of the lateral cycle will inevitably produce a yawing moment.

If such a tandem helicopter is rolled towards starboard (to the right), yawing motion towards port (to the left) will be induced. This characteristic is opposite, unfortunately, to that needed to produce a coordinated turn.

The helicopter gives rise to a number of very distinctive control problems, including the following: its uncontrolled motion is unstable; its control is effected through its major lift generator; it is capable of hovering motion; the pilot has to directly control on its lift force, as well as controlling the motion about its three axes; and its speed range is narrow, the speeds involved not being very high (the upper limit is about 240 knots, i.e. 120 m s^{-1}).

Only the problems involving stability and control of the helicopter are dealt with in this book, and then only briefly. However, for helicopters, more acutely than for fixed wing aircraft, the control and stability characteristics depend very heavily upon the vehicle's distinctive flight dynamics and aerodynamics.

26.1.3 Equations of Motion

Any study of the dynamic response of a helicopter is complicated because each blade of the rotor has its own degrees of freedom, which are in addition to those of the fuselage.

Yet, for small perturbations in the helicopter's motion, a knowledge of the motion of each blade is not required; only the rotor's motion as a physical entity needs to be considered. It is usual to assume that the rotor speed, Ω, is constant. Because such analyses are invariably carried out in a body-fixed axes system and it is assumed that all perturbations are small, the inertia terms can be linearized and the lateral and longitudinal motions may be considered as being essentially uncoupled. It should be remembered, however, that because of the rotation of the rotors, a helicopter does not have lateral symmetry (except for coaxial or side-by-side rotor configurations). There is, consequently, considerable coupling of lateral and longitudinal motions.

For example, consider the roll coupling that can result from yawing motion. However the pedals in the cockpit are moved, a rolling acceleration is experienced because the tail rotor is generally above the roll axis. This can be easily seen from an examination of the equations governing rolling and yawing motion:

$$L = I_{xx}\ddot{\varphi} - I_{xz}\ddot{\psi} \tag{26.1}$$

$$N = I_{xx}\ddot{\psi} - I_{xz}\ddot{\varphi}. \tag{26.2}$$

For a helicopter, if T_{tr} represents the thrust produced by the tail rotor, h represents the height of the hub of the tail rotor above the helicopter's cg and l is the distance aft of the cg at which the tail rotor is located, then:

$$L = hT_{tr} \tag{26.3}$$

$$N = -lT_{tr}. \tag{26.4}$$

It is simple to show that the ratio of rolling to yawing acceleration can be expressed as:

$$\frac{\ddot{\varphi}}{\ddot{\psi}} = \frac{hI_{zz} - lI_{xz}}{hI_{xz} - lI_{xx}}. \tag{26.5}$$

Since $I_{xz} < I_{xx}$ in general, then:

$$\frac{\ddot{\varphi}}{\ddot{\psi}} \simeq -\frac{h}{l}\frac{I_{zz}}{I_{xx}} + \frac{I_{xz}}{I_{xx}}. \tag{26.6}$$

$\frac{I_{xz}}{I_{xx}}$ can take a value in the range 0.1–0.25.

26.1.4 Longitudinal Motion

In wind axes the linearized equations of motion are:

$$m\dot{u} = -mg\theta_F \cos\gamma + \Delta X \tag{26.7}$$

$$m\dot{w} = mV\dot{\theta}_F - mg\theta_F \sin\gamma + \Delta Z \tag{26.8}$$

$$I_{yy}\ddot{\theta}_F = \Delta M \tag{26.9}$$

where ΔX and ΔZ are increments in the aerodynamic forces rising from disturbed flight, ΔM the corresponding increment in pitching moment, γ the angle of climb, and θ_F the pitch attitude of the fuselage. Because it is assumed that the perturbations in u, w and θ_F are small, the increments in the forces and the moment can be written as the first terms of a Taylor Series expansion, i.e.:

$$\Delta X = \frac{\partial X}{\partial u_{sp}}u_{sp} + \frac{\partial X}{\partial w}w + \frac{\partial X}{\partial q}q + \frac{\partial X}{\partial \delta_B}\delta_B + \frac{\partial X}{\partial \delta_{\theta_0}}\delta_{\theta_0} \tag{26.10}$$

where δ_B is the cyclic pitch control term, and δ_{θ_0} the collective pitch control term. The coefficients $\frac{\partial X}{\partial u_{sp}}$, $\frac{\partial X}{\partial w}$, etc. (or in the shorthand X_u, X_w, etc.) are the stability derivatives. Thus:

$$mu_{sp} = X_u u_{sp} + X_w w + X_q q + mg\theta_F \cos\gamma + X_{\delta_B}\delta_B + X_{\delta_{\theta_0}}\delta_{\theta_0} \tag{26.11}$$

$$m\dot{w} = Z_u u_{sp} + Z_w w + Z_q q + mV\dot{\theta}_F - mg\theta_F \sin\gamma + Z_{\delta_B}\delta_B + Z_{\delta_{\theta_0}}\delta_{\theta_0} \tag{26.12}$$

$$I_{yy}\ddot{\theta}_F = M_u u_{sp} + M_w w + M_q q + M_{\dot{w}}\dot{w} + M_{\delta_B}\delta_B + M_{\delta_{\theta_0}}\delta_{\theta_0}. \tag{26.13}$$

The term $M_{\dot{w}}\dot{w}$ is usually included to account for the effect of downwash upon any tailplane that may be fitted.

Because lift is generated by the rotating blades whose tilt angles are considered as the control inputs, it proves to be helpful to employ a non-dimensional form of those equations. The stability derivatives concept is basically similar to the stability derivatives we discussed in the fixed wing aircraft case, with minor care in interpreting them. Let the radius of the rotor blades be denoted by R. The tip speed of any blade is therefore given by ΩR. The blade area is $s\pi R^2$ where the solidity factor, s, of the rotor is given by:

$$s = \frac{bc}{\pi R} \tag{26.14}$$

where b represents the number of blades used in the rotor and c represents the chord of these blades (assuming, of course, that they are all identical).

The reference area, A_{ref}, is given by:

$$A_{ref} \triangleq \pi R^2. \tag{26.15}$$

Following similar logic explained in the fixed wing aircraft case, the non-dimensional stability derivatives, conceptually, can be defined as:

$$x_u = \frac{X_u}{\rho s A_{ref}\Omega R} \tag{26.16}$$

$$x_w = \frac{X_w}{\rho s A_{ref}\Omega R} \tag{26.17}$$

$$x'_q = \frac{X_q}{\rho s A_{ref}\Omega R^2} \tag{26.18}$$

$$z_u = \frac{Z_u}{\rho s A_{ref}\Omega R} \tag{26.19}$$

$$z_w = \frac{Z_w}{\rho s A_{ref}\Omega R} \tag{26.20}$$

$$z'_q = \frac{Z_q}{\rho s A_{ref}\Omega R^2} \tag{26.21}$$

$$m'_u = \frac{M_u}{\rho s A_{ref}\Omega R^2} \tag{26.22}$$

$$m'_{\dot{w}} = \frac{M_{\dot{w}}}{\rho s A_{ref}\Omega R^2} \tag{26.23}$$

$$m'_q = \frac{M_q}{\rho s A_{ref}\Omega R^2} \tag{26.24}$$

$$x_{\delta_B} = \frac{X_{\delta_B}}{\rho s A_{\text{ref}} \Omega^2 R^2} \tag{26.25}$$

$$z_{\delta_B} = \frac{Z_{\delta_B}}{\rho s A_{\text{ref}} \Omega^2 R^2} \tag{26.26}$$

$$m'_{\delta_B} = \frac{M_{\delta_B}}{\rho s A_{\text{ref}} \Omega^2 R^3} \tag{26.27}$$

$$x_{\delta_{\theta 0}} = \frac{X_{\delta_{\theta 0}}}{\rho s A_{\text{ref}} \Omega^2 R^2} \tag{26.28}$$

$$z_{\delta_{\theta 0}} = \frac{Z_{\delta_{\theta 0}}}{\rho s A_{\text{ref}} \Omega^2 R^2} \tag{26.29}$$

$$m'_{\delta_{\theta 0}} = \frac{M_{\delta_{\theta 0}}}{\rho s A_{\text{ref}} \Omega^2 R^3}. \tag{26.30}$$

26.1.5 Lateral Motion

To control lateral motion the following inputs are used: the deflection angle of the lateral cyclic, δ_A and the collective pitch angle of the tail rotor, δ_T. The corresponding equations of motion are:

$$m\dot{v} = Y_v v + Y_p p - mVr + Y_r r + mg\varphi \cos\gamma + mg\psi_F \sin\gamma$$
$$+ Y_{\delta_A} \delta_A + Y_{\delta_T} \delta_T \tag{26.31}$$

$$I_{xx}\dot{p} - I_{xz}\dot{r} = L_v v + L_p p + L_r r + L_{\delta_A} \delta_A + L_{\delta_T} \delta_T \tag{26.32}$$

$$-I_{xz}\dot{p} + I_{zz}\dot{r} = N_v v + N_p p + N_r r + N_{\delta_A} \delta_A + N_{\delta_T} \delta_T. \tag{26.33}$$

I_{xx}, I_{xz}, and I_{zz} are the moments of inertia. The derivatives Y_p and Y_r are usually negligible in helicopter studies. Using the same procedure to non-dimensionalize these equations as that employed with the longitudinal motion produces:

$$i_{xx} = \frac{I_{xx}}{mR^2} \tag{26.34}$$

$$i_{zz} = \frac{I_{zz}}{mR^2} \tag{26.35}$$

$$i_{xz} = \frac{I_{xz}}{mR^2} \tag{26.36}$$

$$\frac{dv}{d\tau} = y_v v + mg\varphi \cos\gamma - \frac{V d\psi_F}{d\tau} + mg\psi_F \sin\gamma + y_{\delta_A} \delta_A + y_{\delta_T} \delta_T \tag{26.37}$$

$$\frac{dp}{d\tau} = l_v v + l_p p + l_r r + \frac{i_{xz}}{i_{xx}} \dot{r} + l_{\delta_A} \delta_A + l_{\delta_T} \delta_T \tag{26.38}$$

$$\frac{dr}{d\tau} = n_v v + n_p p + n_r r + \frac{i_{xz}}{i_{xx}} \dot{p} + n_{\delta_A} \delta_A + n_{\delta_T} \delta_T. \tag{26.39}$$

26.1.6 Static Stability

Static stability is of cardinal importance in the study of helicopter motion since the several equilibrium modes affect each other considerably. For example, any disruption of directional equilibrium will lead to a change in the thrust delivered from the tail rotor, resulting in a corresponding change of the moment of this force (relative to the longitudinal axis, OX) which causes a disruption in the transverse equilibrium of the helicopter. But how does any disruption of directional equilibrium occur in the first place? Suppose the helicopter rotates about the transverse axis, OY, i.e. its longitudinal equilibrium is disrupted. The angle of attack of the main rotor will then change; such a change causes a change in thrust and, consequently, a change in the reactive moment of the main rotor. That change disrupts the directional equilibrium.

The practical significance of this interplay between the balancing forces mean that a helicopter pilot must constantly try to restore the disrupted equilibrium so that controlling (i.e. flying) a helicopter is more complicated and therefore more difficult than flying a fixed wing aircraft. That is why the simple question: Do helicopters possess static stability? requires the examination of a number of factors before an answer can be attempted. Three factors are involved: (1) the static stability properties, if any, of the main rotor; (2) the static stability properties, if any, of the fuselage, and (3) the effect of the tail rotor and any tailplane on any static stability properties.

26.1.7 Static Stability of the Main Rotor

26.1.7.1 Speed

It is assumed that the helicopter is flying straight and level at a speed V. Suppose, in the perturbed flight, the speed is increased by a small amount, ΔV. The flapping motion of the blades therefore increases. As a result, the axis of the cone of the main rotor is deflected aft, from its previous position, by an angle denoted by ε. Such a tilt of the coning axis leads to the development of a force F_x that is in an opposite sense to the direction of flight. As a result of this force, the velocity of the main rotor falls, and hence the helicopter reduces its forward speed.

If it had been assumed that, when the helicopter was flying straight level, the speed has been reduced by an amount ΔV, the cone axis would then have been deflected forward, and the force, F_x, would have developed in the same sense as the direction of flight, thereby causing an increase in the forward speed.

Thus, it can be concluded that with respect to changes in speed, the main rotor is statically stable.

26.1.7.2 Angle of Attack

The helicopter is once more assumed to be flying straight and level with its main rotor at an angle of attack of α_{MR_A}. The thrust delivered by the main rotor passed through the helicopter's cg and hence any moment of the thrust must be zero. Under the influence of a vertical air current, say, the helicopter lowers its nose and, therefore, the angle of attack of the main rotor is reduced by an amount $\Delta\alpha_A$. The vector of thrust is now deflected forward.

A moment, M_T, is established which causes the value of the angle of attack of the main rotor to decrease:

$$M_T = Tl. \tag{26.40}$$

This moment is destabilizing.

If the angle of attack of the main rotor is increased, however, the thrust vector will tilt aft and a nose-up moment, M_T, will be established causing the angle of attack of the main rotor to increase further.

The main rotor is statically unstable, therefore, with respect to fuselage angle of attack. Provided that no translation occurs, a helicopter in hovering motion has neutral stability with respect to any change in attitude.

26.1.7.3 Fuselage Stability

The greatest influence upon the static stability of a helicopter is that of the rotor; the contribution of the fuselage to static stability is not negligible, however. For a single rotor helicopter, for example, the fuselage is statically unstable in all three axes of motion. A small tailplane is sometimes installed at the aft end of the fuselage to improve the static stability of longitudinal motion in straight and level flight. Its influence is practically zero at low speeds and at hover. However, the degree of instability in longitudinal motion can be reduced from the value at hover by increasing forward speed and by reducing the angle of attack until, at negative angles of attack, the fuselage plus tailplane possesses some static stability.

If a helicopter is fitted with a tail rotor it has a profound effect on the fuselage's static stability as if the directional equilibrium is disrupted and the helicopter turns to the right, say, the angle of attack of the blade elements of the tail rotor will increase and, consequently, the thrust from the tail rotor increases by some amount, ΔT. Therefore, the moment of this thrust must also increase thereby restoring equilibrium. In this manner the tail rotor gives the fuselage directional static stability.

If the hub of the main rotor has offset horizontal (lagging) hinges, the hinge moment associated with that offset have a considerable effect on both longitudinal and transverse static stability of that helicopter. The greater the offset of the hinge and the rotational speed of the rotor, the greater the static stability possessed by the helicopter. These same factors also contribute to the increase in damping moment contributed by the main rotor.

26.1.8 Dynamic Stability

Since the flying qualities of a helicopter are markedly different in forward flight and in hovering motion, these two flight regimes are dealt with separately.

26.1.9 Longitudinal Motion

26.1.9.1 Stick-fixed Forward Flight

The pilot's stick being assumed fixed, there are no control inputs $\delta_B, \delta_{\theta_{r0}}, \delta_A$ or δ_T: the dynamic stability properties are determined solely from the coefficient matrix.

For straight and level flight,

$$\gamma \triangleq 0 \tag{26.41}$$

$$V \geq z_q. \tag{26.42}$$

Hence, the corresponding coefficient matrix, A^*_{long}, can be expressed as:

$$A^*_{\text{long}} = \begin{bmatrix} x_u & x_w & x_q & -mg \\ z_u & z_w & V & 0 \\ \tilde{m}_u & \tilde{m}_w & \tilde{m}_q & 0 \\ 0 & 0 & 1 & 0 \end{bmatrix}. \tag{26.43}$$

26.1.9.2 Hovering Motion

When a helicopter hovers, V is zero and, usually, x_w, x_q, m_m and m_w are negligible, i.e. the equations of motion now become:

$$\dot{u}_{\text{sp}} = x_u u_{\text{sp}} - mg\theta_F + x_{\delta_B}\delta_B + x_{\delta_{\theta_0}}\delta_{\theta_0} \tag{26.44}$$

$$\dot{w} = z_u u - z_w w + z_{\delta_B}\delta_B + z_{\delta_{\theta_0}}\delta_{\theta_0} \tag{26.45}$$

$$\dot{q} = m_u u - m_q q + m_{\delta_B}\delta_B + m_{\delta_{\theta_0}}\delta_{\theta_0}. \tag{26.46}$$

Hence, the characteristic polynomial can be shown to be:

$$\Delta_{\text{hover}} = \lambda^3 - (x_u + m_q)\lambda^2 + x_u m_q \lambda + mgm_u \tag{26.47}$$

$$\Delta_{\text{hover}} = (\lambda + p_1)(\lambda^2 + 2\zeta\omega\lambda + \omega^2). \tag{26.48}$$

The factor $(\lambda + p_1)$ corresponds to a stable, subsidence mode, whereas the quadratic factor corresponds to an unstable, oscillatory mode since ζ invariably lies in the range 0 to -1.0. Consequently, the longitudinal dynamics of a helicopter at hover separate into two distinct motions: vertical and longitudinal. It is easy to show that:

$$\frac{w(s)}{\delta_{\theta_0}(s)} = \frac{z_{\delta_{\theta_0}}}{(s - z_w)} = \frac{sh(s)}{\delta_{\theta_0}(s)} \tag{26.49}$$

i.e. the vertical motion of a helicopter at hover is described by a first order linear differential equation, with a time constant given by:

$$T_v = -\frac{1}{z_w}. \tag{26.50}$$

The time-to-half amplitude is typically about 2 s since the value of z_w is typically within the range -0.01 to -0.02.

In many ways, the simplified representation of the vertical motion in response to collective inputs is misleading. The vertical damping, z_w, is not a simple aerodynamic term but is composed of contributions from the fuselage and from the inflow created by the rotor. In hovering motion, the inflow contribution is predominant. The value of z_w, however, which is speed dependent, does have a marked effect on the thrust-to-weight ratio required for helicopter flight. Furthermore, the value of vertical damping required for a particular height response is considerably affected by the response time of the engine(s) driving the rotor. Of considerable importance to any control in helicopters is the nature of the engine response.

The instability of the longitudinal dynamics is a result of coupling of the motion via the pitching moments, which comes about as a result of the change in longitudinal velocity, i.e. M_u (known as speed stability), and the longitudinal component of the gravitational force. For static stability, the requirement is that the constant term of the characteristic polynomial shall be positive, i.e.

$$mgm_u > 0. \tag{26.51}$$

The inequality can be satisfied with a positive value of m_u.

In summary, for a hovering helicopter, the longitudinal dynamics are described by a stable, subsidence mode (a large negative real root due to pitch damping) and a mildly unstable, oscillatory mode (due to the speed stability M_u). A pilot will have good control over the angular acceleration of the helicopter, but poor direct control over translation. Because of the low damping, in hover the control sensitivity is high. This combination of high sensitivity and only indirect control of the translational velocity makes a hovering helicopter prone to pilot-induced oscillations thereby increasing the difficulty of the pilot's task. To aggravate matters, the lateral and longitudinal motions are not decoupled, as supposed, and, for many types of helicopter, a longitudinal cyclic input can result in large corresponding lateral motion. Furthermore, because of the speed stability of its rotor, a helicopter is susceptible to gusts whenever it is hovering and, as a result, its position relative to the ground drifts considerably; this makes the task of station keeping, for which helicopters are universally employed, particularly taxing.

26.1.10 Lateral Motion

26.1.10.1 Hovering Motion
In hovering motion the forward speed is zero. When longitudinal motion in hover is considerable it is found that a number of stability derivatives are either zero or negligible, which leads to a substantial simplification of the equations of motion. However, such simplifications do not occur in lateral motion studies, because the yawing (r) and rolling (p) motions are coupled by virtue of the stability derivatives, l_r and n_p, which have significant values owing to the tail rotor. If, however, it is assumed that the shaft of the tail rotor is on the roll axis, l_r can be considered negligible. Then the characteristic polynomial becomes:

$$(\lambda - n_r)(\lambda^3 - [y_v + l_p]\lambda^2 + y_v l_p \lambda - l_v mg). \tag{26.52}$$

The root ($\lambda = n_r$) means that the yawing motion is stable (since n_r is invariably negative) and independent of sideways and rolling motion. The cubic can be factored into:

$$(\lambda + p_2)(\lambda^2 + 2\zeta_1\omega_1\lambda + \omega_1^2). \tag{26.53}$$

The first factor corresponds to stable rolling, subsidence mode; the quadratic represents an unstable, oscillatory mode. Typically, for the rolling subsidence mode, t_a is less than 0.5 s; the period of the oscillation is about 15–20 s, whereas the time-to-double amplitude is about 20–30 s. The time constant of the yawing mode is about 5 s.

26.1.11 Overview of the Similarities and Differences with Respect to the Fixed Wing Aircraft

In this section, we summarize the similarities and differences of fixed wing aircraft analysis in a conceptual way.

26.1.11.1 Similarities

- The equations of motion development is quite similar to the fixed wing case, as we still use the Newton's laws of motion to develop the equations with the appropriate definitions for the state variables and the control variables. Note that the state variables have similarities with the fixed wing, but the control variables do have considerable conceptual differences.
- Similarly, the notions of static stability and dynamic stability, the development of stability derivatives needed in the equations of motion, the linearization concept and stability in small motions, etc. carry over to this case as well.

26.1.11.2 Differences

- There is significant coupling between all the three degrees of freedom, which does not allow us to do the traditional decoupled analysis we have done for aircraft by considering the longitudinal (pitch) motion separately from the roll/yaw motion. Thus there is no clear cut mode identification such as pure short period and pure phugoid modes. They are all coupled, because speed variations may cause significant pitching moments and so on. Similarly pitching motion may cause significant rolling and yawing moments causing the decoupling between longitudinal motion to lateral/directional motion to be unacceptable.
- Similarly there is not that much stability margin available via the static margin concept. Most helicopters are statically unstable, requiring continuous dependence on the automatic flight control system to keep it stable in both static as well as dynamic situations, especially in the hover mode.
- The pitch bandwidth requirements are significantly different for helicopters and fixed wing aircraft. Let τ_p denote the phase delay parameter. It relates to the rate of change of phase with frequency above the crossover frequency and is a measure of the equivalent time delay between attitude response and pilot control input. The bandwidth, denoted by ω_{bw} is that frequency beyond which closed loop stability is threatened. Thus if we plot the parameter τ_p versus the parameter ω_{bw}, in a planar diagram, for both fixed wing and helicopters, then the range of these parameters for mid-term stability is significantly different with very limited range of ω_{bw} and a corresponding large range in τ_p for helicopters compared to the fixed wing aircraft, pointing to the fact that controlling helicopters is a much tougher and critical task than for a fixed wing aircraft.
- Helicopters have two distinct flight regimes: (i) the hover/low speed regime, (ii) the mid/high speed regime. The hover/low speed regime is unique to a helicopter; no other flight vehicle can so efficiently maneuver in this regime, close to the ground and obstacles, with the pilot having direct control of thrust with quick response times.
- Interestingly, a fixed wing aircraft leaves its tip vortices behind whereas the helicopter rotor blades are forced to operate under the presence of these tip vortices shed by all rotor blades, which in turn translates to considerable demand on control energy expended as well as the complexity in the control logic to account for this.
- Propeller aircraft have axial symmetry but helicopter blades lose that axial symmetry under perturbed conditions.
- In the excellent book by [5], the author presents an interesting and enlightening discussion on Flying Qualities. He categorizes them into two distinct categories, namely (i) handling qualities, reflecting the vehicle's behavior in response to pilot controls

and (ii) ride qualities, reflecting response to the response to external disturbances. In summary, the author concludes that for acceptable flying qualities, in a helicopter, the automatic flight control system plays an extremely crucial role in the sense that, without it, it is almost impossible to achieve acceptable flying qualities.

This in turn prompts the reader (likely the student) to appreciate the importance of learning the art of designing a good automatic flight control system by mastering the contents of both classical, transfer function based control theory as well as the time domain state space based control theory, presented in this book.

26.1.12 Helicopter Tutorial Summary

In this tutorial, we have discussed the dynamics and control issues for a rotor-craft vehicle such as a helicopter in contrast to the fixed wing aircraft in the major part of this book. In this brief tutorial type discussion we attempted to make the reader understand the similarities and differences between rotor-craft vehicle dynamics and control and fixed wing vehicle dynamics and control.

26.2 Tutorial on Quadcopter Dynamics and Control

26.2.1 Quadcopter Tutorial Highlights

In this tutorial, we present a brief overview of another popular flight vehicle that has garnered immense attention in recent years under the umbrella of unmanned aerial vehicles (UAV)s, namely the simple and agile quadcopters. While quadcopters (Figure 26.1) have

Figure 26.1 Quadcopter.

many similarities with the helicopter dynamics and control we discussed in the previous tutorial, a brief overview of the dynamics and control issues specific to quadcopters would be worthwhile and this chapter attempts to do the same in a tutorial fashion.

26.2.2 Unmanned Aerial Systems (UAS) and the role of Quadcopters

Unmanned aerial systems (UASs) with various types of UAVs has emerged as a promising arena for improving mobility and efficiency in aviation in recent times. Among those UAVs, small size quadrotor helicopters (also referred to as drones, in a loose sense) have become extremely popular because of their simplicity, agility, and maneuverability to accomplish various tasks such as inspection, surveillance, package delivery, etc. The Federal Aviation Administration (FAA) approved oil and gas companies (like BP) to inspect oil pipelines for structural damage and leaks [1, 4], the German firm DHL and Amazon have attempted a package delivery system with drones (or quadrotor helicopters) [1, 4], and an Israeli company Bladeworx is attempting to use a drone surveillance system to protect the Jerusalem light rail system from riots and vandalism [1, 4], and there are many other such tasks of societal impact. It is not too far fetched to think of a scenario where the FAA introduces policies to legislate the civil and commercial use of UAS airspace as these types of applications are expected to greatly expand in the very near future.

26.2.3 Dynamics and Control Issues of Quadrotors

A quadrotor schematic is shown in Figure 26.2. They are equipped with four rotors typically arranged either in an X configuration, or in a + configuration, each pair of opposite rotors rotating in clockwise fashion, while the other pair rotate counterclockwise to balance the torque. Control is achieved by varying the motor (rotor) speeds.

26.2.3.1 Mathematical Model and Control Inputs
As far as the mathematical modeling is concerned, the fundamental steps in developing the equations of motion remain the same, conceptually, from those we have already

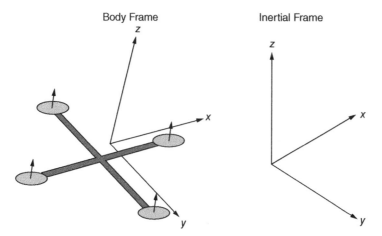

Figure 26.2 Quadcopter body frame and inertial frame.

discussed in this book. In view of the tutorial nature of the coverage of this topic, we thus do not go into the detailed mathematical model development in this chapter. Instead we focus on the issues that make this quadcopter case different from standard fixed wing aircraft. From that point of view, we focus on the control logic part. Note that in a quadcopter, control is achieved essentially by varying the rotor speeds. So let us elaborate on the control inputs needed and how the control law is developed.

The quadrotor is an under-actuated mechanical system, in the sense that there are fewer control variables than the number of degrees of freedom that need to be controlled. The four independent motor thrusts are conceptually the main control variables, where we need to control three attitude rotational degrees of freedom and three translational degrees of freedom. In a quadcopter, its attitude is coupled with the position, i.e. a pitch or roll angular displacement is needed to move it in the (x, y) plane, i.e to move in a forward or backward direction. One of the proposed methods of determining the four virtual control variables that are taken as T_f, the total thrust force, τ_ϕ, the roll torque, τ_θ, the pitch torque, and τ_ψ, the yaw torque. The total thrust force is given by the sum of the thrusts from four rotors, i.e.

$$T_f = f_1 + f_2 + f_3 + f_4. \tag{26.54}$$

A positive roll torque (roll right) is generated by increasing the thrust in one pair of motors (the left motors) and/or decreasing the thrust in the other pair of motors (right motors). Thus, knowing the moment arms of the motors from the cg position, one can determine the roll torque τ_ϕ. Thus we have

$$\tau_\phi = l_1(-f_1 + f_2) + l_2(f_3 - f_4) \tag{26.55}$$

where l_1 and l_2 are the moment arm lengths of the front motors.

Similarly, a positive pitch torque (pitch up) is generated by increasing the thrust in one pair of motors (the front motors) and/or decreasing the thrust in the other pair of motors (back/rear motors). Thus, knowing the moment arms of the motors from the cg position, one can determine the pitch torque τ_θ. Thus we have

$$\tau_\theta = l_3(f_1 + f_2) - l_4(f_3 + f_4) \tag{26.56}$$

where l_3 and l_4 are the moment arm lengths of the back motors.

The drag force on the rotors produces a yawing torque on the quadrotor in the opposite direction of the rotor's rotation. Thus, a positive yaw torque (clockwise from the top view) is generated by increasing the first and third motor speeds and/or decreasing the second and fourth motor speeds, i.e.

$$\tau_\psi = \tau_1 + \tau_2 + \tau_3 + \tau_4. \tag{26.57}$$

Next, we need to analyze the relationship between thrust and the rotor's angular velocity. This relationship varies based on the angle of attack and the free-stream properties of the wind and on the blade flapping and blade geometry. This relationship is difficult to model in an accurate way, but for the small motions such as small angles of attack and low speeds, this relationship is typically approximated and simplified by assuming that the thrust produced by the ith rotor is given by

$$f_i = k_t \Omega_i^2 \tag{26.58}$$

where k_t is a thrust coefficient approximated as a constant.

Similarly, the relationship between the rotor's angular velocity and the counter-torque due to drag on the propeller τ_i, which is also difficult to model, is also approximated and simplified as

$$\tau_i = c_d \Omega_i^2 \tag{26.59}$$

where c_d is assumed to be a constant.

Finally, the relationship between the four virtual control variables and the motor angular velocities can be related by the following matrix relationship, namely

$$
\begin{bmatrix} T_f \\ \tau_\phi \\ \tau_\theta \\ \tau_\psi \end{bmatrix} =
\begin{bmatrix}
k_t & k_t & k_t & k_t \\
-l_1 k_t & l_1 k_t & l_2 k_t & -l_2 k_t \\
l_3 k_t & l_3 k_t & -l_4 k_t & -l_4 k_t \\
c_d & -c_d & c_d & -c_d
\end{bmatrix}
\begin{bmatrix} \Omega_1^2 \\ \Omega_2^2 \\ \Omega_3^2 \\ \Omega_4^2 \end{bmatrix}. \tag{26.60}
$$

Once the controller logic computes the virtual control inputs T_f, τ_ϕ, τ_θ τ_ψ, these are then transformed into the desired angular velocities of the rotors by inverting the above above matrix. This process is labeled as motor mixing in some literature.

Once the above control variable interpretation is understood, the rest of the exercise is to design the control logic and there is an abundance of literature on this depending on the control objective and the performance specifications.

26.2.4 Quadcopter Tutorial Summary

In this section, we have briefly highlighted the important features related to the dynamics and control of quadcopters in a brief, tutorial fashion. The interested reader is encouraged to expand the knowledge base in this area by consulting many recent research papers emerging in this area, few of which are referenced here in this book.

26.3 Tutorial on Missile Dynamics and Control

26.3.1 Missile Tutorial Highlights

In this tutorial, we present a brief tutorial on issues related to missile dynamics, guidance and control [2].

26.3.2 Introduction

Missiles have been effectively used as weapons in many military missions (see Figure 26.3). Typically, when we use the word missile it is understood that it is guided missile. Thus the major discussion on missiles revolves around guidance and control aspects, as they become extremely important. As per the dynamics part, we follow the same mathematical development we outlined before for an aircraft. By using a body axis system the product of inertia term J_{xz} is zero and $I_z = I_y$. Thus for $P = 0$ there is no coupling between the longitudinal and lateral equations. The calculation of the missile transfer functions, conceptually is more complicated than that of an aircraft because of the assumptions we made about the physical properties (such as mass, moments of inertia, cg location, etc) and the flight conditions (such as the Mach number, altitude,

Figure 26.3 Boeing AGM-84L Harpoon Missile. Courtesy of media.defence.gov.

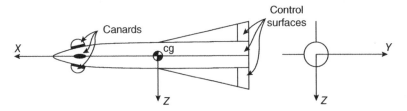

Figure 26.4 An aerodynamic missile and axis system.

etc). Similarly in the computation of all the stability derivatives needed at a given flight condition (i.e. at a given altitude and Mach number (velocity)), in the case of aircraft, we also assume them to be evaluated at constant values of the aerodynamic coefficients. So, strictly speaking, in a missile, these physical properties as well as the flight conditions change relatively rapidly with time, thereby violating some of the assumptions we made in the aircraft case. However, we can still use the same equations of motion we derived for an aircraft for missiles as well, since in majority of the situations the duration of the missile motion is very short and hence we can still assume that the instantaneous values of all these parameters are relatively constant.

In a guided missile (See Figures 26.4 and 26.5), the missile is controlled by commands from the internal guidance system or by commands transmitted to the missile by radio from the ground or a launching vehicle. These guidance commands serve as the desired or reference inputs in a control system. Thus guidance and control are intertwined.

There are various types of guided missiles. Those that are flown in the same manner as manned aircraft, that is missiles that are banked to turn, cruise missiles, and remotely piloted vehicles, which are not of that much interest in this tutorial. We are more focused on aerodynamic missiles, which use aerodynamic lift to control the direction of flight, and ballistic missiles, which are guided during powered flight by deflecting the thrust vector and become free-falling bodies after engine cut-off. One feature of these missiles is that they are roll stabilized. Thus we can assume decoupling between longitudinal

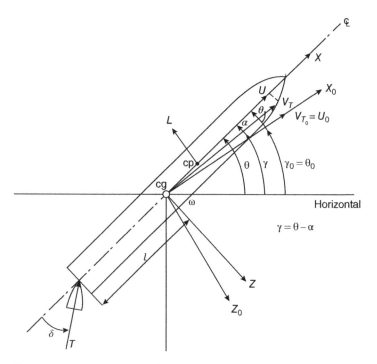

Figure 26.5 Ballistic missile and axis system.

modes and lateral modes. Only from a pure control system design point of view can most of the methods we developed for aircraft case be carried over to the missile control problem. However, as far as the guidance system for ballistic missiles is concerned, the flat earth approximation we made (which amounts to taking the rotating earth as an inertial reference frame) may not hold for computing the guidance commands.

26.3.2.1 Roll Stabilization

For aerodynamic missiles, roll stabilization is accomplished generating the required rolling moment by differential movement of the control surfaces. For ballistic missiles, it is achieved by differential swiveling of small rockets mounted on the side of the missile (e.g. Atlas) or by the main rocket engines if more than one main engine is used.

26.3.2.2 Aerodynamic and Ballistic Missiles

For few types of aerodynamic missiles, control is achieved either by conventional control surfaces with stationary canards or even without canards or completely by canards with no control surfaces on the main lifting surfaces. The typical cruciform missiles have Y and Z axis symmetry, so that controlling pitch axis motion can be done by assuming standard pitch axis short period equations of motion used for the aircraft case.

For ballistic missiles (such as Vanguard) the trajectory of the missile is planned such a way as to make the missile maintain zero angle of attack (see Figure 26.5).

The trajectory of a ballistic-type missile is planned to maintain the missile at a zero angle of attack. This is usually done by programming the pitch attitude or pitch rate to yield a zero-g trajectory.

While some of the assumptions we made in the aircraft case such as:

1. the X and Z axes lie in the place of symmetry, and the origin of the axis system is located at the center of gravity of the missile
2. the perturbations from equilibrium are small

may carry over to the missile case, there are few assumptions we made for aircraft such as:

1. the mass of the missile is constant
2. the missile is a rigid body
3. the Earth is an inertial reference

are not completely valid for the missile case, because the ballistic missile is consuming fuel at a highly rapid rate. However, it is interesting to note that missile control systems designed under the aircraft assumptions have successfully performed, mostly because the duration of the flight times for these missiles are very short and thus the engineering judgment was sufficiently accurate for all practical purposes.

26.3.3 Missile Tutorial Summary

This tutorial briefly covered the basic issues related to the guidance and control of aerodynamic and ballistic missiles.

26.4 Tutorial on Hypersonic Vehicle Dynamics and Control

26.4.1 Hypersonic Vehicle Tutorial Highlights

In this tutorial, we highlight the issues pertaining to dynamics and control of hypersonic vehicles in a brief tutorial fashion [3], see Figure 26.6.

26.4.2 Special Nature of Hypersonic Flight:Hypersonic Flight Stability and Control Issues

Hypersonic vehicles are those which fly at high speeds exceeding Mach numbers of ≥ 5. Naturally, the aerodynamic characteristics of these hypersonic vehicles are significantly different from those of subsonic, transonic and low supersonic vehicles. The major key differences can be summarized as follows:

- Center of pressure (neutral point) position. Unlike subsonic and low supersonic vehicles, for hypersonic vehicles, the center of pressure position (labeled as the neutral point in our previous discussion on aircraft dynamics) does not change for changes in Mach number, angle of attack and altitude. This in turn means that external active control becomes necessary to maintain stability. In other words, essentially, the concept of static stability is absent in this case.
- Robustness is a necessary feature, not simply a desirable feature. The estimation of aerodynamic parameters and the resulting stability derivatives derived from ground tests do not accurately reflect the actual in-flight values making the nominal models themselves not that accurate. Thus robustness to parameters is a necessary feature,

Figure 26.6 NASA X-43a. Pic. courtesy of NASA.

not simply a desirable feature in the design of active controllers for hypersonic vehicles. Thus advanced robust control design methods need to be employed right at the design stage. This aspect of robust control design for aerospace systems has been an active topic of research carried out by this author [7] for a long time and is well documented in the literature. Thus hypersonic vehicle control serves as an incentive to carry out advanced research in this robust control area.

- Importance of nonlinear nature. At hypersonic speeds, the drag and lift forces become nonlinear functions of the angle of attack. At the same time, following a transition phase during transonic and supersonic speeds, the drag and lift coefficients attain constant values at hypersonic speeds. This in turn means that extreme care needs to be taken in interpreting the validity and fidelity of any control design that uses linear models and the need to possibly design inherently nonlinear controllers for hypersonic vehicle models that incorporate the nonlinearities in the model.
- Low lift/drag ratio. Compared to subsonic and supersonic speeds, the maximum value of the lift/drag ratio for hypersonic vehicles is significantly lower. For example, the typical values of L/D for supersonic vehicles is in the range 5–10, whereas it is in the range 1–5 for hypersonic vehicles. How to increase this ratio has been an active topic of research in recent times.
- Heating/thermal issues. Thermal gradients and temperature variations is always a significant issue with hypersonic vehicles. Techniques such as rounding of the nose and other leading edges may be needed to reduce these issues. A means of dissipating heat is a challenging issue. This issue is directly related to the structural integrity issue. Thus an integrated approach to design the airframe taking into account all these

interrelated issues is needed. In particular, issues, such as control structure interactions, vehicle thermal management and others, make the control system design for hypersonic vehicles a truly challenging and exciting task.

26.4.3 Hypersonic Vehicle Tutorial Summary

In summary, it can be seen that a hypersonic vehicle's dynamics and control issues are much more complex and as such more sophisticated dynamic modeling as well as more advanced robust control design techniques have to be used.

Bibliography

1 Nizar Hadi Abbas and Ahmed Ramz Sami. Tuning of pid controllers for quadcopter system using hybrid memory based gravitational search algorithm - particle swarm optimization. *International Journal of Computer Applications(0975-8887)*, 2017.

2 J.H. Blakelock. *Automatic Control of Aircraft and Missiles*. Wiley Interscience, New York, 1991.

3 C.C Coleman and F. Faruqi. On stability and control of hypersonic vehicles. *DSTO-TR-2358*, 2009.

4 David W. Kun and Inseok Hwang. Linear matrix inequality-based nonlinear adaptive robust control of quadcopter. *Journal of Guidance, Control, and Dynamics* Vol. 39, No. 5, May 2016.

5 G.D. Padfield. *Helicopter flight dynamics*, pages 282–287. Blackwell Publishing, Oxford, U.K., 2007.

6 Newman S. Seddon J.M. *Basic Helicopter Aerodynamics. 3rd Edition*. John Wiley, 2011.

7 Rama K Yedavalli. *Robust Control of Uncertain Dynamic Systems: A Linear State Space Approach*. Springer, 2014.

Appendices

Appendix A

Data for Flight Vehicles

A.1 Data for Several Aircraft

Aircraft

1. A-7A
2. A-4D

A.1.1 A-7A

Nominal cruise configuration

Clean airplane

60% fuel

$W = 21,889$ lbs

CG at 30% MGC

$I_x = 13,635$ slug – ft^2

$I_y = 58,966$ slug – ft^2

$I_z = 67,560$ slug – ft^2

$I_{xz} = 2,933$ slug – ft^2

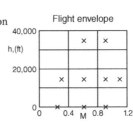

Reference geometry

$$S = 375 \text{ ft}^2$$
$$c = 10.8 \text{ ft}$$
$$b = 38.7 \text{ ft}$$

Flight Dynamics and Control of Aero and Space Vehicles, First Edition. Rama K. Yedavalli.
© 2020 John Wiley & Sons Ltd. Published 2020 by John Wiley & Sons Ltd.

References

1 LTV Vought Aeronautics Div. Rept. No. 2-53310/5R-1981, "A-7A Aerodynamics Data Report", 21 May 1965 (U).
2 LTV Vought Aeronautics Div. Rept. No. 2-53310/5R-5121, Rev. I, "A-7A Estimated Flying Qualities", 20 August 1965 (C).
3 LTV Vought Aeronautics Div., "Updated A-7A Aircraft Lateral-Directional Cruise Device Configuration Data", 25 August 1967.

 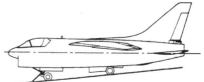

Basic Data Sources
Wind tunnel test and estimates
Some lateral-directional derivatives
Adjusted after flight test.

Table A.1 A-7A flight conditions.

	Flight condition				Flight condition		
	3	4	8		3	4	8
h, ft	0	15,000	35,000	α_0, deg	2.1	13.3	7.5
M_∞	0.9	0.3	0.6	U_0, fps	1,004	309	579
a, fps	1,117	1,058	973.3	W_0, fps	36.8	72.9	76.2
ρ, slft^{-3}	2.378×10^{-3}	1.496×10^{-3}	7.368×10^{-4}	δ_{E_0}, deg	-3.8	-8.8	-5.4
V_V, fps	1,005	317	584	γ_0, deg	0	0	0
q_∞, psf	1,200	75.3	126				

Note: All data is given in fuselage reference axes. Clean, flexible airplane.

Table A.2 A-7A longitudinal dimensional derivatives.

	Flight condition				Flight condition		
	3	4	8		3	4	8
h, ft	0	15,000	35,000	Z_{δ_E}	-318	-23.8	-43.2
M_∞	0.9	0.3	0.6	$M_u + M_{Pu}$	0.00118	0.00183	0.000873
$X_u + X_{Pu}$	-0.0732	0.00501	0.00337	M_α	-40.401	-2.463	-4.152
X_α	-28.542	1.457	8.526	$M_{\dot\alpha}$	-0.372	-0.056	-0.065
X_{δ_E}	11.6	5.63	5.7	M_q	-1.57	-0.340	-0.330
$Z_u + Z_{Pu}$	0.0184	-0.0857	-0.0392	M_{δ_E}	-58.6	-4.52	-8.19
Z_α	$-3,417$	-172.77	-323.54				

Note: All data is given in fuselage reference axes. Clean, flexible airplane.

Table A.3 A-7A lateral directional derivatives.

	Flight condition					Flight condition		
	3	4	8			3	4	8
h, ft	0	15,000	35,000	L'_{δ_A}		−25.2	3.75	7.96
M_∞	0.9	0.3	0.6	L'_{δ_R}		13.2	1.82	3.09
Y_β	−516.57	−38.674	−49.465	N'_β		17.2	0.948	1.38
Y_{δ_A}	−8.613	−0.4755	−1.5593	N'_p		−0.319	−0.031	−0.0799
Y_{δ_R}	62.913	9.732	15.593	N'_r		−1.54	−0.271	−0.247
L'_β	−98.0	−8.79	−14.9	N'_{δ_A}		1.56	0.280	0.652
L'_p	−9.75	−1.38	−1.4	N'_{δ_r}		−11.1	−1.56	−2.54
L'_r	1.38	0.857	0.599					

Note: All data is given in fuselage reference axes. Clean, flexible airplane.

A.1.2 A-4D

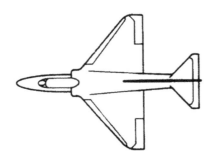

Nominal cruise configuration

Clean airplane

60% fuel

$W = 21,889$ lbs

CG at 30% MGC

$I_x = 13,635$ slug − ft^2

$I_y = 58,966$ slug − ft^2

$I_z = 67,560$ slug − ft^2

$I_{xz} = 2,933$ slug − ft^2

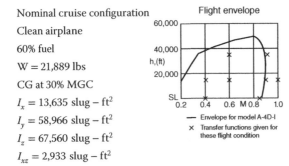

Flight envelope

— Envelope for model A-4D-I
× Transfer functions given for these flight condition

Reference
Geometry

$S = 375$ ft^2

$c = 10.8$ ft

$b = 38.7$ ft

References

1 LTV Vought Aeronautics Div. Rept. No. 2-53310/5R-1981, "A-7A Aerodynamics Data Report", 21 May 1965 (U).

2 LTV Vought Aeronautics Div. Rept. No. 2-53310/5R-5121, Rev. I, "A-7A Estimated Flying Qualities", 20 August 1965 (C).

3 LTV Vought Aeronautics Div., "Updated A-7A Aircraft Lateral-Directional Cruise Device Configuration Data", 25 August 1967.

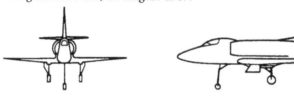

Basic Data Sources
Wind tunnel test and estimates
Some lateral-directional derivatives
Adjusted after flight test.

Table A.4 A-4D flight conditions.

	Flight condition			Flight condition	
	4	7		4	7
h, ft	15,000	35,000	q_∞, psf	301	126
M_∞	0.6	0.6	α_0, deg	3.4	8.8
a, fps	1058	973.3	U_0, fps	634	577
ρ, sl ft^{-3}	1.496×10^{-3}	7.368×10^{-4}	W_0, fps	37.7	89.3
V_V, fps	635	584	γ_0, deg	0	0

Note: All data is given in fuselage reference axes. Clean, flexible airplane.

Table A.5 A-4D longitudinal dimensional derivatives.

	Flight condition			Flight condition	
	4	7		4	7
h, ft	15,000	35,000	Z_{δ_E}	−56.68	−23.037
M_∞	0.6	0.6	$M_u + M_{Pu}$	0.00162	0.001824
$X_u + X_{Pu}$	−0.00938	0.000806	M_α	−12.954	−5.303
X_α	26.797	13.257	$M_{\dot\alpha}$	−0.3524	−0.1577
X_{δ_E}	7.396	6.288	M_q	−1.071	−0.484
$Z_u + Z_{Pu}$	−0.0533	−0.0525	M_{δ_E}	−19.456	−8.096
Z_α	−521.97	−226.242			

Note: All data is given in fuselage reference axes. Clean, flexible airplane.

Table A.6 A-4D lateral directional derivatives.

	Flight condition			Flight condition	
	4	8		4	8
h, ft	15,000	35,000	L'_{δ_A}	21.203	8.170
M_∞	0.6	0.6	L'_{δ_R}	10.398	4.168
Y_β	−144.78	−60.386	N'_β	16.629	6.352
Y_{δ_A}	−2.413	−0.4783	N'_p	−0.02173	−0.02513
Y_{δ_R}	25.133	10.459	N'_r	−0.5144	−0.2468
L'_β	−35.95	−17.557	N'_{δ_A}	1.769	0.5703
L'_p	−1.566	−0.761	N'_{δ_r}	−7.78	−3.16
L'_r	0.812	0.475			

Note: All data is given in fuselage reference axes. Clean, flexible airplane.

A.2 Data for Selected Satellites

Table A.7 lists representative satellite masses with and without propellant for various types of spacecraft. Table A.8 further breaks down the dry mass by the percentage devoted to each subsystem.

Table A.7 Actual mass for selected satellites. The propellant load depends on the satellite design life.

Spacecraft name	Loaded mass (kg)	Propellant mass (kg)	Dry mass (kg)	Propellant mass (%)	Dry mass (%)
Communication satellite					
DSCS II	530.0	54.1	475.9	10.2%	89.8%
DSCS III	1095.9	228.6	867.3	20.9%	79.1%
NATO III	346.1	25.6	320.4	7.4%	92.6%
Intelsat IV	669.2	136.4	532.8	20.4%	79.6%
TDRSS	2150.9	585.3	1565.7	27.2%	72.8%
Average	981.7	166.2	815.6	16.9%	83.1%
Navigation satellite					
GPS Block 1	508.6	29.5	479.1	5.8%	94.2%
GPS Block 2,1	741.4	42.3	699.1	5.7%	94.3%
GPS BLock 2,2	918.6	60.6	858.0	6.6%	93.4%
Average	722.9	44.1	678.7	6.1%	93.9%
Remote sensing satellite					
P80-1	1740.9	36.6	1704.4	2.1%	97.9%
DSP-15	2277.3	162.4	2114.9	7.1%	92.9%

Table A.7 (Continued)

Spacecraft name	Loaded mass (kg)	Propellant mass (kg)	Dry mass (kg)	Propellant mass (%)	Dry mass (%)
DMSP 5D-2	833.6	19.1	814.6	2.3%	97.7%
DMSP 5D-3	1045.5	33.1	1012.3	3.2%	96.8%
Average	1474.3	62.8	1411.6	4.3%	95.7%
		LightSats			
Orsted	60.8	No propulsion	60.8	n/a%	100.0%
Freja	255.9	41.9	214.0	16.4%	83.6%
SAMPEX	160.7	No propulsion	160.7	n/a	100.0%
HETE	125.0	No propulsion	125.0	n/a	100.0%
Clementine	463.0	231.0	232.0	49.9%	50.1%
Pluto Fast Flyby '93	87.4	6.9	80.5	7.9%	92.1%

Table A.8 Mass distribution for selected spacecraft.

Spacecraft name	Percentage of spacecraft dry mass						
	Payload	Structural	Thermal	Power	TT&C	ADCS	Propulsion
	Communication satellite						
DSCS II	23%	23.5%	2.8%	29.3%	7.0%	11.5%	3.0%
DSCS III	32.3%	18.2%	5.6%	27.4%	7.2%	4.4%	4.1%
NATO III	22.1%	19.3%	6.5%	34.7%	7.5%	6.3%	2.4%
Intelsat IV	31.2%	22.3%	5.1%	26.5%	4.3%	7.4%	3.1%
TDRSS	24.6%	28.0%	2.8%	26.4%	4.1%	6.2%	6.9%
Average	27.4%	21.3%	3.6%	31.9%	4.8%	6.9%	3.8%
	Navigation satellite						
GPS Block 1	20.5%	19.9%	8.7%	35.8%	5.8%	6.2%	3.6%
GPS Block 2,1	20.2%	25.1%	9.9%	31.0%	5.2%	5.4%	3.3%
GPS BLock 2,2	23.0%	25.4%	11%	29.4%	3.1%	5.3%	2.7%
Average	21.2%	23.5%	9.9%	32.1%	4.7%	5.6%	3.2%
	Remote sensing satellite						
P80-1	41.1%	19.0%	2.4%	19.95%	5.2%	6.3%	6.1%
DSP-15	36.9%	22.5%	0.5%	26.9%	3.8%	5.5%	2.2%
DMSP 5D-2	29.9%	15.6%	2.8%	21.5%	2.5%	3.1%	7.4%
DMSP 5D-3	30.5%	18.4%	2.9%	29.0%	2.0%	2.9%	8.7%
Average	34.6%	18.9%	2.1%	24.3%	3.4%	4.5%	6.1%

Table A.8 (Continued)

Spacecraft name	Percentage of spacecraft dry mass						
	Payload	Structural	Thermal	Power	TT&C	ADCS	Propulsion
LightSats							
Orsted	21.5%	38.3%	0.8%	15.8%	16.7%	6.8%	none%
Freja	34.1%	22.7%	2.4%	19.05%	8.7%	6.0%	7.0%
SAMPEX	32.5%	23.1%	2.5%	25.0%	10.6%	6.3%	none%
HETE	35.3%	16.0%	1.8%	20.3%	8.5%	18.1%	none%
Pluto Fast Flyby '93	8.7%	18.1%	4.6%	24.1%	23.9%	8.3%	12.3%

This is data for the satellite moment of inertia.

Satellite	Mass (kg)	Moment of inertia (kgm^2)		
	M	I_x	I_y	I_z
LAGEOS	406.97	11.42	10.96	10.96
Firesat II	215	90	60	90
IntelSat IV	595	93	225	702

U.S. space shuttles.

Dimension	Length	System	56.1 m
		Orbiter	37.2 m
	Height	System	23.3 m
		Orbiter	17.3 m
	Wingspan	Orbiter	23.8 m
Weight	Gross at lift off		1,995,840 kg
	Orbiter at landing		84,800 kg
Thrust	Solid rocket boosters (2)		1,305,000 kg each at sea level
	Orbiter main engines (3)		168,750 kg each at sea level
Cargo bay		18 m long, 4.5 m in diameter	

Appendix B

Brief Review of Laplace Transform Theory

B.1 Introduction

Laplace transform theory is very useful in solving linear constant coefficient ordinary differential equations. This is accomplished by first converting the differential equations (in the time domain) to algebraic equations in the Laplace variable (frequency) domain; do all the required algebraic manipulations in the Laplace domain; and then finally converting these Laplace domain equations back into time domain equations via the inverse Laplace transformation technique. Hence in this appendix, we briefly review Laplace transform theory and illustrate its use in solving constant coefficient ordinary differential equations. Elaborate versions of the Laplace transform theory are available in many excellent linear control theory books such as those cited in the text, but in this appendix we present only the very basic version needed and adequate for our purposes.

B.2 Basics of Laplace Transforms

The transformation of a function $f(t)$ to the Laplace domain $F(s)$ is given by the formula

$$F(s) = \mathcal{L}(f(t)) = \int_{0^-}^{\infty} f(t)e^{-st}dt \quad \text{where} \quad s = \sigma + \vec{j}\omega. \tag{B.1}$$

The lower limit notation allows the integral to handle functions that are discontinues at zero because it starts just before zero, which allows it to handle functions such as impulse. To return $F(s)$ from the Laplace domain to the time domain, we use

$$f(t)u(t) = \frac{1}{2\pi j} \int_{\sigma-j\infty}^{\sigma+\vec{j}\infty} F(s)e^{st} \, ds. \tag{B.2}$$

It is important to note that all Laplace transforms start at zero, although sometimes $u(t)$ will be placed next to the function, which is a step function where

$$u(t) = 0 \quad \text{when} \quad t < 0$$
$$u(t) = 1 \quad \text{when} \quad t > 0.$$

Example B.1   Suppose we want the Laplace transform of $f(t) = u(t)$, i.e. of a unit step function.

Solution

Then we would use Equation B.1

$$F(s) = \int_{0^-}^{\infty} f(t)e^{-st}dt$$
$$= \int_0^{\infty} 1e^{-st}dt$$
$$= \frac{-e^{-st}}{s}\Big|_0^{\infty}$$
$$= \frac{-e^{-s\infty}}{s} - \frac{-e^{-s0}}{s}$$
$$= 0 - \left(\frac{-1}{s}\right)$$
$$= \frac{1}{s}.$$

Obviously, considerable effort needs to be spent in carrying out this integration process for getting the Laplace transforms for various given time functions. Fortunately, this was done by our predecessors and thus many good mathematics and control theory textbooks contain tables of a given time function and its corresponding Laplace transformed function. Out of the many time functions, for our present purposes in illustrating the main concepts we consider only very few basic functions that have engineering relevance. Hence Table B.1 shows the most important ones used in linear control system theory. These transforms are very useful in the control theory that is developed in this book and instant familiarity with this table is expected at the undergraduate level to the extent of a suggestion that this table's contents be memorized.

Note that the functions listed in Table B.1 have engineering relevance. They can be thought of as representing frequently occurring input functions with a physical meaning. For example, the unit impulse function δ and the unit step function $u(t)$ are somewhat like the worst case inputs in that they act suddenly. The ramp function t represents a function that grows linearly as a function of time, which is a frequent phenomenon

Table B.1 Laplace transform table for key time functions.

Type	$f(t), t > 0$	$F(s)$
Unit impulse	δ	1
Unit step	$u(t)$	$\dfrac{1}{s}$
Unit ramp	$tu(t)$	$\dfrac{1}{s^2}$
Parabolic	$t^n u(t)$	$\dfrac{factorial n}{s^{n+1}}$
Sine function	$\sin \omega t u(t) t$	$\dfrac{\omega}{\omega^2 + s^2}$
Cosine function	$\cos \omega t u(t)$	$\dfrac{s}{\omega^2 + s^2}$
Exponential function	$e^{-at}u(t)$	$\dfrac{1}{s+a}$

Table B.2 Key Laplace transform theorems.

Name	Theorem
Definition	$f(t) = F(s) = \int_{0^-}^{\infty} f(t)e^{-st}dt$
Linearity theorem 1	$kf(t) = kF(s)$
Linearity theorem 2	$f_1(t) + f_2(t) = F_1(s) + F_2(s)$
Differential	$\dfrac{df(t)}{dt} = sF(s) - f(0^-)$
Differential	$\dfrac{df(t)^n}{dt^n} = s^n F(s) - \sum_{k=1}^{n} s^{n-k} f^{k-1}(0^-)$
Final value	$f(\infty) = \lim_{s \to 0} sF(s)$
Convolution	$f_1(t) * f_2(t) = F_1(s)F_2(s)$

of engineering significance. So are the other few functions covered in the table. The trigonometric functions are very important due to the fact that many periodic, general time functions can be expressed as summation of individual sine and cosine terms and thus if a linear system's response to these functions are known, by superposition principle, it is possible to gain insight into the response to a general time function $f(t)$. The above relationships in the table can also be used to interpret the inverse Laplace transforms. For example, the inverse Laplace transform of $\frac{1}{s+a}$ is $e^{-at}u(t)$ and so on.

Along with the key Laplace transforms listed above, there are several Laplace transform theorems that are very helpful in control theory, which are listed in Table B.2. These theorems play an extremely important role in the utility of Laplace transform theory in solving linear constant coefficient differential equations, as can be seen later.

One of the most important theorems is the differentiation theorem, which helps us to convert a differential equation in the time domain into an algebraic equation in the frequency domain.

Example B.2 Consider the following linear constant coefficient differential equation (with all initial conditions equal to zero) where

$$\frac{d^3y(t)}{dt^3} + 5\frac{d^2y(t)}{dt^2} + 8\frac{dy(t)}{dt} + 2y(t) = \sin 3t.$$

Solution
Using the differentiation theorem and the linearity theorems 1 and 2, we get

$$s^3 Y(s) + 5s^2 Y(s) + 8sY(s) + 2Y(s) = \frac{3}{s^2 + 9}.$$

Hence the function $Y(s)$ is now given by an algebraic expression

$$Y(s) = \frac{3}{(s^2 + 9)(s^3 + 5s^2 + 8s + 2)}.$$

Then we can get the time function $y(t)$ by taking the inverse Laplace transformation of the Laplace function on the right-hand side.

Hence, we now need to learn the process of inverse Laplace transformation, which is briefly reviewed next.

B.3 Inverse Laplace Transformation using the Partial Fraction Expansion Method

Note that, as demonstrated in the above example, a Laplace function $Y(s)$, on the left-hand side, which is the Laplace transform function of a yet to be known time function $y(t)$, is mostly known on the right-hand side, as a ratio of a numerator polynomial in s and denominator polynomial in s i.e.

$$Y(s) = \frac{N(s)}{D(s)}.$$

The way to get the time function $y(t)$ very much depends on the way the denominator polynomial $D(s)$ is factored. Let us consider the simple case of $D(s)$ consisting of simple, distinct first order factors.

Partial Fraction Expansion with Roots that are Real and Distinct

$$F(s) = \frac{N(s)}{(s+a)(s+b)}.$$

The inverse Laplace transformation procedure is best illustrated with the help of an example.

Example B.3 Consider

$$Y(s) = \frac{8}{(s+2)(s+4)}.$$

Solution
First we will use partial fraction expansion to simplify the equation and the one of the key Laplace transforms. So first we expand the equation.

$$Y(s) = \frac{8}{(s+2)(s+4)}$$
$$= \frac{A}{s+2} + \frac{B}{s+4}.$$

Next we have to find the coefficients, A and B. To find A we have to multiple the equation by $s+2$ on both sides to isolate A.

$$F(s) = \frac{8}{(s+4)}$$
$$= A + \frac{B(s+2)}{s+4}.$$

Now letting $s = -2$, which eliminates the B term, giving us $A = 4$. Likewise to find B we multiply the expanded equation by $s + 4$ and we let $s = -4$ like we did with the first term giving us $B = -4$. We thus have a simplified version of the original Laplace function, namely

$$Y(s) = \frac{4}{s+2} + \frac{-4}{s+4}.$$

Now, we use the Laplace transform table content corresponding to $1/(s + a)$ and linearity theorem to obtain the needed time function.

$$y(t) = \text{Laplace inverse of} \left[\frac{4}{s+2} + \frac{-4}{s+4} \right]$$
$$= 4e^{-2t} - 4e^{-4t}.$$

There are many excellent textbooks such as those that elaborate on this inverse Laplace transformation process for various other cases of the expansion of the denominator polynomial $D(s)$ and we refer the reader to that material. The purpose of this brief appendix is to only very briefly convey the overall conceptual procedure of the use of Laplace transform theory to solve linear constant coefficient differential equations. It is expected that these basics in Laplace transform theory are covered elsewhere in a prerequisite mathematics course in the undergraduate curriculum. The best way to learn this process is to solve exercise problems such as those given next.

B.4 Exercises

B.4.1 Exercises on Laplace Transformation

Convert the following functions of time into the Laplace domain functions:

1. $f(t) = te^{4t}$
2. $f(t) = (t + 1)^3$
3. $f(t) = 4t^2 - 5\sin 3t$
4. $f(t) = t\cos t$
5. $f(t) = 4t - 10$
6. te^{10t}
7. $t^3 e^{-2t}$
8. $t\cos 2t$
9. $t\sin 2t$.

Solutions

1. $te^{4t} = \dfrac{1}{(s-4)^2}$

2. $(t+1)^3 = \dfrac{1}{5}(\cos t\sqrt{5})$

3. $4t^2 - 5\sin 3t = \dfrac{8}{s^3} - \dfrac{15}{s^2+9}$

4. $t\cos t = \dfrac{s^2 - 1}{(s^2 + 1)^2}$

5. $4t - 10 = \dfrac{4}{s^2} - \dfrac{10}{s}$

6. $te^{10t} = \dfrac{1}{(s-10)^2}$

7. $t^3 e^{-2t} = \dfrac{1}{(s+2)^4}$

8. $t\cos 2t = \dfrac{s^2 - 4}{(s^2 + 4)^2}$

9. $t\sin 2t = \dfrac{4s}{(s^2 + 4)^2}$.

B.4.2 Exercises on Inverse Laplace Transformation

Determine the inverse Laplace transforms for the given Laplace functions.

1. $\dfrac{1}{s^3}$

2. $\dfrac{1}{s^2} - \dfrac{1}{s} + \dfrac{1}{s-2}$

3. $\dfrac{4s}{4s - 1}$

4. $\dfrac{s}{(s-2)(s-6)(s-3)}$

5. $\dfrac{1}{(s-5)^3}$.

Solutions

1. $\dfrac{1}{s^3} = \dfrac{1}{2}t^2$

2. $\dfrac{1}{s^2} - \dfrac{1}{s} + \dfrac{1}{s-2} = t - 1 + e^{2t}$

3. $\dfrac{4s}{4s - 1} = \cos\dfrac{t}{2}$

4. $\dfrac{s}{(s-2)(s-6)(s-3)} = \dfrac{1}{2}e^{2t} - e^{3t} + \dfrac{1}{2}e^{6t}$

5. $\dfrac{1}{(s-5)^3} = \dfrac{1}{2}t^2 e^{5t}$.

B.4.3 Other Exercises

1. Find the time function, for all positive times, for each of the following

 (a) $Y(s) = \dfrac{s}{s+2}$

 (b) $Y(s) = \dfrac{3s - 5}{s^2 + 4s + 2}$

(c) $Y(s) = \dfrac{10}{s^3 + 25s^2 + 5s}$

(d) $Y(s) = \dfrac{4(s + 1)}{(s + 2)(s + 3)^2}$.

2. Solve the following differential equations using Laplace transform methods.

(a) $\dfrac{dy}{dt} + 4y = 6e^{2t}$ with the initial condition $y(0^-) = 3$

(b) $\dfrac{dy}{dt} + y = 3\cos 2t$ with the initial condition $y(0^-) = 0$

(c) $\dfrac{d^3y}{dt^3} + 5\dfrac{d^2y}{dt^2} + 6\dfrac{dy}{dt} = 0$ with the initial conditions:

$$y(0^-) = 3$$
$$\dot{y} = (0^-) = -2$$
$$\ddot{y} = (0^-) = 7.$$

3. Find the transfer function $\dfrac{Y}{R}(s)$ for the following equation

(a) $\dfrac{d^3y}{dt^3} + 6\dfrac{d^2y}{dt^2} + 2\dfrac{dy}{dt} + 4y = -5\dfrac{d^2r}{dt^2} + 8\dfrac{dr}{dt}$

(b) and the system of equations

$$\dfrac{dx_1}{dt} = -3x_1 + x_2 + 4r$$

$$\dfrac{dx_2}{dt} = -2x_1 - r$$

$$y = x_1 - 2x_2.$$

4. Find the zero-state response (i.e. zero initial conditions) of the following systems. Here, $T(s) \equiv \dfrac{Y}{R}(s)$

(a) $T(s) = \dfrac{4}{s + 3}$ with the input $r(t) = u(t)$

(b) $T(s) = -\dfrac{5s}{s^2 + 4s + 3}$ with the input $r(t) = 6u(t)e^{-2t}$.

5. Find the complete response for the given initial conditions and input $r(t)$

(a) $T(s) = \dfrac{1}{s + 4}$ with the input $r(t) = \delta(t)$ and the initial function $y(0^-) = 0$

(b) $T(s) = \dfrac{s - 5}{s^2 + 3s + 2}$ with the input $r(t) = u(t)$ and the initial conditions

$$y(0^-) = -3$$
$$y(0^-) = 4.$$

Solutions

1. (a) $y(t) = \delta(t) - 2e^{-2t}$
   (b) $y(t) = 5.39e^{-3.41t} - 2.39e^{-0.59t}$

(c) $y(t) = 3(e^{-2t} - e^{-3t}) - [6e^{-2(t-2)} - 6e^{-3(t-2)}]u(t - 2)$

(d) $y(t) = -4e^{-2t} + 4e^{-3t} + 8te^{-3t}$.

2. (a) $y(t) = e^{2t} + 2e^{-4t}$

(b) $y(t) = -\dfrac{3}{5}e^{-t} + \dfrac{3}{5}\cos 2t + \dfrac{6}{5}\sin 2t$

(c) $y(t) = \dfrac{13}{6} + e^{-3t} - \dfrac{1}{2}e^{-2t}$.

3. The solution to the first equation:

$$\dfrac{Y}{R}(s) = \dfrac{-5s^2 + 8s}{s^3 + 6s^2 + 2s + 4}$$

and the solution to the second equation

$$\dfrac{Y}{R}(s) = \dfrac{6s + 21}{s^2 + 3s + 2}.$$

4. (a) $T(t) = \dfrac{4}{3} - \dfrac{4}{3}e^{-3t}$

(b) $T(t) = 15e^{-t} - 60e^{-2t} + 45e^{-3t}$

5. (a) $T(t) = 10e^{-4t}$

(b) $T(t) = -\dfrac{5}{2} + 4e^{-t} - \dfrac{9}{2}e^{-2t}$

Appendix C

A Brief Review of Matrix Theory and Linear Algebra

C.1 Matrix Operations, Properties, and Forms

In this appendix we include briefly some rudimentary material on matrix theory and linear algebra needed for the material in the book. In particular we review some properties of matrices, eigenvalues, singular values, singular vectors and norms of vectors, and matrices. We have used several texts and journal papers (given as references in each chapter of this book) in preparing this material. Hence those references are not repeated here.

Principal Diagonal: consists of the m_{ii} elements of a square matrix M.

Diagonal Matrix: a square matrix in which all elements of the principal diagonal are zero.

Trace: sum of all the elements on the principal diagonal of a square matrix.

$$\text{trace } M = \sum_{i=1}^{n} m_{ii}. \tag{C.1}$$

Determinant: denoted by det[M] or M, definition given in any linear algebra book.

Singular matrix: a square matrix whose determinant is zero.

Minor: the minor M_{ij} of a square matrix M is the determinant formed after the ith row and jth column are deleted from M.

Principal minor: a minor whose diagonal elements are also diagonal elements of the original matrix.

Cofactor: a signed minor given by

$$c_{ij} = (-1)^{i+j} M_{ij}. \tag{C.2}$$

Adjoint matrix: the adjoint of M, denoted by adj[M], is the transpose of the cofactor matrix. The cofactor matrix is formed by replacing each element of M by its cofactor.

Inverse matrix: inverse of M is denoted by $M(-1)$, has the property $MM^{-1} = M^{-1}M = I$, and is given by

$$M^{-1} = \frac{\text{adj}[M]}{|M|}. \tag{C.3}$$

Flight Dynamics and Control of Aero and Space Vehicles, First Edition. Rama K. Yedavalli.
© 2020 John Wiley & Sons Ltd. Published 2020 by John Wiley & Sons Ltd.

Rank of a matrix: the rank r of a matrix M (not necessarily square) is the order of the largest square array contained in M that has a non-zero determinant.

Transpose of a matrix: denoted by M^T, it is the original matrix with its rows and columns interchanged, i.e. $m'_{ij} = m_{ji}$.

Symmetric matrix: a matrix containing only real elements that satisfies $M = M^T$.

Transpose of a product of matrices:

$$(AB)^T = B^T A^T.$$ (C.4)

Inverse of a product of matrices:

$$(AB)^{-1} = B^{-1} A^{-1}.$$ (C.5)

(Complex) conjugate: the conjugate of a scalar $a = \alpha + j\beta$ is $a^* = \alpha - j\beta$. The conjugate of a vector or matrix simply replaces each element of the vector or matrix with its conjugate, denoted by $m*$ or $M*$.

Hermitian matrix: a matrix that satisfies

$$M = M^H = (\overline{M})^T$$ (C.6)

where superscript H stands for Hermitian. The operation of a Hermitian is simply complex conjugate transposition – usually, $*$ is used in place of H.

Unitary matrix: a complex matrix U is unitary if $U^H = U^{-1}$.

Orthogonal matrix: a real matrix R is orthogonal if $R^T = R^{-1}$.

C.1.1 Some Useful Matrix Identities

1. $$[I_n + G_2 G_1 H_2 H_1]^{-1} G_2 G_1 = G_2 [I_m + G_1 H_2 H_1 G_2]^{-1} G_1$$
$$= G_2 G_1 [I_r + H_2 H_1 G_2 G_1]^{-1}$$
$$= G_2 G_1 - G_2 G_1 H_2 [I_p + H_1 G_2 G_1 H_2]^{-1} H_1 G_2 G_1 \quad \text{(C.7)}$$

where G_1 is $(m \times r)$, G_2 is $(n \times m)$, H_1 is $(p \times n)$, and H_2 is $(r \times p)$.
For the following three identities, the dimensions of matrices P, K, and C are: P is $(n \times n)$, K is $(n \times r)$, and C is $(r \times n)$.

2. $$(P^{-1} + KC)^{-1} = P - PK(I + CPK)^{-1}CP$$ (C.8)

3. $$(I + KCP)^{-1} = I - K(I + CPK)^{-1}CP$$ (C.9)

4. $$(I + PKC)^{-1} = I - PK(I + CPK)^{-1}C$$ (C.10)

5. $$\det(I_n + MN) = \det(I_m + NM)$$ (C.11)

where M and N are matrices of appropriate dimensions such that the products MN and NM are square matrices of dimension n and m respectively.

C.2 Linear Independence and Rank

A set of mathematical objects $a_1, a_2, \cdots, a_r$ (specifically, in our case vectors or columns of a matrix) is said to be linearly dependent, if and only if there exists a set of constants $c_1, c_2, \cdots, c_r$, not all zero, such that

$$c_1 a_1 + c_2 a_2 + \cdots + c_r a_r = 0.$$

If no such set of constants exists, the set of objects is said to be linearly independent. Suppose a is a matrix (not necessarily square) with $a_1, a_2, \cdots, a_r$ as its columns

$$A = [a_1 | a_2 | \cdots | a_n].$$

The rank of A, sometimes written rank(A) or $r(A)$ is the largest number of independent columns (or rows) of A. The rank of A cannot be greater than the minimum of the number of columns or rows, but it can be smaller than that minimum. A matrix whose rank is equal to that minimum is said to be of full rank.

A fundamental theorem regarding the rank of a matrix can be stated as follows: The rank of A is the dimension of the largest square matrix with non-zero determinant than can be formed by deleting rows and columns from A.

Thus we can say that the rank of a matrix is the maximum number of linearly independent columns (rows) of the matrix, the test for which is to look for the largest dimension square matrix with non-zero determinant found embedded in the matrix.

Numerical determination of the rank of a matrix is not a trivial problem: if the brute force method of testing is used, a goodly number of determinants must be evaluated. Moreover, some criterion is needed to establish how close to zero a numerically computed determinant must be in order to be declared zero. The basic numerical problem is that rank is not a continuous function of the elements of a matrix; a small change in one of the elements of a matrix can result in a discontinuous change of its rank.

The rank if a product of two matrices cannot exceed the rank of either factor

$$\text{rank } (AB) \leqq \min \, [\text{rank } (A), \,\, \text{rank } (B)]. \tag{C.12}$$

However, if either factor is a non-singular (square) matrix the rank of the product is the rank of the remaining factor:

$$\text{rank } (AB) = \text{rank } (A) \text{ if } B^{-1} \text{ exists}$$

$$\text{rank } (AB) = \text{rank } (B) \text{ if } A^{-1} \text{ exists}. \tag{C.13}$$

C.2.1 Some Properties Related to Determinants

For square matrices A and B, $\det(AB) = \det(A)\det(B)$.

However, $\det(A + B) \neq \det(A) + \det(B)$

$$\det[I_n + MN] = \det[I_m + NM]. \tag{C.14}$$

For a square matrix A, with block diagonal partitioned matrices A_i (i=1,2,$\cdots$ r)

$$\det A = \det[A_1]\det[A_2]\cdots\det[A_r]. \tag{C.15}$$

C.3 Eigenvalues and Eigenvectors

A is an $(n \times n)$ matrix, and v_i is an $(n \times 1)$ vector. The eigenvalue problem is

$$[\lambda_i I - A]v_i = 0. \qquad (C.16)$$

Solution:

$$\det[\lambda_i I - A] = 0 \qquad (C.17)$$

gives the eigenvalues $\lambda_1, \lambda_2, , \lambda_n$. Given λ_i, the non-trivial solution vector v_i of (C.16) is called the eigenvector corresponding to the eigenvalue λ_i. We also refer to $v_1, v_2, , v_n$ as right eigenvectors. These are said to lie in the null space of the matrix $[\lambda_i I - A]$. The eigenvectors obtained from

$$w_i^T[\lambda_i I - A] = 0 \qquad (C.18)$$

are referred to as left eigenvectors. Left and right eigenvectors are orthogonal to each other, that is,

$$w_j^T v_i = \begin{cases} 1 & \text{for } i = j \\ 0 & \text{for } i \neq j. \end{cases}$$

The trace of A, defined as the sum of its diagonal elements, is also the sum of all eigenvalues:

$$\operatorname{tr}(A) = \sum_{i=1}^{n} A_{ii} = \sum_{i=1}^{n} \lambda_i = \lambda_1 + \lambda_2 + \cdots + \lambda_n. \qquad (C.19)$$

The determinant of A is the product of all eigenvalues:

$$\det(A) = \prod_{i=1}^{n} \lambda_i = \lambda_1 \lambda_2 \cdots \lambda_n. \qquad (C.20)$$

If the eigenvalues of A are distinct, then A can be written as

$$A = T \Lambda T^{-1} \qquad (C.21)$$

where Λ is a diagonal matrix containing the eigenvalues. This is called an eigenvector decomposition (EVD). T is called a modal matrix and the Λ matrix is called the Jordan matrix.

The columns of T are the right eigenvectors v_i and the rows of T^{-1} are left eigenvectors w_i^T. Thus

$$T = [v_1 v_2 ... v_n], \ T^{-1} = [w_1^T \, w_2^T ... w_n^T]^T. \qquad (C.22)$$

Note that when the eigenvalues are repeated, whether that matrix can be fully diagonalizable or not (with the similarity transformation by modal matrix), depends on the algebraic multiplicity, denoted by m_i and the geometric multiplicity, denoted by q_i of that repeated eigenvalue. The geometric multiplicity q_i is given by $q_i = n - \operatorname{rank}(A - \lambda_i I)$. If $m_i = q_i$, then pure diagonalization is still possible. If $q_i = 1$, there will be one Jordan block of dimension m_i, with ones on the super diagonal. When $1 < q_i < m_i$, then the pure diagonal form for the Jordan matrix is not possible as there will be q_i Jordan blocks within the overall large Jordan matrix. These details are clearly discussed in [1]. Thus, in summary, the Jordan form for the repeated eigenvalue case needs to be carefully thought about, because it is not always automatically a pure diagonal matrix.

Some properties of eigenvalues:

1. All the eigenvalues of a Hermitian matrix (symmetric in the case of real matrices) are real.
2. All the eigenvalues of a unitary matrix have unit magnitude.
3. If a matrix A is Hermitian (symmetric in the case of real matrices), then the modal matrix T in (C.21) is unitary. The EVD is then

$$A = U\Lambda U^H \tag{C.23}$$

since $U^{-1} = U^H$.
4. If A is Hermitian (symmetric in the case of real matrices), then

$$\min_{x \neq 0} \frac{x^H A x}{x^H x} = \lambda_{\min}(A) \tag{C.24}$$

$$\max_{x \neq 0} \frac{x^H A x}{x^H x} = \lambda_{\max}(A). \tag{C.25}$$

The quantity $\frac{x^H A x}{x^H x}$ is called the Rayleigh quotient. Sometimes we are not interested in the complete solution of the eigenvalue problem (i.e. all the eigenvalues and eigenvectors). We may want an estimate of the first mode. One of the nice properties of Rayleigh's quotient is that it is never smaller than $\lambda_{\min}(A)$. Also, the minimum of the left-hand side of (C.24) is achieved when y is the eigenvector corresponding to $\lambda_{\min}$. Similarly, the maximum is achieved in (C.25) when x is the eigenvector corresponding to $\lambda_{\max}(A)$. Equation (C.24) is particularly useful in the modal analysis of structures represented by finite element models.
5. The eigenvalues of any diagonal matrix (real or complex) are simply the diagonal entries themselves. Similarly the eigenvalues of upper or lower triangular matrices (real or complex) are again simply the diagonal entries themselves.

Some more properties:

1. If A is $(n \times m)$ and B is $(m \times n)$, then

$$AB \text{ is } (n \times n) \text{ and is singular if } n > m. \tag{C.26}$$

2. If A is $(n \times m)$, E is $(m \times p)$, and C is $(p \times n)$, then

$$APC \text{ is } (n \times n) \text{ and is singular if } n > m \text{ or } n > p. \tag{C.27}$$

3. $\qquad A$ is singular *if* $f \lambda_i(A) = 0$ for some i. $\tag{C.28}$

4. $\qquad \lambda(A) = \dfrac{1}{\lambda(A^{-1})} \rightarrow \lambda(A)\lambda(A^{-1}) = 1.$ $\tag{C.29}$

5. $\qquad \lambda(\alpha A) = \alpha \lambda(A); \alpha$ is scalar. $\tag{C.30}$

6. $\qquad \lambda(I + A) = 1 + \lambda(A)$ $\tag{C.31}$

C.4 Definiteness of Matrices

$A_s = \frac{A+A^T}{2}$, is the symmetric part of A

$A_{sk} = \frac{A-A^T}{2}$, is the skew-symmetric part of A.

If all the (real) eigenvalues of matrix A_s are > 0, then A is said to be positive definite.

If all the (real) eigenvalues of matrix A_s are ≥ 0, then A is said to be positive semi-definite.

If all the (real) eigenvalues of $-A_s$ are > 0, then A is said to be negative definite (or if the eigenvalues of A_s are negative).

If all the (real) eigenvalues of $-A_s$ are ≥ 0, then A is said to be negative semi-definite.

If some of the (real) eigenvalues of A_s are positive and some negative, then A is said to be indefinite.

Note that $x^T A x = x^T A_s x + x^T A_{sk} x$ (i.e. $A = A_s + A_{sk}$).

In real quadratic forms:

$x^T A x = x^T A_s x$

(since $x^T A_{sk} x$ is always equal to 0).

Thus the definiteness of A is determined by the definiteness of its symmetric part A_s where λ_i, $i = 1$ to n are the eigenvalues of A_s and Δ_i =determinant of the ith principal minor

$$\Delta_1 = a_{11}, \quad \Delta_2 = \begin{bmatrix} a_{11} & a_{12} \\ a_{12} & a_{22} \end{bmatrix}, \quad \Delta_3 = \begin{vmatrix} a_{11} & a_{12} & a_{13} \\ a_{12} & a_{22} & a_{23} \\ a_{13} & a_{23} & a_{33} \end{vmatrix}, \text{ etc.}$$

where

$$A_s = \begin{bmatrix} a_{11} & a_{12} & a_{13} & \cdots & a_{1n} \\ a_{12} & a_{22} & a_{23} & & \vdots \\ a_{13} & a_{23} & a_{33} & & \vdots \\ \vdots & & & \ddots & \\ a_{1n} & \cdots & \cdots & & a_{nn} \end{bmatrix} \quad \text{(since } A_s \text{ is assumed symmetric, therefore } a_{ij} = a_{ji}\text{).}$$

Corollary: if A_s is ND, then A has negative real part eigenvalues. Similarly if A_s is PD, then A has positive real part eigenvalues. Thus a negative definite matrix is a stable matrix but a stable matrix need not be ND.

Also note that, even though

$A = A_s + A_{sk}$

the eigenvalues of A do not satisfy the linearity property.

i.e. $\lambda_i(A) \neq \lambda_i(A_s) + \lambda_i(A_{sk})$.

However, it is known that $\lambda_i(A)$ lie inside the field of values of A, i.e. in the region in the complex plane, bounded by the real eigenvalues of A_s on the real axis and by the pure imaginary eigenvalues of A_{sk}.

The Definiteness Test conditions are summarized in Table C.1.

Table C.1 Principal minor test for definiteness of matrix A given in terms of A_s (symmetric part of A).

By the definition	the matrix A_s is	if,	or equivalently,
$x^T A x > 0 \ \forall \ x \neq 0$	PD	All $\lambda_i > 0$	All $\Delta_i > 0$
$x^T A x \geq 0 \ \forall \ x \neq 0$	PSD	All $\lambda_i \geq 0$	All $\Delta_i \geq 0$
$x^T A x < 0 \ \forall \ x \neq 0$	ND	All $\lambda_i < 0$	$\Delta_1 < 0, \ \Delta_2 > 0, \ \Delta_3 < 0, \ \Delta_4 > 0$, etc.
$x^T A x \leq 0 \ \forall \ x \neq 0$	NSD	All $\lambda_i \leq 0$	$\Delta_1 \leq 0, \ \Delta_2 \geq 0, \ \Delta_3 \leq 0, \ \Delta_4 \geq 0$, etc.
$x^T A x > 0$ for some x $x^T A x < 0$ for other $x \neq 0$	Indefinite	Some $\lambda_i > 0$ Some $\lambda_i < 0$	None of the above

C.5 Singular Values

Let us first define inner product and norms of vectors.

Inner product: the inner product is also called a scalar (or dot) product since it yields a scalar function. The inner product of complex vectors x and y is defined by:

$$< x, y > = (x^*)^T y = y^T x^* = x_1 * y_1 + x_2 * y_2 + \dots + x_n * y_n = \sum_{i=1}^{n} x_i * y_i \qquad (C.32)$$

where $(\cdot)^*$ indicates the complex conjugate of the vector in parenthesis. If x and y are real, then

$$< x, y > = \sum_{i=1}^{n} x_i y_i = x_1 y_1 + x_2 y_2 + \dots + x_n y_n. \qquad (C.33)$$

Note that the inner product of two real vectors of same dimension $< x, y >$ can also be written as

$$< x, y > = \text{Trace } [xy^T]. \qquad (C.34)$$

Note that when x and y are complex $< x, y > = x^T y^*$. However, when x and y are real,

$$< x, y > = x^T y = y^T x = < y, x > . \qquad (C.35)$$

Norm or length of a vector: the length of a vector x is called the Euclidean norm and is (also known as the l_2 norm)

$$\|x\|_E = \|x\|_2 = \sqrt{< x, x >} = \sqrt{x_1^2 + x_2^2 + \dots + x_n^2}. \qquad (C.36)$$

The definition of a spectral norm or l_2 norm of a matrix is given by

$$\|A\|_2 = \max_{x \neq 0} \frac{\|Ax\|_2}{\|x\|_2} \qquad \text{where} \quad A \in C^{m \times n}. \qquad (C.37)$$

It turns out that

$$\|A\|_2 = \max_i \sqrt{\lambda_i(A^H A)}, \quad i = 1, 2, \dots, n$$

$$= \max_i \sqrt{\lambda_i(AA^H)}, \quad i = 1, 2, \dots, m. \qquad (C.38)$$

Note that $A^H A$ and $A A^H$ are Hermitian and positive semi-definite and hence eigenvalues of $A^H A$ and $A A^H$ are always real and non-negative. If A is non-singular, $A^H A$ is positive definite, and the eigenvalues of $A^H A$ and $A A^H$ are all positive.

We now introduce the notion of singular values of complex matrices. These are denoted by the symbol σ. If $A \in C^{n \times n}$, then

$$\sigma_i(A) = \sqrt{\lambda_i(A^H A)} = \sqrt{\lambda_i(A A^H)} \geq 0, \quad i = 1, 2, ..., n \tag{C.39}$$

and they are all non-negative since $A^H A$ and $A A^H$ are Hermitian.

If A is non-square, i.e, $A \in C^{m \times n}$, then

$$\sigma_i(A) = \sqrt{\lambda_i(A^H A)} = \sqrt{\lambda_i(A A^H)} \tag{C.40}$$

for $1 \leq i \leq k$, where $k =$ number of singular values $= \min(m, n)$ and $\sigma_1(A) \geq \sigma_2(A) \geq ... \geq \sigma_k(A)$.

$$\sigma_{max}(A) = \max_{x \neq 0} \frac{\|Ax\|_2}{\|x\|_2} = \|A\|_2 \tag{C.41}$$

$$\sigma_{min}(A) = \min_{x \neq 0} \frac{\|Ax\|_2}{\|x\|_2} = \frac{1}{\|A^{-1}\|_2} \tag{C.42}$$

provided A^{-1} exists. Thus the maximum singular value of A, $\sigma_{max}(A)$ is simply the spectral norm of A. The spectral norm of A^{-1} is the inverse of $\sigma_{min}(A)$, the minimum singular value of A. The spectral norm is also known as the l_2 norm. Usually we write $\bar{\sigma}(A)$ and $\underline{\sigma}(A)$ to indicate $\sigma_{max}(A)$ and $\sigma_{min}(A)$.

It follows that

$$\sigma_{max}(A^{-1}) = \|A^{-1}\|_2 = \frac{1}{\sigma_{min}(A)} \tag{C.43}$$

$$\sigma_{max}(A^{-1}) = \frac{1}{\|A\|_2} = \frac{1}{\sigma_{max}(A)} \tag{C.44}$$

$$\sigma_{min}(A) = 0 \quad \text{if } A \text{ is singular.} \tag{C.45}$$

Let us now introduce the singular value decomposition (SVD). Given any $(n \times n)$ complex matrix A, there exist unitary matrices U and V such that

$$A = U \Sigma V^H = \sum_{i=1}^{n} \sigma_i(A) u_i v_i^H \tag{C.46}$$

where Σ is a diagonal matrix containing the singular values $\sigma_i(A)$ arranged in descending order, u_i are the column vectors of U, i.e.

$$U = [u_1, u_2, ..., u_n] \tag{C.47}$$

and v_i are the column vectors of V, i.e.

$$V = [v_1, v_2, ..., v_n]. \tag{C.48}$$

The v_i are called the right singular vectors of A or the right eigenvectors of $A^H A$ because

$$A^H A v_i = \sigma_i^2(A) v_i. \tag{C.49}$$

The u_i are called the left singular vectors of A or the left eigenvectors of $A^H A$ because

$$u_i^H A^H A = \sigma_i^2(A) u_i^H. \tag{C.50}$$

For completeness let us also state the SVD for non-square matrices. If A is an $(m \times n)$ complex matrix, then the SVD of A is given by:

$$A = U\Sigma V^H = \sum_{i=1}^{K} \sigma_i(A)u_i v_i^H \tag{C.51}$$

where

$$U = [u_1, u_2, ..., u_m] \tag{C.52}$$
$$V = [v_1, v_2, ..., v_n] \tag{C.53}$$

and Σ contains a diagonal non-negative definite matrix Σ_1 of singular values arranged in descending order in the form

$$\Sigma = \begin{bmatrix} \Sigma_1 \\ \cdots \\ 0 \end{bmatrix} \text{ if } m \geq n$$

$$= \begin{bmatrix} \Sigma_1 & \vdots & 0 \end{bmatrix} \text{ if } m \leq n \tag{C.54}$$

where

$$\Sigma_1 = \begin{bmatrix} \sigma_1 & 0\cdots & & 0 \\ 0 & \sigma_2 & \cdots & 0 \\ \vdots & \vdots & \ddots & \vdots \\ 0 & 0 & \cdots & \sigma_p \end{bmatrix}$$

and

$$\sigma_1 \geq \sigma_2 \geq \cdots \geq \sigma_p \geq 0, \quad p = \min\{m, n\}.$$

Let us digress momentarily now and point out an important property of unitary matrices. Recall that a complex matrix A is defined to be unitary if $A^H = A^{-1}$. Then $AA^H = AA^{-1} = 1$. Therefore, $\lambda_i(AA^H) = 1$ for all i, and

$$\|A\|_2 = \bar{\sigma}(A) = \underline{\sigma}(A) = 1. \tag{C.55}$$

Therefore, the (l_2) norm of a unitary matrix is unity. Thus, unitary matrices are norm invariant (if we multiply any matrix by a unitary matrix, it will not change the norm of that matrix).

Finally, the condition number of a matrix is given by

$$\text{cond}(A) = \frac{\underline{\sigma}(A)}{\bar{\sigma}(A)}. \tag{C.56}$$

If the condition number of a matrix is close to zero, it indicates the ill-conditioning of that matrix, which implies inversion of A may produce erroneous results.

C.5.1 Some useful singular value properties

1.   If $A, E \in C^{m \times m}$, and $\det(A + E) > 0$, then $\bar{\sigma}(E) < \underline{\sigma}(A)$.   (C.57)

2. $\sigma_i(\alpha A) = |\alpha|\sigma_i(A), \quad \alpha \in C, A \in C^{m \times n}$.   (C.58)

3. $$\bar{\sigma}(A + B) \leq \bar{\sigma}(A) + \bar{\sigma}(B), \quad A, B \in C^{m \times n}.$$ (C.59)

4. $$\bar{\sigma}(AB) \leq \bar{\sigma}(A)\bar{\sigma}(B), \quad A \in C^{m \times k}, \quad B \in C^{k \times n}.$$ (C.60)

5. $$\underline{\sigma}(AB) \geq \underline{\sigma}(A)\underline{\sigma}(B), \quad A \in C^{m \times k}, \quad B \in C^{k \times n}.$$ (C.61)

6. $$|\underline{\sigma}(A) - \underline{\sigma}(B)| \leq \bar{\sigma}(A - B), \quad A, B \in C^{m \times n}.$$ (C.62)

7. $$\underline{\sigma}(A) - 1 \leq \underline{\sigma}(I + A) \leq \underline{\sigma}(A) + 1, \quad A \in C^{n \times n}.$$ (C.63)

8. $$\underline{\sigma}(A) \leq |\lambda_i(A)| \leq \bar{\sigma}(A), \quad A \in C^{n \times n}.$$ (C.64)

9. $$\underline{\sigma}(A) - \bar{\sigma}(B) \leq \underline{\sigma}(A + B) \leq \underline{\sigma}(A) + \bar{\sigma}(B), \quad A, B \in C^{m \times n}.$$ (C.65)

10. $$|\underline{\sigma}(A) - \underline{\sigma}(B)| \leq \bar{\sigma}(A + B), \quad A, B \in C^{m \times n}.$$ (C.66)

11. $$\underline{\sigma}(A) - \bar{\sigma}(E) \leq \underline{\sigma}(A - B) \leq \underline{\sigma}(A) + \bar{\sigma}(B).$$ (C.67)

12. $$\text{rank}(A) = \text{the number of non-zero singular values of } A.$$ (C.68)

13. $$\sigma_i(A^H) = \sigma_i(A), A \in C^{m \times n}$$ (C.69)

C.5.2 Some Useful Results in Singular Value and Eigenvalue Decompositions

Consider the matrix $A \in C^{m \times n}$.

Property 1:

$$\sigma_{\max}(A) \triangleq \|A\|_s = \|A\|_2$$
$$= [\max_i \lambda_i (A^T A)]^{1/2} = [\max_i \lambda_i (AA^T)]^{1/2}.$$ (C.70)

Property 2: if A is square and $\sigma_{\min}(A) > 0$, then A^{-1} exists and

$$\sigma_{\min}(A) = \frac{1}{\sigma_{\max}(A^{-1})}.$$ (C.71)

Property 3: the standard norm properties, namely

(a) $\|A\|_2 = \sigma_{\max}(A) > 0$ for $A \neq 0$ and $= 0$ only show $A \equiv 0$. (C.72)

(b) $\|kA\|_2 = \sigma_{\max}(kA) = |k|\sigma_{\max}(A)$. (C.73)

(c) $\|A + B\|_2 = \sigma_{\max}(A + B) \leq \|A\|_2 + \|B\|_2 = \sigma_{\max}(A) + \sigma_{\max}(B)$ (C.74)

(triangle inequality).

Property 4: special to $\|A\|_2$ (Schwartz inequality)

$$\text{i.e. } \|AB\|_2 \leq \|A\|_2 \|B\|_2 \tag{C.75}$$

$$\text{i.e. } \sigma_{max}(AB) \leq \sigma_{max}(A)\sigma_{max}(B). \tag{C.76}$$

Property 5: $\sigma_{min}(A)\sigma_{min}(B) \leq \sigma_{min}(AB).$
It is also known that:

$$\sigma_{min}(A) \leq |\lambda(A)|_{min} \leq |\lambda_i(A)|_{max} = \rho(A) \leq \sigma_{max}(A). \tag{C.77}$$

Note that: $\lambda(A + B) \neq \lambda(A) + \lambda(B)$
$\lambda(AB) \neq \lambda(A)\lambda(B)$.

Result 1: given the matrix A is non-singular, then the matrix $(A + E)$ is non-singular if

$$\sigma_{max}(E) < \sigma_{min}(A).$$

Result 2: if A is stable and A_s is negative definite then $A_s + E_s$ is negative definite and
hence $A + E$ is stable if

$$\sigma_{max}(E_s) \leq \sigma_{max}(E) \leq \sigma_{min}(A_s).$$

Result 3: if A_s is negative definite, $A_s + E_s$ is negative definite if

$$\rho[(E_s(F_s)^{-1})_s] < 1$$
$$\text{or } \sigma_{max}[(E_s(F_s)^{-1})_s] < 1$$

because for a symmetric matrix $|\lambda(.)_s|_{max} = \rho[(.)_s] = \sigma_{max}[(.)_s].$
Result 4: for any given square matrix A

$$\rho(|A|) = \rho(A_m) \geq \rho(A)$$
$$\sigma_{max}(A_m) \geq \sigma_{max}(A)/$$

Result 5: for any two given square non-negative matrices A_1 and A_2 such that $A_{1ij} \geq A_{2ij}$
for all i, j then,

$$\rho(A_1) \geq \rho(A_2)$$
$$\sigma_{max}(A_1) \geq \sigma_{max}(A_2).$$

C.6   Vector Norms

A vector norm of x is a non-negative number denoted $\|x\|$, associated with x, satisfying:

(a) $\|x\| > 0$ for $x \neq 0$, and $\|x\| = 0$ precisely when $x = 0$.
(b) $\|kx\| = |k|\|x\|$ for any scalar k.
(c) $\|x + y\| \leq \|x\| + \|y\|$ (the triangle inequality).

The third condition is called the triangle inequality because it is a generalization of the fact that the length of any side of a triangle is less than or equal to the sum of the lengths of the other two sides.

We state that each of the following quantities defines a vector norm.

$$\|x\|_1 = |x_1| + |x_2| + \dots + |x_n| \tag{C.78}$$
$$\|x\|_2 = (|x_1|^2 + |x_2|^2 + \dots + |x_n|^2)^{1/2} \tag{C.79}$$
$$\|x\|_\infty = \max_i |x_i|. \tag{C.80}$$

The only difficult point in proving that these are actually norms lies in proving that $\|\cdot\|_2$ satisfies the triangle inequality. To do this we use the Hermitian transpose x^H of a vector; this arises naturally since $\|x\|_2 = (x^H x)^{1/2}$.

We note that

$$\|x\|_\infty \le \|x\|_1 \le \|x\|_\infty$$
$$\|x\|_\infty \le \|x\|_2 \le \sqrt{n}\|x\|_\infty.$$

From the Schwartz inequality applied to vectors with elements $|x_i|$ and 1 respectively, we see that $\|x\|_1 \le \sqrt{n}\|x\|_2$. Also, by inspection, $\|x\|_2^2 \le \|x\|_1^2$. Hence

$$\frac{1}{\sqrt{n}}\|x\|_2 \le \|x\|_\infty \le \|x\|_2$$

$$\|x\|_2 \le \|x\|_1 \le \sqrt{n}\|x\|_2$$

$$\frac{1}{n}\|x\|_1 \le \|x\|_\infty \le \|x\|_1. \tag{C.81}$$

Let A be an $m \times n$ matrix, and let A be the linear transformation $A(x) = Ax$ defined from C^n to C^m by A. By the norms $\|A\|_1$, $\|A\|_2$, $\|A\|_\infty$ we mean the corresponding norms of A induced by using the appropriate vector norm in both the domain C^n and the range C^m. That is

$$\|A\|_1 = \max_{x \ne 0} \left\{ \frac{\|Ax\|_1}{\|x\|_1} \right\} \tag{C.82}$$

$$\|A\|_2 = \max_{x \ne 0} \left\{ \frac{\|Ax\|_2}{\|x\|_2} \right\} \tag{C.83}$$

$$\|A\|_\infty = \max_{x \ne 0} \left\{ \frac{\|Ax\|_\infty}{\|x\|_\infty} \right\}. \tag{C.84}$$

Let A be an $m \times n$ matrix. Then:

(i) $\qquad \|A\|_1 = \max_j \sum_{i=1}^{m} |a_{ij}|. \tag{C.85}$

(ii) $\qquad \|A\|_\infty = \max_i \sum_{j=1}^{n} |a_{ij}|. \tag{C.86}$

(iii) $\qquad \|A\|_2 = [\text{maximum eigenvalue of } A^H A]^{1/2}$

$\qquad\qquad = \text{maximum singular value of } A. \tag{C.87}$

Since we can compare vector norms we can easily deduce comparisons for operator norms. For example, if A is $m \times n$, using (C.81) find that

$$\|Ax\|_1 \leq m\|Ax\|_\infty \leq m\|A\|_\infty\|x\|_\infty \leq m\|A\|_\infty\|x\|_1 \tag{C.88}$$

so that $\|A\|_1 \leq m\|A\|_\infty$. By similar arguments we obtain

$$\frac{1}{\sqrt{m}}\|A\|_2 \leq \|A\|_\infty \leq \sqrt{n}\|A\|_2$$

$$\frac{1}{\sqrt{n}}\|A\|_2 \leq \|A\|_1 \leq m^{1/2}\|A\|_2$$

$$\frac{1}{n}\|A\|_\infty \leq \|A\|_1 \leq m\|A\|_\infty. \tag{C.89}$$

C.7 Simultaneous Linear Equations

C.7.1 Introduction

Sets of simultaneous linear equations are ubiquitous in all fields of engineering. The control systems engineer is particularly interested in these systems of equations because many problems including, but not limited to, optimization, pole placement, and stability are dependent upon the solution(s) to these systems. In this appendix we discuss the basic theory to deal with systems of linear equations and briefly review the nature of the solutions under various conditions.

C.7.2 Problem Statement and Conditions for Solutions

Consider the set of simultaneous linear algebraic equations

$$\begin{aligned}
a_{11}x_1 + a_{12}x_2 + \cdots + a_{1n}x_n &= y_1 \\
a_{21}x_1 + a_{22}x_2 + \cdots + a_{2n}x_n &= y_2 \\
&\vdots \\
a_{m1}x_1 + a_{m2}x_2 + \cdots + a_{mn}x_n &= y_m.
\end{aligned} \tag{C.90}$$

Since programs such as MATLAB can easily handle large-order matrices, it is often convenient to place the system of equations in matrix matrix notation as such:

$$Ax = y. \tag{C.91}$$

Note here that A is of dimensions $m \times n$; x is of dimensions $n \times 1$; and y is of dimensions $m \times 1$. Any vector, say x_1, which satisfies all m of these equations is called a solution to the system of linear equations. Note also that this system of equations can have one (unique) solution, infinite solutions, or no solutions depending on the nature of the data in the known or given quantities, namely the matrix A and the vector y.

Following the treatment given in [1], to determine the nature of the solution without excessive computation, we form the augmented matrix W. The augmented matrix is nothing more than the coefficient matrix A with the y vector tagged along as an additional column, i.e. $W = [A|y]$. Then, comparing the rank of the coefficient matrix A, denoted by r_A, with that of the augmented matrix W, denoted by r_W we can shed considerable light on the nature of the solution to the given system, as follows:

1. If $r_W \neq r_A$, no solution exist. The equations are inconsistent, i.e. the equations in the system contradict one another.
2. If $r_W = r_A$, at least one solution exists.
   (a) If $r_W = r_A = n$ there is a unique solution for x.
   (b) If $r_W = r_A < n$, then there is an infinite set of solution vectors.

Example C.1   Given the augmented matrix $W = \begin{bmatrix} 1 & 0 & 1 & -32 \\ 0 & 1 & 0 & -1 \\ 1 & 0 & 1 & -12 \end{bmatrix}$, determine the nature of the solution(s) to the system.

Solution
We count the number of linearly independent rows in the W matrix to obtain $r_W = 3$. Next we recognize that the coefficient matrix A is embedded inside the augmented matrix W; more specifically, it consists of the first three rows and columns of W. Since rows one and three are repeated (i.e. linearly dependent), $r_A = 2$. Consulting the guidelines above, we can concur that no solutions exist (because $r_A \neq r_W$).

Example C.2   Given the augmented matrix $W = \begin{bmatrix} 1 & 1 & 1 & -32 \\ 0 & 1 & 0 & -1 \\ 1 & 0 & 1 & -12 \end{bmatrix}$, determine the nature of the solution(s) to the system.

Solution
Following the same steps as above, we obtain $r_W = 3$. We then obtain the rank of the A matrix: since row three is a linear combination of rows one and two ($R_3 = R_1 - R_2$), $r_A = 2$. Again, we say there are no solutions to this system of equations because $r_W \neq r_A$.

Example C.3   Given the augmented matrix $W = \begin{bmatrix} 1 & 1 & 1 & -32 \\ 0 & 1 & 0 & -1 \end{bmatrix}$, determine the nature of the solution(s) to the system.

Solution
Here, both W and A are composed of two linearly independent rows. Therefore, $r_W = r_A = n = 2$. Following the guidelines given earlier in the section, we therefore decide this system has one unique solution vector.

Example C.4   Given the augmented matrix $W = \begin{bmatrix} 1 & 1 & 1 & -32 \\ 0 & 1 & 0 & -1 \\ 1 & 1 & 0 & 4 \\ 1 & 0 & 1 & -31 \end{bmatrix}$, determine the nature of the solution(s) to the system

Solution
Notice that in both W and A, row four is a linear combination of rows one and two ($R_4 = R_1 - R_2$). Therefore, $r_W = r_A = 3 < n$. Therefore we conclude that this system has infinitely many solutions.

C.8 Exercises

Exercise C.1. Given the set of simultaneous equations

$$\begin{bmatrix} 1 & 3 & 2 \\ 2 & 5 & 3 \\ 3 & 7 & 4 \\ 4 & 9 & 5 \end{bmatrix}\begin{bmatrix} x_1 \\ x_2 \\ x_3 \end{bmatrix} = \begin{bmatrix} 1 \\ 0 \\ -1 \\ 1 \end{bmatrix}$$

investigate the nature of the solution.

Exercise C.2. Given the system of equations

$$\begin{bmatrix} 1 & 0 & 1 \\ 0 & 1 & 0 \\ 2 & 2 & 2 \end{bmatrix}\begin{bmatrix} x_1 \\ x_2 \\ x_3 \end{bmatrix} = \begin{bmatrix} 0 \\ 0 \\ 0 \end{bmatrix}$$

investigate the nature of the solutions.

Exercise C.3. Find the determinant, transpose, inverse, trace, rank, eigenvalues, and normalized eigenvectors of the matrix

$$A = \begin{bmatrix} 1 & 1 \\ 1 & 2 \end{bmatrix}$$

Exercise C.4. Find the rank of the matrix

$$A = \begin{bmatrix} 3 & 2 & 9 \\ 1 & 0 & 3 \\ 2 & 1 & 6 \end{bmatrix}$$

and if appropriate find the inverse.

Exercise C.5. Find the eigenvalues and eigenvectors of

$$A = \begin{bmatrix} 4 & -2 & 0 \\ 1 & 2 & 0 \\ 0 & 0 & 6 \end{bmatrix}$$

and the corresponding Jordan matrix and modal matrix.

Exercise C.6. Determine the definiteness of the following matrices

(a) $A = \begin{bmatrix} -6 & 2 \\ 2 & -1 \end{bmatrix}$

(b) $A = \begin{bmatrix} 4 & -3 \\ 3 & -1 \end{bmatrix}$

(c) $A = \begin{bmatrix} 6 & -2 \\ 2 & 1 \end{bmatrix}$

(d) $A = \begin{bmatrix} 13 & 4 & -13 \\ 4 & 22 & -4 \\ -13 & -4 & 13 \end{bmatrix}$

(e) $A = \begin{bmatrix} -1 & 3 & 0 & 0 \\ 3 & -9 & 0 & 0 \\ 0 & 0 & -6 & 2 \\ 0 & 0 & 2 & -1 \end{bmatrix}$

(f) $A = \begin{bmatrix} 8 & 2 & -5 \\ 2 & 11 & -2 \\ -5 & -2 & 8 \end{bmatrix}$.

Exercise C.7. Find the eigenvalues and eigenvectors and then use a similarity transformation to diagonalize

$$A = \begin{bmatrix} 0 & 1 \\ -3 & -4 \end{bmatrix}.$$

Exercise C.8. Consider the eigenvalue–eigenvector problem for

$$A = \begin{bmatrix} 1 & 2 \\ -2 & -3 \end{bmatrix}.$$

Exercise C.9. Find the eigenvalues, eigenvectors, and the Jordan form for

$$A = \begin{bmatrix} 1 & 1 \\ 1 & 1 \end{bmatrix}.$$

Exercise C.10. Find the eigenvalues and Jordan form of

$$A = \begin{bmatrix} 1 & 0 & 0 & -3 \\ 0 & 1 & -3 & 0 \\ -0.5 & -3 & 1 & 0.5 \\ -3 & 0 & 0 & 1 \end{bmatrix}$$

given that the characteristic equation is

$$\lambda^4 - 4\lambda^3 - 12\lambda^2 + 32\lambda + 64 = 0$$

whose roots are found to be $\lambda_t = -2, -2, 4, 4$.

Exercise C.10. Let

$$A = \begin{bmatrix} 3 & 1 & 0 & 0 & 0 & 0 & 0 \\ 0 & 3 & 0 & 0 & 0 & 0 & 0 \\ 0 & 0 & 3 & 0 & 0 & 0 & 0 \\ 0 & 0 & 0 & 4 & 1 & 0 & 0 \\ 0 & 0 & 0 & 0 & 4 & 0 & 0 \\ 0 & 0 & 0 & 0 & 0 & 4 & 1 \\ 0 & 0 & 0 & 0 & 0 & 0 & 4 \end{bmatrix}.$$

(a) What are the eigenvalues of A?
(b) How many linearly independent eigenvectors does A have?
(c) How many generalized eigenvectors?

Exercise C.12. Find the eigenvalues and Jordan form of

$$A = \begin{bmatrix} 4 & 2 & 1 \\ 0 & 6 & 1 \\ 0 & -4 & 2 \end{bmatrix}.$$

Bibliography

1 W.L. Brogan. *Modern Control Theory*. Prentice Hall, 1974.

Appendix D

Useful MATLAB Commands

D.1 Author Supplied Matlab Routine for Formation of Fuller Matrices

```
1   function [Aoutput,Sign_Aoutput]=TILLYP(A,MTHD)
2   Siz_A=size(A);
3
4   if (MTHD == 1)   %Tilde
5      ci=1;ri=1;
6      for p=2:Siz_A(1)
7      for q=1:p-1
8        for r=2:Siz_A(2)
9        for s=1:r-1
10         if (r==q)
11            Aoutput(ri,ci)= -A(p,s);
12         elseif (r~=p & s==q)
13            Aoutput(ri,ci)= A(p,r);
14         elseif (r==p & s==q)
15            Aoutput(ri,ci)= A(p,p)+A(q,q);
16         elseif (r==p & s~=q)
17            Aoutput(ri,ci)= A(q,s);
18         elseif (s==p)
19            Aoutput(ri,ci)= -A(q,r);
20         else
21            At(ri,ci)= 0;
22         end
23         ci=ci+1;
24         if (r==Siz_A(2) & s==r-1)
25            ci=1;
26         end
27        end
28        end
29        ri=ri+1;
30      end
31      end
32      Ac=Aoutput;
33      As=Aoutput;
34   elseif (MTHD == 2)   %Lyapunov
35      ci=1;ri=1;
36      for p=1:Siz_A(1)
37      for q=1:p
38        for r=1:Siz_A(2)
```

Flight Dynamics and Control of Aero and Space Vehicles, First Edition. Rama K. Yedavalli.
© 2020 John Wiley & Sons Ltd. Published 2020 by John Wiley & Sons Ltd.

```
39      for s=1:r
40        if (p > q)
41          if (r==q & s<q)
42            Aoutput(ri,ci)= A(p,s);
43          elseif (r>=q & r ~=p & s==q)
44            Aoutput(ri,ci)= A(p,r);
45          elseif (r==p & s==q)
46            Aoutput(ri,ci)= A(p,p)+A(q,q);
47          elseif (r==p & s<=p & s~=q)
48            Aoutput(ri,ci)= A(q,s);
49          elseif (r>p & s==p)
50            Aoutput(ri,ci)= A(q,r);
51          else
52            Aoutput(ri,ci)= 0;
53          end
54        elseif (p == q)
55          if (r == p & s < p)
56            Aoutput(ri,ci)=2*A(p,s);
57          elseif (r == p & s == p)
58            Aoutput(ri,ci)=2*A(p,p);
59          elseif (r > p & s == p)
60            Aoutput(ri,ci)=2*A(p,r);
61          else
62            Aoutput(ri,ci)= 0;
63          end
64        end
65        ci=ci+1;
66        if (r==Siz_A(2) & s==r)
67          ci=1;
68        end
69      end
70      end
71      ri=ri+1;
72    end
73    end
74    Ac=Aoutput;
75    As=Aoutput;
76  end
77  end
```

D.2 Available Standard Matlab Commands

Table D.1 Useful Matlab Commands.

Matlab commands	Description
acker	Pole placement design for single-input systems
c2d	Convert model from continuous to discrete time
canon	State space canonical realization
care	Continuous time algebraic Riccati equation solution
compan	Compute the companion matrix corresponding to the polynomial
cond	Condition number with respect to inversion
cross	Vector cross product
ctrb	Form the controllability matrix
d2c	Convert model from discrete to continuous time
dare	Solve discrete time algebraic Riccati equations (DAREs)
det	Determinants of matrix
dlqr	Linear quadratic (LQ) state feedback regulator for discrete time state space system
dlyap	Solve discrete time Lyapunov equations
eig	Eigenvalues and eigenvectors
expm	Matrix exponential
eye	Identity matrix
initial	Initial condition response of state space model
inv	Inverse matrix
kron	Kronecker tensor product
lqr	Linear quadratic regulator (LQR) design
lsim	Simulate time response of dynamic system to arbitrary inputs
lyap	Continuous Lyapunov equation solution
norm	Norm of linear model
obsv	Observability matrix
pinv	Moore–Penrose pseudoinverse
place	Pole placement design
poly	Polynomial with specified roots or characteristic polynomial
rank	Rank of matrix
reside	Partial fraction expansion (partial fraction decomposition)
roots	Polynomial roots
ss	Create state space model, convert to state space model
ss2ss	State coordinate transformation for state space model
ss2tf	Convert state space representation to transfer function
svd	Singular value decomposition
tf2ss	Convert transfer function filter parameters to state space form
trace	Sum of diagonal elements

Index

a

Acceleration
  absolute   20, 21
  control system   257–259
  due to gravity   26, 27, 58, 156, 259, 266
  lateral   134, 266
  normal   127, 139, 252
  relative   6, 24
Ackermann's gain formula   373–375
active control   104, 146, 155–158,
    445–449, 468, 469
Active stabilization of satellites   143
Actuators
  hydraulic   433, 437
  limiting/saturation   434
Adjustable systems   209–210
Adverse yaw   107, 134
Aerodynamic
  forces   6, 33, 36, 41, 54, 78, 99, 125, 138,
    146, 452, 454
  moments   32, 36, 125
Aerodynamic center   82, 86, 87, 91, 127
Aerodynamic data   89, 93, 112, 114, 128,
    130, 138
Aerodynamic torques   144–146
Aileron
  angle (definition)   105
  effectiveness   134
  rudder interconnect   107–111
Aircraft
  altitude-speed (flight) envelope   124,
    125, 128, 259, 440
  dynamic stability and control   3, 77, 111,
    117–140, 155–162
  electromechanical controls   253

flight control systems (AFCS)   99, 138,
    251, 252, 263, 265, 267, 433–442,
    451–462
  static stability and control   3, 77–114,
    117–140, 143
  transport aircraft model   111, 253
Airfoil   79, 80, 83, 85, 100
Algebraic equations   188, 479, 481, 499
Algebraic Riccati equation (ARE)   388,
    389, 393
Altitude
  -hold autopilot   264–265
  vs. Mach number plot   162
Angle of attack
  feedback   258
  rate of change of   87, 91
  sensor   252, 256, 257
  trim value of   79, 80, 87, 89, 95
Angles of departure and arrival   221–223
Angular acceleration   460
Angular momentum   23, 24, 29–31, 33, 47,
    48, 50, 51, 59, 60, 144, 149–151, 155,
    160, 272, 277, 424, 445, 453
Angular velocity
  components   10, 18, 30, 31, 32, 42, 47,
    54, 58, 160, 305
  vector   10, 18, 32, 36, 47, 48, 51, 54, 58,
    160, 305
Approximations   5, 6, 8, 20, 25, 83, 117,
    128, 130, 135, 138, 234, 248
  in modeling dynamic systems   5–21
Arbitrary pole (eigenvalue) placement   397
Aryabhata satellite   269
Aspect ratio   91, 132, 256, 452
Asymptotic stability   325, 327, 336, 340

Flight Dynamics and Control of Aero and Space Vehicles, First Edition. Rama K. Yedavalli.
© 2020 John Wiley & Sons Ltd. Published 2020 by John Wiley & Sons Ltd.